D1415805

Torrey's Morphogenesis of the Vertebrates

Torrey's Morphogenesis of the Vertebrates

Fifth Edition

Alan Feduccia
UNIVERSITY OF NORTH CAROLINA
AT CHAPEL HILL

Edward McCrady
UNIVERSITY OF NORTH CAROLINA
AT GREENSBORO

John Wiley & Sons, Inc.
New York   Chichester   Brisbane   Toronto   Singapore

*The copyright for the drawings prepared by
Lois M. Darling has been assigned to the artist.*

ACQUISITIONS EDITOR / Sally Cheney
PRODUCTION MANAGER / Katharine Rubin
DESIGNER / Kevin Murphy
PRODUCTION SUPERVISOR / Micheline Frederick
MANUFACTURING MANAGER / Denis Clarke
COPY EDITOR / Marjorie Shustak
MANUFACTURING MANAGER / Catherine Faduska

Preface

Given the long tradition of *Morphogenesis* as a text for comparative embryology and anatomy, the task of producing a fifth edition was particularly onerous. Many changes have taken place in college curricula, and the embryology/anatomy offerings have often shifted away from the more scholarly and conceptual approaches exemplified by this text to courses oriented to the practical aspects of memorizing human structure. Fortunately, many schools, including ours, have not been so short-sighted, and still offer traditional courses in the evolution and development of morphology. Thus, the reception accorded the previous edition was most welcome. Yet because of the excessive demands of modern curricula, the trend has continued to combine anatomy and embryology into a single-semester course. The study of embryology is merely the changing anatomy of the embryo, and the causal analysis of its development is treated in courses designated as developmental biology or experimental embryology, which treat only the mechanisms relating to the development of the embryo and which have little time to deal with the complex and fascinating changes in the embryo through ontogeny, much less the phylogenetic history of the adult form.

As the editions of *Morphogenesis* evolved over the years, the book became more and more inclusive, and, like many texts, became larger and larger with each edition. In contrast to the previous editions, therefore, we eliminated from this edition some material that was no longer desirable for a text of this nature, without altering any important facet of study. (In the previous edition, for example, there were over 100 new glossary entries and 40 new illustrations; in this edition, while some new illustrations have been incorporated, the overall length of the text has been shortened by one-tenth.)

The reader will find the fifth edition to be a somewhat shortened, refined, and updated version of *Morphogenesis*, with many changes, deletions, and additions, some new illustrations, and completely updated selected readings at the end of the text.

We are indebted to the editorial staff at John Wiley & Sons for their labor and expertise in producing this text. Particular thanks are due to Sally Cheney, Biology Editor. In addition, we thank Micheline Frederick, Production Supervisor, Marjorie Shustak, Senior Copy Editor, and Josephine Della Peruta, our copyeditor. We also thank Sally McCrady for editorial assistance.

University of North Carolina   **Alan Feduccia**
Chapel Hill and Greensboro, 1990   **Edward McCrady**

Contents

Part One
Panorama

Chapter One

History of the Human Body

THE CONTINUITY OF LIFE

The essential substance of which all cells are composed is known as *protoplasm,* a complex system of organic and inorganic constituents that is the "physical basis of life." Not only are all organisms composed of cells and protoplasm (and its products), but all the activities that collectively constitute life and by way of which we distinguish the living from the nonliving are associated with it. Conversely, protoplasm exists solely in the form of organisms and any consideration of it apart from living things is a meaningless abstraction. By the same token, the word protoplasm does not describe a homogeneous something which, like water, possesses distinctive chemical properties; on the contrary, the term is a pervasive one that describes all the per-

mutations of form and function in living systems.

The task of the biologist has been, and still is, to understand this remarkable material in all its variety. The magnitude of the task is suggested both by the physical and chemical intricacy of protoplasm and by the kinds of things it does. That is, life is more than a structural complication. It is the distinctive behavior of which this physiochemical structure is capable that makes it the vehicle of life. Paramount in this behavior is the capacity for *reproduction,* without which life would disappear from the earth because of two important conditions.

First, all organisms and the various parts thereof have a limited life span. Although an organism has a measure of ability to cope with the ravages of aging and use on its parts, sooner or later the forces of age gain the upper hand and death of the individual ensues. Thus the persistence of life upon the earth calls for replacement of the mortal individual. Second, although there is no absolute reason to assume that living systems of the simplest kind may not originate from nonliving matter on the earth at present, neither is there positive evidence that such "spontaneous generation" of new life occurs nowadays or ever has occurred since the original inception of life countless millions of years ago. In fact, it is generally held that the physical composition and chemical composition of the earth's surface and atmosphere are no longer amenable to the de novo creation of life. Hence it follows that all present-day organisms have been derived from preexisting organisms and those, in turn, from previous forms, and so on back to the beginning of life.

THE PRINCIPLE OF EVOLUTION

It has become increasingly clear in recent years that the basis of reproductive behavior lies in the ability of that entity we call the *gene* to duplicate itself. In fact, we may envisage the origin of life itself in the assembly out of nonliving ingredients of a molecule capable of reproducing itself. But in addition to the mere(!) capacity for duplication, the primeval gene also possessed the ability to reproduce its chemical variations, thus providing for an increased variety of genes. Consequently, we may conceive of the original genes as having assumed the role of centers of chemical activity for the construction of organized systems of chemical substances that constitute the cells and tissues of organisms as we usually think of them. From these simple beginnings, because of the ability of genes to reproduce their variations and to imprint these variations on their host cells, there arose the present multitudinous variety of living things.

What we are saying is essentially this: In contrast to the relatively simple structure and limited variety of the first living things, present-day plants and animals exhibit a bewildering array of kind and form. And there has been, so far as we know, no spontaneous creation of new form; new organisms have been derived only from preexisting organisms. Therefore it follows that, as new generations of organisms have arisen down through the ages, changes have occurred. If this were not so, present-day living things would differ in no way from the original living things. This is the essence of the principle of *evolution*—that all organisms have arisen from common ancestry through a gradual process of change and diversification.

There are no biologists who presently dissent with the principle of evolution. That this was not always so is revealed by 200 years of history in which the evolutionary mode of thought struggled to establish itself. Yet, even after evolution as a basic principle became a conviction of the scientific mind, the mechanism of the evolutionary process remained at issue and it continues to be debated and actively investigated to this day. The immediate point is to emphasize the distinction between the *fact* of evolution, the evidence for which with respect to the vertebrates will be marshaled in the coming chapters, and evolutionary *theory* designed to explain how the described changes may have come about.

THE MECHANISM OF EVOLUTION

The modern-day concept of the mechanism of evolution had its beginning in the work of Charles Darwin (1809–1882). It is beyond our province to record the theoretical speculations of Darwin's predecessors and the detailed observational data upon which Darwin's own conclusions were based. Suffice it to say that Darwin realized that all organisms produce far more offspring than can possibly survive, so that some elimination of progeny is inevitable. Moreover, in any population of cross-breeding organisms, variations (anatomical, physiological, and behavioral) in individuals occur. This means that some of the progeny will be better *adapted* than others to cope with their environment. These favored individuals will be more likely to survive and produce offspring than will the less fit members of the population. In other words, *natural selection* occurs. As Darwin saw it, that is, the factors restricting the unlimited reproduction of members of a species would have a lesser effect on those best fitted to their environment and, by the same token, some members of a given population, thanks to the variations among them, would be better adapted to cope with a modified environment or to extend their range so as to exploit a new environment. Such naturally selected, better adapted individuals would leave more offspring than their less favored associates and thus, over many generations, distinctly new and different species would emerge. The key to success, of course, was *variation* and Darwin himself, because of the limitations of the knowledge of his times, was never able to provide an explanation of the origin of variation. That had to await the development of modern genetic theory. So let us consider variation in this setting.

One need only look at any population of sexually reproducing organisms to become aware of the differences between individuals. These variations are, to be sure, in some measure the result of differences in nutrition, effects of disease, and other environmental factors. But they are in far greater measure the result of differences in genetic constitution. Although all the individuals in a population have essentially the same numbers and kinds of genes in their chromosomes, genetic analyses reveal that the genes may occur in many alternative forms. The geneticist expresses this by saying that among the individuals of a population two or more *alleles* occur at a large proportion of the gene loci. The sum total of the different alleles in a population is, therefore, considerable and is said to con-

stitute the *gene pool* of the population. Since the variations residing in the gene pool provide the potential for evolutionary change, it is essential to inquire into the source of these variations.

Genes are known to consist of *deoxyribonucleic acid* (*DNA*), a gigantic organic molecule that is composed of two long, twisted chains of structural units known as nucleotides. Each nucleotide unit is composed of the sugar deoxyribose, phosphoric acid, and a nitrogenous base. There are actually four different kinds of nitrogenous bases: adenine, thymine, cytosine, and guanine, commonly designated by the letters A, T, C, and G; hence there are four different kinds of nucleotides. Because the nitrogenous bases in the nucleotides opposite each other in the double-ribboned DNA molecule can fit together only in two combinations—A with T and C with G—only four kinds of double units are possible along the length of the DNA molecule: A–T, T–A, C–G, and G–C. The order of these base pairs varies, however, and the specificity of any part of the DNA molecule depends upon this order. In effect, a triplet of nucleotide pairs provides a code that determines the position of a particular amino acid in a molecule of protein under synthesis. Any change in the code of nucleotide pairs in the DNA molecule will result in a corresponding change in the protein it codes and, by extension, in the cells and even the organisms of which the protein is a part. Such an alteration of the DNA molecule constitutes *mutation.* It is through mutation, and only through mutation, that the number of alleles at a gene locus is increased, thereby enlarging the gene pool. Mutations are the sole source of *new* genetic variability and therefore lie at the root of evolutionary potential; however, certain phenomena of *recombination* of the chromosomes themselves add important ingredients.

The sexually reproducing individual is itself a product of a single cell that results from the union of two cells, the ovum and spermatozoon. Thus, each of its cells contains two sets of chromosomes and genes, respectively maternal and paternal in origin. But with the production of the next generation of egg and sperm cells, homologous chromosomes pair up and one member of each pair is ultimately passed on to each of the reproductive cells. In other words, maturation of the reproductive cells brings a halving of the original number of chromosomes. Since, in this process, the original pairs of chromosomes segregate in a completely random fashion, a chromosomal basis for the Mendelian rules of segregation and independent assortment is provided. Subsequent union of the reproductive cells, fertilization, then provides for a new and unique combination of genetic propensity.

It is essential to reemphasize that mutations are the only source of new kinds of genes. It is this genetic variation within the gene pool that is the raw material of evolution.

To analyze all of the possible components of natural selection would call for a large-scale treatise on evolution and population genetics. Suffice it to speak only in generalities. Any variant in the gene pool that enables organisms to cope more successfully with the circumstances of their environment or with a changed environment has positive adaptive value. Individuals best adapted to their environment will gradually dominate the population, and their more poorly adapted colleagues will gradually be eliminated; or individuals with unique adaptations may invade a new environment and set up a new population distinct from the original. In the chapters to come, we shall look to certain conspicuous adaptations that have furthered the major steps of vertebrate evolution.

HISTORY OF THE HUMAN BODY

With the principle of the continuity of life in mind, it follows that all living things, past and present and including humans, are re-

lated. Any two organisms, however great the structural differences between them, are joined by intermediate links, and, ideally, we should be able to plot the ancestral course of any species along this connecting chain. That this can rarely if ever be done completely derives from apparent discontinuities in the chain resulting from past extinction of many forms that are largely unknown to us today.

The closeness of relationship between two organisms is measured by the number of structural features that they possess in common. Thus, John Doe is most like his brothers and sisters and parents; that is, they are his nearest relatives. He has fewer points of identity with, is more distantly related to, his aunts and uncles and cousins. There is established in the same fashion John Doe's position in relation to all other humans, to primates, to mammals, and to vertebrates. In other words, a human being is a vertebrate and is related in some degree to all other kinds of vertebrates. But because humans, compared with other vertebrates, are a relatively recent arrival on the evolutionary scene, we may find in these earlier established forms indications of the evolutionary pathway that has led to humans. The history of these vertebrate relatives is, to a degree, human history.

The succession of forms that culminates in any given structural entity constitutes the evolutionary history, or *phylogeny,* of that entity. By common consent the concept of phylogeny is applied either to the total body form or any part thereof. In other words, we may speak of the phylogeny of the human or the phylogeny of the excretory system or the phylogeny of a single organ such as the heart. The relationships and lines of descent constituting a given phylogenetic history are established through comparison of anatomical features, *comparative anatomy.* When such comparative study involves data from extinct organisms, available only as fossils, it is designated *paleontology.* But sometimes the likenesses and unlikenesses of parts of organisms are not self-evident and it is necessary to turn to another line of evidence. We refer to the sequence of transformations presented during the embryonic life of the individual, for the real nature of an adult part is often revealed by the developmental events that bring it into being. This developmental history of the individual or its parts is known as *ontogeny.* Now it is obvious to everyone that the newly hatched bird or the newborn human infant is not fully formed; some measure of development follows hatching or birth. That is, the term *ontogeny,* "origin of the individual," refers to the totality of developmental operations. For that more restricted segment of ontogeny that precedes hatching or birth we employ the term *embryogeny,* "origin of the embryo." The study of an embryo is designated as *embryology.*

Not only is understanding of adult form furthered by revelation of the embryonic history of that form, but comparative embryological studies have demonstrated that the more closely animals are related, the more nearly alike are their ontogenetic histories. Ontogeny thus provides an important tool for the establishment of the relationships on which the lines of evolutionary descent are based.

THE RECAPITULATION PRINCIPLE

Often known as the father of modern embryology, Karl Ernst von Baer (1792–1876) in 1828 published a book on embryology in which he first described some general principles that became known as von Baer's laws. These "laws" state that the more general characteristics of any group of animals appear earlier in the embryo than the more specialized characteristics, and that the embryo of a higher species may resemble the embryo of a lower species, but not its adult form. In 1866, Ernst Haeckel (1834–1919) published a theory he termed the "biogenetic law," but that subsequently became

known as the recapitulation principle. According to Haeckel's theory, the successive stages of individual development (ontogeny) correspond to successive adult ancestors in the line of evolutionary descent (phylogeny). The scheme may be thought of as working something like this: if, for instance, a series of ontogenetic stages *a-b-c* produces an adult fish, the addition of new steps, making the ontogeny *a-b-c-d-e,* would produce an amphibian, and so on through reptile to mammal. Accordingly, a developing mammal would first be a fish, then an amphibian, and then a reptile before it became a mammal; it would, in other words, pass through the adult stages of its ancestors. As Haeckel put it, *ontogeny recapitulates phylogeny!* In its application, the principle was made to work in two directions: on the one hand, it provided a causal explanation for the order of ontogenetic events; on the other, the events of embryonic development were employed to help establish phylogenetic relationships.

Obviously, if the idea of recapitulation was to be embraced, as it was wholeheartedly, and was to persist, as it does in some measure to this very day, it had to embody an element of truth. This derives from the fact that a developing embryo proceeds toward its goal of final form by an indirect route. Along the way it exhibits conditions that are indeed *suggestive* of those possessed by its ancestors. Comparing, for example, a human embryo with an adult shark (Figure 1-1), we note in the embryo such features as a tail, a series of gill-like pouches in the neck region, and a layout of blood vessels—all of which are fishlike in general appearance. Closer scrutiny, however, would reveal that the likenesses between embryo human and adult fish are superficial at best. In fact, if we expand the comparison to encompass a broad developmental series of vertebrates (Figure 1-2), a different set of impressions begins to emerge. Inspection of the early embryonic stages of vertebrates pictured in Figure 1-2 reveals that they really look very little like fully formed fishes,

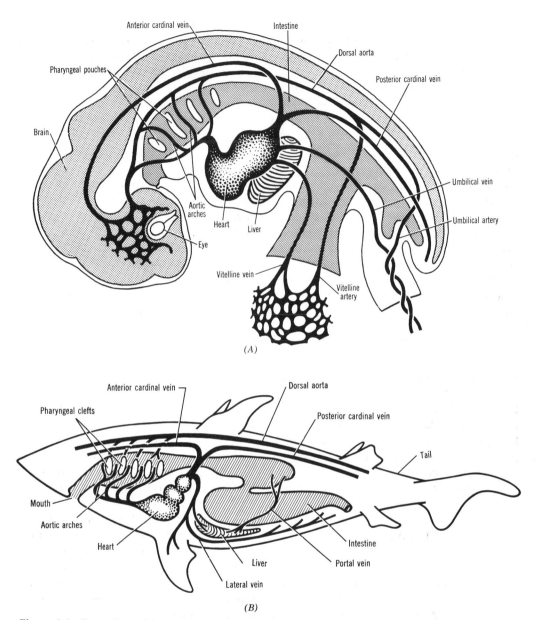

Figure 1-1 Comparison of the structure of a human embryo (*A*) with that of an adult fish (*B*).

amphibia, and reptiles. On the contrary, it is only at the beginning of development that a real likeness between these distinct categories of vertebrates obtains. As development proceeds, the embryos of these different animals become more and more dissimilar. The more distantly related two animals are, the earlier these differences begin to manifest themselves; conversely, the developmental pathways of more closely related forms run parallel for a longer time (see the mammalian embryos in Figure 1-2).

It was observations of this kind that had led Karl von Baer, many years before the rise of evolution theory, to make two major points which generalized embryonic development in vertebrates. Restated: (1) An-

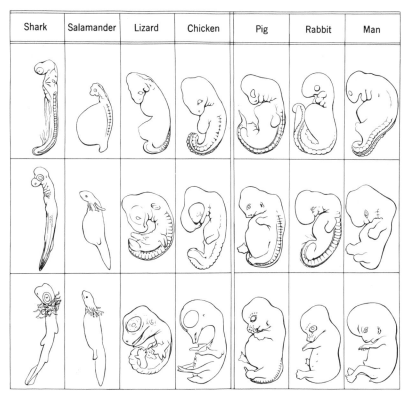

Shark	Salamander	Lizard	Chicken	Pig	Rabbit	Man

Figure 1-2 Comparative anatomy of representative vertebrate embryos at equivalent developmental stages, illustrating von Baer's principles. (See text for interpretation.)

imals are more similar at early stages of their development than they are when fully formed, and this resemblance becomes less as they grow older. (2) Therefore, instead of passing through the adult stages of other animals during its ontogeny, a developing animal moves away from them.

STRUCTURAL AND FUNCTIONAL SIMILARITY

The links in the evolutionary chain can be established only by comparing a part in one organism with a *corresponding* part in another—"corresponding" in the sense that the two are genetically related. The problem is to distinguish between the correspondences resulting from inheritance from a common ancestry and those superficial resemblances that have no community of

origin. Truly equivalent parts, in the sense of having had a common ancestry and regardless of structure and function, are said to be *homologous,* or are *homologues.* Parts having similar functions, whether homologues or not, are said to be *analogous,* or are *analogues.*

Homologies are often of a self-evident kind that reveal an easily recognized evolution of form and function. Take, for example, the front fin of a type of fish believed to have been the ancestral form from which all the terrestrial vertebrates stemmed (Figures 1-3A and 2-23). It contains the basic skeletal elements which with modest modification could have provided the skeleton for the walking limbs of the first Amphibia (Figure 1-3B). Once established in the Amphibia, this pattern was passed down through diverse descendants. Although showing many variations to accommodate

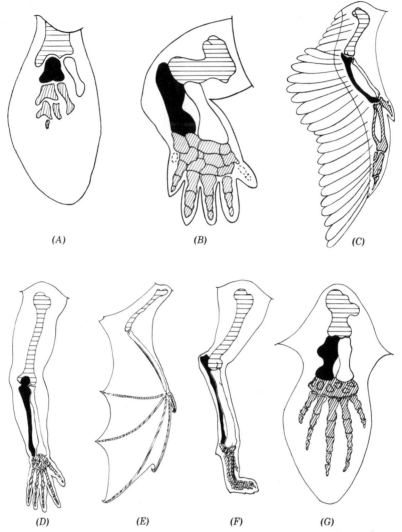

Figure 1-3 Homologous bones in the forelimbs: (*A*) crossopterygian fish (*Eusthenopteron*); (*B*) primitive amphibian (*Eryops*); (*C*) bird; (*D*) human; (*E*) bat; (*F*) dog; (*G*) whale. Crosshatching = humerus, white = radius, black = ulna, diagonal hatching = wrist and hand.

to functions of running, flying, swimming, and such (Figures 1-3*C–G*), the basic structural theme is clearly retained; hence all these limbs are homologous.

Sometimes, however, the evidence for homology is much more subtle. To illustrate, certain small bones in the middle ear of a mammal such as the human have been established as the homologues of particular jaw elements of fishes (Figure 1-4). Verifi-

cation of this identity not only calls for the data of comparative anatomy and paleontology, but requires embryological analysis and an assessment of the relationships of adjacent parts such as muscles and nerves. Much of the story of vertebrate morphogenesis, the history of the human body, is found in the more obscure homologies of this kind.

The wings of insects and birds, although

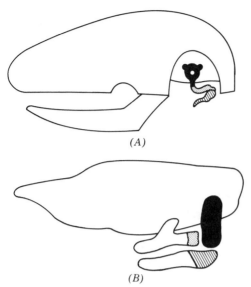

(A)

(B)

Figure 1-4 Diagram illustrating the bones of the middle ear of a mammal (*A*) and their homologous counterparts in the jaws of fishes (*B*).

they serve the same purpose, are not homologous; they are constructed on entirely different plans, have unlike embryonic histories, and have evolved wholly independently. They are thus analogous. *Homology* is correspondence between structures of different organisms resulting from their inheritance of these structures from the same ancestry. *Analogy* refers to similarity of function of parts of different animals. However, we shall not set up one against the other. Parts may be both homologous and analogous, as, for example, the forelimbs of a bat and a bird, both of which, with the same fundamental type of structure, are subservient to flight. In other cases parts may be analogous but not homologous, as with the wings of insects and birds. And, as illustrated by the already mentioned ear and jaw bones, parts may be homologous and not analogous.

Convergence or convergent evolution is common among vertebrates and refers to similar attributes in organisms from two or more lineages that did not share common ancestry. Convergence generally results from the evolutionary acquisiton of similar adaptations in different lineages to cope with similar environments. The remarkable similarity of the flippers of an ichthyosaur (an extinct marine reptile) and porpoise (marine mammal) has no basis in common ancestry and is therefore an example of convergent evolution.

Chapter Two

Ancestry and History of the Vertebrates

The point has been made that in some measure all vertebrates are related to one another, and just as we can say that a given vertebrate type has been derived from another vertebrate type, we should also be able to point out the invertebrate ancestor of the vertebrates.

What are the qualifications that must have been possessed by this invertebrate ancestor? Put in other terms, the fact that a diversity of forms such as fishes, frogs, and humans are lumped together in one category, the Vertebrata, suggests that they possess certain features in common and that these are features that the invertebrate ancestor, with a minimum of modification, must have been able to supply. The immediate task, then, is to identify those characters that distinguish the vertebrates.

THE VERTEBRATE BODY PLAN

The vertebrates as a group constitute only one category (subphylum) out of three included in a still larger category (phylum), the Chordata. In other words, all vertebrates are chordates but not all chordates are vertebrates. The vertebrates share certain features with a number of nonvertebrated forms. Specifically, there are three structures possessed by all chordates at some period of their life that serve to distinguish them from all other animal phyla (Figure 2-1). The first is the *notochord,* the structure from which the phylum Chordata derives its name. It is a pliant rod, composed of a distinctive vesicular tissue, which lies along the dorsal midline and provides the axis of support for the body. The notochord

13

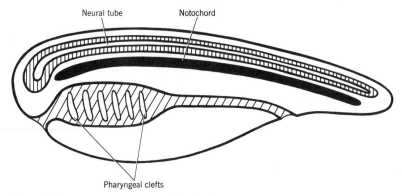

Neural tube   Notochord

Pharyngeal clefts

Figure 2-1 Diagnostic characters of a chordate.

is present in the embryos of all chordates and in some forms it persists throughout life. In most chordates, however, the notochord has only a temporary existence, ultimately being replaced by the "backbone," or vertebral column. The second is the *dorsal neural tube.* This, the central nervous system of chordates, lies just above the notochord and is hollow throughout its length. The third is a set of *pharyngeal clefts* (slits), passageways from the pharyngeal portion of the digestive tract to the exterior. In aquatic chordates the pharyngeal clefts are ordinarily carried over to adult life and, through the device of filaments formed on the walls of the clefts, serve as respiratory organs (gills). In terrestrial chordates the clefts are present only as transient structures during early embryonic life, often failing to open to the outside, and are subsequently adapted to nonrespiratory ends.

A fourth characteristic of chordates, although perhaps not as salient a feature as that mentioned earlier, is the possession of a *subpharyngeal gland,* located ventrally in the pharynx, that is capable of binding iodine and other related substances. In primitive chordates it is termed simply a subpharyngeal gland or *endostyle* (see Figure 2-5*A* and Figures 2-11*B, C*); in vertebrates its homologue is the thyroid gland.

In addition to these exclusively diagnostic characters, most chordates also exhibit

the following features: (1) A portion of the body extends rearward of the terminal opening of the digestive tract as a *tail.* (2) A *liver* lies ventral to the digestive tract, and *kidneys* (with a distinctive kind of tubular unit) rest dorsal to the digestive tract. (3) There is a *heart* ventral to the digestive tract, pumping blood through the body via a closed system of vessels; the blood flows forward from the heart to the dorsal side, where it courses posteriorly, and then back to the heart. (4) An *internal skeleton* (over and beyond the notochord) provides support and protection for body parts.

Three other features are shown by chordates, but these are shared with other animal phyla and are thus not exclusively diagnostic. (1) The chordate body is *bilaterally symmetrical,* which is to say that the right and left sides are mirror images of each other. (2) The serial repetition of such structures as nerves, blood vessels, muscles, and certain other parts signifies *segmentation,* or *metamerism.* (3) All chordates have a true body cavity, or *coelem.*

Applying the preceding diagnosis of chordate organization to the vertebrates specifically, there emerges this picture of vertebrate form (Figure 2-2). Along with members of several other animal groups, the vertebrates are bilaterally symmetrical, segmented, and coelomate. As chordates, they possess a dorsal neural tube, notochord, and pharyngeal clefts. Added to

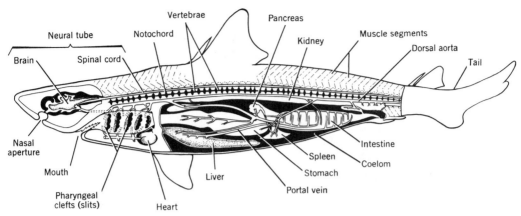

Figure 2-2 Diagnostic characters of a vertebrate, exemplified by a shark.

these features are a tail, heart and liver ventral to the gut, closed blood system, dorsal kidneys, and an internal skeleton. It is from the last-named feature that the vertebrates derive their name; the skeletal parts protecting and supporting the neural tube behind the brain consist of a segmental series of structures known as *vertebrae*.

CLASSIFICATION OF THE CHORDATES

All of the non-vertebrate chordates are commonly referred to as the protochordates. They include a diverse assemblage, including hemichordates, urochordates, and cephalochordates. Those exhibiting the least number of chordate features are often collectively classified in the subphylum Hemichordata, and the animals included herein may not be closely related at all. Because of their heterogeneity, and because some lack a hollow dorsal nervous system and the structure (stomocord) is conventionally but doubtfully interpreted as a notochord, their rank as a distinct phylum seems more appropriate.

Phylum Hemichordata

(*Figure 2-3*)
Solitary or colonial marine, wormlike forms with the body and coelom divided into three parts.

Class Pterobranchia
(*Figure 2-3A*)
Aggregated or colonial animals living in a secreted encasement. The first body division of the larva secretes the case, the second provides the mouth and a pair of tentacle-bearing arms (*lophophores*), and the third provides most other parts of the body. A single pair of pharyngeal clefts occurs in two of the three genera, none in the third.

Class Enteropneusta
(*Figures 2-3B, C*)
Wormlike animals termed acorn worms with numerous gill slits (used in filtering particulate food matter out of the water) and a terminal anus. The most anterior body division consists of a tonguelike proboscis used in burrowing; the second division is the "collar" containing the mouth; the third contains the pharynx with numerous clefts, the long intestine, and the gonads.

Phylum Chordata

(*Figure 2-1*)
Animals with a notochord, pharyngeal clefts, and a dorsal neural tube.

Subphylum Urochordata

(*Figures 2-4 and 2-5*)
The urochordates (Greek *oura*, "tail," indicating the position of the notochord in the larvae) are a group of marine forms, commonly known as tunicates, ascidians, or

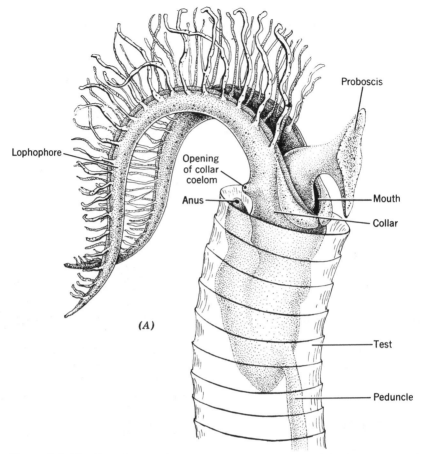

Figure 2-3 The Hemichordata. (*A*) Diagram of a representative pterobranch. (After Delage and Hérouard.) (*B*) Model of *Dolichoglossus,* a representative enteropneust. (Courtesy American Museum of Natural History.) (*C*) Detail of head region of an enteropneust. (After Dawydoff.)

"sea squirts." These are colonial or solitary marine organisms, the adults of which are baglike in shape (Figure 2-4). They are composed of an enclosed barrel-shaped net that constitutes the pharyngeal region and is utilized in filtering food particles out of the sea water. They are often called "sea squirts" for that reason, but are more commonly termed tunicates because of their tough covering or tunic, which is composed of a cellulose-like material termed *tunicin.* Tunicates range in size from microscopic to several inches. The tadpolelike larvae (Figure 2-5) are free swimming and are utilized for dispersal. The larvae exhibit a well-de-veloped notochord and dorsal nerve cord in the tail region. Recent studies utilizing the electron microscope have shown that the striated muscle in the tail region is identical to that of vertebrates. In addition the larvae exhibit a well-developed pharyngeal basket for filter feeding. Upon metamorphosis the larvae attach themselves to the substrate utilizing a special adhesive organ, the notochord and nerve cord degenerate, and they become visceral, filter-feeding organisms composed of little more than the pharyngeal basket. In both the tadpole larva and the adult, food particles are filtered from sea water, and water that has been

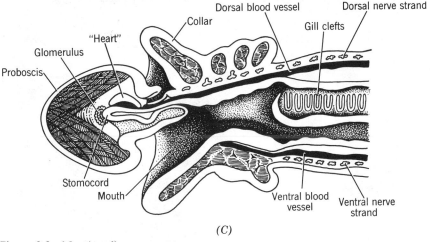

Figure 2-3 (*Continued*)

strained passes into an atrium and out the excurrent siphon; an endostyle along the ventral surface of the pharynx secretes a mucoid substance that aids in the entrapment of food particles. The agglutinated food "band" then passes into the intestine for digestion.

Subphylum Cephalochordata
(*Figure 2-6*)

Cephalochordates comprise a small number of marine species, all of which are placed in two genera, but are commonly referred to as *lancelets* or "*amphioxus*" (Greek *amphi,* "both," and *oxys,* "sharp"), an old generic name now superseded by *Branchiostoma.* The aberrant extension of the notochord into the head region (presumably as an adaptation for burrowing into the sand) gives the group its name, Cephalochordata. Lancelets are small (one to several inches), semitranslucent, fishlike forms that occur in coastal waters where for the most part they remain buried in the sand with their head regions protruding. Filter feeding is accomplished in a manner quite similar to that described for the urochordates. Food particles enter the mouth with sea water, and the pharynx (branchial basket) serves to filter the particles which then pass out into an *atrium* and out an atriopore. The food particles are agglutinated by a mucoid secretion from the endostyle (located beneath the pharynx) and the food strings are carried dorsally up the pharynx in "conveyer belt" fashion by cilia. There a food band is formed that enters the intestine where digestive processes take place.

The lancelets or amphioxi are the living protochordates that most clearly resemble the vertebrates and have therefore received

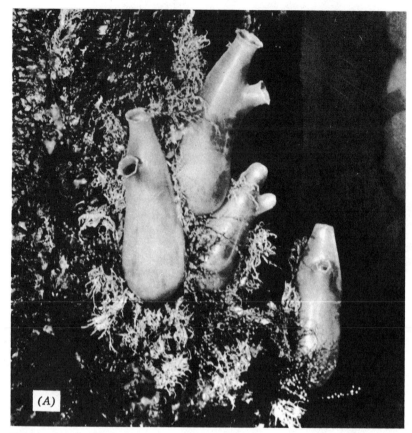

Figure 2-4   The Urochordata. (*A*) *Ciona,* a sea squirt of tunicate. Group on wharf piling. (Courtesy American Museum of Natural History.) (*B*) Diagram of the internal anatomy of a tunicate. Arrows show intake of water, its passage through numerous slits in pharynx for food filtration and aeration, and route through intestine.

considerable attention from comparative morphologists. They are the only proto-chordates to possess metamerically arranged muscle segments, or *myomeres* (Greek *myos,* "muscle," and *meras,* "part"), a condition known as *myomerism* which presages the common condition in vertebrates. There are no true paired fins in the lancelets, rather a pair of ventrolateral tissue extensions known as *metapleural folds* that extend from the anterior region of the pharynx to the atriopore.

Subphylum Vertebrata
(*Figure 2-2*)
Chordates with a vertebral column.

ORIGIN OF THE VERTEBRATES

With this background of chordate classification and the general organization of the vertebrates, we are now prepared to return to the problem posed at the beginning of this chapter: the origin of the vertebrates. This is actually a double problem that calls for answers to two questions. First, from which nonchordate group did the original chordates stem? Second, what is the evolutionary relationship of the vertebrates to the other two subphyla of chordates?

With respect to the first question, almost every major nonchordate phylum other than the Mollusca has at one time or another

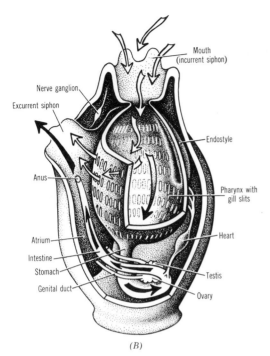

Nerve ganglion

Excurrent siphon

Mouth (incurrent siphon)

Endostyle

Anus

Pharynx with gill slits

Heart

Atrium

Intestine

Stomach

Genital duct

Testis

Ovary

(B)

Figure 2-4 *(Continued)*

been proposed as a possible progenitor of the chordate line. Generally speaking, the approach has been to consider forms such as annelids and arthropods that possess segmented bodies, a ventral nerve cord, and a dorsal heart. The idea has been that by turning such an animal over, thereby bringing the nerve cord uppermost and the heart ventral, the basic plan of a chordate would be approached. But this simple expedient does not completely achieve the result. The source of the notochord and pharyngeal clefts is left unaccounted for, and there are other alterations of structural detail, such as the change of a solid nerve cord to a hollow tube, for which only highly speculative solutions have been offered. More likely clues have come from certain facts pertaining to the Echinodermata (starfishes, sea urchins, etc.) and Hemichordata.

At first sight the Echinodermata would appear to be unlikely candidates for chordate progenitors. The highly specialized, radially symmetrical adult echinoderm,

however, is preceded by a bilaterally symmetrical larva of quite different design (Figure 2-7A). The Hemichordata also have a larval stage remarkably like this (Figure 2-7B); in fact, this hemichordate larva, when first discovered, was thought to be an echinoderm. The similarity of these larvae is strongly indicative of a close relationship between echinoderms and hemichordates. That the chordates in turn have an affinity with these two groups is suggested by facts such as these: In all three—echinoderms, hemichordates, and chordates—there are many similarities of early embryogeny, notably in the manner of development of the coelom. And the three groups also share certain qualities of biochemical makeup. A general affinity of echinoderms, hemichordates, and chordates therefore seems likely.

The lines of succession of these three phyla are not so easily resolved. Great differences of opinion exist, but a view that is coming to be accepted more and more looks to the origin of the echinoderms, hemichordates, and chordates in a common stock. One can only speculate as to the architecture of this hypothetical ancestor. It may have been a form not unlike the larvae of present-day echinoderms and hemichordates, or it may have resembled a pterobranch hemichordate. In any event, the thought is that the echinoderms represented one line of evolution from such a stock and a second line led to the hemichordates and chordates. This is not to say that the chordates came from the hemichordates (a view held by some authorities) or vice versa. It seems preferable to consider these two phyla as representing divergent lines from a common starting point. This, then, is the tentative answer to one portion of the two-part question of the origin of the vertebrates: common ancestry of chordates and hemichordates from a line that traces back to a stock that also produced the echinoderms.

As to the second question—the relationships of the Urochordata, Cephalochordata,

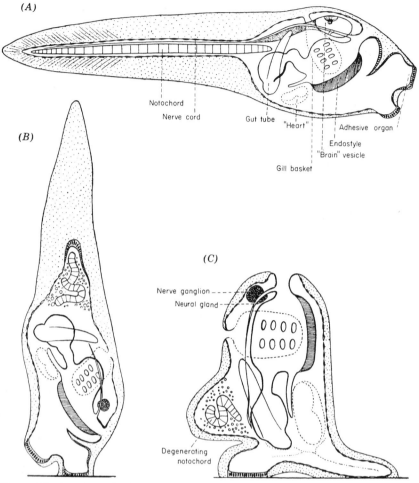

Figure 2-5 Diagrammatic illustration showing the metamorphosis of a solitary tunicate. (*A*) The free-swimming "tadpole" larva. (*B*) The larva has attached itself to the bottom using its adhesive organ, and tail degeneration begins. (*C*) The sessile transforming adult form is little more than a barrel-shaped "pharynx" for filtering particulate matter from sea water. Except for remnants of the notochord, most of the "somatic" structures have disappeared. (From Romer, *The Vertebrate Body*, 4th ed., after Dawydoff, courtesy W.B. Saunders Co.).

and Vertebrata—the picture is somewhat clearer. It should be appreciated that present-day protochordates (a term encompassing Urochordata and Cephalochordata) are specialized survivors of a long evolutionary process. They are not primitive animals "attempting to achieve vertebrate status"; they should be interpreted and evaluated as the terminals of, rather than landmarks along, an ancient evolutionary pathway. Consequently, many possible an-

swers, often framed in highly sophisticated argument, have been proposed.

In all these answers, a main focus of attention has been the *larval* urochordate (Figure 2-5). The motile larvae of the urochordates are elongate and bilaterally symmetrical; more significantly, they possess a dorsal neural tube and a notochord. The usual destiny of these larvae is transformation to sessile adults. It may have happened, however, that the motile larva of some se-

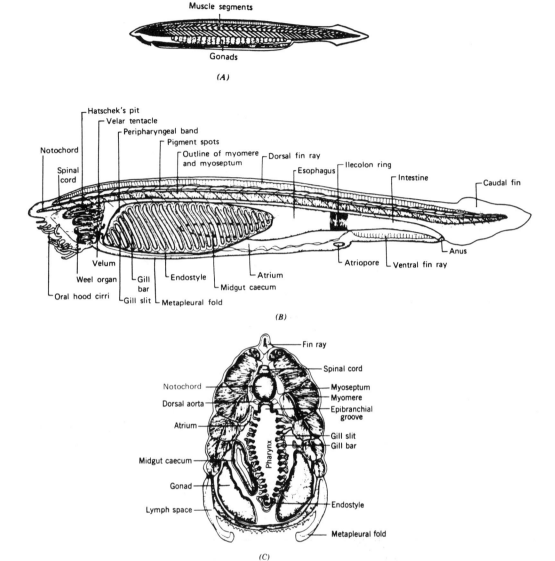

Figure 2-6 The Cephalochordata. (*A*) External morphology of *Amphioxus.* (*B*) Schema of internal anatomy. (*C*) Cross section of *Amphioxus* through the pharynx. (*A*, modified after Romer; *B* and *C*, from Feduccia, *Structure and Evolution of Vertebrates,* courtesy W.W. Norton & Company.)

dentary tunicate sloughed its adulthood, so to speak, and became capable of its own reproduction. (The term *neoteny* is used to describe the phenomenon of reproduction at a larval stage of organization.) In this way the larva would have become an independent form and a potential source of the two lines (Cephalochordata and Vertebrata) of free-swimming chordates. In fact, this is the view that is held by some investigators— the view that a neotenous, free-swimming larva first arose within the Urochordata, which in turn gave rise to *Amphioxus* and the vertebrates.

It is of great interest here to note that one group of tunicates has abandoned completely the adult sessile stage, and attains sexual maturity as tadpole larvae. These

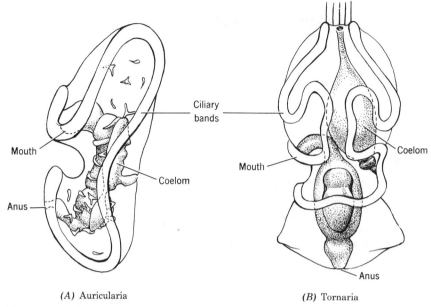

(A) Auricularia   *(B)* Tornaria

Figure 2-7   Echinoderm and hemichordate larvae. (*A*) Auricularian larva of a sea cucumber. (*B*) Tornarian larva of an enteropneust. (Drawn from Ziegler models.)

neotenic urochordates, referred to as the *Larvacea,* are thus important living models for the neotenic larval theory.

Many investigators have thought it likely that the urochordates themselves, as well as the cephalochordates and the vertebrates, arose from a primitive "prechordate" stock consisting of free-living, relatively long-lived larvae. Again, the larval type remains hypothetical. Nevertheless, this prototype may be supposed to have had pharyngeal clefts, a notochord, and a dorsal neural tube, the possession of which would give it the potentiality to evolve either in the urochordate direction of short-lived larvae and sessile adults as we know them today, or toward a prolongation of larval life and the neotenous, independent activity that favors the emergence of cephalochordates and vertebrates.

Even if this argument is accepted, there remains the issue of the relationship of the Cephalochordata to the Vertebrata. One view has been that *Amphioxus* is a degenerate derivative of an early vertebrate stock.

Conversely, the vertebrates have been said to have stemmed from a cephalochordate stock. The tendency today, however, is to regard the Cephalochordata and Vertebrata as having had a common origin from the stock that also produced the urochordate line, although all three groups share some morphological characters, the cephalochordates and vertebrates resemble each other more closely than they resemble the urochordates.

The recent discovery of a middle Cambrian age (530 million years old), well-preserved cephalochordate fossil named *Pikaia* (for nearby mount Pika) from the famous Burgess Shale of western Canada, has given credence to the idea that, indeed, the cephalochordates are primitive chordates. Not only do they possess features that one would presume to be present in the ancestral vertebrates, but they occur earlier in time than the first vertebrates, examples of which are first known from the latter part of the Cambrian Period, considerably later in time than the Burgess Shale *Pikaia.* There is no longer any reason to doubt that the cephalochordates occupy a position as probable vertebrate ancestors.

Figure 2-8 summarizes the story. Two evolutionary lines diverged from a common source, as yet hypothetical but possibly of a form similar to the present-day tornarian larva of hemichordates. The chordate progenitor presumably consisted of a free-swimming larva whose life became prolonged and which became capable of reproduction (neoteny) and in some unexplained fashion acquired pharyngeal clefts, a notochord, and a neural tube. From this similarly hypothetical stock, one line led to the Urochordata and another to the early diverging sublines of the Cephalochordata and Vertebrata.

VERTEBRATE CLASSIFICATION AND EVOLUTIONARY TRENDS

Before beginning a detailed study of the evolution of the various lines of vertebrates, it will be profitable to offer the reader an overview of the classification and evolution of the major groups of vertebrates. Perhaps the best starting place is to subdivide the various vertebrates into convenient groupings known as classes; this hierarchy of the taxonomic system illustrates the most distinctive assemblages. We may begin with the most advanced of the vertebrate classes, the mammals (class Mammalia) and the birds (class Aves). Both groups exhibit great advancement in their attainment of high and constant metabolic rates, thus making possible high levels of activity. Because most of their body heat is derived from within the body, from internal thermogenesis, both groups are termed *endotherms* (warm blooded). This is in contrast to all of the other classes of vertebrates that derive most of their body heat from outside their bodies, from the ambient environment; these groups are termed *ectotherms*

(cold blooded).[1] Of course, there are shades of gray between true endothermy and true ectothermy, but the two major divisions generally hold true. Mammals are warm-blooded, advanced tetrapods that have a covering of hair for insulation, and nourish their young by lactation. Birds are best characterized by their possession of feathers and a myriad of structural adaptations for flight. The other two classes of tetrapod vertebrates are the classes containing the reptiles (class Reptilia), and the amphibians (class Amphibia). Reptiles are primarily land-dwelling forms that lack many of the sophisticated features of the birds and mammals. Now restricted to only a few groups (turtles, crocodiles, snakes, and lizards), they once enjoyed the dominant role as terrestrial tetrapods during the Mesozoic Era, and some may have even been warm blooded. The amphibians, from which the reptiles sprang, are for the most part semiaquatic forms that are represented today primarily by frogs, toads, and salamanders, but once showed a tremendous evolutionary adaptive radiation. All of the previously mentioned classes of vertebrates may for convenience be grouped as a superclass Tetrapoda (vertebrates with four legs).

The remaining classes of vertebrates comprise the various groups of fishes; because they are all built around a common blueprint, they may be grouped for convenience as a superclass Pisces. There are various classifications of the fishes, but they seem most naturally treated as four classes. First to appear in the geologic strata is the

---

[1]In the past the term *poikilotherm* has been used to designate "cold-blooded" animals and *homoiotherm,* "warm-blooded" animals. These terms are more vague than the somewhat equivalent ectotherm and endotherm used here, as they refer simply to the relative constancy of body temperature (*see* the Glossary), rather than the actual source of body heat. Thus, some reptiles warm themselves from the sun behaviorally and become effectively homoiothermous during the day.

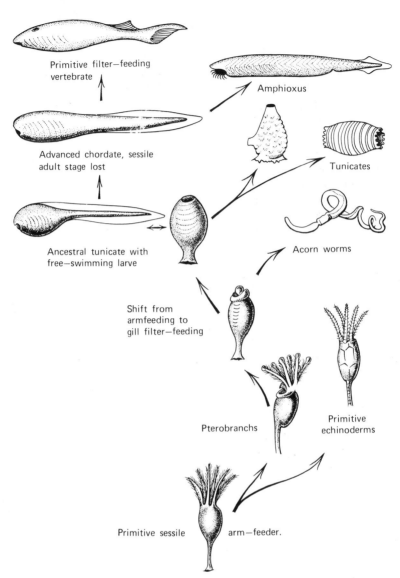

Primitive filter—feeding
vertebrate

Amphioxus

Advanced chordate, sessile
adult stage lost

Tunicates

Ancestral tunicate with
free—swimming larve

Acorn worms

Shift from
armfeeding to
gill filter—feeding

Pterobranchs

Primitive
echinoderms

Primitive sessile   arm—feeder.

Figure 2-8 Family tree suggesting the possible mode of early evolution of vertebrates. The echinoderms may have arisen from animals not too dissimilar to the pterobranchs; the acorn worms, from pterobranch descendants that had evolved a gill system but were little more advanced in other regards. Tunicates represent a stage in which, in the adult, the gill apparatus has become highly evolved for feeding, but the important point is the development in some tunicates of a free-swimming larva with advanced features of notochord and nerve cord and free-swimming habits. In further progress to amphioxus and the vertebrates the old sessile adult stage has been abandoned, and it is the larval type that has initiated the advance. (From Romer, *The Vertebrate Story,* courtesy University of Chicago Press.)

jawless class Agnatha, now extinct except for the lampreys and hagfishes. The first of the armored, jawed fishes belong to the class Placodermi, and the modern cartilaginous fishes, the sharks, skates, and rays, are perhaps best treated as the class Chondrichthyes. The advanced bony fishes, which constitute nearly half of the living species of vertebrates, are placed in the class Osteichthyes.

Still two other means of subdividing the vertebrates emphasize the development of jaws and the development of a special type of egg. Thus, the jawless fishes may be termed the Agnatha and all jawed vertebrates, the Gnathostomata. Reptiles, birds, and mammals differ from all other vertebrates in the type of egg they possess (this is discussed in great detail in Chapter 9). Consequently, these three classes may be termed the Amniota, and all other groups, the Anamniota.

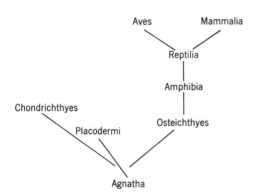

A very simple phylogeny of the vertebrate classes is illustrated above; a complete panorama of vertebrate phylogeny is illustrated in Figure 2-9.

A classification of vertebrates is presented at the end of this chapter. It is an abbreviated one intended as a summary for the student and adapted to the needs of this chapter and the remainder of this course. Only those categories deemed useful are given and common names are provided when possible. An asterisk preceding a given taxon indicates that it is extinct.

THE GEOLOGIC RECORD

In the search for the ancestry of chordates and vertebrates we were concerned primarily with comparisons of living invertebrates (echinoderms) and living hemichordates, urochordates, cephalochordates, and vertebrates. Because they are soft-bodied forms, hemichordates and protochordates have left behind few fossils; however, in tracing the history of vertebrates the study of fossils becomes of utmost importance, as vertebrates, with their skeletons of bone, have left behind a remarkable record in the rocks of the earth. In studying these fossil forms, and in tracing the history of the various groups of vertebrates, it is necessary to have some understanding of the geologic record, a record of earth history that covers at least several billion years. Fossils of importance, however, occur only during three divisions of geologic time that cover slightly over a half billion years; these major divisions are termed *eras.* Eras are then subdivided into *periods* of shorter duration, and the periods into *epochs* of even shorter duration. Table 2-1 illustrates the major divisions of geologic time and the notable geologic and biotic events that have occurred in each.

ANCESTRY OF THE VERTEBRATES

Class Agnatha (Jawless Fishes). The first vertebrates to appear in the fossil record were already well-developed armored fishes that lacked jaws. No doubt preceded by a long geologic history, the armored, jawless fishes, collectively known as the *ostracoderms,* first appeared in marine sediments of late Cambrian age, diversified during the Ordovician and Silurian Periods, and declined by Devonian time with the appearance of more advanced fishes. The ostracoderms (Figure 2-10) were primarily small to medium-sized fishes that were protected by a heavy, dermally derived armor that acted as an exoskeleton. There were no

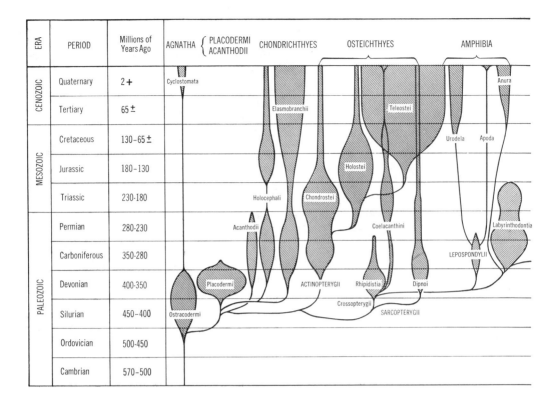

ERA	PERIOD	Millions of Years Ago
CENOZOIC	Quaternary	2+
CENOZOIC	Tertiary	65±
MESOZOIC	Cretaceous	130–65±
MESOZOIC	Jurassic	180–130
MESOZOIC	Triassic	230–180
PALEOZOIC	Permian	280–230
PALEOZOIC	Carboniferous	350–280
PALEOZOIC	Devonian	400–350
PALEOZOIC	Silurian	450–400
PALEOZOIC	Ordovician	500–450
PALEOZOIC	Cambrian	570–500

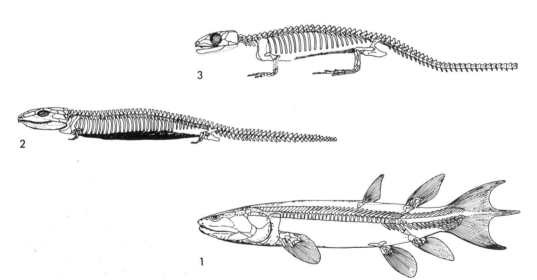

Figure 2-9 From fish to mammal. A series of skeletons in approximately true phylogenetic sequence from a rhipidistian crossopterygian to a placental mammal. (1) Crossopterygian, *Eusthenopteron;* (2) *Pholidogaster,* an early labyrinthodont tending in a reptilian direction; (3) *Hylonomus,* one of the oldest and most primitive of known reptiles; (4) *Sphenacodon,* a Permian pelycosaur representing the group from which therapsids were derived; (5) *Lycaenops,* a generalized therapsid with improved four-footed locomotion; (6) the three shrew *Tupaia,* a generalized placental mammal. (Redrawn from an illustration by and with the permission of the late Professor A.S. Romer.)

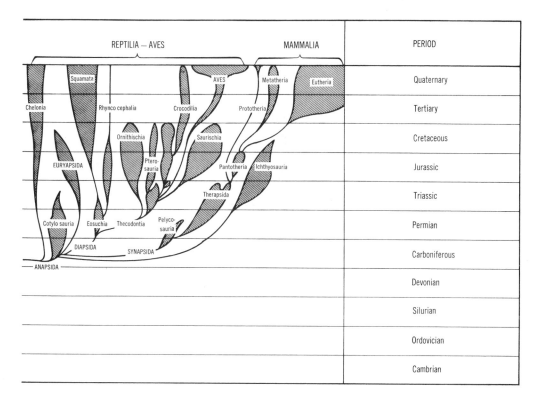

	REPTILIA — AVES			MAMMALIA		PERIOD
	Squamata	AVES		Metatheria	Eutheria	Quaternary
Chelonia	Rhynco cephalia	Crocodilia		Prototheria		Tertiary
	Ornithischia	Saurischia				Cretaceous
EURYAPSIDA	Ptero-sauria		Pantotheria	Ichthyosauria		Jurassic
		Therapsida				Triassic
Cotylo sauria	Eosuchia	Thecodontia	Pelyco-sauria			Permian
DIAPSIDA	SYNAPSIDA					Carboniferous
ANAPSIDA						Devonian
						Silurian
						Ordovician
						Cambrian

Figure 2-9 (*Continued*)

TABLE 2-1

SCHEMA OF GEOLOGIC TIME AND THE MAJOR FEATURES OF BIOTIC AND
GEOLOGIC EVOLUTION

Era (with approximate duration)	Period	Time since Beginning of Period (millions of years)	Epoch	Major Biotic and Geologic Events
Paleozoic (approximately 340 million years)	Cambrian	570		Most phyla of invertebrates present (probably all); trilobites and brachiopods dominant; cephalochordates present; origin of first fishes (Agnatha) toward end of period; diversification of marine algae; continents generally low; little known of period
	Ordovician	500		Diversification of jawless fishes (Agnatha); brachiopods and cephalopods dominant; early evolution of terrestrial plant communities; continents very low lying, with the most extensive inland seaways of geologic time
	Silurian	450		First jawed fishes appear; ostracoderms dominant; first terrestrial arthropods appear on land; club mosses, horsetails and ferns dominant; water scorpions (eurypterids) major marine predators; continents generally flat with extensive seaways
	Devonian (age of fishes)	400		Origin of amphibians; all major fish groups present; sarcopterygians, sharks abundant; ostracoderms become extinct; first winged insects; diversification in terrestrial plants; extensive inland seaways and many freshwater basins present
	Carboniferous	350		Origin of reptiles; amphibians adaptively radiate into many specialized types; extensive coal swamp forest, low-lying continents, and tropical and subtropical conditions; large terrestrial arthropods present; sarcopterygians dominant
	Permian	280		Reptiles adaptively radiate and replace amphibians as the dominant group; first mammallike reptiles appear; modern insect orders develop; chondrosteans become more dominant than sarcopterygians; single land mass (Pangaea) begins to split; elevated continents and extensive glaciation in Southern Hemisphere causing cool temperatures; Appalachian Revolution at end of period

TABLE 2-1
(*Continued*)

Era (with approximate duration)	Period	Time since Beginning of Period (millions of years)	Epoch	Major Biotic and Geologic Events
Mesozoic (age of reptiles; approximately 165 million years)	Triassic	230		Mammals originate; mammallike reptiles dominant; reptiles adaptively radiate; first dinosaurs, plesiosaurs, ichthyosaurs, pterosaurs, and turtles present; gymnosperms (cycads and conifers) dominant; extinction of primitive groups of amphibians; continents mainly exposed; rising temperatures with much aridity
	Jurassic	180		Reptiles (especially dinosaurs) dominant on land and in the seas; birds evolve; holostean fishes dominant; gymnosperms continue dominance; tropical and subtropical conditions with moderate continental elevation and seaways beginning to cover continents
	Cretaceous	130		Extinction of large land and marine reptiles at end of period; teleost fishes abundant; extinction of many invertebrates including the dominant ammonites; archaic mammals abundant; angiosperm plants begin their dominance; gymnosperms decline; tropical and subtropical throughout entire period, but cool at close with the beginning of the Rocky Mountain or Laramide Revolution. Oceanic seaways widespread across continents throughout the period; South America and Africa split apart toward end of period
Cenozoic (age of mammals; approximately 65 + million years)			Paleocene	Archaic mammals predominate; evolution of many groups of birds; culmination of late Cretaceous mountain building; cool at beginning, but warmer and dry toward the end; subtropical plants
			Eocene	Many modern orders and suborders of mammals and birds present; tropical and subtropical forests dominant; lemuroid primates and first horses present; erosion of mountains

TABLE 2-1
(*Continued*)

Era (with approximate duration)	Period	Time since Beginning of Period (millions of years)	Epoch	Major Biotic and Geologic Events
	Tertiary	65±	Oligocene	Modern families of mammals present; some extinction of primitive types; generally warm and humid, but probably cooler in certain regions
			Miocene	Appearance of modern subfamilies of mammals and all families of birds present; development of grassloads and many grazing mammals; moderate seasonal climates
			Pliocene	Many modern genera of mammals and birds appear; evolution of humans; generally warm with continued rise of continents, volcanic activity, and disappearance of many marine embayments
	Quaternary	2+	Pleistocene (Ice Age)	Modern species emerge; repeated cold and warm periods corresponding to four extensive periods of glaciation in the Northern Hemisphere and corresponding interglacials; extensive extinction of mammals and birds; humans enter the New World across the Bering Land Bridge
			Recent	End of the last ice age; warmer; modern humans ascend through stone, bronze, and iron ages

paired fins, but many had paired "stabilizers" on either side of the trunk region and flaps in back of the head. Most of the ostracoderms were probably bottomdwellers that were perhaps filter feeders or grazers, though some may have foraged in a variety of manners. It is of interest that bone was so well developed in the ostracoderms; even the internal head skeleton was a bony structure. Many reasons have been given for the early evolution of a bony external armor. First, the waters in which most ostracoderms occur were teeming with giant water scorpions, known as eurypterids, and these voracious invertebrates certainly preyed heavily upon the small ostracoderms. Less convincing arguments are that the armor developed to prevent water loss or for the storage of calcium ions. The ancestral ostracoderms may have been cartilaginous because there is no fossil record leading up to the abrupt appearance of these highly developed jawless fishes.

With the advent of the Osteichthyes the ostracoderms became extinct in Devonian times, but are survived by two eellike jawless fishes, collectively known as the *cyclostomes*. These include the living lampreys and hagfishes (Figure 2-11*A*). In both the lampreys and hagfishes the dermal

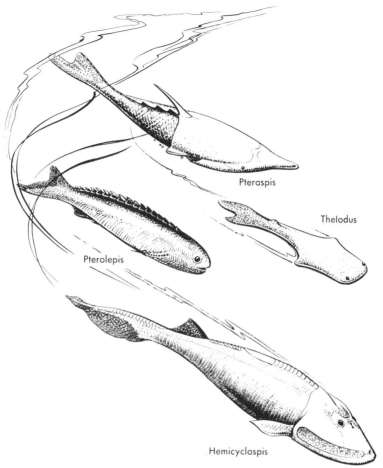

Pteraspis

Thelodus

Pterolepis

Hemicyclaspis

Figure 2-10 Representatives of four orders of ostracoderms from the Silurian and Devonian periods, all drawn to the same scale. (From Colbert, *Evolution of the Vertebrates,* John Wiley & Sons. Prepared by Lois M. Darling.)

armor has been completely lost and is replaced by a slimy skin devoid of scales. Lampreys consist of both freshwater and marine species, and one has become a permanent resident in the Great Lakes. Perhaps the best known lamprey is the marine lamprey (*Petromyzon*), and its life history is typical of the group. Lampreys are highly predaceous and utilize their rounded suction mouths to attach to other fishes; they then bore into the body wall with their rasping tongue, and suck the blood while secreting an anticoagulant. When the breeding season occurs, lampreys migrate long distances up freshwater streams to spawn. Fertilization

is external, and upon hatching, a creature quite unlike the adult emerges that may undergo metamorphosis and maturation for up to 7 years before returning to the sea. So unlike the adult is the immature form that when first discovered it was given the distinction of a different genus, and is still called an *ammocoetes* larva for that reason. The ammocoetes larvae (Figures 2-11*B, C*) are quite similar superficially to the lancelets (Cephalochordata) and feed in much the same manner. They burrow into the substrate and filter particulate matter from the water, but the water exits via external gill slits rather than through an atrium and

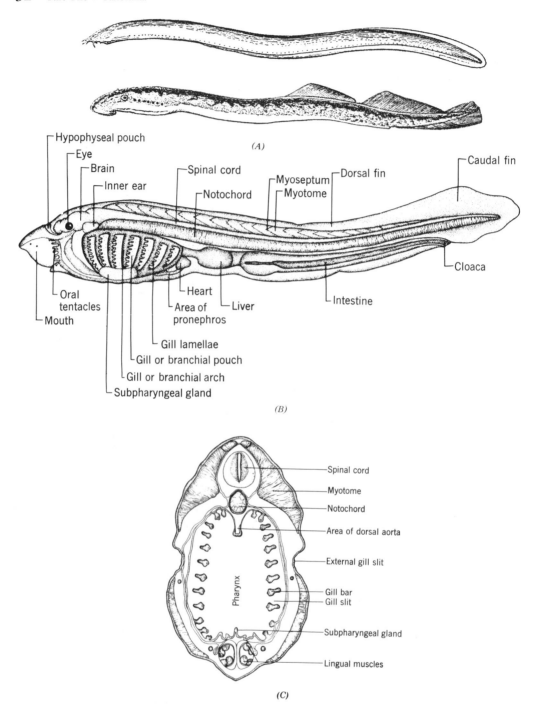

Figure 2-11   The Cyclostomata. (*A*) The hagfish, *Myxine* (upper), and the lamprey, *Petromyzon* (lower). (*B*) Morphology of the ammocoetes larva of the lamprey. (*C*) Cross section through the pharynx of the ammocoetes larva. (*A* through *C*, from Romer and Parsons, after Dean; *D* and *E*, from Feduccia, *Structure and Evolution of Vertebrates,* courtesy of W.W. Norton and Company.)

atriopore. Of course, there are many other differences, but the superficial similarities are quite reminiscent of vertebrate ancestry. Hagfishes (*Myxine*) are strictly marine and make their living by boring into dead or moribund fishes; they lay eggs but development is direct. Hagfishes and lampreys are degenerate forms that have persisted into the modern fauna probably because of their peculiar habits for survival. Both have skeletons composed entirely of cartilage, but aside from that the resemblances are more superficial, and some authors have suggested that they arose from different ostracoderm ancestors.

Figure 2-12 compares the upper and lower head regions of a living lamprey and an extinct ostracoderm. Note that in both forms there is a single nostril (nasohypophyseal) opening located in the midline on top of the head, and immediately posterior is a pineal or median eye, in addition to the paired lateral eyes. Both have a mouth without jaws, but in the ostracoderm the gill openings are ventral; in the lamprey they are lateral along the pharynx.

A new era in the history of fishes opened with the advent of jaws, and opened to the vertebrates opportunities for exploitation of innumerable ecological niches and thus expanded their evolutionary potential immeasurably.

Another major advance with these jawed fishes was the evolution of functional paired appendages.

Because there was an enormous hiatus in the fossil record from the actual first vertebrates to the ostracoderms, there is equally a major gap in the paleontological record between the class Agnatha and the first of the Gnathostomata ("jaw mouths"). Like the ostracoderms before them, the first jawed fishes did not spring forth full fledged; there must have been preliminary steps. The only explanation for the paucity of data seems to lie in the virtual absence of the continental geological strata that might be expected to provide the record.

Class Placodermi. The ostracoderms reached the apex of their evolutionary development and diversification during the Silurian Period, but at the end of the Silurian there appeared the first jawed fishes, or

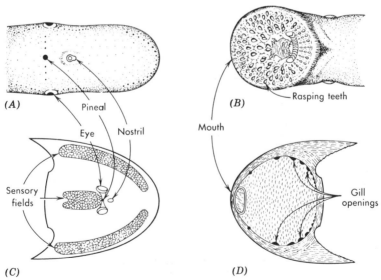

Figure 2-12 Comparison of the upper and lower surfaces of the head regions of a lamprey (*A* and *B*) and the Devonian ostracoderm, *Cephalaspis* (*C* and *D*). (From Colbert, *Evolution of the Vertebrate*, John Wiley & Sons.)

placoderms ("plate skin"), which were to become very important during the next period, the Devonian, and grow extinct by the end of the Paleozoic Era. The Placodermi (Figure 2-13) came in a variety of sizes and shapes, but all had plated armor, at least on the anterior part of the body and the head region. Some featured heavy armor over the fore portion of the body, which was divided into "head" and "thorax" sections connected by ball-and-socket joints. All had paired fins of one kind or another and all had some kind of jaws. Placoderms ranged in size from rather small to up to more than 3 meters. The Acanthodii were once though to be a group of placoderms but are now known to represent a group of true bony fishes; they are discussed with the Osteichthyes. Placoderms represent the only class of vertebrates to have become completely extinct and to have no representatives in the modern fauna.

Class Chondrichthyes (Cartilaginous Fishes). Although gaps in the paleontological record again make for uncertainty, it is generally agreed that the placoderms are related to the Chondrichthyes (fishes with cartilaginous skeletons) rather than to other groups of later fishes. The Chondrichthyes are all derived from radiating lines emerging during the Devonian Period and constitute a side branch of vertebrate evolution from which no group of higher vertebrates evolved. The Chondrichthyes flourished through Carboniferous and Permian times but, by the end of the Paleozoic Era, many of their evolutionary lines had died out. Nevertheless, these fishes have continued to the present day in modest variety. A subclass Elasmobranchii encompasses the sharks, skates, and rays (Figure 2-14), and is characterized by peculiar placoid scales. Modified placoid scales give rise not only to the teeth of elasmobranchs, but also to accessories such as fin spines, claspers, and even the stings of sting rays. The sharks are placed in a separate order Selachii, the skates in the order Batoidea. The subclass Holocephali is represented by the so-called "chimaera" or ratfish (Figure 2-15). Chimaeras are rather rare oceanic forms that

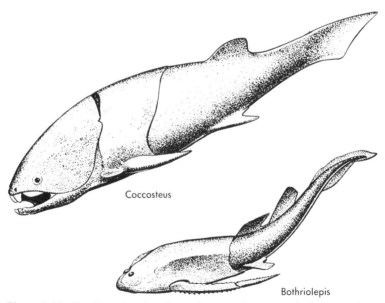

Coccosteus

Bothriolepis

Figure 2-13 Two Devonian placoderms, drawn to the same scale, about one-third natural size. (From Colbert, *Evolution of the Vertebrates,* John Wiley & Sons. Prepared by Lois M. Darling.)

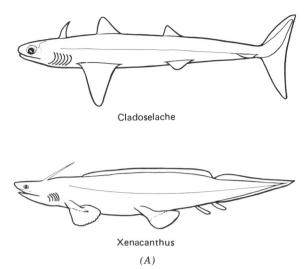

Cladoselache

Xenacanthus

(A)

Figure 2-14 (A) The extinct genus *Cladoselache,* primitive sharks of the order Cladoselachii, and *Xenacanthus,* a member of the order Pleuracanthodii (freshwater forms). (B) Living sharks of the order Selachii. The basking shark is a filter feeder and represents the world's second largest fish, the largest being the whale shark. (C) A living skate, guitarfish, ray, and manta, all members of the order Batoidea. (A, from Nelson, *Fishes of the World,* John Wiley & Sons; B and C, from Jordan and Evermann, *The Fishes of North and Middle America,* Pt. IV.)

lack the placoid scales characteristic of elasmobranchs. In addition, all of the gill slits open into a common opening covered by an operculum. Some authorities feel that they may represent a separate line of evolution from the placoderms independent of the elasmobranchs.

Since the cartilaginous fishes are off the main highway of vertebrate evolution, we need not consider them further in the present context other than to point out that the all-cartilage skeleton, once interpreted as being a primitive character, is now generally (although not universally) believed to be a result of specialization. After all, the placoderms from which they stemmed featured external bony armor and variable amounts of internal bone.

The Devonian Period also saw the rise of all major groups of advanced bony fishes.

Class Osteichthyes (Bony Fishes). The class Osteichthyes is by far the most diverse and numerous of all the groups of fishes. It appears that by Silurian time much of the major evolution in the fish groups was occurring in fresh waters, but unfortunately most of the continental deposits are of marine sediments. Bony fishes appear to be allied by their common possession of salient features of the skull, a bony operculum that forms a common opening for all of the external gill slits, a pectoral girdle attached to the skull by a series of bones, and various types of scales covering the body.

Although bony fishes underwent their tremendous adaptive radiation beginning in early Devonian time, along with their chondrichthian cousins, the first bony fishes appeared earlier, well back into the Silurian, as a group of minnow-sized fishes known as the subclass Acanthodii (Figure 2-16). The acanthodians, once placed with the placoderms, are now thought to represent an initial radiation of bony fishes. They are characterized by a bony skeleton and a covering of diamond-shaped scales. Paired fins were well developed, each supported by a strong

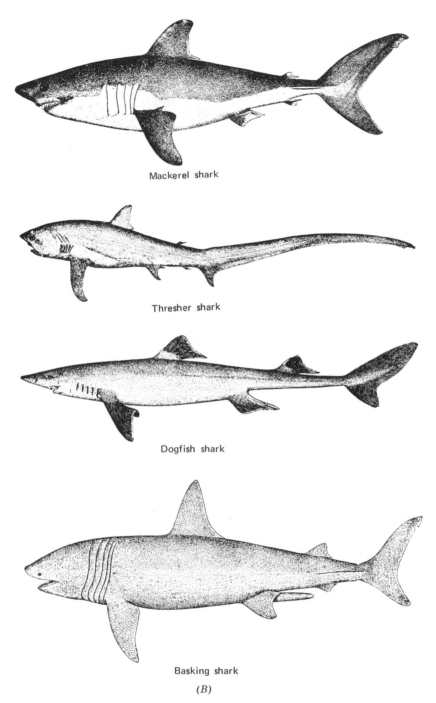

Mackerel shark

Thresher shark

Dogfish shark

Basking shark

(B)

Figure 2-14   (*Continued*)

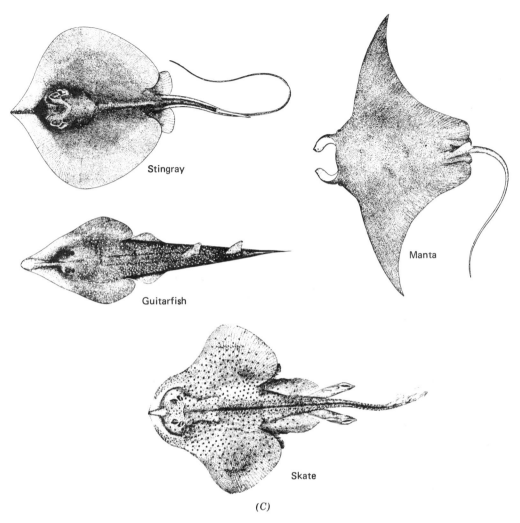

Stingray

Guitarfish

Manta

Skate

(C)

Figure 2-14 *(Continued)*

spine, but in addition to the usual two pairs, as many as five accessory paired spines were present. The name "spiny shark" was used previously to characterize this group. Unlike the placoderms whose jaws show only shearing edges serving as teeth, some acanthodians featured true teeth, especially on the lower jaw. And, like modern fishes, the gill region was covered on each side by a bony lid or operculum. This enigmatic group, which became extinct by Permian time, shows little affinity with either the sharks or placoderms, and certain features of the skull and branchial region ally them clearly with the true Osteichthyes; however, they are far too specialized to be on the mainstream of bony fish evolution. They are thus placed as a separate subclass of the Osteichthyes and are considered as a interesting side group. Figure 2-17 illustrates the evolutionary lines of the primitive fishes.

As early as the middle of the Devonian Period the bony fishes were the dominant fish group in freshwater, and by later Paleozoic they had diversified into a number of marine lines. By the close of the Triassic Period, however, bony fishes were dominant in marine waters, and the marine environment became the focus of their evolutionary adaptive radiation. Lungs were

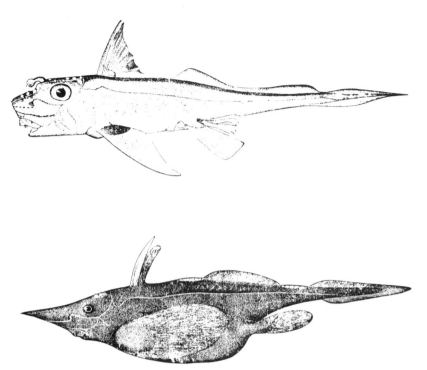

Figure 2-15 Two species of living holocephalian fishes or chimaeras. (From Jordan and Evermann.)

apparently present in the primitive lines of bony fish as either air-breathing organs or as hydrostatic organs (swim bladders), and presumably served as an adaptation for living in hot freshwater situations of Devonian time where low oxygen tension in the water necessitated breathing aerial oxygen. Later, with the invasion of the marine environment the organs were lost in many forms. There were numerous evolutionary lineages of bony fishes, but here it is only of interest to consider the major lines, particularly those leading to the tetrapods (four-footed vertebrates).

By early Devonian the bony fishes (apart from the acanthodians) had already split into two major evolutionary lines, the subclasses Actinopterygii (ray-finned fishes) and the Sarcopterygii (lobe-finned fishes). Figure 2-18 illustrates this basic bifurcation; note that the sharks (Chondrichthyes) appear and begin to diversify at approximately the same time. Although the actinopterygians or ray fins represent by far the vast

Figure 2-16 The acanthodian *Climatius*, about natural size. Acanthodians were primitive, bony jawed fishes thought to be related to the Osteichthyes. Note the operculum behind the head region. (From Colbert, *Evolution of the Vertebrates*, John Wiley & Sons.)

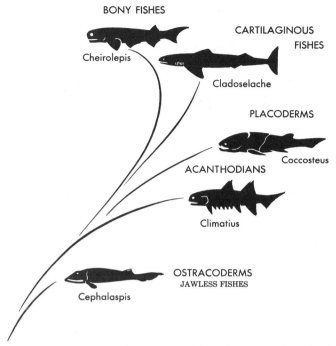

Figure 2-17 Evolution of the primitive fishes. There are really no fossil links between any of the gnathostome fishes and the agnathostomes, or between any of the bony fish lines and the placoderms. Note the early appearance of the bony fishes as the acanthodians. (From Colbert, *Evolution of the Vertebrates,* John Wiley & Sons. Prepared by Lois M. Darling.)

majority of not only living fishes, but also living vertebrates, they are not on the direct line leading to the tetrapods and will only be considered briefly here.

The ray-finned fishes are so named because the fins are supported by bony rays that radiate from the base of the fin, unlike the lobe-finned fishes, which have a fleshy fin. The subclass Actinopterygii is classified into three distinctive superorders: the Chondrostei (paleoniscoid fishes), the Holostei, and the Teleostei. These three superorders are considered representative of three stages in the evolution of modern bony fishes in the order indicated. The three groups differ in a number of important anatomical features, including the nature of the pectoral girdle, the conformation of certain skull bones, the types of scales, the degree of ossification of the skeleton, and many other minor characteristics. The an-

cient paleoniscoids or Chondrostei were the major representatives of the actinopterygians during the Triassic when they were quite abundant; they are survived today by only a few forms, the sturgeons (*Scaphyrhynchus*), the paddlefish or "spoonbill" (*Polyodon*) of the Mississippi River (a filter feeder), and the Nile bichir (*Polypterus*), a form considered by some not to be of chondrostean affinity. The bichir has a pair of simple lungs and fleshy fins, for which reason it was considered at one time a lobefin. Chondrosteans typically have the uptilted sharklike heterocercal tail. The Holostei were dominant during the Jurassic and Cretaceous Periods and are survived today only by the gars (*Lepisosteus*) and the bowfin (*Amia*). They show considerable changes from the ancestral paleoniscoids, but before the close of the Cretaceous the most advanced bony fish

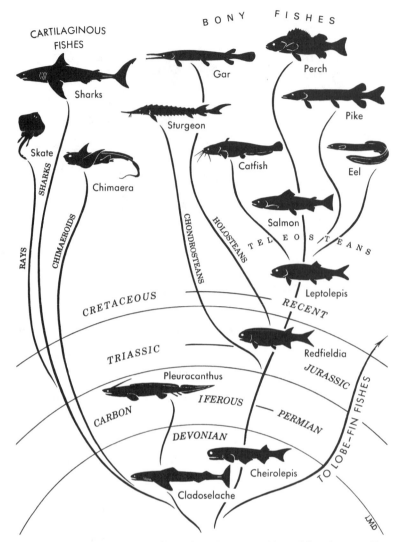

Figure 2-18   Major evolutionary lines of cartilaginous and bony fishes. (From Colbert, *Evolution of the Vertebrates,* John Wiley & Sons. Prepared by Lois M. Darling.)

type, the Teleostei, had evolved from the Holostei and were the dominant fish group in the late Mesozoic oceans. Teleosts exhibit considerable reduction in the skull bones over their predecessors, a reduction in the scales, and many other advancements. There are more than 20,000 species of living teleosts that are arranged in approximately 30 orders. Figure 2-19 shows the changes in structure from the early paleoniscoids to the modern teleosts. Note that in the teleosts the tail is symmetrical

dorsally and ventrally (homocercal) as opposed to the uptilted or heterocercal tail of the chondrosteans (also typical of sharks). In the most advanced teleosts the pelvic fins have actually migrated anterior to the pectorals.

More focal to the actual history of tetrapods, however, is the other major subclass of bony fishes, the Sarcopterygii or lobe-finned fishes (formerly termed the Choanichthyes: nostril + fishes). These fleshy finned fishes contain two orders, the Cros-

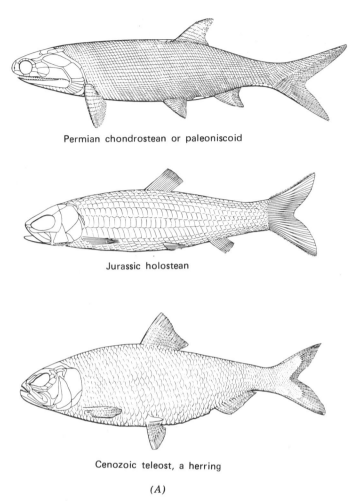

Permian chondrostean or paleoniscoid

Jurassic holostean

Cenozoic teleost, a herring

(A)

Figure 2-19 (*A*) Three major changes in the evolution of ray-finned fishes or actinopterygians. (*B*) The three living chondrosteans (*Polypterus* is a living representative of the primitive paleoniscoids with modified fin structure). (*C*) Living holostean fishes. (*D*) Some representatives of the advanced teleost fishes showing the great diversity in their morphology. (*A*, from Colbert, *Evolution of the Vertebrates,* John Wiley & Sons; *Polypterus,* from Romer, after Dean; and others, from Jordan and Evermann.)

sopterygii, from which the tetrapods descend, and the Dipnoi, or lungfishes, a group not ancestral to any other vertebrate group. Sarcopterygians were particularly abundant during the Devonian Period when the crossopterygians were the dominant predators of freshwaters. Both crossopterygians and lungfishes had lungs, internal nares, fleshy fins, and a special type of scale, the cosmoid scale.

The lungfishes or Dipnoi are represented by three surviving genera, one each on the continents of South America, Africa, and Australia. They are of interest for a number of reasons—notwithstanding the fact that they are "cousins" rather than the actual ancestors of the tetrapods. First, they are termed "primary" freshwater fishes, a term applied to fish that are unable to tolerate the least salinity in the water. Thus, their disjunct distribution seemed enigmatic until our recent understanding of drifting continents; they evolved before the splitup of

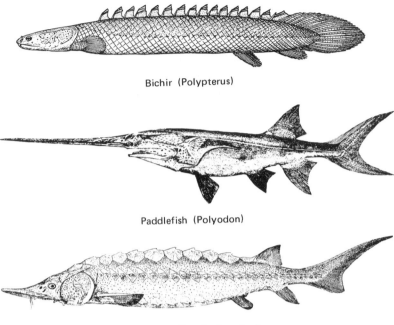

Bichir (Polypterus)

Paddlefish (Polyodon)

Sturgeon (Scaphyrhynchus)

(B)

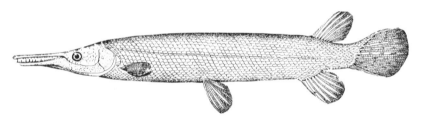

Gar Pike (Lepisosteus)

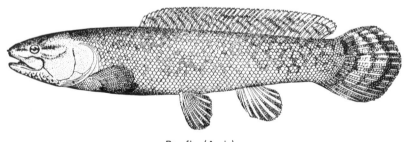

Bowfin (Amia)

(C)

Figure 2-19   (*Continued*)

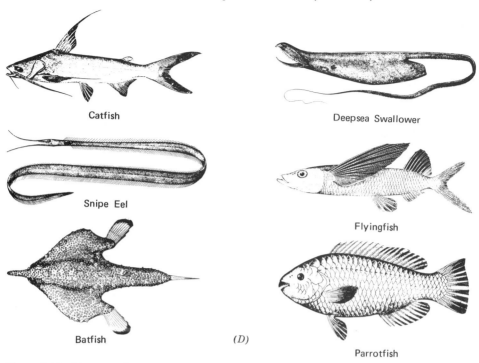

Catfish

Deepsea Swallower

Snipe Eel

Flyingfish

Batfish   (D)

Parrotfish

Figure 2-19 (*Continued*)

the continents and are now simply residents of their ancestral homes. Lungfishes are also of interest in that they live in regions of stagnant water, where seasonal drought is common. The Australian form can withstand these conditions by air breathing; the others actually bury themselves in the mud and undergo a period of dormancy (*aestivation*) until the pools again are filled with water. Figure 2-20 compares ancient and recent lungfishes.

Although lungfishes are of interest, it is the other order, the Crossopterygii (unfortunately there is no common name), that is of particular interest to us here since they represent the actual ancestors of the tetrapods. The typical crossopterygians, termed rhipidistians (suborder Rhipidistii), which were the dominant freshwater predaceous fishes of the Devonian, became rare in the following period, the Carboniferous, and were extinct by the end of the Paleozoic. However, a strange sideline of crossopterygian evolution was the suborder

Coelacanthini (Figure 2-21), a group that evolved in Mesozoic seas and became highly specialized in the head region. The last fossils are known from marine sediments of the Cretaceous; however, in 1939 a strange fish caught off the coast of South Africa proved to be a living coelacanth—it was named *Latimeria.* Since that time a number of specimens have been recovered from the depths of the Indian Ocean. Lungfish and coelacanths are highly specialized sidelines of the Sarcopterygii and are obviously not ancestral to the tetrapods; however, in the rhipidistians one sees an almost perfect transitory form between fish and amphibian. The skull is very similar in its bony anatomy to that of the first amphibians; there are all of the precursor bony elements in the fleshy fins for the tetrapod hand; internal nares and lungs are present; and even the teeth show strange labyrinthine infoldings of the dentine that characterize the first amphibians and give them the name *labyrinthodonts.* Figure 2-22 il-

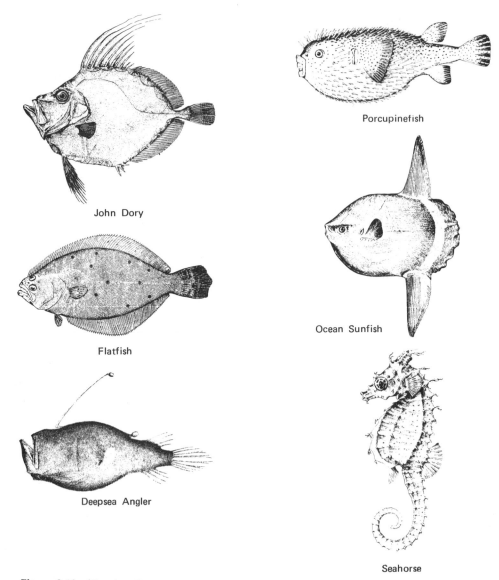

John Dory

Flatfish

Deepsea Angler

Porcupinefish

Ocean Sunfish

Seahorse

Figure 2-19   (*Continued*)

lustrates the major adaptive lines of the lobe-fins and the amphibians; the rhipidistians *Osteolepis* and *Eusthenopteron* lead almost directly to the first amphibian, the upper Devonian labyrinthodont *Ichthyostega*. Labyrinthodonts (Figure 2-23) flourished through the Carboniferous and Permian periods before extinction in the Triassic. The labyrinthodonts are significant not only because they show a linkage to the crossop-terygian fishes, but certain of them also suggest transition to the first reptiles.

A persisting puzzle, however, attends the three surviving orders of Amphibia (Figure 2-24), which are all placed in the subclass Lissamphibia. The Anura (frogs and toads), the Urodela (salamanders and newts), and the Apoda (limbless amphibians, or caecilians) are so different as to pose questions of their true nature and origin. One view

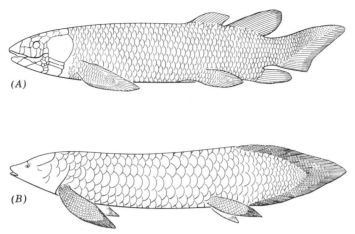

(A)

(B)

Figure 2-20 Ancient and recent lungfishes. (*A*) Devonian lungfish (*Dipterus*) from Europe. (*B*) Modern Australian lungfish (*Neoceratodus*). (From Colbert, *Evolution of the Vertebrates*, John Wiley & Sons.)

holds that the Anura stemmed from some labyrinthodont, whereas the Urodela and Apoda came from a group of small Carboniferous and Permian forms known as the Lepospondyli. Another view traces the ancestry of all three orders to the lepospondyls.

TRANSITION FROM AQUATIC TO AMPHIBIOUS LIFE

Many biologists have pondered the question of the transition from a fully aquatic existence to an amphibious one—Why should fish leave the water to assume an amphibious mode of life, half in water and half on land? Any of several explanations may pertain, but all may have been in part responsible for the change in habitats. It is thought that the end of the Devonian was a period of seasonal drought, and under these conditions ponds in which the rhipidistians dwelled would have been contracting and drying up. With their already developed fleshy limbs, lungs, and other "preadaptations," these fish would have found it no great task to meander over the

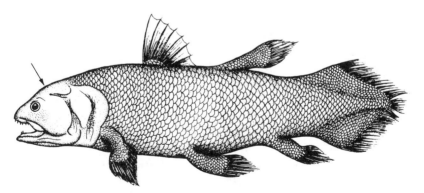

Figure 2-21 *Latimeria*, a living marine coelacanth, and the only living crossopterygian fish. The arrow shows the position of the spiracle. (From Feduccia, *Structure and Evolution of Vertebrates*, W.W. Norton & Co., after Smith.)

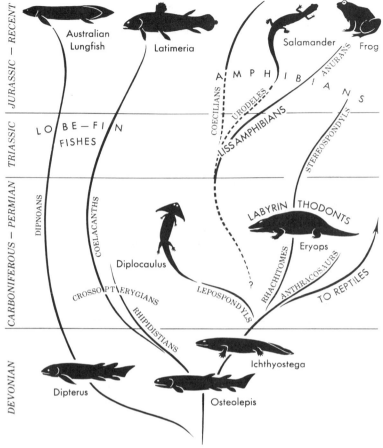

Figure 2-22 Evolution of the lobefin fishes and the amphibians. (From Colbert, *Evolution of the Vertebrates,* John Wiley & Sons. Prepared by Lois M. Darling.)

land temporarily in search of new sources of water. This "seasonal drought" hypothesis is perhaps the most widely endorsed. Yet, there are other feasible theories. A "predation" hypothesis points to the fact that the young of many modern freshwater teleosts remain in shallow water to avoid predation by the adults of their own and other species. One could imagine young rhipidistians not only remaining in shallow water but perhaps also wandering out onto mud flats to avoid predation. Once out, either by this or the previous hypothesis, they would have been faced with a new food supply in the form of arthropods that had already made the transition to land in the previous geologic period. And, even without seasonal

drought or predation by adults, the transition can perhaps most parsimoniously be explained by a "dispersal" or "overpopulation" hypothesis, by which, even under humid conditions, young individuals would find it advantageous to disperse to less crowded ponds to breed, thus increasing the probability of successful reproduction. A number of modern teleosts—perhaps the most well known is the walking catfish (*Clarias*) of southeastern Asia—commonly exit ponds under extremely humid conditions. Thus, the transition from an aquatic to an amphibious mode of existence is easily explained by any one or a combination of possible happenings.

The humid environment of the Carbon-

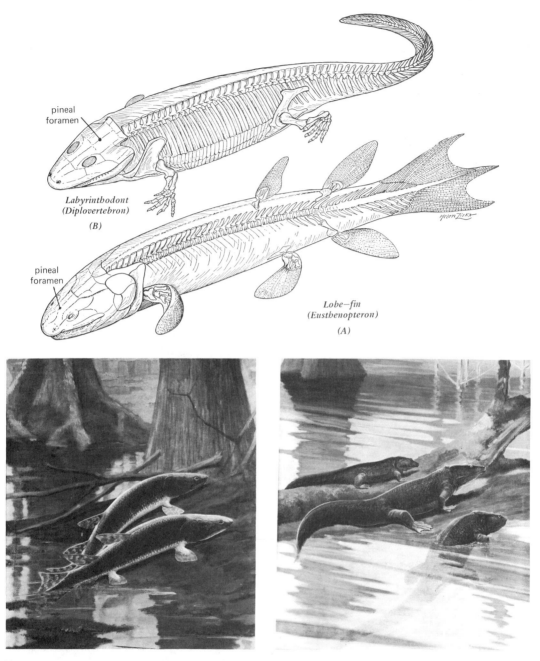

pineal
foramen

Labyrinthodont
(Diplovertebron)
(B)

pineal
foramen

Lobe—fin
(Eusthenopteron)
(A)

Figure 2-23 *Upper,* Devonian rhipidistian crossopterygian, *Eusthenopteron* (*A*), and a Carboniferous labyrinthodont amphibian, *Diplovertebron* (*B*). The amphibian exhibits many advancements over the crossopterygian, especially in the loss of the median fins, interlocking vertebrae, the evolution of actual limbs from paddles, the development of strong ribs, increase in size and development of the scapulocoracoid, loss of the posttemporal bar that connected the shoulder girdle with the skull, and great increase in the size of the pelvis. *Lower,* An artist's reconstruction of the rhipidistian *Eusthenopteron* and the labyrinthodont *Diplovertebron,* illustrating their probable habitats. (*Upper,* from Gregory, *Evolution Emerging,* Courtesy Macmillan Co., New York; *lower,* Courtesy of the American Museum of Natural History.)

Figure 2-24 Modern Amphibia: (*A*) bullfrog (Anura); (*B*) red salamander (Urodela); (*C*) *Ichthyophis*, a burrowing, limbless amphibian (Apoda). (Courtesy American Museum of Natural History.)

iferous Period, the period that produced the world's great deposits of coal from the plant debris of those lush times, was highly favorable to the amphibians. But the coming of the Permian brought changes in the climate; inland waters receded, swamps dried up, high mountain ranges arose, and the lush vegetation gave way to hardy plants capable of surviving great extremes of temperature. In these new circumstances the amphibian line began to degenerate. Some survived by exploiting the limited resources of water; others succeeded in bypassing the water requirements for embryonic development and began to deposit eggs that could incubate on dry land, an innovation ranking in importance with the appearance of jaws.

TRANSITION FROM AQUATIC TO TERRESTRIAL DEVELOPMENT

While Carboniferous amphibians were attaining a higher degree of perfection in adaptations for living on land, there remained one great impasse for the full transition to land—the development of a land egg. Amphibians, for the most part, are bound to the water, at least during the breeding season. Their eggs must be deposited in the water, or damp places on land, and they

have an extended larval stage. It was thus in the line leading to the first reptiles, the anthracosaurian labyrinthodonts, that the invention of the land egg developed and it is at that point that the class Reptilia begins and finds its definition. The "invention" of the self-contained egg, complete with its own food and water supply and sealed in a waterproof covering, opened up the land for evolutionary exploitation. The transition was a gradual one indeed, for it is often difficult to decide whether some fossilized skeletons are those of advanced amphibians or primitive reptiles. But those clearly judged to be reptiles constitute a group of so-called "stem reptiles," the order Cotylosauria of the subclass Anapsida. *Seymouria* (Figure 2-25) is a small tetrapod from the Permian of Texas that exhibits a strange combination of reptilian and am-

Figure 2-25   Painting of *Seymouria,* a structural intermediate between amphibians and reptiles. (Courtesy American Museum of Natural History.)

phibian characteristics. This form is an almost perfect structural intermediate between the amphibians and reptiles, yet it occurs far too late to have been an actual ancestor; the first reptiles occur in the Carboniferous. Thus, *Seymouria* is a good example of a "temporal relict" or a persistent structural ancestor, and illustrates what the actual ancestor must have resembled. Was *Seymouria* an anthracosaurian amphibian or a stem reptile? The ultimate answer, of course, would depend on whether *Seymouria* deposited an amniotic egg on land or returned to the water to lay a typical amphibian egg. There is no direct paleontological evidence to support either view; *Seymouria,* like the rhipidistians *Osteolepis* and *Eusthenopteron,* confirms the Darwinian expectation of intermediate forms linking major groups of organisms.

EVOLUTION OF REPTILES

The Permian was a period of profound changes in the earth's crust. There was major mountain building toward the end of the period, glaciation in the Southern Hemisphere, and a general elevation in the continents. These changes led to the demise of many ancient groups of amphibians; the reptiles took over as the dominant group of vertebrates and, from their cotylosaurian beginnings, rapidly differentiated into many branches.

Classification of the diversity of reptiles has posed a continuing problem, but there is an empirical scheme that has considerable usefulness. This is a system based on the number and kinds of openings in the temporal region of the skull (Figure 2-26). Like that of their labyrinthodont ancestors, the skull of the cotylosaurs was a solid structure, pierced only by the openings for the nostrils, eyes, and pineal gland. As the reptiles evolved, additional openings appeared in the temporal area behind the eyes. In some there is a single opening high up on each side of the roof behind the eyes; in

others there is a single opening lower down on each side; in still others there are two openings on each side. The following is a listing of the subclasses of reptiles based on this system.

Subclass Anapsida
(*Figure 2-26*)
No temporal openings in the skull.

Subclass Synapsida
(*Figure 2-26*)
A single lateral temporal opening on each side, bounded above by the postorbital and squamosal bones.

Subclass Euryapsida
(*Figure 2-26*)
A single dorsolateral opening on each side, bounded below by the squamosal and postorbital bones.

Subclasses Lepidosauria and Archosauria (Diapsida)
(*Figure 2-26*)
Two openings on each side, separated by the squamosal and postorbital bones.

We shall not undertake to trace the full succession of reptilian types. It is sufficient for our purposes to suggest only the more notable lines of radiation (Figure 2-27). The anapsid line led to the Chelonia (turtles), who soon reached their present-day form and came down through the ages virtually unchanged. Other branches led to the swimming reptiles: the flippered euryapsidan plesiosaurs (Figure 2-28) and the fishlike ichthyosaurs (Figure 2-29), convergent with the modern porpoises of the class Mammalia. There were many other minor groups of marine reptiles during the Mesozoic.

The first subclass of the diapsid reptiles is the Lepidosauria (Figures 2-26 and 2-30); it is first represented by an ancestral group of lizardlike forms, the order Eosuchia, all of which are extinct. Arising in the early Mesozoic from eosuchian stock is the order Rhynchocephalia, another group of lizardlike forms that, though never prominent, is today survived by the New Zealand tuatara

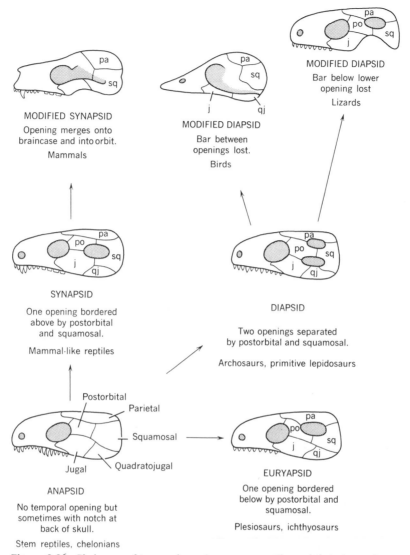

Figure 2-26 Phylogeny of temporal openings among reptiles and their descendants. (From Hildebrand, *Analysis of Vertebrate Structure,* John Wiley & Sons.)

(*Sphenodon*). The tuatara has a well-developed pineal eye on top of its head in the middorsal line. In addition to the Rhynchocephalia there arose still another group from eosuchian stock, the order Squamata, which includes the snakes and lizards. The snakes are placed in a suborder Serpentes; snakes are derived from lizard ancestors of the suborder Lacertilia. In addition to giving rise to the snakes, the ancient lizards (very much like the modern Old World varanid lizards) gave rise to a very successful group of late Cretaceous, giant marine lizards known as the mosasaurs.

RULING REPTILES (ARCHOSAURS)

The other major line of the diapsid type of skull led to the basal stock of the archosaurs, the order Thecodontia, (Figure 2-31) so

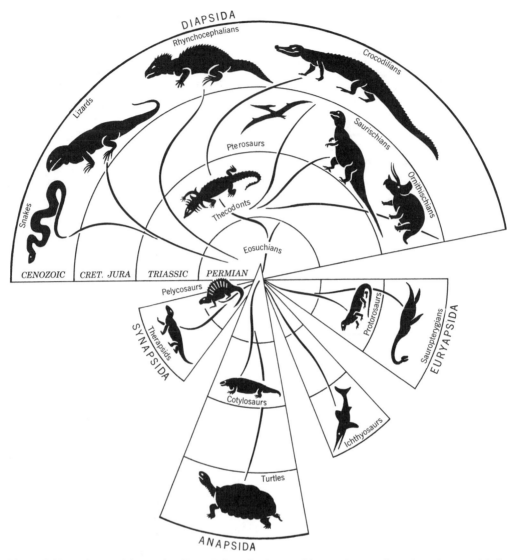

Figure 2-27 Radiation of the reptiles. (From Colbert, *Evolution of the Vertebrates,* John Wiley & Sons. Modified from a drawing by Lois M. Darling.)

named because the teeth are set in sockets. The thecodonts were relatively small, agile reptiles, many of which walked on their hind legs (bipedal) and used their arms for grasping. Although their existence was confined to the Triassic Period, they had the distinction of being forerunners of the great array of reptiles that followed in succeeding periods. These ruling reptiles, known as the archosaurs, give the Mesozoic Era the vernacular name "the age of reptiles." Aside

from giving rise to the archosaurs, thecodonts also had several side lineages, some of which came to resemble the crocodiles.

One of the most interesting groups derived from the thecodonts are the pterosaurs of the order Pterosauria (Figure 2-32). These flying reptiles, quite unrelated to the birds, had a wing or patagium of skin that extended across the hand in batlike fashion. Much debate has centered around their potential for flight, but most authorities be-

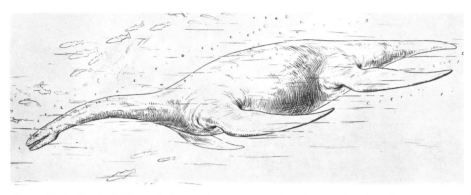

Figure 2-28 *Cryptocleidus,* a plesiosauroid euryapsid reptile. (Courtesy American Museum of Natural History.)

lieve that they were primarily gliders, perhaps capable of some propelled flight. Pterosaurs ranged in size from a small bird to a large airplane. *Pteranodon,* a late Cretaceous form, had a wing spread of approximately 7 meters; but a recently discovered Texas pterosaur may have had a wing spread perhaps as great as 16 meters. Pterodactyls were contemporaneous with birds but became extinct with the close of the Cretaceous Period.

The crocodiles and alligators, order

Figure 2-29 Painting of *Ichthyosaurus.* (Courtesy American Museum of Natural History.)

Figure 2-30 Living survivors of the various lines of diapsid evolution. (*A*) The rhynchocephalian *Sphenodon*. (*B*) Hognose snake and (*C*) collared lizard, squamate reptiles. (*D*) A crocodilian. (Courtesy American Museum of Natural History.)

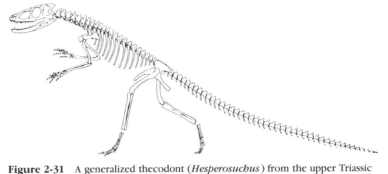

Figure 2-31   A generalized thecodont (*Hesperosuchus*) from the upper Triassic of Arizona. This reptile was slightly over a meter in length. (From Colbert, *Evolution of the Vertebrates*, John Wiley & Sons.)

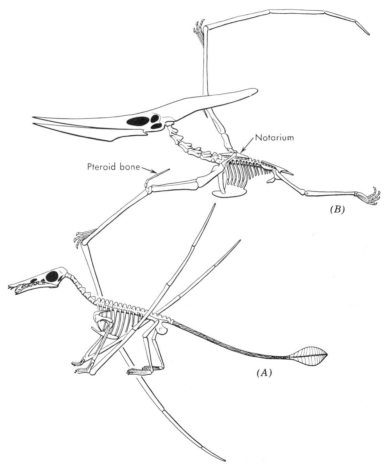

Figure 2-32   Flying reptiles or pterosaurs. (*A*) *Ramphorhynchus* of Jurassic age, with a wing spread of slightly less than a meter. (*B*) *Pteranodon* of Cretaceous age, with a wing spread of about 7 m. The pteroid bone in the wrist of *Pteranodon* was a support for the wing membrane or patagium; the notarium (fused thoracic vertebrae) served to attach the upper edge of the scapula to the backbone, thus making a strong base for the great wing. Note the loss of the tail by Cretaceous time and the greatly increased keel on the sternum that served for the origin of the flight muscles. (From Colbert, *Evolution of the Vertebrates*, John Wiley & Sons.)

Crocodilia (Figure 2-30), were also an off-shoot of the thecodonts, and paralleled some of the bizarre thecodont groups at least superficially. They were a very successful group, and by the end of the Cretaceous some forms had attained a length of approximately 17 meters and no doubt fed on small dinosaurs. The crocodiles and alligators are the only archosaurs to survive into the recent fauna.

The most dramatic of the thecodont derivatives, and the forms that most typically characterize the "age of reptiles" are various groups of the dinosaurs (terrible lizards); these forms, some no larger than a chicken, first appeared in the Triassic, and by Cretaceous time had evolved into bizarre and often highly adaptive monsters of great size. The dinosaurs bifurcated early into two major lineages, the orders Saurischia (meaning "reptile hips") and Ornithischia (meaning "bird hips," because of superficial resemblance to that of birds). (See Figures 2-33 and 2-34.)

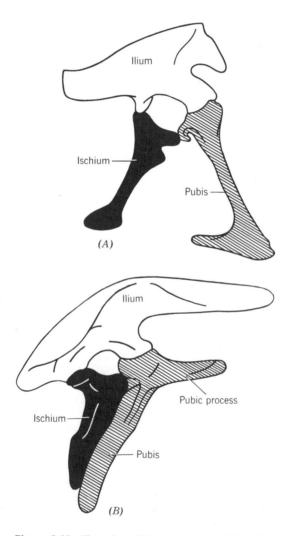

Figure 2-33 The pelvis of the two orders of dinosaurs. (*A*) Saurischian pelvis, with pubis directed forward. (*B*) Ornithischian pelvis, with pubis parallel to ischium and pubic process directed forward. (From Colbert, *Evolution of the Vertebrates,* John Wiley & Sons.)

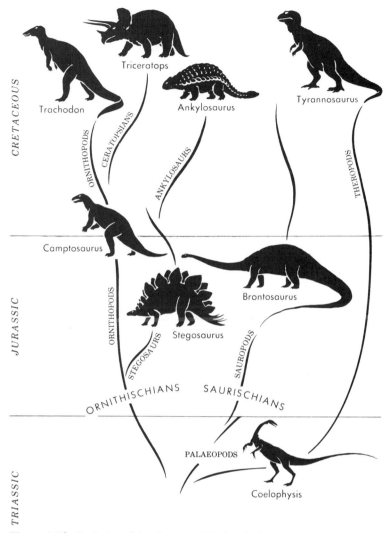

Figure 2-34 Evolution of the dinosaurs. The line leading to the Theropoda split into the large carnosaurs and the small coelurosaurs. (From Colbert, *Evolution of the Vertebrates,* John Wiley & Sons. Prepared by Lois M. Darling.)

Most of the ornithischians returned to a four-footed pose, although some remained bipedal. They were all plant eaters and they showed a much greater variety of form than the saurischians. Some of nature's most bizarre designs appeared in the ornithischian dynasty. There was, for instance, *Stegosaurus,* a ponderous beast 6.5 meters in length, weighing at least 10 tons, having powerful legs, and bearing a parapet of heavy, triangular plates down its back. *Stegosaurus* had a pelvic enlargement of the spinal cord ap-

proximately 20 times the size of its brain that no doubt functioned for motor activity of the enormous hind limbs. Others, like the ankylosaurs, were veritable walking fortresses, armed with an assortment of plates and spikes. The culmination of defensive armament was exhibited by a family of horned ornithischians. *Triceratops,* for example, was 6.5 meters long, stood 2.5 meters high and, as its name implies, carried three horns projecting from a flaring frill of bone that encased the skull.

Another common group of ornithischians were the duckbill dinosaurs such as *Trachodon*. These are among the most commonly preserved of all the dinosaurs because of their aquatic habits.

But the saurischians were the rulers of their day: some of their herbivores were the largest land animals that ever lived; their carnivores were truly terrors. Take *Apatosaurus* (*Brontosaurus*), a huge herbivore measuring nearly 23 meters and weighing 30 tons; or *Allosaurus,* a monster carnivore more than 10 meters long, armed with rows of knifelike teeth and prehensile claws; or the tyrant of them all, *Tyrannosaurus rex,* a towering machine of destruction at least 16 meters long. However, early in the evolution of the saurischians there was not only a basic bifurcation between the carnivorous forms (Theropods) and the large herbivores (Sauropods), but also a split in the carnivorous line into the large forms, such as *Tyrannosaurus,* and a group of small carnivores known as the coelurosaurs (Figure 2-35). They diversified greatly and one form became superficially similar to the ostrich through convergent evolution; it was called *Struthiomimus* (ostrich mimic). Small Jurassic coelurosaurs such as the chicken-sized *Comsognathus* are thought by some to be near the ancestry of birds. It is particularly among the coelurosaurs that some paleontologists have found evidence of larger brain size and increased activity.

The extinction of the dinosaurs of the land, the swimming reptiles of the sea, and the flying reptiles of the air at the end of the Mesozoic Era poses an enigma of evolution. Why should creatures so powerful and apparently so well adapted to their ecological niches vanish almost simultaneously, leaving only relatively inconsequential descendants—turtles, snakes, crocodiles, and small lizards—to survive to the present? Several answers have been proposed: climatic change, disease, cosmic radiation, the rise of flowering plants to which the herbivores could not adjust, and so on. However, recent discoveries of large concentrations of iridium (a metal of the platinum group) in sediments worldwide at the Cretaceous/Tertiary boundary seem explainable only by the impact of an extraterrestrial body such as a meteorite, approximately 10 kilometers in diameter. Such an impact would have created a crater some 100 kilometers in diameter and would have thrown debris up to 60 times its mass into the upper atmosphere, thus creating a dense cloud for several years and blocking sunlight. Such an event would have had devastating effects, and the extinctions that

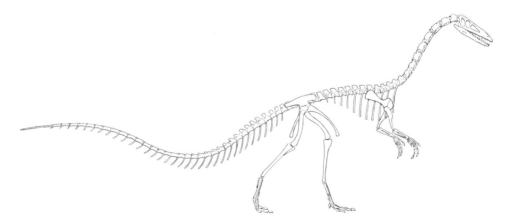

Figure 2-35 Skeleton of the North American upper Triassic theropod *Coelophysis,* illustrating the structure in an early and generally primitive coelurosaurian dinosaur. It is about 3 m in length. (From Colbert, *Evolution of the Vertebrates,* John Wiley & Sons.)

characterized the close of the age of reptiles would have surely occurred.

Nevertheless, the record of terrestrial animals indicates that the basic cause of this massive extinction was in the hands of geologic events. We do know that the late Cretaceous was a time of extensive mountain building, the American Rockies were among the ranges formed. Thus, the low lying landscape suddenly emerged and much of the formerly swampy and lagoonal regions where the dinosaurs ranged was drained and disappeared. Inland seaways suddenly disappeared and with them the habitat for the myriad of marine reptiles. In addition, there was concomitant drastic climatic change, and new plants emerged as dominant types. As the herbivorous dinosaurs, so highly specialized, began to dwindle, so also did the carnivores that were dependent on them. One can imagine that with 100 million years of warm, moderate climates into which the dinosaurs could adaptively radiate and specialize, only minor changes would have been terribly disruptive and extinction would occur. Once the base of the ecologic pyramid was perturbed, then the entire structure would soon collapse. Perhaps the extraterrestrial event expedited the inevitable.

With the great reptiles gone, the mammals, which emerged in the late Triassic and lived obscurely in the shadow of the dinosaurs for the next 100 million years, began to expand rapidly. Birds had begun their evolution in the Jurassic, but remained in obscurity until late Cretaceous and Tertiary time.

BIRDS (CLASS AVES)

Birds have frequently been termed "glorified reptiles," and although there is some truth in this cliché, birds are in reality finely tuned metabolic machines, adapted in almost every feature for the extremely high physiological demands of flight. Because all birds either fly or are derived from flying

ancestors, they constitute a group of vertebrates far more uniform structurally than any other. In fact, so highly evolved are modern birds that before intermediate fossil stages were discovered, the apparent "gap" between living birds and living reptiles was used as evidence against the theory of evolution. Later, however, perfectly intermediate fossils were found.

Because of their extremely light bones (often hollow or pneumatized) avian fossils are not often preserved, and unlike the glorious fossil record of reptiles that has permitted us to understand clearly their evolutionary meanderings, the history of birds is pieced together, where possible, by a few lucky acts of preservation. Most important has been the discovery of six specimens of the oldest known bird, *Archaeopteryx lithographica* (Figure 2-36), from fine-grained limestone sediments that were laid down in late Jurassic time in Bavarian lagoons. The specimens of *Archaeopteryx* ("ancient wing") were buried in calcareous muds that later formed lithographic limestone. These limestones were once mined for the purpose of making lithographs; now they are mined for tile. Indeed, had it not been for the extremely fine preservation of these fossils, and hence the preservation of feather impressions, *Archaeopteryx* would surely have been classified as a small saurischian (coelurosaurian) dinosaur. The fifth specimen, initially uncovered in 1951, was for 20 years incorrectly identified as a small coelurosaurian dinosaur, *Compsognathus!* However, a number of paleontologists still believe that birds were derived earlier in time, from thecodonts. It is clear that the first birds were indeed "glorified reptiles," little removed from their reptilian ancestry. The crow-sized *Archaeopteryx* possessed a slightly modified diapsid skull with an altered temporal region and a long undifferentiated tooth row on both upper and lower jaws. In addition, and unlike modern birds, there were three well-developed "fingers" on each wing with strongly decurved claws,

Figure 2-36   Fossilized skeleton of the first known bird, *Archaeopteryx*. (From Colbert, *Evolution of the Vertebrates*, drawn from the Berlin specimen, John Wiley & Sons.)

probably to aid in climbing up trees. There was a long reptilian tail with a pair of tail features attached to each vertebra; in modern birds most of the tail vertebrae are lost and the remaining fused into a single *pygostyle* to which all of the tail feathers attach.

ARBOREAL VERSUS CURSORIAL ORIGIN OF FLIGHT

Two major theories have been invoked to explain the origin of flight and endothermy

in birds. Briefly, the cursorial (running) theory pictures the ancestral bird or protoavis as a running bipedal reptile that attains endothermy; feathers evolve from reptilian scales for insulation. Once present, feathers could be modified to form flight feathers, perhaps to aid in gliding across the landscape to escape predation. In other words, endothermy would have to have evolved in coelurosaurian dinosaurs for some reason, and feathers would have been a preadaptation for flight; that is, feathers would have first evolved for a function other than flight. The arboreal theory pictures the protoavis as primarily a tree-dwelling form that developed feathers first from reptilian scales on the posterior aspect of the wings. Any feathers in that region would be beneficial in developing an airfoil in jumping, and later parachuting, gliding, and finally active flight. Most paleontologists favor the arboreal theory of bird origin. The protoavis was an active animal, both on the ground and in the trees, like so many modern lizards. It could have jumped and glided from limb to limb, been active on the ground, and then used its claws to aid in climbing back up into the trees. Of course, the question of the origin of feathers remains enigmatic; it must be explained in terms of intermediate, microevolutionary events, each of which would be adaptive at a given moment. All we really have are the six specimens of *Archaeopteryx* and lots of speculation. What of the feathers of *Achaeopteryx?*

The flight feathers, primaries and secondaries, are essentially identical to those of modern birds; in other words, they have remained unchanged for approximately 140 million years. This must certainly indicate an aerodynamic function for the wing. It appears as though *Archaeopteryx* could at least glide, if not engage in some active flight. *Archaeopteryx* alone constitutes the avian subclass *Archaeornithes.*

NEORNITHES

Avian fossils recovered from sediments only slightly younger than *Archaeopteryx,* in early Cretaceous time, exhibit most of the major flight adaptations of modern birds. By the late Cretaceous Period a number of highly adapted diving birds existed that closely (although convergently) resembled modern loons; they are known as *Hesperornis* and *Baptornis.* These forms possessed reptilian teeth like *Archaeopteryx,* as did a gull-like form, *Ichthyornis;* however, in almost all other features they were modern birds. These three toothed forms are placed in their own superorder, *Odontognathae.* No other modern birds possess

teeth, and all are most properly treated as a single superorder, Neognathae (Figures 2-37 and 2-38).

Some authors recognize still another subdivision of birds termed the superorder Palaeognathae. This group, often referred to as the *ratites* (Latin *ratis,* "raft," referring to the flat unkeeled sternum) includes the ostrich of Africa, the cassowary and emu of Australia, the kiwi and extinct moas of New Zealand, and the rheas of South America. It should be noted here that the vast majority of birds have a keeled sternum for the attachment of flight muscles and are termed *carinates* (Latin *carina,* "keel of a ship").

More than half of the nearly 9000 species of living birds belong to the single order of songbirds, Passeriformes. These are the most highly evolved of the class and have exhibited massive recent transformation at the species level. The remainder of modern birds come in a great variety of styles, ranging from the wading and diving forms to the hawks, owls, doves, and so on. The penguins of the Southern Hemisphere, like the auks of the Northern Hemisphere, are wing-propelled divers, and still have a keeled sternum for the origin of flight muscles; they literally fly through the water. Both groups descended from different flying oceanic birds and are therefore convergent.

MAMMALS (CLASS MAMMALIA)

The mention of mammals requires us to backtrack to the one still unconsidered class of reptiles, the Synapsida, a group that diverged from the other reptiles almost at the base of their evolutionary advent. Thus, mammals and their antecedents are rather remotely related to any of the existing orders of reptiles. The group leading to mammals that goes farthest back in time is the order Pelycosauria, which appeared in the late Carboniferous and radiated in the early Permian (Figure 2-39). In many structural features they resembled their immediate ancestors, the cotylosaurs, but in others,

SONG BIRDS
Robin

FOREST BIRDS
Owl

OCEANIC BIRDS
Albatross

UPLAND BIRDS
Pheasant

ADAPTATIONS TO VARIED HABITATS

WADING AND SHORE BIRDS
Plover

FLIGHTLESS GROUND BIRDS
Ostrich

Archaeopteryx

AQUATIC BIRDS
Penguin

Figure 2-37 Some major adaptive lines among the birds. (From Colbert, *Evolution of the Vertebrates,* John Wiley & Sons. Prepared by Lois M. Darling.)

such as the temporal region of the skull, they show a clear line leading toward the mammalian condition. Pelycosaurs came in a variety of morphological forms, both herbivores and carnivores. Some had enormous extensions of the dorsal vertebral spines that give the appearance of a sail, and are thought to have functioned in temperature regulation. In some one sees the beginning of the differentiation of the tooth row away from the simple homogeneous dentition of their forebears. Pelycosaurs had the usual sprawled out posture of typical reptiles, but in one line leading away from them in the Permian the posture was more erect, the body more elevated, and the length of the stride was increased. Thus, in late Permian and early Triassic time the order Therapsida

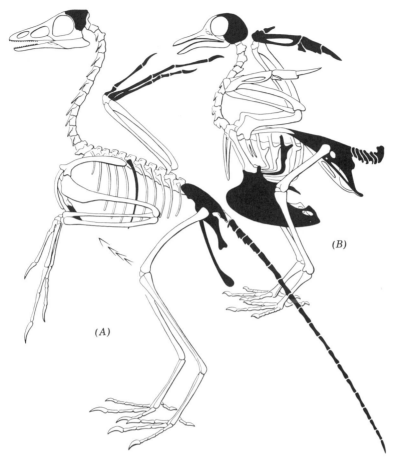

Figure 2-38 Comparison of adaptations in the skeletons of the earliest known bird, *Archaeopteryx* (*A*), and a modern pigeon (*B*). Comparable regions of the skeleton (braincase, hand, sternum, rib, pelvis, tail) are shaded black. In modern birds the tail is reduced to several vertebrae and a fusion of several into the terminal pygostyle, the braincase is expanded and the bones of the head fused, wing bones are coalesced, the sacrum is fused into a single solid structure, the ribs are rigidly bound by uncinate processes, and the large sternum is keeled for the attachment of the major flight muscles. All are adaptations associated with flight. (From Colbert, *Evolution of the Vertebrates,* modified after Heilmann, John Wiley & Sons.)

or mammallike reptiles (Figure 2-40) became a successful group, exhibiting both herbivorous and carnivorous adaptive radiations. However, it was primarily in the carnivores that one sees the progression toward the mammalian grade of organization. There were many diverse lines of mammallike reptiles, some of which very closely approached the mammalian structural condition, but apparently only one actually gave rise to the mammals.

Therapsids enjoyed a period of great success during the early Triassic, but by the end of that period other reptile groups had become dominant, particularly the dinosaurs, and the therapsids could not adequately compete and vanished by the close of the Triassic. Before their departure they evolved rather small forms that left behind as their legacy the mammals. These small Mesozoic mammals had evolved at about the same time as the dinosaurs, but like the

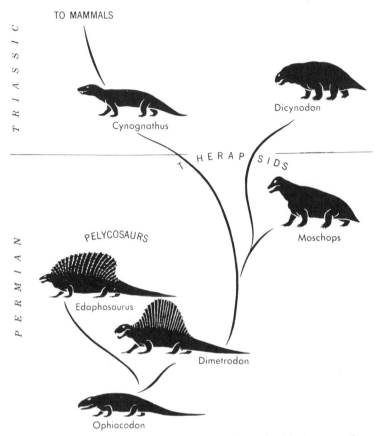

Figure 2-39 Evolution of synapsid or mammallike reptiles. The line going from *Cynognathus* (cynodonts) leads to the mammals. (From Colbert, *Evolution of the Vertebrates,* John Wiley & Sons. Prepared by Lois M. Darling.)

therapsids found themselves unable to actively compete with the giants of the Mesozoic. Obviously, during periods of warm, moderate climates, like those of the Jurassic and the Cretaceous, there is no great advantage during the day in being an endotherm. The mammals remained subdominant throughout the Mesozoic, and it was not until the close of the Cretaceous and the extinction of the dinosaurs that they emerged to take over as the dominant type of tetrapod; the Cenozoic Era is thus named "the age of mammals." It would appear that mammals probably occupied the nocturnal habitus throughout the Mesozoic, and perhaps endothermy would have been advantageous in coping with the fluctuating temperatures of the Mesozoic nights. It should be remembered that the mammallike reptiles had evolved earlier, during the Permian, when major changes in the earth's surface were occurring and rather dramatic climatic change occurred; during that period endothermy would have also been advantageous.

WHAT IS A MAMMAL?

When comparing modern mammals with modern reptiles, one encounters no problem in immediately differentiating the groups. Mammals are endotherms insulated by hair; have a muscular diaphragm for

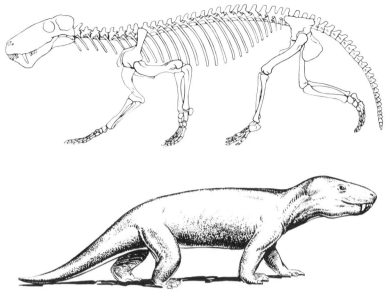

Figure 2-40 Therapsids or mammallike reptiles. *Top,* Skeleton of *Lycaenops,* illustrating mammallike characters and pose in an advanced therapsid. *Bottom,* Reconstruction of the dog-sized *Cynognathus,* a cynodont therapsid on the main line to the mammals. (From Colbert, *Evolution of the Vertebrates,* John Wiley & Sons. *Bottom,* prepared by Lois M. Darling.)

breathing, a four-chambered heart, a single aorta, and two occipital condyles; and nourish the young by lactation. In addition, in all but the egg-laying mammals there is a functional placenta that serves to nourish the fetus and remove waste products from it. However, when one examines the fossil record carefully one finds many therapsids (some not on the direct line to mammals) that are perfect intermediates between reptiles and mammals, thus rendering these simple definitions based only on living forms invalid. For example, it is likely that many of the advanced therapsids were endothermic; in some there is evidence for a diaphragm, and some form of insulation could have existed. Their habits may have been quite similar because the locomotor apparatus as well as the differentiation of teeth was quite similar to that of primitive mammals. So, what are the major differences between therapsids and mammals? Most, of course, must necessarily be associated with the skeleton. In mammals there is a single

bone in the lower jaw, the dentary; three ear ossicles in the middle ear, the malleus, incus, and stapes; and a jaw joint between the dentary and squamosal bones. In reptiles, there is more than one bone in the lower jaw; a single ear ossicle, the stapes or columella (as in birds and amphibians); and a jaw joint between the articular (lower) and quadrate (upper) bones. The reasons for these differences are discussed in Chapter 11. It should, however, be pointed out that even with these salient features separating the two groups, there are fossil therapsids so intermediate that the definitions do not hold very well. Nonetheless, the success of mammals resides in their intelligence, which owes to their extensive development of the cerebral cortex; what they lacked in brawn they made up for in brain. Activity, implemented by an even body temperature and directed by intelligence, was the key to their survival and proliferation. Once the dinosaurs passed from the scene the mammals took over as the dominant group, and are today represented by three major groups: prototherians, metatherians, and eutherians.

PROTOTHERIA

The sole living prototherians are contained in a single order *Monotremata* that contains as living species the duckbill platypus and spiny anteaters of the Australian and New Guinea regions (Figure 2-41). Monotremes are quite close to their therapsid ancestry and they lay large reptilian eggs that are incubated in a nest or in a temporary pouch in the abdomen. There are no nipples for the mammary glands (modified sweat glands), and the milk simply seeps out onto tufts of hair. In addition, the testes are abdominal, there is a cloaca that receives both excretory and reproductive products, and the middle ear bones, the malleus and incus, are rather large and re-

semble the reptilian jaw bones from which they evolved. Indeed, in the monotremes we see living fossils, mammals derived very early from a different line than that which gave rise to the marsupials and placentals.

THERIA

The subclass Theria appears to have had an origin different from that of the Prototheria, and is distinguished from the latter by the fact that all species bear young that are nourished from well-formed nipples from the mammary glands (Figures 2-42 and 2-43). The infraclass Metatheria, distinguished by the presence of an abdominal pouch or "marsupium," contains only a single order of living forms, the order *Marsupialia.* Marsupials are known in the fossil record as far back as mid-Cretaceous time and were surely cosmopolitan in distribution; today their stronghold is the Australian

Platypus

Echidna

Figure 2-41 The two living monotremes or egg-laying mammals. *Top,* Duckbill platypus of Australia. *Bottom,* Spiny anteater or echidna of Australia and New Guinea. (From Colbert, *Evolution of the Vertebrates,* John Wiley & Sons. Prepared by Lois M. Darling.)

Figure 2-42 Adaptive radiation among the recent marsupials of the Australian region. (From Colbert, *Evolution of the Vertebrates,* John Wiley & Sons. Prepared by Lois M. Darling.)

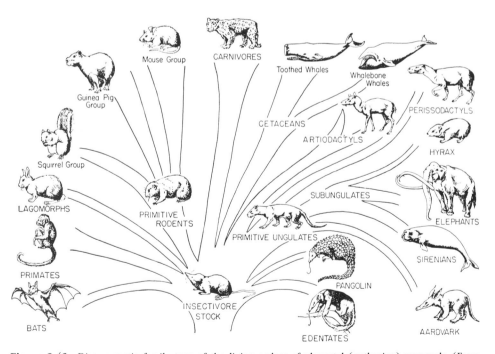

Figure 2-43 Diagrammatic family tree of the living orders of placental (eutherian) mammals. (From Romer and Parsons, *The Vertebrate Body,* 5th ed., courtesy W.B. Saunders, Philadelphia.)

region, with a few genera in the New World. In the Tertiary, however, there was an adaptive radiation of marsupials in South America somewhat equivalent to that in Australia. After fertilization the fetus comes into contact with the uterine living of the mother to form a rather poorly developed placental contact. The newborn is very poorly developed and emerges from the female to find its way to the pouch or marsupium and attaches firmly to one of the nipples therein. In the safety of the pouch, it undergoes a second period of maturation, and even after disavowing the nipple attachment, it may have an affinity for the refuge of the pouch for some time. The myriad of marsupials, both herbivorous and carnivorous, appear to have evolved primarily in isolation in Australia and South America, and illustrate remarkable examples of convergent evolution with the placentals.

The infraclass Eutheria comprises the "true" placental mammals; they underwent an enormous adaptive radiation during the Cenozoic Era, and today are represented by forms that constitute about 15 or more orders (Figure 2-43). With the exception of the insectivores and primates that appear in late Cretaceous, most of the remaining orders have representatives back into Eocene or Paleocene time. Unlike metatherians, the eutherians give birth to relatively well-developed young, there is a very well-developed placenta and a high degree of parental care.

The great diversification of the placental lineages, once begun, took the form of a tremendous adaptive radiation that left little behind in the way of specific branching subdivisions—instead there was a very complicated profussion of many small complicated phylogenetic lines leading in all directions. However, we do know that the mammals had remained obscure during the reign of the dinosaurs, and were all small forms that no doubt fed on a variety of grubs and insects. These ancestral placentals were probably in structure and habits not so unlike the members of the order Insectivora

(see Figure 2-43) including the living shrews, hedgehogs, moles, and so on, which in their shy, nocturnal habits still give us a glimpse, albeit hazy, of placental ancestry. Perhaps the living mammals least modified structurally from the ancestral condition are the tree shrews (*Tupaia*), now confined to the Oriental Region. These forms have often been thought of as being close to the actual ancestry of primates. Another insectivore often placed in a separate order is the flying lemur (*Galeopithecus*) of the East Indies. The flying lemur is really a glider; however, a group of true flying mammals directly derived from insectivore stock is the order Chiroptera or bats.

Aside from primates, among the more interesting living orders is the Edentata, including forms with simple teeth such as the sloths, armadilos, and anteaters. An enormous order is the Rodentia, which includes a vast myriad of squirrels, rats, mice, guinea pigs, and so forth, and their allies the Lagomorpha, the hares and rabbits. The term ungulate is applied to living placentals that are herbivorous and possess hoofs. However, a number of primitive lines of subungulates possessed claws. One line appears to have given rise to the living African order Hyracoidea, the hyrax (cony of the Bible) that appears in many features to be the living form most closely resembling the order Proboscidea, the elephants. Also thought to be within this general grouping is the order Sirenia, the sea cows and manatees, that have returned to the aquatic mode of life. The most spectacular of the aquatic mammals are in the order Cetacea, the whales and porpoises (or dolphins). Whales come in two varieties: the toothed forms that are carnivorous and the enormous plankton filter feeders that strain particulate matter from the sea, utilizing their epidermally derived strainers known as baleen. All modern "carnivores" are placed in a single order, the Carnivora, and are thought to be derived from a primitive creodont stock. These forms include the dogs, cats, racoons, bears, and such marine forms as the seals and their allies. The true ungulates, primitively derived from clawed ungulates that lost the claws as running adaptations, include two orders, the Perissodactyla or odd-toed hoofed mammals, such as the horses, tapirs, and rhinoceroses, and the order Artiodactyla, or even-toed hoofed mammals, such as the cattle, sheep, pigs, hippos, camels, deer, giraffes, and antelopes.

Archaic mammals were predominant throughout the Paleocene Epoch, and by Eocene time many of the modern orders and suborders of mammals had appeared. However, it seems that birds got a slight jump on the mammals as successful carnivores and evolved some enormous types such as *Diatryma* (Figure 2-44) which had a skull similar in size to that of a horse, but

Figure 2-44 The giant, flightless "terror cranes" (*Diatryma*) became the bipedal carnivores after the reign of the dinosaurs and before the mammals developed their sophisticated carnivorous lines. The "terror cranes" are known from the Paleocene and Eocene of North America and Europe, a time when there was a trans-Atlantic land connection between the two continents. *Diatryma* stood nearly $2\frac{1}{2}$ m tall and had a head the size of that of a modern horse.

these bipeds disappeared by the end of the Eocene. By Oligocene time the modern families are present, by Miocene time the modern subfamilies, and by the Pliocene many of the modern genera had appeared in much the same form as they exist at present.

Toward the close of the Cenozoic was the beginning of the Ice Age or Pleistocene Epoch, a period of several million years characterized by the spread of glaciers over much of the Northern Hemisphere. There were four glacial advances and four retreats or interglacial periods. It was during the Pleistocene that humans began their major migrations and crossed the Bering Land Bridge into the New World. However, toward the close of the Ice Age the climate apparently became quite severe and many of the then living mammals and birds became extinct. Nowhere is a better record of the life of Pleistocene mammals than at the Rancho La Brea tar pits, now within the city of Los Angeles. Figure 2-45 shows an Imperial Mammoth sinking slowly into the tar pits during the Pleistocene; this small trap provides us with a most spectacular picture of the North American Ice Age. Figure 2-46 illustrates the variety of mammal life present in the region during the Pleistocene.

PRIMATES (ORDER PRIMATES)

The order Primates (Figure 2-47), the order to which we belong, was an early offshoot of insectivore stock; in fact, certain ancient fossils are often difficult to assign to one or the other order. However, the primitive primates are generally thought to have been arboreal, and this habit is believed to have been largely responsible for the development of many traits that we think of as being characteristic of primates. Arboreal life requires great agility in climbing and in grasping; the development of the hand no doubt resulted. Acute vision and coordination are also necessary and, perhaps most important,

Figure 2-45   North American Pleistocene imperial mammoth trapped in the Rancho La Brea tar pits. (Courtesy Dr. D.P. Whistler and the Museum of Natural History of Los Angeles County.)

a high level of development of the brain is required—the most salient feature of primate evolution. Although tree shrews (*Tupaia*) may resemble the ancestral primate condition in many features, the arboreal lemurs or lemuroids are the most primitive forms acknowledged to be true primates; though common in early Cenozoic time throughout much of the world, they are now confined (not much changed from their original condition) on the island refugium of Madagascar. A living analogy of a slightly later stage in primate evolution is illustrated by the arboreal tarsier (*Tarsius*), an East Indian form. Two distinctive lines of monkeys appear to have diverged early from a tarsierlike ancestral stock, a New World group and an Old World group. The great apes are, of course, the group most closely approaching humans in many anatomical similarities, though not on the mainstream of human evolution. On the other hand, human ancestors are now coming to light, and the fossil australopithecines or "man apes" of Africa appear to link humans to their primate ancestry.

CLASSIFICATION OF THE VERTEBRATES

Class *Agnatha:* jawless fishes (agnathostomes).
  *Subclass *Ostracodermi:* armored, jawless fishes, including cephalaspids and a number of other orders.[2]
  Subclass *Cyclostomata:* cyclostomes or round-mouth fishes (lampreys and hagfishes).
*Class *Placodermi:* jawed (gnathostome) fishes with armor, the placoderms.
Class *Chondrichthyes:* cartilaginous fishes.

[2]An asterisk indicates that the taxon is extinct.

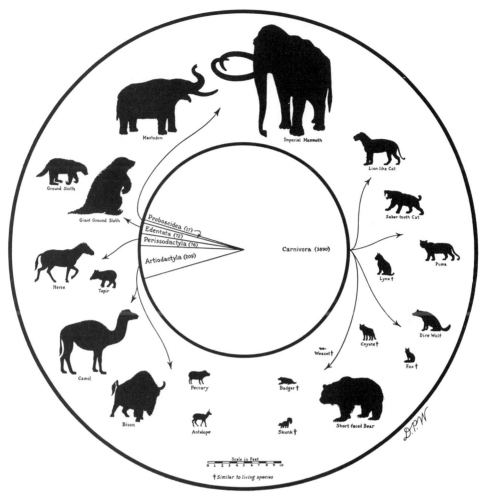

Figure 2-46 Diversity of Pleistocene mammals in North America as indicated by those recovered from the Rancho La Brea tar pits. Note the preponderance of predatory forms, no doubt trapped while attempting to capture prey animals already in the tar. This chart does not include insectivores, bats, rodents, and rabbits. (Courtesy Dr. D.P. Whistler and the Museum of Natural History of Los Angeles County.)

Subclass *Elasmobranchii:* elasmobranchs, including the extinct cladoselachians and pleuracanths, and the sharks and allies.

 Order *Selachii:* sharks and allies.
 Order *Batoidea:* skates and rays.
Subclass *Holocephali:* chimaeras or ratfishes.

Class *Osteichthyes:* bony fishes.
 Subclass *Acanthodii: acanthodians or "spiny sharks" (little known early bony fishes).

Subclass *Actinopterygii:* ray-finned fishes.

 Superorder *Chondrostei:* ancient paleoniscoids or chondrosteans, including the living sturgeons, paddlefish, and Nile bichirs.
 Superorder *Holostei:* holosteans, including the living gars and bowfins.
 Superorder *Teleostei:* modern bony fishes or teleosts.
Subclass *Sarcopterygii:* lobe-finned fishes or sarcopterygians.

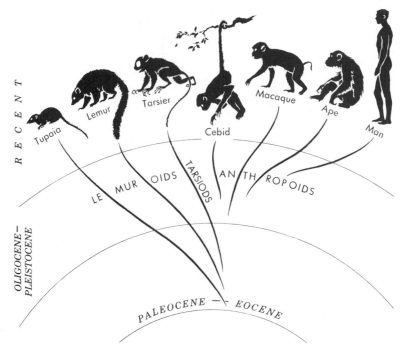

Figure 2-47 Evolution and relationships of the major groups of primates. *Tupaia* represents in a general way the type of primitive placental (either insectivore or primate) from which the basic stock of the primates, was derived. (From Colbert, *Evolution of the Vertebrates,* John Wiley & Sons. Prepared by Lois M. Darling.)

Order *Crossopterygii:* predaceous lobefins.

> ***Suborder Rhipidistii:*** crossopterygians ancestral to tetrapods.

> **Suborder *Coelacanthini:*** marine crossopterygians, including the living *Latimeria.*

> **Order *Dipnoi:*** lungfishes (three living forms).

Class *Amphibia:* amphibians.

> ***Subclass Labyrinthodontia:*** primitive labyrinthodont amphibians.

>> ***Order Ichthyostegalia:*** the earliest amphibians from the Devonian with a fishlike tail.

>> ***Order Anthracosauria:*** amphibians ancestral to the reptiles.

> ***Subclass Lepospondyli:*** common Paleozoic amphibians in varied forms.

> **Subclass *Lissamphibia:*** the living amphibians—salamanders, frogs, and caecilians.

Order *Anura:* frogs and toads.
Order *Urodela:* salamanders.
Order *Apoda:* caecilians or apodans.

Class *Reptilia:* reptiles.

> **Subclass *Anapsida:*** stem reptiles and turtles.

>> ***Order Cotylosauria:*** stem reptiles.

>> **Order *Chelonia:*** turtles.

> **Subclass *Lepidosauria:*** diapsid reptiles, including lizards, snakes, and rhynchocephalians.

>> ***Order Eosuchia:*** ancestral diapsids of the Permian and Triassic.

>> **Order *Rhynchocephalia:*** the living tuatara (*Sphenodon*) of New Zealand and related fossils.

>> **Order *Squamata:*** diapsids with reduced temporal arches.

>>> **Suborder *Lacertilia:*** lizards, including primitive and diverse

forms, some limbless, and the Cretaceous marine mosasaurs.

Suborder *Serpentes:* snakes (derived from lizards).

Subclass *Archosauria:* "ruling reptiles"; diapsid skulls often with other fenestrae; many with bipedal locomotion.

*Order *Thecodontia:* small bipedal diapsids of the Triassic ancestral to the archosaurs; some became similar to crocodiles.

Order *Crocodilia:* crocodiles and alligators.

*Order *Pterosauria:* pterosaurs or flying reptiles with membranous wings.

*Order *Ornithischia:* herbivorous dinosaurs with pubis parallel to ischium and a prepubic process directed forward; includes the suborders Ornithopoda with duckbills or hydrasaurs, stegosaurs, ankylosaurs, and ceratopsians.

*Order *Saurischia:* very large herbivorous and small and large carnivorous dinosaurs with a triradiate pelvis, with pubis directed forward.

*Suborder *Sauropodomorpha:* the largest quadrapedal herbivorous dinosaurs, including *Apatosaurus* and *Diplodocus.*

*Suborder *Theropoda:* carnivorous, bipedal dinosaurs; large forms are termed carnosaurs, and small forms coelurosaurs (probably ancestral to birds).

*Subclass *Euryapsida* (also called *Parapsida* or *Synaptosauria*): A possibly heterogeneous assemblage of forms with a single temporal opening whose classification is highly debated; includes many marine reptiles such as the plesiosaurs (order *Sauropterygia*), and the ichthyosaurs (order *Ichthyosauria*).

*Subclass *Synapsida:* mammallike reptiles with a single temporal opening; ancestral to the mammals.

*Order *Pelycosauria:* primitive late Carboniferous and Permian reptiles somewhat similar to the stem reptiles.

*Order *Therapsida:* mammallike reptiles of the late Permian and Triassic; ancestral to the mammals.

Class *Aves:* birds.

*Subclass *Archaeornithes:* ancestral birds with teeth and a skeleton similar to small coelurosaurian dinosaurs; represented by *Archaeopteryx* of the Jurassic.

Subclass *Neornithes:* modern birds, all but *Archaeopteryx.*

*Superorder *Odontognathae:* Cretaceous toothed birds.

Superorder *Palaeognathae:* ratites, including the flightless ostrich, rhea, emu, and kiwi. May not be a natural group and is often included in the following superorder.

Superorder *Neognathae:* all other modern birds.

Class *Mammalia:* mammals.

Subclass *Prototheria:* egg-laying mammals or monotremes; includes the duckbill platypus and spiny anteaters of Australia and New Guinea.

Order *Monotremata:* living egg-laying mammals.

Subclass *Theria:* normal mammals that give birth to their young; includes the extinct order *Pantotheria* that was ancestral to the higher mammals.

Infraclass *Metatheria:* marsupials or pouched mammals; young born at an immature stage of development; primarily Australian and South American.

Order *Marsupialia:* living marsupials.

Infraclass *Eutheria:* "placental" or higher mammals.

Order *Insectivora:* primitive eutherians thought to be ancestral to a number of other orders; includes the living shrews, hedgehogs, and moles; flying lemurs of the order *Dermoptera* are often placed as a suborder of the insectivores.

Order *Chiroptera:* bats.

Order *Primates:* lemurs, tarsiers, monkeys, apes, and humans.

Order *Edentata:* sloths, armadillos, American anteaters.

Order *Rodentia:* gnawing mammals including squirrels, rats, mice, and guinea pigs.

Order *Lagomorpha:* hares and rabbits.

Order *Hyracoidea:* conies or hyrax of Africa; this and the following two orders are probably related.

Order *Proboscidea:* elephants.

Order *Sirenia:* sea cows, manatees, and dugongs; aquatic offshoots of ungulate stock.

Order *Cetacea:* whales, dolphins or porpoises.

Order *Perissodactyla:* odd-toed hoofed mammals, such as horses, tapirs, and rhinoceroses.

Order *Artiodactyla:* even-toed hoofed mammals, such as cattle, sheep, pigs, camels, antelopes, deer, and giraffes.

Order *Carnivora:* typical carnivores with teeth adapted for flesh eating; includes dogs, cats, bears, and the marine carnivores such as the seals and walruses.

Suborder *Fissipedia:* terrestrial carnivores.

Suborder *Pinnipedia:* marine carnivores; seals, sea lions, and walruses.

Chapter Three

A Preview of Embryogeny

Although the methods and data of comparative anatomy and paleontology have clearly marked the routes of changing vertebrate form from fish to mammal, it is equally clear that an understanding of the causal mechanisms underlying evolution will not come from such an approach alone. Behind every adult structure and form there lies an embryonic history, and behind this history, in turn, lie the hereditary factors, genes, that control and direct the transformations that constitute any given embryogeny.

Because, as we have previously noted, all animals are in some degree related, it follows that there are many similarities among their individual patterns of development. It is possible, in fact, to sketch the early stages in development in a broad outline that applies roughly to all classes of vertebrates.

Such an outline—a preview of embryogeny, if you will—can suggest the nature of the events that will be given closer scrutiny later on. It is convenient to present the outline in the form of a series of steps, keeping in mind, however, that development does not proceed by halting, circumscribed intervals but is a gradual operation wherein one stage grades imperceptibly into the next.

MATURATION OF THE GAMETES

A new individual is inaugurated in a single cell (zygote) that results from the union of a male gamete (spermatozoon) with a female gamete (ovum or egg). Although embryogeny in the strict sense may be thought of as beginning at this point, the gametes themselves have a developmental history of great importance. This is the period during which the number of chromosomes in both egg cell and sperm cell is reduced to the haploid set. For the spermatozoon it is also the time when it takes on its distinctive form and structure; for the egg it is the period when it acquires its nutritive materials known as "yolk" and also assumes an internal organization that will play a very important role in later development.

FERTILIZATION

The term *fertilization* describes the actual union of ovum and spermatozoon. It also encompasses two other very important events: the activation of the egg, which causes it to begin dividing, that is, to initiate development proper; and the union of the haploid nuclei of the two gametes, resulting in restoration of the diploid chromosome number. As we shall see later, these two events are essentially distinct, for activation of the egg can occur without union of the nuclei.

CLEAVAGE

Upon activation, the egg cell enters the period of cleavage or segmentation. This con-sists of a series of mitotic cell divisions by which the egg is segmented into smaller and smaller parts. There are many different patterns of cleavage, and we shall later explore the factors responsible for these patterns. As a result of cleavage, a hollow, usually spherical mass of cells known as the *blastula* is formed.

GASTRULATION

As the cells continue to multiply, the various regions of the blastula become folded and otherwise moved about in such a way as to build up an embryo (known at this time as a *gastrula*) consisting of three distinct layers. These are the so-called *primary germ layers.* The outermost one is called *ectoderm,* the innermost one *endoderm,* and the one in the middle *mesoderm.* The movements by which these layers are brought into position collectively consti-tute gastrulation.

GENERAL ORGANIZATION OF AN EMBRYO

At the stage of development just described, the embryo consists essentially of a hollow cylinder whose wall exhibits three strata of cells (Figure 3-1). The outer stratum, as

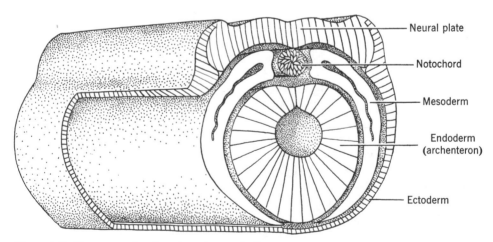

Figure 3-1 Stereogram of early embryonic organization.

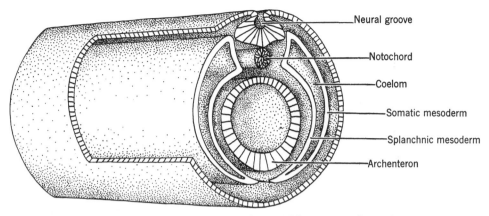

Figure 3-2 Stereogram of embryonic organization showing differentiation of mesoderm and neural ectoderm.

noted, is ectoderm and represents the embryonic skin; the innermost stratum, endoderm, is the future lining of the digestive tube, or gut, and the space that it encompasses is the cavity of the digestive tube (*archenteron*). The mesoderm, then, lies between the embryonic skin and embryonic gut. Because the form acquired by the early embryo foreshadows and thus makes more understandable the organization of the adult, its further progression will be followed (Figures 3-1 to 3-3).

Neural Tube. The ectoderm along the mid-dorsal axis of the embryo becomes thicker than the remaining ectoderm. Then the margins of this thicker plate elevate and its center is depressed, forming a shallow trough along the length of the body. Later the margins of the trough meet dorsally in the median line, forming a tube that subsequently separates from the parent ectoderm. This is the *neural tube* whose anterior portion will give rise to the brain, the remainder becoming the spinal cord.

The Neural Crest. Arising as a wedge of cells between the two dorsal edges of the neural folds, the neural crest separates into two halves which, with the establishment of the neural tube, come to lie on each side of the neural tube between it and the overlying epidermis. Major derivatives of the neural crest are the ganglia of the cranial and spinal nerves. But many neural crest cells migrate from the region of the neural tube and give rise to many different tissues in all parts of the body. They contribute, for instance, to the sympathetic nervous system; sensory ganglia of cranial nerves V, VII,

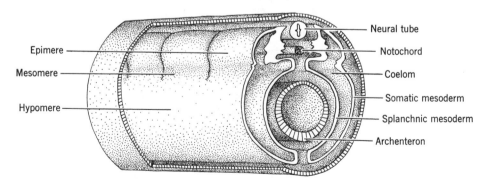

Figure 3-3 Stereogram of embryonic organization showing later differentiation of mesoderm and neural tube.

IX, and X; adrenal medulla; pigment cells; teeth; and some parts of the skull, and many of these features represent innovations of vertebrates over their cephalochordate ancestors.

Notochord and Mesoderm. The notochord and mesoderm develop coincidently. Initially the mesoderm lies wedgelike between the roof of the gut (with which it is intimately associated) and the overlying ectoderm. Progressively, then, it spreads ventrolaterad. In the meantime, the median-dorsal region of the mesoderm segregates from that which is more lateral. It is this median strip that rounds up and forms the *notochord* lying between the gut roof and the developing neural tube above. The lateral mesoderm continues to spread, and eventually the spreading sheets on either side meet in the ventral midline. At the same time, these mesodermal sheets split into two layers with a cavity between. One layer (*splanchnic mesoderm*) lies next to the gut, the other (*somatic mesoderm*) next to

the ectoderm; the space between the two layers is the *coelom*. It should be emphasized at this point that the coelom is not just any space in the body. It is exclusively a cavity bounded by an epithelium of mesodermal origin.

Not only does the mesoderm split into two layers, but on each side of the embryonic body the mesoderm becomes subdivided horizontally into three distinct regions. Most dorsal and flanking the neural tube and notochord is the *epimere;* most ventral and flanking the gut is the *hypomere;* intermediate to epimere and hypomere is a narrow *mesomere.* Almost hand in hand with this dorsoventral subdivision of the mesoderm comes the first indication of the basic segmental character of the vertebrates. This takes the form of a series of transverse divisions of the epimere and mesomere. The hypomere does not become segmented. Still to be noted are the several derivations of each epimeric block (Figure 3-4). Its lateral wall is known as the *dermatome;* its dorsomedian wall is the

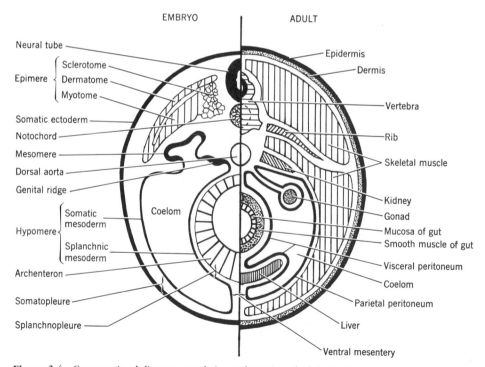

Figure 3-4 Cross-sectional diagram correlating embryonic and adult structure.

sclerotome; its ventromedian wall is the forerunner of the *myotome.* These parts are all components or derivatives of the epimere.

EMBRYO AND ADULT

As already mentioned, the structure of the adult is foreshadowed by the organization of the embryo. Figure 3-4 is intended to correlate adult and embryonic structure. The left side of the diagram pictures an embryo as seen in a transverse section at trunk level; the right side shows the adult in similar transverse section. (It will be appreciated that the embryo is drawn on a larger scale than the adult.) The drawing of the embryo also incorporates certain additional features. Those parts of the coelom incorporated in the epimere and mesomere are relatively reduced, whereas that in the hypomere is greatly expanded. It is also customary to refer to splanchnic mesoderm and adjacent gut endoderm as *splanchnopleure* and to somatic mesoderm and adjacent body (somatic) ectoderm as *somatopleure.*

The outline plan pictured in Figure 3-5 not only supplements Figure 3-4, but also provides a preview of the pathways of development that will be taken up individually in the chapters that follow. It will be noted that all pathways lead back to one of the three primary germ layers, that is, the origin of every organ is identified with a given germ layer. This correlation between organ and primary germ layer was originally established for the chick embryo. It was subsequently extended to encompass all vertebrate development and became the foundation on which comparative embryology rests. It means that in the primary germ layers we see a common ancestry, a common starting point for the development of all classes of vertebrates.

The preceding statement of the germ layer concept requires some qualification. In the first place, when we speak of a particular organ or part being derived from a given germ layer, we have in mind only the intrinsic functional portion of that organ, for there is no structure in the animal body that is the product of a single germ layer exclusively. To illustrate, the intestine is considered to be endodermal in origin, yet only its secreting and absorbing interior lining is so derived. The muscles, peritoneum, connective tissue, blood vessels, and nerves, which constitute the bulk of the digestive tube, are derived from mesoderm and ectoderm.

Second, modern observational and experimental evidence clearly demonstrates that the organ–germ layer homology is far less rigid or specific than we have been prone to believe. Although a particular germ layer may normally and ordinarily give rise to a specific part, transplantation experiments have revealed that in certain circumstances the cells of one germ layer may provide structures that are usually considered the exclusive products of an entirely different germ layer. Furthermore, even in normal development there are instances in which the cells of one germ layer give rise to structures customarily formed by those of other germ layers. The developing tail of vertebrates is illustrative. The muscles of the tail are a product of ectoderm, whereas muscles in general are mesodermal in character. Similarly, the skeleton is ordinarily derived from mesodermal connective tissue, yet the gill skeleton is produced by neural crest cells originally associated with the nervous system.

We may now ask whether, in the light of such facts, the germ layer doctrine retains any significance. The answer must still be in the affirmative. The various embryonic cells, whatever the range of their individual potentialities, do, as groups, assume certain spatial relations that we describe in terms of germ layers, and there is a high degree of homology between these layers and the products provided by them. At the very least, the concept is useful as a descriptive scheme of classification and identification.

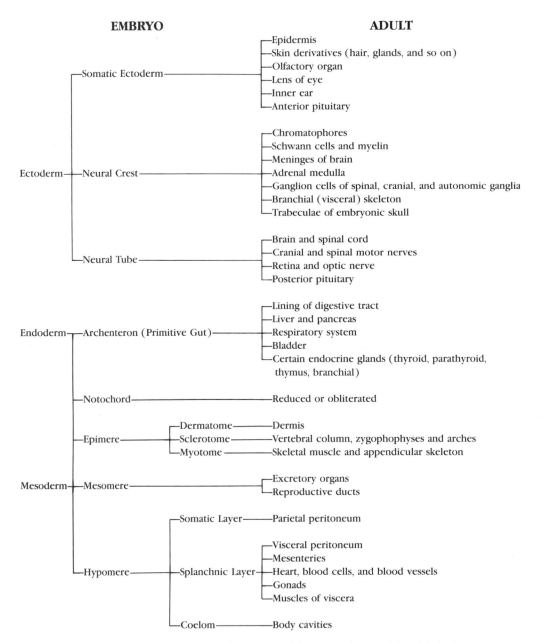

Figure 3-5   Chart showing the embryonic differentiation of the principal parts of the adult body.

THE COMPONENTS OF DEVELOPMENT

Even the brief sketch just provided should suggest that what has been described as development is an orderly, progressive sequence of changes, bringing a steady increase in complexity. From the relatively small and simple there emerge the relatively large and complex. But "development" is a pervasive term that encompasses and describes a wide array of operations. For analytical purposes, therefore, it is helpful to break the whole into its component parts. Provided it is understood that the breakdown is artificial—that one component has no real meaning divorced from the others—development may be said to consist of the following processes: *growth, morphogenesis,* and *differentiation.*

Growth. Growth, by definition, simply means developmental increase in mass. It results from synthesis of new protoplasm, both the cytoplasmic and nucleic, and cytoplasmic products such as the fibers of connective tissue and the matrix of bone and cartilage. Because the physiological economy of the individual is based on the cell and because there are minimum requirements of cell surface–volume ratio, increase in mass is usually accompanied by cell division, which is a distinguishing characteristic of growth. Development per se is inaugurated by division of the fertilized egg, and cell division continues throughout ontogeny and, to a variable degree, throughout the life of the individual. This is not to say, however, that cell division is limitless and uniform in rate throughout the embryo. To the contrary, multiplicative growth exhibits many nuances that contribute to the increasing heterogeneity of embryonic organization. Different parts of the embryo seldom grow at the same rate. Certain organs or tissues may build up rapidly, whereas others do not. Also, the rate of growth of parts may change, so that the relative proportions of various structures are altered with increasing age. Generally speaking, multiplicative growth is rapid early in development and then tapers off until it is ultimately confined to a few sites.

Such variations serve to point up a number of questions, the complete answers to which remain elusive. What determines local rates of growth, so as to create varying and changing proportions among parts? What factors set the size limits on organs and organisms? What is the nature of the stimuli that produce the renewed growth seen in tissue repair (wound healing) and replacement of lost parts (regeneration)?

The answers are needed not only because they may provide an understanding of the mechanisms of growth regulation during embryogeny, but also because the phylogenetic history of vertebrates records many instances in which the forces of natural selection have exploited changing proportional dimensions and size relationships between wholes and parts.

Morphogenesis. Morphogenesis literally means the generation of new form. Because the emergence of a new shape may be a consequence of many kinds of operations, the term may be and often is used in a variety of contexts. Here it will be employed to designate the organized movements of multiplying cells and the attendant physical reshaping of areas.

Individual cells or groups of cells may undergo *migration* to new positions, thus setting up new spatial relationships with other parts of the developing embryo. Migration (frequently results in *aggregation,* producing sheets, cords, and masses. A sheet of cells may undergo *delamination* (splitting) into two or more layers; a cord or mass may hollow out, that is, undergo *cavitation.* Finally, there may be *folding,* which may consist of an elevation of a local area, invagination (inpocketing), evagination (outpocketing), or bending.

Differentiation. Like the term development, differentiation is a pervasive expression that encompasses and describes a host of operations responsible for increasing diversification of form and function. That is, it refers to the events by which cells and other parts become different from one another and also different from what they were originally.

These emerging differences manifest in a number of ways. First, there is *morphological* differentiation. As they multiply, individual cells and groups of cells become structurally different from other cells and groups of cells. For example, from a common starting point in generalized ectoderm, nerve cells and epidermal cells acquire distinguishing features of size, shape, and internal architecture. Second, there is *behavioral* differentiation. Although all cells exhibit common basic attributes, such as metabolism and irritability, special functional capabilities are eventually superposed on these general properties. Thus nerve cells come to transport electrical disturbances, muscles to contract, gland cells to secrete special products, and so on. Third, and perhaps most important, there is *chemical* differentiation. Consider, for example, the egg itself, whatever its vertebrate class. From the beginning it has a complex chemical composition that is often reflected in a measure of pattern and spatial arrangement; and as the egg cleaves and progressively becomes converted to blastula, gastrula, and definitive embryo, individual cells and areas of cells become biochemically different from one another.

The elaboration and assortment of chemical differences lie at the root of the morphological and behavioral differences that ultimately emerge; the acquisition of special biochemical properties by the cells antedates the appearance of a recognizable neural tube, heart, or whatever else. Putting it somewhat differently, there comes a time when a given region of an egg or embryo takes on those chemical qualities that dictate the course of development to be followed by that region. The process by which this fixity of developmental prospect is acquired is called *determination*. Differentiation, therefore, might be redefined as follows: the series of events by which materials become chemically determined and then assume distinctive form and function.

Part Two

Developmental Preliminaries

Chapter Four

The Gametes

In all vertebrates reproduction is carried out by sexual means and involves the utilization of specialized *germ cells* or *gametes.* There are two types of these, the sperm cell or *spermatozoon* and the egg cell or *ovum,* produced by male and female, respectively. Although the beginning of a new individual is usually dated from the time when the sperm unites with the egg, these two types of cells have important developmental histories all their own.

PRIMORDIAL GERM CELLS

According to the doctrine of the "continuity of the germ plasm," the germ cells are not literally produced in the parental body, but have a common origin with that body in an earlier germinal material. The multi-cellular vertebrate is in a sense a dual organism in which the germ cells are early set aside and lead a life that is semi-independent of the organism proper. Moreover, this germinal material is passed on without interruption from one generation to the next; it is a potentially immortal substance residing temporarily in each of a succession of mortal individuals. So goes the doctrine in theory. But what of the evidence of the early segregation of germ substance?

Some of the clearest facts derive from specific invertebrates in which the future sex cells have been identified with certain of the early cleavage cells. Among the vertebrates, however, there are only two known instances of early segregation of germinal material. In one of the teleosts, some of the cells in the fifth cleavage stage have been equated with future germ cells. In the

85

egg of the frog, a cytoplasmic substance in the yolk near the vegetal pole (p. 94) has been traced to the roof of the gut, the site where definitive germ cells ultimately are recognizable. Otherwise in vertebrates, identification of the primordial germ cells comes somewhat later. Even so, when first recognizable, they are external to the prospective gonads, often even external to the embryo proper (Figure 4-1).

Whether these extragonadal cells are truly the forerunners of the gametes has long been a moot point. Some investigators have held that the so-called primordial germ cells are merely peculiar "endodermal wandering cells" having nothing to do with reproduction and that gamete formation is a matter of differentiation of somatic cells within the gonads. Others have conceded that the primordial germ cells eventually make their way to the gonads, where they then form gametes, but that they are supplemented by additional cells that arise within the gonads. Most recent data clearly favor the view that the primordial germ cells are the sole progenitors of the gametes.

1. In the very young embryo chick, the primordial cells occupy a crescentic area in a position anterior to the embryo proper. If this area is excised or destroyed by cautery or radiation, sterile gonads result in the later embryo. If the circulatory system of such an embryo is connected to a normal embryo of the same age, however, the gonads of the damaged embryo will be populated with germ cells from the normal embryo.

2. An experiment with toad embryos takes advantage of the fact that the primordial germ cells initially lie in the endoderm of the gut wall. When the germ cell containing endoderm of one species of toad is replaced by a graft of endoderm from another species whose eggs are distinctive in size and color, host females reared to adulthood lay eggs characteristic of the donor species.

3. Unlike anuran amphibians, the primordial germ cells of urodeles originally lie in the prospective lateral plate mesoderm (hypomere). In experiments with the Mexican axolotl, germ cell bearing

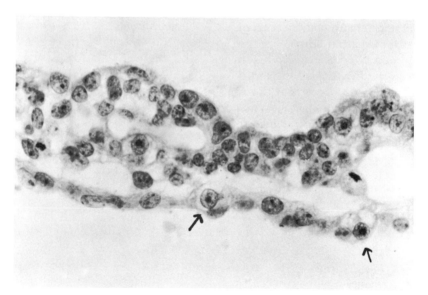

Figure 4-1 Primordial germ cells (indicated by arrows) in the yolk sac (splanchnopleure) of a young human embryo. (From Witschi, *Contributions to Embryology,* Vol. 32, Carnegie Institution of Washington.)

mesoderm from embryos carrying a dominant gene for dark pigmentation is grafted in place of mesoderm in a genetically albino host. When mature hosts are subsequently mated with white animals, the resulting progeny are pigmented like the donors. When, in a variation of this experiment, gastrula ectoderm is grafted unilaterally in place of mesoderm, none of the hosts produces progeny that are genetically like the donors of the ectoderm, indicating that somatic tissue cannot produce substitute germ cells. The conclusion that the primordial germ cells, and only these cells, can be the forerunners of the gametes seems inevitable.

The means by which the primordial germ cells reach their final abode in the gonads is not fully understood. They may possibly migrate under their own power, following cell surface signals such as fibronectin in the dorsal mesentery and peritoneum into the region of the developing gonads. Or, as demonstrated in the chick embryo, they may be wafted passively through the blood vessels. In this connection, in a recent experiment, the primordial germ cells of a turkey were injected into the vascular system of a young chick embryo whose own germ cells had previously been immobilized by radiation. The gonads of the chick were subsequently populated exclusively by turkey germ cells, recognizable by virtue of characteristic nuclear and cytoplasmic features. This experiment demonstrates not only the extragonadal source of germ cells but their transport via the circulatory system. The germ cells in birds and reptiles insert themselves into the extraembryonic circulation by forcing themselves between the endothelial cells of two capillaries, a process termed *diapedesis.*

GEMETOGENESIS

Whatever the means, the primordial germ cells do reach the gonads and there give rise to the definitive gametes. After a period of quiescence in the early embryo, these cells multiply rapidly and continue to do so until the individual reaches sexual maturity. They may now be referred to collectively as gonial cells; separately, they are *oogonia,* if they are destined to form ova, and *spermatogonia,* if they will form spermatozoa. The former lie in the ovaries where they are surrounded by special cells known as *follicle cells* (Figures 4-14 and 9-2); the latter grow and mature in the walls of the *seminiferous tubules* of the testis (Figure 4-10).

There are, of course, two basic functions with which eggs and sperm cells are concerned: (1) they serve to bring together for the new individual the hereditary material provided by its parents; (2) they furnish the material substance out of which development of the new individual is initiated.

The totality of preparatory events is termed *gametogenesis,* origin of the gametes. *Spermatogenesis* refers specifically to origin of the spermatozoon, *oogenesis* to origin of the egg. The key event in either case is the following: whereas the nucleus of a gonial cell, like an ordinary body cell, contains two of each kind of chromosome and thus two of each kind of gene, in the gamete nucleus these are reduced to one representative of each kind. The reduction is accomplished by a sequence of two divisions termed the *meiotic* or maturation divisions (*meiosis*). The matured gamete nuclei, containing only one of each sort of chromosome and gene, are said to be *haploid,* and the original condition is said to be *diploid.*

Remember that the sexually reproducing individual is itself a product of a single cell that results from the union of two gametes. Each of its cells, including the gonial cells, therefore contains two sets of chromosomes and genes, respectively maternal and paternal in origin. As a gonial cell prepares for meiosis, each chromosome in the cell is replicated and then somehow finds the one homologous to it that came from the other

Figure 4-2 A gametogonium at the beginning of maturation. Each chromosome thread of paternal origin makes side-by-side contact (synapsis) with the like maternal thread. (This figure and Figures 4-3 through 4-8 are from Muller et al. *Genetics, Medicine, and Man,* courtesy Cornell University Press.)

parent. The members of each pair synapse together in lengthwise contact. If, for instance, there are six chromosomes, *A-B-C-A'-B'-C',* in the gonial cell, when pairing is completed there are three doublets, *AA'-BB'-CC'* (Figure 4-2). This pairing is known as *synapsis* and it is so exact that each gene in a chromosome lies next to its corresponding gene in the homologous chromosome.

The replication of DNA during prophase of meiosis I doubles the amount of DNA in the nucleus. That is, for each original double-stranded DNA molecule there are now two double-stranded molecules, and therefor each of the original pairs of genes and chromosomes is represented by two pairs; every doublet of homologous chromo-

Figure 4-4 Diagrammatic detail of one tetrad to show the interchange of chromosomal material as a consequence of crossing over.

somes has become a complex of four strands, called a *tetrad* (Figure 4-3).

It is commonly found, as an accompaniment of the twisting about each other of the four strands that constitute a tetrad, that two of the four, one from each double chromosome, have broken at identical points and have joined together again in such a way as to have interchanged corresponding segments of maternal and paternal material (Figure 4-4). By this process, known as *crossing over,* new combinations of maternal and paternal genes are assembled in the chromosomes. The stage is now set for the meiotic division.

The chromosomes in the interlocked groups of fours become arranged across the "equatorial plane" of the division spindle (Figure 4-5). The maternal and paternal homologues of each tetrad are then dragged apart and guided to the opposite poles of the cell. The cell body itself divides (Figure 4-6) and the chromosomes of the two daughter cells, now in pairs (*dyads*) instead of fours, again line up in the equatorial plane. The process of chromosome sepa-

Figure 4-3 Continuation of maturation. Each chromosome of a pair is now duplicated, so that the threads are in groups of four, called tetrads.

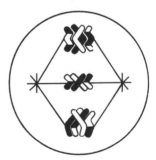

Figure 4-5 Metaphase of first meiotic division, showing tetrads lying in equatorial plane and about to become separated into two sets of pairs (dyads).

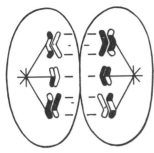

Figure 4-6 First cell division with each cell containing a set of dyads.

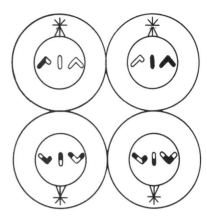

Figure 4-8 Second cell division, resulting in four cells, each containing just one complete set of chromosomes, that is, the haploid number of chromosomes.

ration is repeated, now with one member of each pair being dragged to each pole (Figure 4-7) because in this division, the centremeres divide as if in a normal mitotic division. Division of each cell body again follows (Figure 4-8). By these two divisions, then, four cells are formed, each with a nucleus that contains one chromosome derived from each original group of four chromosomes. Thus each nucleus now has just one complete set of chromosomes (haploid number) in place of the two sets (diploid number) possessed by the original gonial cell.

It is purely a matter of chance just which member of each group of four chromosomes a given nucleus comes to have. The randomness of this assortment is the source of the Mendelian rules of segregation. When this is added to the phenomenon of crossing over, mentioned earlier, and to the incidence of gene mutation, we have the mechanism for the production of variation that is so essential for evolution.

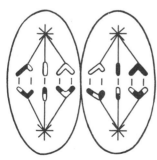

Figure 4-7 Start of second meiotic division with members of dyads being pulled apart.

SPERMATOGENESIS AND THE STRUCTURE OF THE SPERMATOZOON

In the development of a spermatozoon (Figure 4-9), any given spermatogonium enters a growth period, during which mitotic divisions serve to maintain the supply of gonial cells and produce cells that enter the meiotic process. Such cells are designated *primary spermatocytes.* Cytoplasmic division is incomplete throughout these gonial and spermatogonial divisions, producing clusters of cells in the same stage all connected by cytoplasmic bridges. The first meiotic division produces a *secondary spermatocyte;* equivalently, the first meiotic division of the oocyte produces a *secondary oocyte* from a *primary oocyte* (see Figure 4-9), although in the case of oogenesis, no cytoplasmic bridges are retained between cells. The meiotic divisions occur fairly early in the history of the spermatocyte. Because there are two of these divisions, one spermatocyte that starts the process will eventually form four cells. These are termed *spermatids,* which then transform into characteristic spermatozoa. All steps occur in the walls of the seminiferous tubules (Figure 4-10). Spermatogonia lie near the outer boundaries of the tubules; later stages are found at successively more central levels.

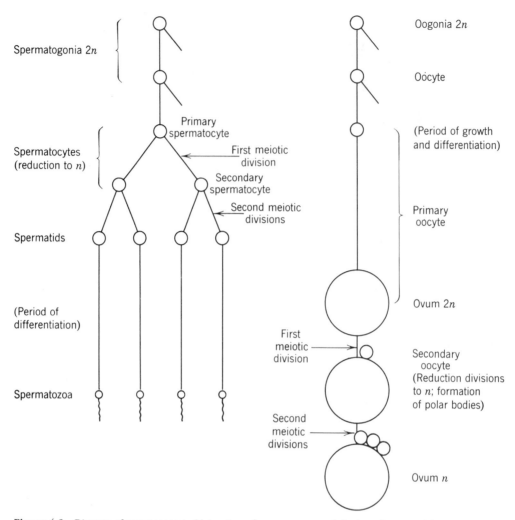

Figure 4-9   Diagram of gametogenesis. Maturation of spermatozoa on left, that of ova on right.

As multiplication of spermatogonia and spermatocytes proceeds, there is a progressive displacement of the more mature, earlier generation cells so as to move them centrally and bring mature spermatozoa to border directly upon the cavities of the tubules.

The spermatozoa of different kinds of animals exhibit a great variety of form. Those of vertebrates, however, are all built to a common plan (Figure 4-11), which features two main parts—*head* and *tail*. The tail, in turn, is subdivided into four regions—*neck,*

middle piece, principal piece (tail), and *end piece.*

The head exhibits a diversity of shapes in different groups of vertebrates. It may be spheroidal (teleosts), rod- or lance-shaped (Amphibia), spirally twisted (passerine birds), spoon-shaped (humans and many other mammals), or hooked (mouse and rat). Whatever its shape, the head serves two functions: *genetic* and *activating.* The genetic function is embodied in the *nucleus,* which consists of a dense concentration of DNA, the material of the genes.

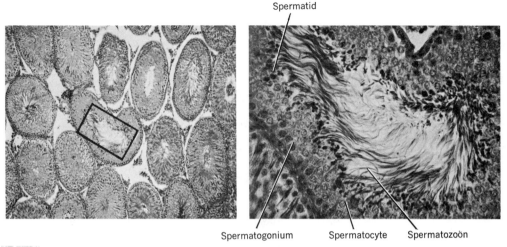

Figure 4-10 Seminiferous tubules of a 60-day old rat. *Left,* 75×. *Right,* area in rectangle magnified 340×. (Photos by Torrey.)

Capping the forward end of the nucleus is the *acrosome,* which, as we shall see (Chapter 5), plays an important role in binding sperm to egg envelopes, as well as in activating the egg in fertilization.

Electron microscopy has revealed a highly complex architecture in the middle piece. The most internal structures are two proteinaceous *central fibers* that are surrounded by a ring of nine double *inner fibers.* This array of axial filaments is presumably anchored in the *centriole* of the neck in the same way that the fibers of a conventional cilium or flagellum of a protozoon, say, are connected to basal granules. Surrounding this internal axial system is a second row of nine more massive *outer fibers,* which apparently originate in the so-called *connecting pieces* within the neck. Still more external are *mitochondria,* arranged spirally around the rings of filaments. All of these parts, as well as those in other regions of the sperm cell, are encased by the *sperm plasma membrane.* In the principal piece and the end piece, the fiber system is reduced to the axial complex of two central fibers surrounded by the ring of nine inner fibers.

A spermatozoon is clearly a highly specialized type of cell. Not only is the greater part of its total volume represented by nuclear material, but such cytoplasm as does occur is dedicated to the propulsion system. This system is essentially a counterpart of that found in a cilium or flagellum—a ring of nine double fibers surrounding a core of two single fibers—and is powered by energy initiated by mitochondria. The sperm cell is also totally devoid of stored food and protective envelopes, which characterize female gametes, and its size is only a fraction of that of even the smallest egg cell. Thus, a spermatozoon may be looked upon as hardly more than a motile nucleus whose destiny is to carry the paternal genetic information to the ovum and to activate the ovum.

OOGENESIS AND THE STRUCTURE OF THE OVUM

Maturation of the ovum is a somewhat more complicated operation, involving as it does not only the production of the haploid nucleus, but the acquisition of food reserves and preliminary organization of the cytoplasm from which the embryo will be

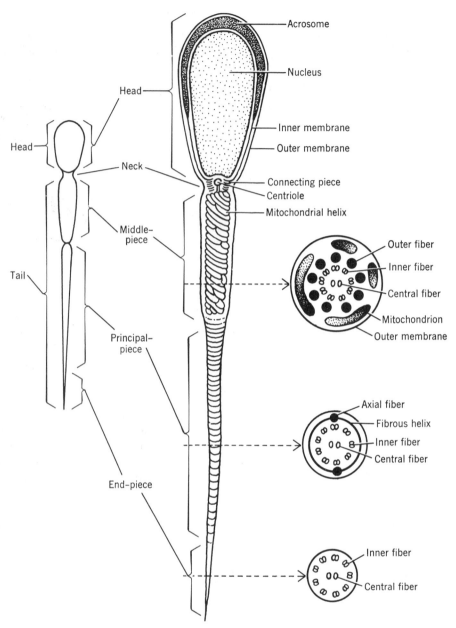

Figure 4-11 Structure of a spermatozoon.

formed. Much of this is accomplished during a period of growth and differentiation, introduced between the oocyte stage and the start of the maturation divisions (Figure 4-9). What usually happens is that the first meiotic division advances to late prophase and then is arrested in a suspended state, while the nucleus and cytoplasm become

involved in major synthetic activities. As a result of these activities, not only does the oocyte increase greatly in size and volume, but important qualitative changes also occur. These elaborations of quantity and variety of oocyte content can be traced to both internal and external sources.

Internally, the nucleus becomes decidedly enlarged, while the chromosomes in-crease in length by a process of decon-densation. The chromosomes may also ex-hibit (notably in the amphibian oocyte) numerous paired loops that project trans-versely from the main chromosomal axes. The shaggy appearance given to the chro-mosomes by these loops has led to their being called *lampbrush chromosomes* (Fig-ure 4-12). Although lampbrush chromo-somes have not been identified in the oocytes of all vertebrates, the consensus is that they are phylogenetically widespread and throughout evolution have played a fundamental role in oogenesis. It is believed that the stretching and looping of the chro-mosomes favor the essential activity in which the DNA-composed genes now en-gage, namely, the synthesis of messenger and ribosomal RNA which moves to the cy-toplasm and there promotes an intense syn-thesis of ribosomes and, to a more limited extent, protein.

Although some of this synthesis is based on materials already present in the oocyte, the bulk of the raw material is obtained from outside. Some may come from the ovarian cells that surround the oocyte, but most of the materials appear to originate at more remote sites in the body of the female and are transferred to the ovary by way of the bloodstream. The ovarian cells that sur-round the growing oocyte thus either pro-vide materials directly or act as a trans-ferring system for substances produced elsewhere and conveyed via the circulation. To some extent, the materials brought in are "readymade" proteins or other organic substances; others are simpler chemical building blocks, which are then assembled within the oocyte. In either case, there is a buildup of new cytoplasm and of variable amounts of food reserves that are collec-tively referred to as "yolk."

The synthesis of proteins encoded by messenger RNA results in an architecture

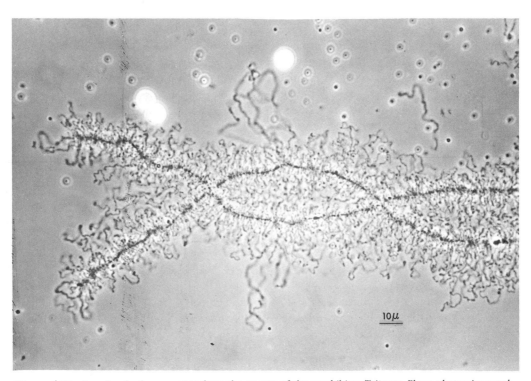

Figure 4-12 Lampbrush chromosomes from the oocyte of the amphibian, *Triturus.* Phase photomicrograph. (Courtesy Professor Gall.)

that is distinctive for each species of egg. Often this architecture manifests itself by local differences in density or pigmentation, but more commonly it resides at a more subtle biochemical level. The important thing is that the cytoplasm of the fully grown oocyte is by no means a homogeneous, undetermined mass; it has a qualitative organization that can greatly affect the developmental events to come. When, in subsequent chapters, we turn to the early development of representative chordates and vertebrates, we shall consider some of the specifics of egg organization in greater detail.

That which is termed yolk is a complex of variably assembled components, rather than a defined chemical substance. The principal components are proteins, phospholipids, and fats, in different combinations, depending on the species of egg. Although in some species yolk may possibly be synthesized by the oocyte itself, it ordinarily is produced in the liver, and is transported in a soluble form via the bloodstream to the ovary. There it is transferred by the ovarian cells that surround the oocyte and is deposited as platelets or granules. In the eggs of some forms (*Amphioxus,* tunicates, and eutherian mammals) the amount of yolk laid down is small, and it is scattered fairly evenly throughout the cytoplasm. Such eggs are referred to as being *microlecithal.* Other ova (Amphibia, Dipnoi, and Petromyzontia) acquire a greater amount of yolk and are described as *mesolecithal.* Still others (Myxinoidea, Chondrichthyes, Osteichthyes, reptiles, and birds) contain enormous food reserves and are designated *macrolecithal.* In the eggs of the meso- and macrolecithal types, the yolk is concentrated more in one hemisphere than in the other, leading to the designation *telolecithal.* Because of the uneven distribution of yolk, such an egg is said to have a *vegetal pole,* where the concentration of yolk is greatest, and an *animal pole,* where it is smallest. In fact, in macrolecithal eggs the

cytoplasm is confined to a small cap at the animal pole whereas the bulk of the total egg volume consists of yolk. Although its primary role is nutritional, we shall see later that the amount and distribution of inert yolk exercise an important influence on patterns of cleavage and the orderly movements of cells and tissues during gastrulation.

We have said that the maturation divisions are temporarily suspended while the egg substance is modeled in the manner just described. For many invertebrates these divisions will not be resumed until the egg is stimulated by a spermatozoon in fertilization. In most vertebrates, however, the oocyte now completes its first division, only to stop once more in the middle of the second where it remains until fertilized. It is only rarely that both maturation divisions are completed prior to fertilization. In any event, the interelated growth period leads to an alteration of nucleocytoplasmic proportions in the divisions that follow. Accordingly, at the first division there is an asymmetrical fractioning of the cytoplasm (and yolk), resulting in one cell substantially of the same size as the original and a small *polar body* (Figures 4-9 and 4-13). This is repeated in the second maturation division, when a second polar body is produced, while the first one may divide. Therefore, the maturation divisions, instead of producing four ova from one oocyte, give rise to one definitive ovum, containing practically all the cytoplasm from the starting oocyte, and two or three polar bodies that ordinarily degenerate.

EGG MEMBRANES

The egg cell, like all other cells, is bounded by a *plasma membrane* whose thickness is well below the limits of resolution of the light microscope. It is an integral part of the living cell and as such plays a major role in the passage of metabolites to and from the egg and, as we shall see, is also involved in

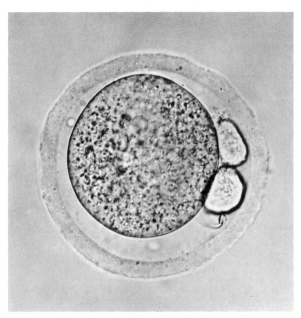

Figure 4-13 Ovum of the mouse, showing two polar bodies. (From Lewis and Wright, *Contributions to Embryology,* Vol. 25, Carnegie Institution.)

fertilization. In addition to the plasma membrane, the eggs of most animals come to be surrounded by one or more accessory envelopes, which are classified with reference to their origin. Primary membranes are those formed by the ovum itself secondary membranes are produced by the immediately surrounding cells of the ovary; tertiary membranes are laid down by the oviducts or special glands associated with the oviducts.

Of the primary envelopes, the *vitelline membrane,* is the most constant (Figure 4-14). It is usually thin and transparent and is closely applied to the surface of the underlying plasma membrane. Less constant is the *zona radiata,* so called because of its striated appearance. It has a wide distribution among vertebrate eggs, notably those of the shark, certain bony fishes, amphibians, and reptiles. In birds a zona radiata is not self-evident, but in mammals it is usually conspicuous (Figure 4-14) and is referred to as the *zona pellucida.*

Secondary envelopes are the chitinous shells surrounding the eggs of cyclostomes and the thin membranes immediately outside the vitelline membrane of the frog's egg. There are no secondary membranes about the eggs of urodeles, reptiles, and birds. Nor is there a true membranous envelope around the mammalian egg, but the cells constituting the *ovarian follicle* (Figure 4-14), within which the mammalian egg lies, are ovarian in origin. The structure and development of the ovarian follicle will be given special attention in a later consideration of the mammalian reproductive cycle (Chapter 9).

Tertiary membranes are formed after the egg has left the ovary. Among vertebrates, the albumen and horny capsule surrounding many elasmobranch eggs, the gelatinous coverings of amphibian eggs, and the albumen and shell membranes and calcareous shell of reptilian and avian eggs (Figure 4-15) are outstanding examples. Tertiary membranes are lacking in mammals, with the exception of the Prototheria.

Two requirements must be fulfilled if an

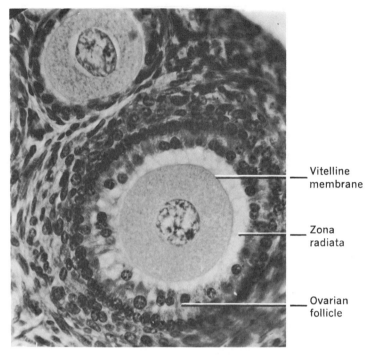

Figure 4-14 Ovum of the ground squirrel, *Citellus*, showing the vitelline membrane, zona radiata, and follicular envelope. (Photo by Torrey.)

egg is to function successfully. One is the provision of organic food adequate to keep the new individual going until it can feed itself or establish other means of nutrition. The other is an adequate water supply. Eutherian and metatherian mammals have solved both problems, or bypassed them if you will, through the device of uterine development, but for all other vertebrates these requirements have had to be met in other ways. The food problem has been handled in whole or in part by the buildup of

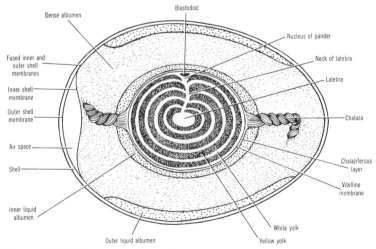

Figure 4-15 Diagrammatic longitudinal section of a hen's egg.

yolk reserves in the egg, although the elaboration of supplementary devices exemplified by the uterine nourishment of shark embryos may be noted. So long as vertebrates lived and bred in the water, there was no water problem. But the colonization of dry land required the establishment of water-conserving envelopes within which development could proceed. Although for most Amphibia a return to water for breeding is the normal event, many of them employ remarkable devices to ensure a water supply. For instance, some deposit their eggs in a stable foam that holds water for long periods; some surround the eggs with an oviducal jelly that becomes saturated with water; some store in their urinary bladders a very large quantity of water which is used to moisten the eggs. It is in reptiles and birds that the egg becomes self-contained, fully laden with yolk and surrounded by albumen and a waterproof shell.

Chapter Five

Fertilization

Although, as we observed in Chapter 4, important developmental events occur as an accompaniment of gametogenesis, the occasion of fertilization is the point from which the start of a new individual is conventionally dated. By definition, fertilization means the union of ovum and spermatozoon. This union involves many complex events, in which the sperm and egg are cooperating partners. These events obviously will not take place unless sperm cell and egg cell get into proximity in the first place and, having done so, make physical contact. Moreover, after first contact the sperm is confronted with the barriers of assorted membranes which must be penetrated to reach the interior of the egg. All of these problems—coming together, contact, and barrier penetration—require some exami-

nation prior to a consideration of the events of fertilization proper.

THE COMING TOGETHER OF THE GAMETES

All the events leading up to and encompassed in fertilization take place in a fluid medium. For most water-dwelling animals this simply means the aquatic environment itself. The customary procedure is for eggs and sperm simultaneously to be deposited externally in the surrounding water. That this is not an entirely haphazard arrangement is revealed by the many patterns of mating behavior that ensure that male and female will be in the vicinity of each other when the gametes are released. There is some evidence, too, that the female may release a chemical substance that stimulates the male to shed its gametes. This evidence, however, derives primarily from certain invertebrates.

In some aquatic vertebrates and all terrestrial ones the liquid medium is supplied by both male and female. Sperm are suspended in fluids provided by glands that are associated with the male reproductive ducts, and eggs lie in fluid within the female reproductive passages. This calls for a mechanism by which sperm cells, in their liquid medium, are transferred directly to the female reproductive tract, where fertilization is consummated. To this end, the vertebrates utilize a variety of intromittent (copulatory) organs (see Chapter 15).

The mere deposition of eggs and sperms in the general vicinity of each other within a liquid medium does not ensure their meeting, even though, at first thought, the possession of a propelling tail by the spermatozoon would seem to provide adequately for this. Presumably, a spermatozoon would simply swim until a chance meeting with an ovum was effected. True, this motility may facilitate the meeting of the gametes, but many investigators have explored the possibility that after their random movements have brought them within effective range, spermatozoa may be attracted by chemical substances released by the eggs. This phenomenon, which may be termed "chemotaxis," has, however, never been adequately demonstrated in any vertebrate. On the other hand, it has been shown in the case of some invertebrates that an increase in the motility of spermatozoa does occur under the influence of a substance emanating from the egg. Similarly, in mammals the liquid media provided by male and female, particularly the latter, are known to activate the spermatozoa. The obvious conclusion here is that anything that causes the sperm to swim more vigorously will lead to the greater likelihood of egg-and-sperm contact. Interestingly enough, however, in all those vertebrates

in which fertilization is internal, that is to say, in which spermatozoa are deposited in the female tract in an act of copulation, the subsequent movement of the sperm cells from the site of deposition to the site of fertilization usually depends little on the active swimming of the spermatozoa themselves. Rather, spermatozoa tend to be transported passively by muscular contractions of the female tract. This is certainly true for the many species of mammals that have been investigated, and it is not unlikely in other classes of vertebrates as well. In addition, it has been demonstrated in certain reptiles and birds that in the midst of the cilia, whose general effective beat is such as to maintain a posteriorly flowing stream of fluid in the female tract, a narrow band beats in reverse to produce a restricted "upward" stream, which carries spermatozoa forward to meet downcoming eggs.

CONTACT AND BARRIER PENETRATION

However they attain proximity, and admittedly there is much yet to be learned about it, ovum and spermatozoon do make contact.

The bulk of our knowledge on the contact interactions between eggs and sperm derives from studies on invertebrates, notably sea urchins, and also mollusks and annelids. Among vertebrates, the mouse has been shown to possess a specific glycoprotein in the structure of the zona pellucida that reacts with and binds the outer membrane of the sperm head. The zona glycoprotein acts as a receptor for the bindin protein of the sperm cell membrane.

Many sperm types must penetrate secondary or tertiary envelopes before contacting the egg's vitelline membrane. In certain invertebrates the egg envelopes are broken down by a lytic (dissolving) enzyme released by the acrosome of the sperm cell. The mammalian egg, when released from the ovary, is commonly encased in a layer of cells of ovarian origin, termed the *corona radiata*. These cells are held together by an adhesive known as hyaluronic acid. The corona is the barrier through which the spermatozoon must first penetrate, and to do so, its acrosome produces an enzyme, *hyaluronidase*, which serves to dissolve the adhesive and disperse the cells.

ACROSOME REACTION

The breaching of the membranous barrier(s) is not only mediated by lytic agents provided by the acrosome of the spermatozoon; the acrosome itself undergoes morphological changes. In recent years, thanks to the advent of electron microscopy, detailed observations of these changes have been made on the spermatozoa of a number of species. Most of these have been echinoderms and annelid worms, but included is a hemichordate, *Saccoglossus*, which we may employ as an illustrative case.

The architecture of the inactive acrosomal region of *Saccoglossus* is diagrammed in Figure 5-1. That which is termed the *acrosomal vesicle* is bounded by an *acrosomal membrane* and contains an *acrosomal granule*. The granule is surrounded in large part by fine, grainy material except at the apex where an *apical space* (arrow) lies

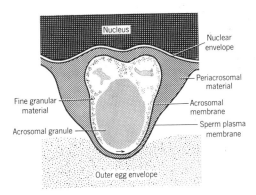

Figure 5-1 Inactive acrosomal region of spermatozoon of *Saccoglossus*. (From Colwin and Colwin, *J. Cell Biol.*, Vol. 19.)

between the granule and membrane. Periacrosomal material containing proteolytic enzymes lies between the acrosomal membrane and the sperm plasma membrane and also between the acrosomal membrane and the nuclear envelope.

As the spermatozoon makes its initial contact with the vitelline envelope, the first step in the sequence of changes pictured in Figure 5-2 occurs. First, the apical portion of the acrosomal membrane disappears so as to open the apical space, and around the rim of the opening the acrosomal membrane and sperm plasma membrane become continuous (Figure 5-2A). At the same time, the acrosomal granule begins to disintegrate, as does also that portion of the egg envelope to which it has become exposed, presumably signifying the release of lytic enzymes. Also, a conspicuous indentation appears in that region of the acrosomal membrane abutting the nucleus (arrow in Figure 5-2A). The indentation continues to deepen, thus creating a steadily elongating *acrosomal filament* that traverses the egg envelopes (jelly and vitelline membrane) and finally approaches the plasma membrane of the ovum (Figures 5-2B, C).

From studies on a number of invertebrates it is now known that the acrosomal filament is produced by an explosive polymerization of actin. This is one of a number of examples of actinlike proteins (similar to actin in muscle) serving a cytoskeletal function and producing changes in cell shape. The acrosomes of many different kinds of sperm have been shown to produce a specific protein, *bindin*, which reacts with bindin receptors (glycoprotein) on the egg vitelline membrane on contact, thus attaching the acrosome tip of the sperm to the membrane. Sperm of several species have been shown to require a period of exposure to compounds in the female reproductive tract before the bindin reaction can occur. Such preparation is a major part of what is known as sperm capacitation. Studies of the acrosomal reaction in rabbit sperm show similar events

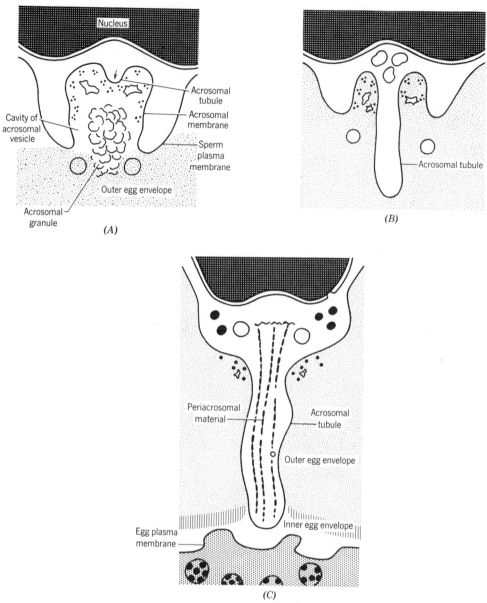

Figure 5-2 Sequence of acrosomal changes in *Saccoglossus,* leading to initial contact with plasma membrane of egg. (From Colwin and Colwin, *J. Cell Biol.,* Vol. 19.)

with one exception: no acrosome filament is formed.

ZYGOTE FORMATION

The initial definitive contact between the two gametes occurs with the meeting of the sperm and egg plasma membranes. These two membranes now fuse (Figure 5-3A). Although the two gametes are still distinct, because their plasma membranes are now continuous, they are essentially components of a single cell, the *zygote.* Accordingly, the contents of the formerly separate cells are now continuous. Egg cytoplasm

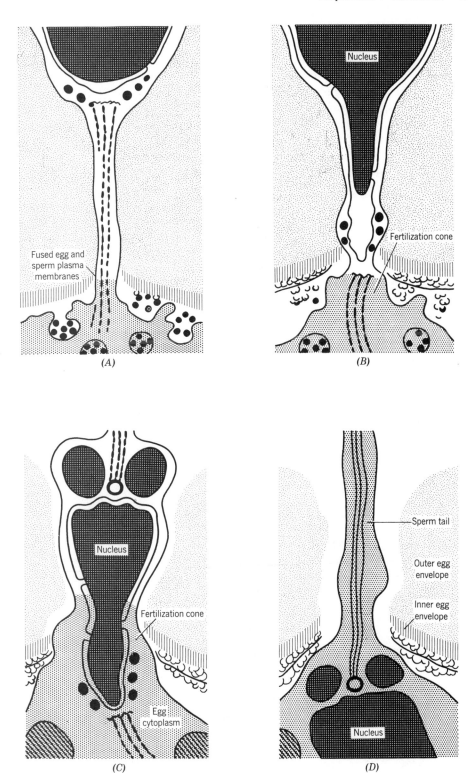

Figure 5-3 Zygote formation in *Saccoglossus*. (From Colwin and Colwin, *J. Cell Biol.*, Vol. 19.)

gradually intrudes upward underneath the sperm-contributed part of the now common plasma membrane and, as a "fertilization cone," engulfs the sperm nucleus, middle piece, and tail (Figures 5-3*B–D*). Meanwhile, the sperm nucleus becomes very much attenuated and, as the fertilization cone subsequently recedes, the sperm nucleus and its associated axial structures enter the egg cytoplasm (Figure 5-4). No such fertilization cone appears in vertebrate eggs.

ACTIVATION OF THE OVUM

The initial contact of the acrosome with the plasma membrane of the ovum triggers a train of reactions that bring the ovum out of its relatively quiescent state. Even as the sperm is completing its entrance into the egg cytoplasm, the egg itself responds with a variety of reactions. These reactions are not alike in kind for all types of eggs and, unfortunately, the events in vertebrate eggs are less well known than those in eggs of invertebrates. There are, however, some general events that commonly take place.

First, the egg surface undergoes a physiochemical change which, in some instances, results in the formation of a membrane that blocks the entrance of later-arriving sperm. A wave of changed membrane potential has been shown to follow contact between the acrosome and the egg cell membrane in sea urchin eggs, resulting in electrical charge differences similar to those in a neuron during nerve impulse transmission. This change in membrane potential sweeps over the surface of the egg in a few hundred milliseconds, immediately preventing any other sperm from fertilizing the egg, and initiating all of the later events in egg activation. It is known as the *fast block* to polyspermy.

The eggs of most vertebrates have many small vesicles, the *cortical granules,* in the peripheral cytoplasm just inside the cell membrane immediately following the fast block event. Ca^{2+} ions are released into the cytoplasm of several kinds of eggs, including those of teleosts, and, in another re-

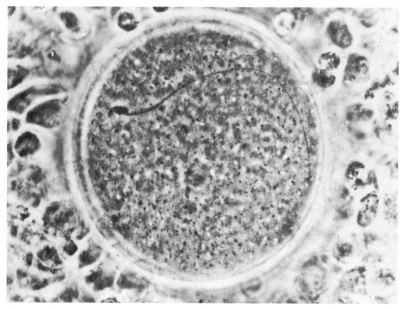

Figure 5-4   Live fertilized mouse egg, showing the entire spermatozoon in the cytoplasm. (From *Studies on Fertility,* Vol. 6, 1954. Courtesy Professor A.S. Parkes.)

markable parallel to events in neurons, this increase in Ca^{2+} ions causes the movement of the cortical granules toward the cell membrane, beginning at the point of sperm contact. The vesicles fuse with the membrane on contact, discharging their contents into the space between the vitelline membrane and the egg cell membrane. Enzymes in the cortical granule discharge cause a shift of water into the subvitelline space, thus elevating the membrane off the egg cell surface. Others catalyze cross linkage of molecules in the vitelline membrane, causing it to harden in this new configuration. The result of this alteration of the membrane is that all its bindin receptors are inactivated, releasing any other sperm that had attached to it and preventing other sperm from adhering to it. This reaction takes from one to several minutes and constitutes the *slow block* to polyspermy.

In mammals, the equivalent to the vitelline membrane is the *zona pellucida,* which is altered in similar fashion to prevent polyspermy, although it has the further role of preventing premature implantation as well. Even in cases where such a membrane does not appear, the surface changes may still be such as to prevent the entrance of more than one sperm. But this seems to be true only in microlecithal eggs. In the heavily yolked eggs of reptiles, birds, and some Amphibia the entrance of more than one sperm is a normal occurrence. Only one sperm nucleus will ultimately unite with the egg nucleus; the accessory sperm will be shunted into the yolk where they ultimately degenerate.

Second, a common reaction of the ovum to sperm penetration is a rearrangement of its internal constituents. We shall explore this in some detail when we turn later to the beginnings of development in tunicates and amphibians. For the present it need only be added that one possible consequence of this rearrangement is the establishment of the plane of bilateral symmetry of the embryo to come.

Third, the visible structural changes of the kinds just mentioned are accompanied by alterations in the physiological properties of the egg substance. These include such things as changes in permeability of the egg surface and alterations in the rate and by-products of metabolism, especially protein synthesis and oxygen consumption, as well as calcium release from the endoplasmic reticulum of the egg, which is accompanied by an increase in pH. In response to the stimulus of the sperm, the ovum becomes a truly dynamic entity and embarks on embryogeny through a series of physiochemical transformations that lead without break into the period of cleavage.

UNION OF THE HAPLOID NUCLEI

If activation of the ovum may be looked upon as serving the immediate end of inaugurating ontogeny, the union of the two haploid nuclei serves the long-range interests of evolution. This, as one writer has put it, is the culminating benefit that comes from the "luxury" of sexual reproduction. Not only is the diploid number of chromosomes restored for the organism to be, but that diploid set represents a new combination of hereditary prospects on which the evolutionary mechanism may capitalize.

It will be recalled that in most vertebrate eggs oogenesis comes to a standstill after the first meiotic division and is resumed only after ovulation or sperm fusion with the egg cell membrane (or through the agency of some artificial stimulus). Entrance of the spermatozoon serves to incite the second maturation division (Figure 5-5*A*). With the production of the second polar body, the egg nucleus (*female pronucleus*) is ready for union with the *male pronucleus* provided by the sperm head. The male pronucleus always moves toward the female pronucleus; the latter in turn often moves to meet the male. A microtubular array extends from the centriole of

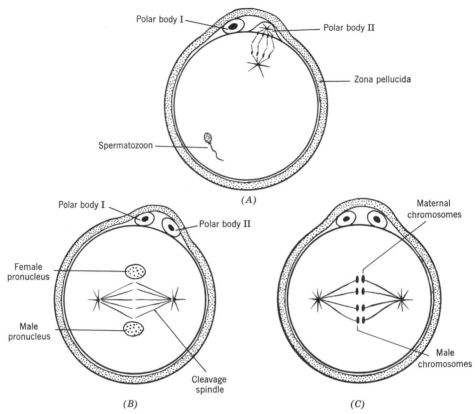

Figure 5-5   Semidiagrammatic figures of stages in the events of fertilization. (*A*) First polar body formed and second in process of forming in consequence of entrance of spermatozoon. (*B*) Second polar body formed and pronuclei approaching each other. (*C*) Maternal and paternal chromosomes arranged on spindle in preparation for the first cleavage.

the male pronucleus to the female pronucleus, facilitating the movement toward each other in some species. In many groups, however, no sperm centrioles can be found after spermiogenesis is complete. In a few animal types, the pronuclei now literally fuse. Ordinarily, however, in mammals and other vertebrates, each pronucleus loses its membrane, and concomitantly its chromatin resolves into the haploid set of chromosomes. The two sets of chromosomes then arrange themselves across the spindle (Figure 5-5*C*). This arrangement, marking readiness for the first cleavage division, represents completion of the process of fertil-

ization. Figure 5-6 summarizes the events of fertilization as seen in the mouse.

PARTHENOGENESIS

The facts of natural and artificial parthenogenesis indicate that the activation and nuclear phases of fertilization are essentially distinct and that development does not absolutely require a union of egg and sperm nuclei. In many animals, notably the insects, development may occur normally without fertilization; and even in those forms in which fertilization ordinarily is required,

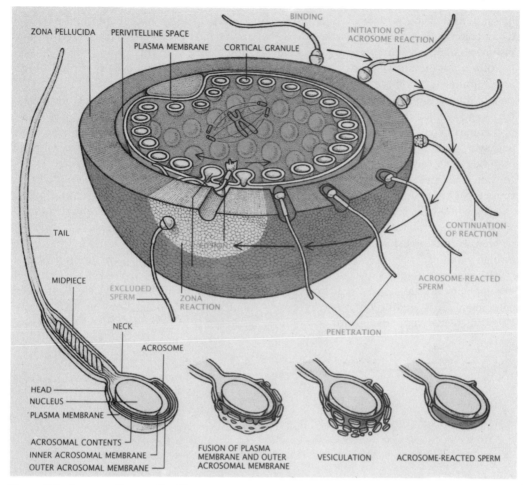

Figure 5-6 Summary diagram of fertilization events as seen in the mouse. Sperm structure and the acrosome reaction are shown at the bottom, and sperm penetration of the zona and the subsequent cortical reaction above. Modification of zone proteins by the cortical reaction causes extra sperm to be released from outer surface of the zona. (From P.M. Wassarman, Fertilization in Animals, *Sci. Amer., 259* (1988), p. 81.)

the sperm cell may be replaced in part or in whole. To illustrate, spermatozoa may still be effective as activating agents after treatment with X rays, ultraviolet light, or chemical substances, which put their nuclear equipment out of action. Or the spermatozoon as a whole may be dispensed with, for the eggs of almost every animal group, including mammals, can be started on a course of development, often to the completion of a new individual, by a host of chemical and physical agents. This strongly suggests that the ovum possesses in itself all the capacities to form an embryo, needing only some agent to "pull the trigger," so to speak. The spermatozoon is the usual releasing agent, but many substitutes can function in its place.

As we have already noted, in normal development the spindle for the first cleavage division is assembled around centrioles brought in by the sperm. What is the source of the centrioles in parthenogenesis when no sperm centriole is available? The answer varies with the animal form. In some forms the egg centrioles that remain after the last maturation division, ordinarily destined to degenerate, are retained along with the division spindle itself, and this spindle system is utilized for the cleavages that are to follow. Another solution is exemplified by the case of the parthenogenetically developing frog's egg. The easiest way to incite development artificially in the frog is to stab the egg with a very small, sharp needle, a method, however, that is effective only if the needle carries some foreign protein into the egg. The protein provides a center around which the cleavage spindle arises. In other instances cleavage spindles appear to arise spontaneously without benefit of foreign proteins.

Although parthogenesis occurs naturally in some vertebrates, such as some birds and amphibians, it is routine only in very rare cases, such as whip-tailed lizards, *Cnemi-dophorus uniparens,* all the adults of which are females. Recent experiments have shown that elements of both male and female genomes are necessary for full development in mammals.

As to chromosome numbers, one would expect parthenogenetic individuals to be haploid, and this indeed is often the case. There is, however, a great tendency for the diploid chromosome number to be restored. In fact, advanced or complete development occurs only if regulation to diploidy is accomplished; haploid individuals generally succumb in early embryonic stages. Regulation to diploidy may take place in a number of ways, the most common being fusion of haploid cleavage nuclei prior to cell division, or fusion of the egg nucleus with a polar body nucleus before cleavage begins.

Chapter Six

Cleavage and Gastrulation: General Considerations

Activation of the egg, whether through the intervention of a spermatozoon or through some parthenogenetic agent, is followed by a period during which the egg divides and subdivides to produce an increasing number of cells. This is the period of *cleavage,* or segmentation. Cell division goes on, of course, throughout the life of the individual, embryo and adult, but initially cell multiplication is the principal visible morphogenetic event. Until other events begin to overshadow cell division per se, development may be said to be in the cleavage phase, which is arbitrarily considered to end with the establishment of a cluster of cells termed the *blastula.*

As the cells continue to multiply, the various regions of the blastula are redistributed in such a way as to create an embryo consisting of three layers, the *primary germ*

layers. The embryo at this time is called the *gastrula.* The assorted kinds of movements by which a blastula is converted to a gastrula collectively constitute *gastrulation.*

CLEAVAGE

A broad view of cleavage shows that ideally there occurs a regular rhythm of mitotic divisions by which the original ovum is converted to two cells, then the two to four, the four to eight, and so on; the individual cells so produced are termed *blastomeres.* Generally speaking, the geometrical relationships of the blastomeres are established in accordance with the following principles: (1) The blastomeres tend to divide into equal parts. (2) Each new plane of division tends to intersect the preceding one at a

109

right angle. (3) As the number of blastomeres increases, the size of the blastomeres decreases; that is, there is no growth of any of the blastomeres during these mitotic cycles. The G_1 and G_2 phases of the cell cycle immediately before and after DNA synthesis are eliminated.

The first two principles derive from the manner in which the mitotic spindle orients itself with respect to the cytoplasm of the cell. The typical position of the mitotic figure is within the center of the cell. So long as nothing displaces the spindle from this central location, the cells that result from division will be equal in size. It should also be kept in mind that because the area of constriction that divides the cytoplasm (cytokinesis) is organized by the centrioles of the cell, the plane of cell division is always at a right angle to the long axis of the spindle. Therefore, the direction of the division plane in reference to the whole cell is determined by those forces responsible for the orientation of the mitotic figure. Ordinarily in a homogeneous spheroidal egg, the spindle will be oriented transversely to the polar axis, and thus the first division will be vertical (Figure 6-1A). The spindles for the second set of divisions will then line up parallel to the first division plane, so that the ensuing divisions are again vertical but at right angles to the first (Figure 6-1B). With each quadrant as long as the entire egg, but no more than half its diameter, the spindles for the third set of divisions will now lie parallel to the polar axis, and the divisions themselves will be horizontal and thus at right angles to the previous two (Figure 6-1C).

A simple but illuminating experiment serves to point to the dependence of cleavage plane on spindle orientation. If an egg undergoes cleavage while compressed between two glass plates, the spindles always orient to the long axes of the flattened blastomeres and all the cleavages are vertical. The result is a flat plate of cells rather than a spherical cluster.

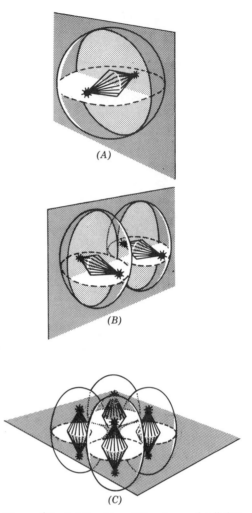

(A)

(B)

(C)

Figure 6-1 Relationships of the planes of cell division to the orientation of the spindles. Explanation in text.

The third principle is only a formal way of stating that during the period of segmentation there is no increase in cytoplasmic mass—the embryo does not grow. The fertilized ovum is a cell in which there is a relatively large amount of cytoplasm in proportion to nucleoplasm, and a principal "reason" for cleavage is to redress this ratio. With each cell division there is a synthesis of additional DNA and protein to provide for the increased number of nuclei. The synthesis of nuclear material occurs at the expense of cytoplasmic reserves, so that actually nuclear gain is balanced by cytoplasmic loss. At the conclusion of the cleavage period, then, the cytoplasm of the original egg, somewhat decreased in amount, is "packaged" in numerous separate units rather than one, and each cytoplasmic package has its own full complement of genes. The multiplication of packages is accompanied by a large increase in cell surfaces, that is, an adjustment in surface–volume ratios, and regional cytoplasmic differences in the original egg may be allocated to individual cells or groups of cells.

If the geometric principles outlined above operated literally, the cleavage patterns of all egg types would necessarily be alike. Cleavage patterns, however, show great variety.

One factor is the amount and distribution of yolk in the egg (see Chapter 4). For example, if yolk is fairly abundant and unevenly distributed, we find that the mitotic spindles in the original egg and later blastomeres are displaced from the center of the cells toward the less yolky ends and that, upon division, blastomeres of unequal size result. Moreover, cytoplasmic division proceeds much more rapidly in less yolky blastomeres. The inertness of yolk introduces an obstruction to the forces generated by the microfilament apparatus responsible for cytoplasmic division. In a given period blastomeres that are heavily laden with yolk will, therefore, divide less often than those that have smaller amounts of yolk. In fact, where the store of yolk is very large, as in the eggs of reptiles and birds, the heavily laden region will not divide at all; cleavage is confined to a small area at the animal pole where alone there is any appreciable amount of cytoplasm. In this reference, cleavage types may therefore be classified as follows:

1. *Holoblastic (Total) Cleavage.* The whole egg divides as do the successive blastomeres (frogs, primitive osteichthyes).

   a. Equal. In microlecithal eggs; the blastomeres are equal in size (mammals).

   b. Unequal. In mesolecithal eggs; the blastomeres are unequal in size (amphibians).

2. *Partial Cleavage.* Part of the egg remains undivided; in moderately macrolecithal eggs (teleost fish).

3. *Meroblastic Cleavage.* Division in only a small area at the animal pole of the egg; in extremely macrolecithal eggs (reptiles, birds, and primitive mammals).

The variety found in cleavage patterns does not derive entirely from the yolk factor; other aspects of cytoplasmic organization play an even more important role. Even in eggs containing minute amounts of yolk, where we might expect a simple expression of the geometric principles of cleavage, differences in blastomere size and geometric relationships appear sooner or later. The microlecithal eggs of certain invertebrates, although not our real concern, provide especially illuminating illustrations of cleavage patterns that reflect cytoplasmic architecture. In annelids and mollusks, for instance, the blastomeres may be unequal in size from the very start. Moreover, the spindles are alternately tilted obliquely to the right and to the left, so that successive generations of blastomeres are oriented in a twisted fashion. The rotational movement of blastomeres resulting from this tilting has lead to the descriptive term *spiral cleavage.* Another example is provided by the echinoderms (Figure 6-2). The first two divisions are regular, resulting in four blastomeres of equal size. But the third, horizontal division produces an upper quartet of cells slightly smaller than the lower quartet, and from this time onward the divisions in the animal and vegetative halves take entirely different courses.

Thus, the eggs of all animal types incor-

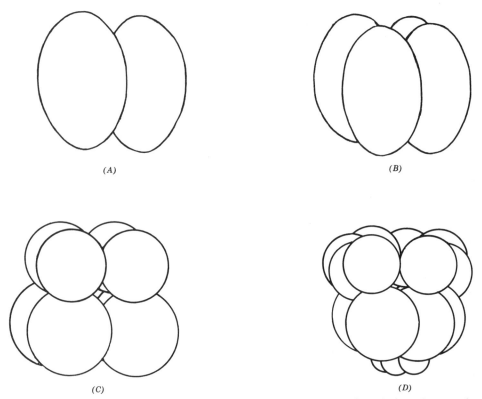

Figure 6-2   Cleavage of the egg of an echinoderm. The first two divisions (*A* and *B*) produce equal-sized blastomeres. The third division (*C*) produces an upper quartet of blastomeres smaller than the lower quartet. (*D*) Sixteen-cell stage, showing small blastomeres at vegetative pole.

porate an organizational pattern, and this pattern plays a far more important role in conditioning cleavage than does whatever amount of yolk may be present. So as we turn shortly to cleavage in specific chordate and vertebrate groups, we should keep in mind that an egg is more than a homogeneous mass that divides under the guidance of purely mechanical rules, and be alert to manifestations of cytoplasmic organization.

It is interesting to note that similarities of cleavage patterns in separate groups of animals may provide evidence of evolutionary relationships. For example, the eggs of annelids and mollusks exhibit spiral cleavage, and for this and other reasons these two phyla of animals are believed to be closely related. The similarities of cleavage patterns in reptiles and birds and later developmental similarities in mammals also point to close ancestral relationships.

GASTRULATION

The period of cleavage has been said to end with the establishment of the blastula. Because the production of the numerous blastomeres during cleavage shows many variations in patterns, it follows that the arrangement of the cells of the blastula, that is, the form of the blastula, will likewise show many variations. Some of the more conspicuous of these "styles" of blastulas will be characterized in subsequent considerations of specific animal groups. For the present it need only be emphasized that, despite the variations in their gross form, all blastulas possess in common regions of distinctive developmental potential, which are destined to be moved about in such a way as to construct a stratified embryo, the gastrula.

Gastrulation, the complex of movements that creates a gastrula, exhibits as many variations as does blastulation. But when we come to examine specific cases of gastrulation, certain general features applying to them all will emerge. Reduced to their essentials, the form-establishing (formative or morphogenetic) movements consist of various kinds of rearrangements of groups and areas of cells. These rearrangements include invagination, migration, spreading, and stretching to different degrees, depending on the animal type. Whatever the specific expressions, however, the cardinal feature of gastrulation is the integration of the operations involved. Although gastrulation may be broken down into a number of particular movements for convenience of analysis and description, the particulars are meaningful only in relation to one another and to the whole process.

Analyses of the morphogenetic movements of gastrulation are remarkably complete for a wide variety of organisms, but only at the descriptive level. That is, much is known of "what" takes place. The same cannot be said for "how." What we should like to understand is the nature of the forces responsible for the movement and guidance of areas of cells from one place to another. Although much has been learned about cell movements in gastrulation, its overall control and coordination remain unexplained.

Chapter Seven

Cleavage and Germ Layer Formation in Ascidians, *Amphioxus* and Amphibians

We now examine ways in which these crucial early developmental events occur in several types of chordates.

Whatever the status of the ascidians (Urochordata) and *Amphioxus* (Cephalochordata) with respect to the origin of the vertebrates (see Chapter 2), there are features of their early embryogeny that make them admirable subjects for an introduction to the beginnings of vertebrate development.

In both *Amphioxus* and ascidians, the egg is relatively small, averaging from 0.1 to 0.2 millimeter in diameter, and is microlecithal. More importantly, we see in the eggs of these forms a clear demonstration of a significant fact commented on earlier (p. 111, Chapter 6), namely, that an egg is more than a mere homogeneous mass of featureless cytoplasm out of which a diver-sity of parts will emerge; it possesses an architecture that in considerable measure presages the parts of the embryo to come. So it is in both urochordates and cephalochordates, but because circumstances are more sharply defined in the ascidian egg, it makes the better illustration. Let us consider specifically the simple, noncolonial form, *Styela,* whose eggs feature a natural pattern of pigmentation and distribution of yolk and other inclusions. This pattern permits a ready following of the developmental displacements of the cytoplasm.

The immature egg of *Styela* (Figure 7-1A) exhibits a yolk-free peripheral layer of orange-yellow cytoplasm and a central area, slate gray in color, through which yolk granules are scattered. The egg nucleus, termed the *germinal vesicle* at this time, is quite large and rests within the central area, close

115

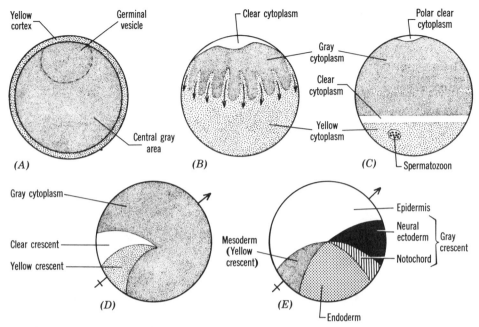

Figure 7-1 Cytoplasmic movements in the egg of the tunicate *Styela,* on the occasion of fertilization. (*A*) Unfertilized egg. (*B*) Few moments after penetration by spermatozoon. Yellow cortical cytoplasm and bulk of clear plasm from germinal vesicle stream toward vegetative pole. (*C*) Stratification of cytoplasmic areas. Layer of clear cytoplasm between yellow cap below and gray cytoplasm above. (*D*) Side view of slightly later stage, showing crescent of clear plasm immediately above yellow crescent. (*E*) Prospective organ regions immediately prior to first cleavage. White = epidermis, fine stipple (yellow crescent) = mesoderm, coarse stipple = endoderm, black = neural ectoderm, lined = notochord. Neural ectoderm and notochord constitute gray crescent. Arrow indicates anteroposterior axis. (After Conklin, *J. Acad. Nat. Sci. Vol. 13.*)

beneath the yellow peripheral layer. The position occupied by the germinal vesicle serves to identify the animal pole of the egg. The side opposite the germinal vesicle, therefore, is the vegetal pole.

When the egg is released from the ovary, the nuclear membrane of the germinal vesicle dissolves, releasing a clear residual plasm, and the spindle for the first maturation division is formed. But the maturation divisions do not take place until the ovum is penetrated by the sperm. Entrance of the sperm, which usually occurs at the vegetative pole, not only incites the maturation divisions but sets off movements within the cytoplasm (Figure 7-1*B*). The cortical layer of yellow cytoplasm streams rapidly toward the vegetal pole where it collects as a cap (Figure 7-1*C*). Simultaneously, all but a

small portion of the clear plasm provided by the germinal vesicle also flows to the lower pole where it accumulates between the yellow cap and the gray yolk-laden cytoplasm which now occupies the upper two thirds of the egg (Figure 7-1*C*).

After the clear and yellow substances have collected at the vegetal pole, the sperm nucleus starts to move toward that side of the egg that future development shows to be posterior. In the meantime, the egg will have completed its maturation divisions and the egg nucleus descends to meet the sperm nucleus just below the equator. As the sperm nucleus moves centroposteriorly, it appears to draw the layers of yellow and clear protoplasm along with it. The yellow substance, which remains largely at the surface, then assumes the form of a crescent, with its broad dimension directed posteroventrally and its arms extended a quarter of the way around each side of the egg. The clear substance extends more deeply into the center of the egg, but it also forms a crescent immediately above the yellow crescent (Figure 7-1*D*).

A further sorting out of cytoplasmic areas ensues as the egg prepares to begin cleavage. The substance composing the clear crescent gradually diffuses into the animal hemisphere, and the gray yolky plasm becomes confined to the vegetative region. Coincidentally, the gray plasm differentiates into two areas: a dark gray mass at the vegetative pole and a lighter gray crescent immediately opposite the yellow crescent (Figure 7-1*E*). At this time, therefore, four general regions may be distinguished in the egg: an upper hemisphere of light protoplasm which originated in the germinal vesicle, a yellow crescent posteroventrally, a gray crescent anterodorsally, and a vegetative area of dark gray yolky substance. Because of the natural differences in pigmentation, the developmental fate of these regions has been traced with considerable precision, and we may view the areas as the components of a "map" of prospective organ regions. Accordingly, the upper clear plasm represents prospective epidermal ectoderm, the yellow crescent is prospective mesoderm, and the dark gray yolky plasm is prospective endoderm. The gray crescent actually encompasses two developmental prospects; the upper portion becomes the neural ectoderm, whereas the lower portion becomes the lower notochord (Figure 7-1*E*).

That these prospective organ regions actually possess biochemical differences foreshadowing the parts they are to produce is demonstrated by centrifuging eggs before they begin cleaving. Such treatment serves to shuffle the regions so that they come to occupy abnormal positions. An embryo will still develop from a centrifuged egg, but its tissues and rudiments of organs have a chaotic arrangement. It is an orderly chaos,

however, in the sense that each part develops in the position where its cytoplasmic precursor came to lie as the result of centrifugation.

The first cleavage plane of *Styela* passes through the polar axis and bisects the mesodermal and chordaneural crescents (Figure 7-2*A*). Because it has been determined that the first cleavage plane corresponds to the median axis of the embryo to come, it follows that the first two blastomeres represent the material source of the right and left halves of the embryo and that each contains half of each prospective organ region. The second set of divisions is at a right angle to the first, cutting off two slightly smaller blastomeres at the posterior side and two larger ones in front (Figure 7-2*B*). This results in the prospective mesoderm coming to lie in the two posterior cells and the chordaneural material in the two anterior cells. Division of each of the four blastomeres is in the horizontal plane, producing an eight-cell stage that consists of an upper (apical) tier of slightly smaller blastomeres and four larger basal cells below (Figure 7-2*C*). The third set of cleavages serves to parcel out the prospective organ materials in such a way that the prospective mesoderm lies in the posterobasal pair of blastomeres, the chorda in the anterobasal blastomeres, and the neural ectoderm in both anteroapical and basal cells. The two posteroapical cells and parts of the anteroapical ones provide the epidermal ectoderm; the endoderm lies in the four basal cells.

It should be obvious from the foregoing that although cleavage in *Styela* is holoblastic, the resulting blastomeres are only approximately equal in size. Small amount of yolk to the contrary, division is not absolutely equal but reflects the cytoplasmic architecture of the egg. Cleavage serves to parcel into individual blastomeres areas of cytoplasm of distinctive biochemical composition and thus of precise developmental potentiality. That this parceling is real indeed has been demonstrated by experi-

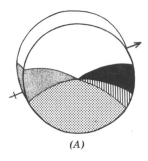

(A)

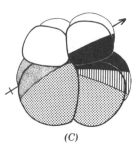

(B)

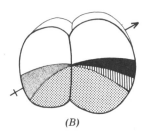

(C)

Figure 7-2   First three cleavages in *Styela*. (*A*) Two-cell stage, side view. Mesodermal and chordaneural crescents bisected. (*B*) Four-cell stage, side view. Chordaneural crescent shared by two larger anterior blastomeres; mesodermal crescent in two smaller posterior blastomeres. (*C*) Eight-cell stage, side view, consisting of four smaller apical blastomeres and four larger basal blastomeres. Symbols indicate distribution of prospective organ regions in the eight cells. White = epidermis, fine stipple = mesoderm, coarse stipple = endoderm, black = neural ectoderm, lined = notochord. Arrow indicates anteroposterior axis.

ments in which certain blastomeres are injured so as to prohibit them from further developmental participation. If, for instance, the four right-hand blastomeres of

an eight-cell stage (Figure 7-2*C*) are injured, the left-hand blastomeres produce only a left half-embryo; or if the four rear blastomeres are removed, the four in front produce only those parts they are normally destined to produce, namely, epidermis, neural tube, notochord, and endoderm.

Although the pigment pattern in the egg of *Amphioxus* is of another character and is far less evident, the conditions just described for *Styela* obtain in *Amphioxus* as well. Thus, at the eight-cell stage of the latter, we see a pattern of blastomeres and prospective organ regions of essentially the same form (Figure 7-3*A*). From here on,

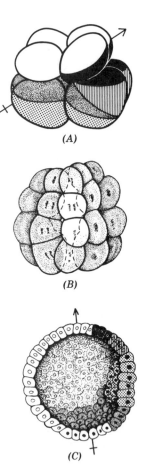

(A)

(B)

(C)

Figure 7-3   Cleavage stages and blastula of *Amphioxus*. (*A*) Eight-cell stage corresponding to that of *Styela*. Symbols are the same as in Figure 7-2. Arrow indicates future longitudinal axis. (*B*) Thirty-two-cell stage. (*C*) Hemisection of blastula.

events are a little clearer in *Amphioxus* than in *Styela,* so the remainder of the description pertains to *Amphioxus.*

The fourth set of divisions is again polar, with the plane of division in each cell tending to parallel that of the first division, but not precisely. If the geometrical rules of cleavage were to apply literally, the fifth series should also be vertical and at right angles to the fourth. Actually, however, the fifth series is horizontal, resulting in 32 cells arranged in four tiers and with a modest gradation in size from larger basal cells to smaller apical cells. (Figure 7-3*B*). The sixth cleavage planes are again approximately meridional and, as in earlier divisions, the blastomeres divide synchronously. But beginning with the seventh series, the synchrony of the divisions declines, so that the progression in numbers of blastomeres becomes arithmetical rather than geometrical. Meanwhile, viscous fluid accumulates in the center of the mass of cells. This serves to push all the blastomeres outward, so that they become arranged in a single layer enclosing a central cavity. It is this hollow sphere that we term the *blastula* whose liquid-filled cavity is the *blastocoel* (Figure 7-3*C*).

Because of natural size differences among its cells or through artificial devices for marking or otherwise identifying cells, it is possible to map the developmental prospects of various areas of the blastula. Such a map is found to be not unlike that which prevails at the eight-cell stage. Accordingly, the region corresponding to the vegetative pole of the original egg is identified as prospective endoderm, and is flanked by prospective mesoderm posteroventrally and chordaneural material anterodorsally. The remainder represents prospective epidermal ectoderm (Figure 7-3*C*).

Gastrulation, by which the blastula is converted to a stratified *gastrula,* is accompanied by continuing and rapid cell multiplication, especially in the areas of prospective ectoderm and mesoderm. More importantly, however, entire areas undergo

movement, and it is on these morphogenetic movements that we shall center attention. Because there is no clear understanding of the forces behind these movements and the role, if any, played by cell multiplication, we must be content with a pure description of events.

The gastrulation process begins with a flattening of the area involving the prospective endoderm. This endodermal plate then gradually invaginates, or folds inwardly, into the blastocoel (Figures 7-4*A, B*). The new cavity created by the invagination is the *gastrocoel* (*archenteron*), which opens to the exterior by the *blastopore.* The circular rim of the blastopore is termed the *lip;* thus with reference to the future orientation of the embryo, the prospective mesoderm lies in the ventral lip of the blastopore and the prospective notochord lies in its dorsal lip. As the endodermal plate continues to move inward, its dorsal portion proceeds more rapidly, leading to an accompanying early inflection, or turning under, of the dorsal lip (Figures 7-4*B, C*). This serves to bring the prospective notochordal material to the interior and beneath the prospective neural ectoderm (Figure 7-4*D*). Continued invagination of the endoderm results in its coming to meet the opposite epidermal wall with an ensuing elimination of the blastocoel. By the same token, the gastrocoel is greatly deepened, and in the process the prospective mesoderm in the ventral lip is also brought into the interior (Figure 7-4*D*). At the same time, the gastrula exhibits a general elongation that serves to stretch out the neural ectoderm and the notochord along the longitudinal axis (Figure 7-4*D*). One effect of the reorientation of the notochord from a transverse to a longitudinal band is the steady pulling of the mesoderm from the other side in a dorsoanterior direction (Figure 7-4*E*). With still further elongation of the gastrula, there is created an internal primitive gut tube containing a notochordal strand dorsally, a strip of mesoderm on either side of that, and a major component

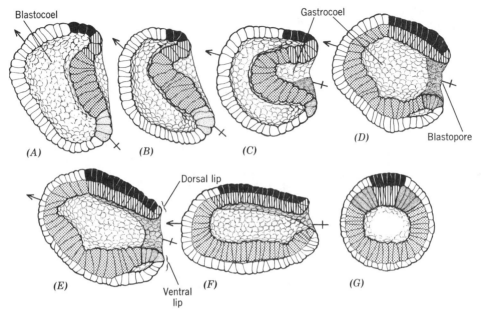

Figure 7-4 Gastrulation in *Amphioxus*. (*A* through *F*) Longitudinal hemisections. (*G*) Transverse section through the midbody of *F*. Arrows indicate anteroposterior axes. White = epidermis, fine stipple = mesoderm, coarse stipple = endoderm, black = neural ectoderm, lined = notochord.

of endoderm. Externally, neural ectoderm lies in the dorsal midline above the notochord, and epidermal ectoderm occupies the remainder (Figure 7-4*F*). Described somewhat differently, the gastrula now consists of two layers, an outer *ectoderm,* neural and epidermal, and an inner mass of cells encompassing prospective notochord, mesoderm, and endoderm (Figure 7-4*G*). The next phase of development, then, involves the segregation of the components of these inner cells so as to establish definitive endodermal and mesodermal layers. This is accompanied by the creation of the notochord and the inauguration of the neural tube.

As the gastrula continues to elongate, the blastopore constricts. Meanwhile, the elongated strip of neural ectoderm, which may now be referred to as the *neural plate,* thickens somewhat, and its center sinks downward to form a trough or groove (Figure 7-5*C*). Concomitantly, the ectoderm just above the ventral lip grows up and over the much reduced blastopore as a curved

ridge whose arms extend forward above the neural plate (Figures 7-5*A, B*). The free edges of these ectodermal arms steadily come together much as the component halves of a "zipper" might be drawn together, thus roofing over the neural groove (Figures 7-5*B, C*). But as this process goes on, the neural plate proper becomes detached from the general ectoderm, its groove continues to deepen, and finally its dorsolateral margins roll together in the midline to create a *neural tube* (Figures 7-5*D, E*).

While all of this is occurring, the dorsolateral regions of the prospective endoderm work steadily inward from each side toward the dorsal midline. One consequence of this is an upward buckling of the prospective notochord; another is an outward bulging and then a complete undercutting of mesodermal sacs on each side (Figures 7-5*C, D*). This, as least, is what appears to happen, but there is no experimental evidence to indicate whether the notochord and mesoderm are "pushed out"

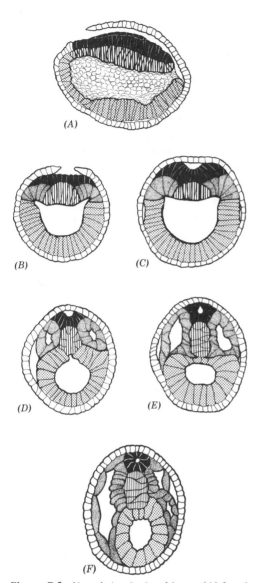

(A)

(B)   (C)

(D)   (E)

(F)

Figure 7-5 Neurulation in *Amphioxus*. (*A*) Longitudinal hemisection. (*B* through *F*) Transverse sections. Explanation in text. Symbols are the same as in Figure 7-4.

by the forces of convergence within the endoderm, or the endoderm only closes in the gap resulting from the active withdrawal of the notochord and mesoderm. Be that as it may, the meeting of the margins of endoderm creates an intestinal primordium composed entirely of endoderm, above which the definitive *notochord* be-

comes pinched off, as are a pair of longitudinal sacs of *mesoderm* (Figures 7-5*D, E*). Subsequently, then, the long horizontal sacs of mesoderm become sliced transversely into a series of pouches (Figure 7-5*E*).

The details of the later history of the *Amphioxus* embryo need not be followed further. It should be noted, first, that from the very beginning the mesodermal pouches are hollow, which is to say that they contain a *coelom, Amphioxus* thus stands in contrast to the vertebrates where, as we shall soon, see, the mesoderm is initially solid and becomes split into two layers with an intervening coelom. Vertebrates, that is, feature a *schizocoel,* a coelom formed by the splitting of an originally solid layer, whereas *Amphioxus* features an *enterocoel,* in reference to the fact that the coelomic spaces trace their origin to the cavity of the primitive gut. In each case, the end result is the same: the mesoderm comes to consist of an outer *somatic* layer associated with the epidermal ectoderm, and an inner *splanchnic* layer associated with the endodermal gut, with the coelom lying between the two layers. The further history of the mesoderm follows the general pattern already outlined in our preview of embryogeny (Chapter 3).

The fully mature egg of a frog of salamander is typically a large spherical cell, measuring some 2.0 to 3.0 millimeters in diameter. With respect to yolk content, amphibian eggs are of the *mesolecithal* type, that is, they contain a moderate quantity of yolk—more than the eggs of ascidians and *Amphioxus,* but much less than those of birds and reptiles. The yolk is present in the form of small platelets or ovoid granules, and, although scattered throughout the cytoplasm, shows a gradation in concentration from one pole to the other. Accordingly, amphibian eggs are said to be moderately *telolecithal,* with the greatest concentration of yolk located at the *vegetal* pole and the least concentration at the *animal* pole. Aside from the distribution of yolk, there is less visible indication of cytoplasmic organization in amphibian than in ascidian eggs.

The most notable features are the presence of a relatively yolk-free cortical layer whose animal half is more heavily pigmented than the vegetal half, and a clear area of cytoplasm surrounding the germinal vesicle which lies near the animal pole. The surface of the cortex has also been shown to be in the nature of a rigid, extensible membrane. It has been shown that this surface coat plays an important role in the later shifting of materials in gastrulation.

Prior to fertilization, the egg is radially symmetrical about an axis passing through its poles. The sperm normally enters the egg somewhere in the animal hemisphere, in an area between 30° and 50° from the animal pole. The open entry point will eventually be part of the ventral side of the body, near the anterior end. Thus, open entry confers dorsoventral symmetry on the egg. This symmetry is a consequence of internal rearrangements that appear to be triggered by the entering sperm cell. The first obvious result of sperm penetration is a slight lifting of the vitelline membrane, which permits the egg to revolve so that the yolk-laden vegetal hemisphere lies downward. As the sperm nucleus then proceeds along the path that will bring it into fusion with the egg nucleus, it appears to be accompanied by a flow of cytoplasm in the same direction, possibly in a manner akin to that which we have seen in ascidians. In any event, this flow is accompanied by a rotation of the cortical cytoplasm relative to the deeper cytoplasm parallel to the midsagittal plane. This movement of the cortical cytoplasm reveals an area of cytoplasm approximately 180° from the sperm entry point, which is very lightly pigmented. In frogs and some urodeles, the resulting depigmentation of this area is of such a degree as to create a visible crescent-shaped region known, because of its color, as the *gray crescent* (Figure 7-6). If the gray crescent is marked by a suitable dye so that its subsequent maneuvers can be followed, it is found in part to be the forerunner of the notochord. Its position, therefore, corresponds roughly to the chordaneural crescent of the ascidian egg and as such marks the dorsal side of the embryo to come. Similar marking tech-

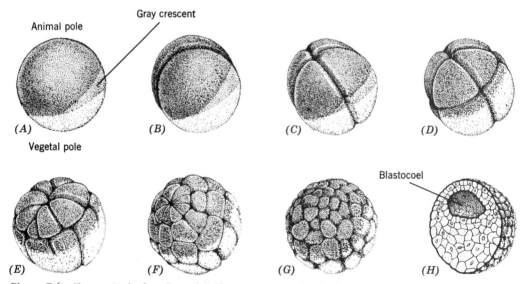

Figure 7-6 Cleavage in the frog, *Rana.* (*A*) The gray crescent shortly after entrance of the spermatozoon. (*B*) Two-cell stage, side view. The plane of division is considered to have bisected the gray crescent and thus corresponds to the future median axis. (*C*) Four-cell stage. Blastomeres still equal in size. (*D*) Eight-cell stage, showing initial inequality of blastomeres. (*E*) Twelve- to sixteen-cell stage. (*F*) Late cleavage. (*G*) Early blastula. (*H*) Diagrammatic hemisection of blastula. (Drawn from Ziegler models.)

niques also demonstrate that the gray crescent is destined to be distributed to the right and left sides of the embryo. In other words, a plane bisecting the crescent will correspond to the median axis of the future embryo. Commonly, although not invariably, the first cleavage plane coincides with the prospective median axis; thus each of the first two blastomeres contains half the gray crescent and represents one side of the future embryo. Recent experiments with *Xenopus* eggs show that rotation of the cortical cytoplasm is the main determinant of the location of the axes of the embryo, and that the rotation must be complete so that normal gastrulation can occur after cleavage.

Clearly, the gray crescent, or its equivalent in those species in which a visible crescent is not evident, is an important manifestation of early embryonic organization. If the first division happens to be such that one blastomere contains all the crescent and the other none and these blastomeres are then separated, the blastomere containing the crescent will produce a complete embryo, whereas the other will form only a few abortive parts. When this experimental result is coupled with the earlier observation that the crescent is destined to furnish the notochord, one sees in the crescent the forerunner of the "organizer" that is to play a major role in the assembly of the embryo.

In accordance with the generalizations submitted in Chapter 6, a mitotic spindle tends to orient so that its long axis coincides with that of the cytoplasmic mass within which it lies. Because the greater concentration of cytoplasm of the amphibian egg occupies the upper third of the animal hemisphere, the first cleavage spindle is displaced toward the animal pole with its axis at a right angle to the polar axis. And because the plane of cell division is always perpendicular to the spindle axis, it follows that the first cleavage is meridional, that is, it passes through the poles (Figure 7-6*B*). These mechanical rules, however, determine only that the first cleavage must be

meridional; they do not determine which meridian will separate the first two blastomeres. In most amphibians, the first cleavage plane coincides with the plane of bilateral symmetry running through the middle of the gray crescent and the sperm entry point. This suggests that the first cleavage plane is in some measure adjusted to a preexisting internal organization—an organization that is manifested by the position of the gray crescent with respect to the site of sperm penetration.

Division of the egg is complete, and thus the first cleavage may technically be described as *holoblastic*. Furthermore, the first two blastomeres are equal in size. The second set of divisions is likewise holoblastic and equal, and the planes are again meridional, but at right angles to the first plane (Figure 7-6*C*). But with the third set of divisions, the unequally distributed yolk makes its influence felt. The spindles orient parallel to the polar axis and, displaced as they are toward the animal pole, the four blastomeres not only cleave latitudinally, but unequally. The eight-cell stage, therefore, consits of four smaller animal cells and four larger vegetal cells (Figure 7-6*D*). From this time on all divisions will be unequal. So, except for the first two divisions, amphibian cleavage may be characterized as holoblastic-unequal. The planes of the fourth cleavage are in general meridional (Figure 7-6*E*), but henceforth the planes become more and more variable. The yolk also begins to make its presence known in another way: the more heavily laden vegetal blastomeres divide more slowly and with lesser frequency, so that the differences in sizes in blastomeres from vegetal to animal pole become steadily greater (Figures 7-6*F, G*).

Early in the cleavage process, a small space, the blastocoel, begins to appear within the cell cluster. Initiated as a consequence of the curvature of the inner surfaces of the blastomeres, the blastocoel rapidly increases in size and, with the more rapid multiplication of the animal cells, shifts more and more toward the animal pole. Meanwhile, it becomes infiltered by water and fluid secreted by the surrounding cells. The culmination of cleavage per se is a hollow ball, the *blastula,* featuring a relatively thin, multilayered roof of animal cells covering the fluid-filled blastocoel that is floored by large yolky blastomeres (Figure 7-6*H*). The blastocoel space is a regular feature of development in higher animals, and permits the cell movements of gastrulation to occur. It also plays a role in separating the prospective neural ectoderm cells from the endoderm until the mesoderm mantle comes between them.

It will be understood that cell multiplication goes on throughout embryogeny, in fact, in some measure throughout the life of the individual, but with the introduction of new morphogenetic events, cleavage may be said to have ended, and development enters a new phase. This is a phase during which the materials constituting the blastula are redistributed in such a way as to create a stratified embryo termed the *gastrula.* The events by which the blastula is converted to a gastrula constitute *gastrulation,* and these, to repeat, are essentially movements of areas of cells.

Unlike the Ascidians whose natural color differences facilitate the following of materials as they undergo translocation, amphibian blastulas provide little more than the distinguishing pigment of the animal hemisphere. Many investigators have added stains to the outside of the amphibian blastula, especially vital dyes, which are retained by cells for considerable periods and in no way interfere with normal processes. Specifically, a small block of agar stained with a dye such as Nile Blue Sulfate or Neutral Red is pressed against a chosen area of the blastula for a short period (Figure 7-7*A*). The dye diffuses from the block into the cells, which will retain their distinctive coloration for several days. By so marking several areas simultaneously, it is possible to follow the movements of gastrulation by continuous observation. Moreover, the per-

sistence of the dye makes it possible to establish the ultimate locations of the marked materials in the later embryo (Figure 7-7*B*). This, in turn, enables one to project the pattern of parts back onto the blastula and thus construct a map of prospective organ regions. It should be appreciated, of course, that such a map does not necessarily imply that the blastula is a mosaic of areas that differ from one another in biochemical architecture or otherwise; it represents only a description of what particular areas are

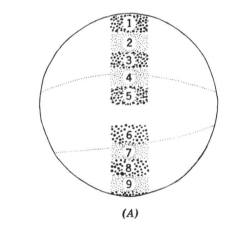

(A)

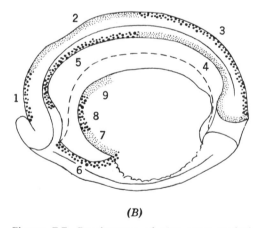

(B)

Figure 7-7 Development of nine areas marked by vital dyes during gastrulation of a urodele. Fine dots = neutral red, heavy dots = Nile Blue Sulfate. (*A*) Numbered areas along the dorsal midline of a blastula. (*B*) Positions occupied by marked areas in early neurula stage. (After Vogt.)

destined to become, not what they are. And the particular areas marked are all on the surface of the blastula, providing no information about the prospects of deeper cells.

Prospective organ region maps have been worked out for a number of species of amphibians. They differ somewhat in details, but the fundamental pattern is the same for all and is highly reminiscent of that observed in ascidian and *Amphioxus*. Let us then consider the case of a representative anuran (Figure 7-8). Roughly, the entire animal hemisphere is prospective ectoderm, with the neural ectoderm largely on the future dorsal side, and the general epidermal ectoderm on the anteroventral side. The lowermost part of the vegetal hemisphere will provide the endoderm of the gut and its derivatives. Between the prospective ectoderm and endoderm there lies a band of prospective mesoderm, broadest dorsally and tapering off ventrally on either side. This is the area that corresponds to the gray crescent of the original egg. Its broad middorsal region becomes the notochord, followed on each side by the prospective somites and lateral and tail mesoderm. Other known details, both at the surface and within the interior, need not concern us. But one other general feature bears on the displacements that we are about to discuss: the areas that will form the principal axial organs have their longest dimensions transverse to the median plane.

In other words, an outstanding feature of gastrulation will be the wheeling into the midline of materials originally at right angles to the long axis.

The basic movement is one of *spreading,* or *stretching,* and most of the technical jargon that has evolved to describe gastrulation movements relates to variations on this theme. Observations of the behavior of cells isolated in culture suggest that it is their inherent nature to elongate and flatten. But during gastrulation cells perform not as individuals but as societies, with the result that entire areas exhibit oriented spreading. Gastrulation is inaugurated by the deformation of certain prospective endodermal cells in a circumscribed area below the equator of the blastula. These cells assume the elongate shape of a bottle and move toward the interior (Figure 7-9). Their steadily elongating necks remain attached to the surface of the blastula, however, so that, as the bulky cell bodies move inward, a pull exerted along their attenuated necks creates an indentation at the surface. With continued multiplication and attenuation of the bottle cells, the indentation, *invagination,* deepens. A more significant force drawing the blastopore cells inward is provided by the spreading sheet of prospective mesoderm cells dorsal to the blastopore. Several layers of these cells interdigitate to form one layer, thus spreading the sheet forward. Because the invagination will be expanded to form the cavernous archenteron, the original indentation is appropriately termed the *blastopore,* the opening of

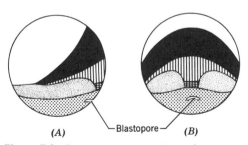

(A)   Blastopore   **(B)**

Figure 7-8 Prospective organ regions of an anuran amphibian at the beginning of gastrulation. (*A*) Side view. (*B*) Dorsoposterior view. White = epidermis, black = neural ectoderm, lined = notochord, cross lined = prechordal plate, fine stipple = mesoderm, coarse stipple = endoderm.

Figure 7-9 Schematized section through an amphibian gastrula showing elongate bottle cells. (After Holtfreter.)

the archenteron to the exterior. We have already seen that the region of the animal pole is the future anterior end of the embryo; so the blastopore marks the posterior end. And because the prospective notochord and neural ectoderm identify the dorsal side of the embryo, the area immediately above the blastopore may be spoken of as the "dorsal lip" of the blastopore. Gradually the blastoporal invagination extends circulolaterally on each side, so that the blastopore becomes crescentic, then horseshoe-shaped, and finally circular (Figure 7-10). Accordingly, the materials bounding the blastopore on the sides represent its "lateral lips" and the material on the ventral side, the "ventral lip." Obviously, however, the terms dorsal, lateral, and ventral refer only to the topography of a continuous circular lip.

Although at the start the blastopore lies wholly within the pigment-free prospective endoderm, by the time it completes the full circle the darkly pigmented tissue of the animal hemisphere is found to have moved into its lips. Thus, except for the exposed mass of light-colored yolky cells, the "yolk plug" within the blastopore, the entire embryo is covered by material originally confined to the animal hemisphere alone. This is a way of saying that as an accompaniment of the maturation of the blastopore, there is a steady streaming of peripheral materials toward its lips. Of equal significance, the materials that converge on the blastopore do not simply pile up, but move through the blastoporal lips to the interior and are redistributed. That is, the substance originally in the lips of the blastopore "turns the corner"—undergoes *involution*—and moves to the interior, to be followed by other materials that move in to replace it. The lips of the blastopore are therefore not anatomical entities but only the constantly changing structural expression of the turning of surface tissues toward the inside. Let us then look at the external and internal streaming movements of the cell sheets involved (Figures 8-5 and 8-6).

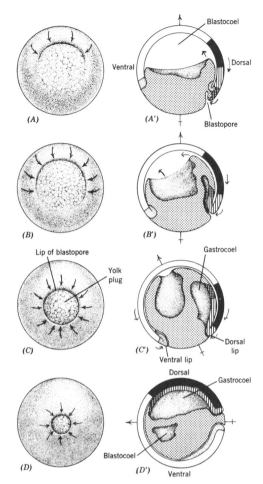

Figure 7-10 Germ layer formation in an amphibian. (*A* through *D*) Surface views looking toward the blastopore. (*A'* through *D'*) Sagittal sections. Rotation of the embryo is indicated by the change in orientation of the anteroposterior axis and the dorsal and ventral sides. Arrows indicate direction of morphogenetic movements. Symbols for prospective organ regions are the same as for Figure 8-3.

Initiation of the blastopore is accompanied by a flow of surface material toward it. As already noted, the blastopore originates within the prospective endoderm; thus the substance originally abutting the dorsal lip will provide the forepart of the gut. Reaching the interior by involution around the dorsal lip, this pharyngeal endoderm moves steadily forward toward the head region. The cell movement is an active migration along a network of fibronectin

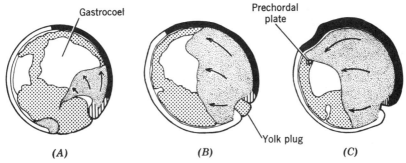

Figure 7-11 Internal streaming of the mesodermal mantle during the later phases of gastrulation of a urodele. Fine stipple = mesoderm, coarse-stipple = endoderm. (Based on figures by Nelsen.)

protein molecules secreted by the cells of the roof of the blastocoel. The cell membranes of the invaginating layers attach to the fibronectin material and pull themselves forward along it. Meanwhile, the prospective pharynx is trailed by the prechordal mesoderm which, in turn, is followed by the anterior part of the prospective notochord. As these materials flow inward around the dorsal lip, they become considerably narrowed and, in the process of forward flow, greatly elongated. Because the totality of the prospective chordamesoderm in the blastula is laid out as a broad band transverse to the long axis of the future embryo, the posterior part of the notochord, the somites, and lateral mesoderm (hypomere) will move along pathways that ultimately bring them into longitudinal orientation. At the surface, the flow of these more laterally disposed materials is along radial lines that converge on the blastopore. Looked at somewhat differently, the steady transition of the blastopore from crescent to horseshoe to circle is a manifestation of the orderly wheeling in of the mesodermal areas from the sides and bottom. The rear of the notochord converges on the dorsolateral lips of the blastopore, followed by the prospective somites and finally the lateral mesoderm. Once inside, these materials take different courses. The notochordal and somitic mesoderm streams upward and stretches out along the longitudinal axis; the lateral and ventral mesoderm spreads out broadly in a forward and later downward direction between the ectoderm and endoderm. It should be understood, of course, that although we have spoken separately of the movements of prospective prechordal mesoderm, notochord, somites, and hypomere, the entire mesodermal mantle moves as a unit. Looked at as a unit, the flow at the surface is along radices converging on the circlar blastopore; within the interior there is a concomitant streaming of the dorsal and dorsolateral components toward and along the dorsal midline, and a fanlike, divergent spreading of the hypomere along the sides.

Throughout gastrulation the embryo retains its spherical shape and a uniform size. This means that as soon as the prospective pharyngeal endoderm and the mesodermal mantle move out of the surface, their places are taken by ectoderm. To accomplish this, both neural and epidermal areas exhibit a pronounced thinning and spreading. With respect to the neural ectoderm, there is a concomitant reorientation comparable to that observed in the axial mesoderm, namely, a convergence upon and stretching out along the median axis, so that material originally disposed transversely to the long axis comes to lie parallel to it. As for the epidermal ectoderm, it tends to fan out much as does the hypomere beneath.

It is in consequence of the removal of endodermal and mesodermal material from the surface and the compensating spreading of the ectoderm that ectoderm finally arrives at the circular lip of the blastopore; that is, the yolk plug comes to be bounded by ectoderm. Meanwhile, the forward movement of the early invaginated pharyngeal endoderm has been supplemented by an internal spreading of the yolk-laden general endoderm in a roughly anteroventral direction. Several consequences derive from this. It serves, first, to deepen steadily the archenteron and, coincidentally, to reduce and eventually eliminate the blastocoel. Second, as the endodermal mass accumulates on the future ventral side, the center of gravity is shifted and the embryo rotates so as to bring its dorsal side uppermost. Third, the protruding yolk plug gradually withdraws to the interior, and, as it does so, the diameter of the blastopore steadily contracts until at the end of gastrulation, the blastopore is recognizable only as a narrow slit.

The events that we have described so far create a spherical gastrula clothed externally by ectoderm and containing endodermal and mesodermal components internally. The part of the ectoderm that lies lengthwise as a broad, middorsal band is the forerunner of the nervous system; the remainder is potential epidermis. Within the interior, the archenteron represents the lumen of the gut to come. For the present, the walls and floor of the primitive gut, or *archenteron,* are composed of endoderm, but in most amphibians its roof is of mesoderm, specifically the middorsal mesoderm, which will provide the prechordal plate and the notochord. In considerable measure, then, the internal situation is comparable to that seen previously in *Amphioxus:* a composite inner layer involving both endoderm and mesoderm. There thus remains to be considered the segregation of definitive endoderm and mesoderm and the creation of a purely endodermal archenteron. These events actually have their beginning at an earlier time in gastrulation, and their completion overlaps the first steps in organogenesis.

We must remember that prior to gastrulation the prospective mesoderm and prospective endoderm are continuous on the surface of the blastula. They remain continuous so long as the mesoderm pursues its convergent flow toward the blastopore. But it should also be recalled that whereas the mesoderm enters the interior by involution around the lips of the blastopore, the bulk of the endoderm moves inward by mass flow without either involution or invagination. Only the prospective pharynx follows the involution pattern. These different styles of flow foreshadow the completely divergent pathways that will be taken by endoderm and mesoderm within the interior. As soon as the mesodermal components reach the inside, they separate from the endoderm (except in the pharyngeal region); thus the mesodermal mantle acquires a free margin on each side. The middle of the mesoderm spreads forward, but its free edges fan out ventrolaterally. Meanwhile, the massive endoderm that has glided onto the ventral side becomes depressed along its middorsal surface. The depression gradually deepens and, as it does so, its bounding margins steadily elevate. In other words, the margins of mesoderm and endoderm exhibit countermovements: as the free edge of the mesodermal mantle moves forward and downward on either

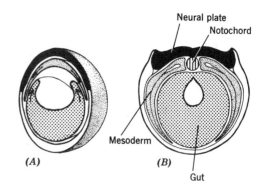

Figure 7-12 Diagrams of neurulation in a urodele. (*A*) Countermovements of margins of mesodermal mantle and endodermal ridges of gut. (*B*) Early differentiation of notochord, mesoderm, and neural plate.

side, and endodermal ridges flanking the trough in the floor move upward (Figure 7-12*A*). With the meeting of the endodermal ridges in the dorsal midline, a tubular, endodermal gut is created. Concomitantly, that mesoderm in the dorsal midline is set off as a definitive notochord, separate from the prospective epimere, mesomere, and hypomere on each side; and the overlying neural ectoderm exhibits the thickening and marginal elevations anticipating the neural tube (Figure 7-12*B*). Recently, marking experiments have shown that some amphibians, notably *Xenopus,* separate the involuting mesoderm from the endoderm at the site of the dorsal lips of the blastopore. This produces an archenteron initially completely lined with endoderm, and apparently avoids the process of the elevation of the labial walls of endoderm.

Chapter Eight
Avian Cleavage and Germ Layer Formation

THE OVUM AND FERTILIZATION

The actual ovum of a bird (the domestic chicken, for example) is the part of the "egg" that is commonly called the "yolk." Food resources make up the bulk of the ovum; the cytoplasm of the cell is confined to a small plate, termed the *blastodisc,* which is approximately 3.0 millimeters in diameter and occupies the animal pole (see Figure 4-15). The first maturation division and the extrusion of the first polar body occur just prior to ovulation. Completion of maturation and formation of the second polar body take place after sperm penetrate the ovum. As we have noted elsewhere (Chapter 5), the mechanism for barring the entrance of more than one spermatozoon does not operate in the highly yolky eggs of birds, reptiles, and some amphibians.

In birds, *polyspermy* prevails. Some 3 to 5 spermatozoa may enter the chicken's ovum, and in pigeons the number may range from 12 to 25. Only one of these, however, consummates fertilization by fusing with the egg nucleus; the remainder are shunted into the yolk where they form subsidiary nuclei whose purpose is not entirely clear. In any event, they shortly degenerate.

After fertilization, the ovum passes down the oviduct, acquiring en route layers of albumen, shell membranes, and shell (Figure 4-15). The oviducal journey consumes 15 to 20 hours, during which time cleavage takes place and germ layer formation is initiated. Much difficulty has attended the making of complete and accurate observations on the developmental events of these early hours. Because initial embryogenesis occurs while the ovum is traversing the oviduct, properly spaced stages are hard to acquire. Also, the young blastodisc is fragile and subject to distortion by the conventional histological methods of fixation, staining, and sectioning.

CLEAVAGE AND BLASTULA

Because the mass of inert yolk is very great, that is, because a macrolecithal condition prevails, the yolk itself does not become subdivided. Cleavage is confined to the blastodisc and thus is described as being *meroblastic* (partial), in contrast to the holoblastic (total) cleavage seen in eggs containing moderate or small amounts of yolk. Another factor conditioning the cleavage pattern is the orientation of the early generations of the mitotic spindles. The fact that the cytoplasmic blastodisc has the form of a shallow, biconvex plate means that for a time the long axes of the spindles lie parallel to the longest dimension of the cytoplasm and the divisions are therefore meridional.

The first division appears as an irregular furrow extending across the central portion of the blastodisc. The furrow not only falls short of the margin of the disc, but also fails, to cut completely through the depth of the disc (Figures 8-1A, a). This results in two incompletely separated blastomeres whose cytoplasmic contents are continuous around the ends of and beneath the furrow. It has yet to be established whether the first furrow coincides with the long axis of the prospective embryo.

The second cleavage division, takes place at right angles to the first (Figure 8-1B). The furrows may intersect the first one at the same central point, in which case they form a continuous furrow, or they may be off center. Also, like the first cleavage furrow, the second set of furrows neither reach the margin of the blastodisc nor cut entirely through its depth.

A third set of vertical furrows, roughly at right angles to the second, but showing

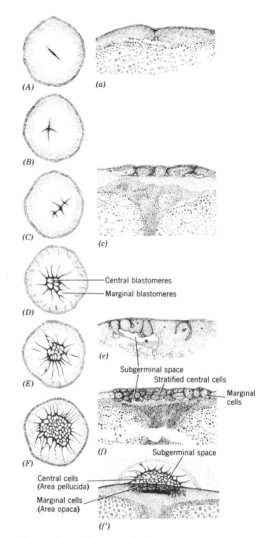

Figure 8-1 Cleavage of the germinal disc of the ovum of the domestic chicken. Left-hand figures are surface views; right-hand figures are transverse sections. (*A,a*) 2-cell stage. (*B*) 4-cell stage. (*C,c*) 8-cell stage. (*D*) 16-cell stage. (*E,e*) 32-cell stage. (*F,f*) 154-cell stage. (*f'*) Diagrammatic hemisection of blastoderm of 154 cells. (After Patterson, *J. Morphol.*, Vol. 21.)

considerable variability, produce eight incompletely separated blastomeres (Figure 8-1C). From here on the pattern of cleavage becomes very irregular. As seen from the surface, two groups of blastomeres become set off: (1) completely bounded *central blastomeres* formed by the cutting off of the inner ends of the original incomplete cells, and (2) incompletely bounded *marginal blastomeres* produced by furrows running in a radial direction (Figure 8-1*D*). Additional central cells are created by subdivision of the original ones and by cutting off the central ends of the marginal cells, and at the same time the marginal cells are multiplied by further peripheral, radial divisions that ultimately reach the margin of the disc (Figure 8-1*E, F*).

Shifting attention from the surface to the deeper levels of the blastodisc, the third set of furrows, although originally shallow and vertical like the first two, soon curve so as to run horizontally and parallel to the surface and thus undercut the cytoplasm and separate it from the yolk (Figure 8-1*c*). This process is repeated as the central cells increase in number, so that the area of central blastomeres becomes free from the underlying yolk while the marginal blastomeres remain in contact with the yolk. Two other concomitant events also occur: (1) the yolk beneath the central area liquefies (Figure 8-1*e*), and (2) divisions of the blastomeres in

planes parallel to the surface bring a stratification of cells (Figure 8-1*f*). The result is a compact plate, several cells thick, which is called the *blastoderm*. The freeing of the central blastomeres from the yolk creates the *subgerminal space,* which is filled with the aforementioned liquefying yolk. At this stage of development it is customary to abandon the terms *central* and *marginal* cells and refer instead to two general areas in the blastoderm: the *area pellucida,* beneath which lies the subgerminal space, and the *area opaca* (*opaqua*), still in contact with the yolk (Figure 8-1*f′*).

Closer scrutiny of the blastoderm reveals two other features: (1) one portion of the blastoderm, known by its future development to represent the posterior end, is somewhat thicker because the cells are more numerous; (2) the cells throughout vary in size and, although the bulk of the yolk is external to the blastoderm, contain some yolk—the larger ones more, the smaller ones less (Figure 8-2*A*). There ensues a segregation in which the larger yolk-

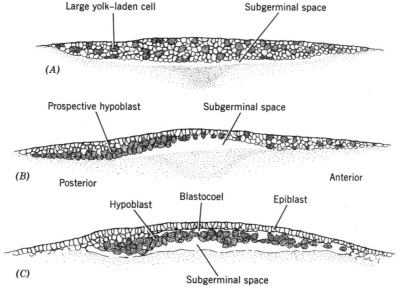

Figure 8-2 Longitudinal sections through blastoderms of successive ages showing origin of the hypoblast. (*A*) Larger yolk-laden cells intermixed with smaller cells. (*B*) Accumulation of larger cells in subsurface position, notably at the rear. (*C*) Organization of larger cells to form hypoblastic layer, concomitantly with rise of blastocoelic space and thinning of epiblast. (Based on data from Pasteels and Peter.)

rich cells gradually accumulate at the undersurface of the blastoderm, leaving the smaller yolk-poor cells at the surface (Figure 8-2*B*). The yolk-poor cells align themselves into an epithelium, which is largely only a single cell thick. The yolk-rich cells, however, do not for some time become organized in a coherent sheet. At the rear, especially, they remain piled up two or three deep; toward the front they tend to be scattered and isolated.

In the posterior portion of the area pellucida, where the larger cells are most numerous, a number of irregular horizontal slits appear between the two categories of cells. These thin cavities gradually expand and coalesce. In consequence, there is created a narrow space that demarcates *epiblast* and *hypoblast* (Figure 8-2*C*). But for the present a definitive hypoblast exists only at the rear; forward, the number of hypoblastic cells remains too few to organize a complete layer. Thus, from rear to front the hypoblast frays out in scattered islands of cells (Figure 8-2*C*). Further differentiation of the hypoblast overlaps other events, which are still to be considered. The hypoblast steadily spreads forward and laterad and in so doing becomes transformed into a very thin epithelium. The factors that cause this spreading are (*a*) continued multiplication of the cells of the hypoblast that were formed earlier, especially at the rear of the blastoderm; (*b*) further contributions from the epiblast by cells moving downward via the later-developing primitive streak; and (*c*) inherent tendencies of the cells to flatten and spread when in contact with the undersurface of the epiblast.

A point of considerable theoretical importance is the interpretation of the subgerminal space and the later-appearing cleft between the epiblast and hypoblast. If the hypoblast (endoderm) arises through the undertucking of the free posterior edge of the blastoderm, the subgerminal space becomes the equivalent of the blastocoel into which the prospective endoderm moves. After all, it is into the blastocoel of the em-

bryo of an amphibian or *Amphioxus* that the endoderm invaginates or flows, and the avian situation might be considered comparable. However, initiation of the hypoblast is in large measure a consequence of splitting, or delamination. Accordingly, it is the narrow cleft between the epiblast and hypoblast that should correspond to the *blastocoel.* Thus, the bilaminar blastoderm whose origin and form we have just considered may be said to represent the *blastula,* although a very much flattened one compared with that of *Amphioxus* or an amphibian (Figure 8-3). The epiblast represents the animal half of the blastula, the hypoblast the vegetal half, and the intervening space the blastocoel. But unlike Amphibia and *Amphioxus,* a large part of the endoderm in birds is precociously set off and, except for its continued elaboration in the manner described earlier, undergoes no translocation during gastrulation.

If we allow for considerable variation in the status of the endoderm (hypoblast), the blastula as we have just described it represents the state of affairs at the time the egg is laid, that is, at the zero hour of incubation. With the beginning of incubation, development enters a new phase.

GASTRULATION

In the discussion of amphibian gastrulation it was emphasized that as a consequence of so-called formative or morphogenetic movements, given areas of material assumed new relative positions and became established as the primary germ layers. It was also pointed out that by employment of the technique of vital staining, not only could the movements of these areas be followed, but their ultimate developmental prospects could be ascertained. By projection of the fates of areas back to the original blastula, there could be constructed what is known as a prospective organ-region or *fate map*. Utilization of the same technique, plus the additional devices of marking parts

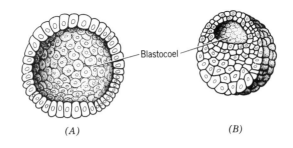

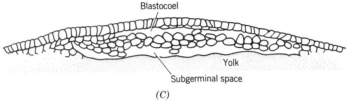

Figure 8-3   Comparative anatomy of the blastulae of (*A*) Amphioxus, (*B*) an amphibian, and (*C*) a bird.

of the blastoderm with carbon particles or tagging with radioactive thymidine, has provided a comparable understanding of the avian embryo.

We have seen that at the beginning of gastrulation the hypoblast (endoderm) is already present in part. We now encounter a major difference in the fates of the embryonic cells of the avian embryo and those of the amphibian. Many of the cells of the blastoderm will be involved in the construction of the organs necessary for the support of the embryo before hatching, but will not be part of the body of the embryo. These organs will ultimately be lost or consumed by the time of hatching. They are said to be *extraembryonic,* in contrast to the *embryonic* tissues. Their primordia can be located on the blastoderm fate map, making it significantly different from that of the frog. The main concern, then, is with the upper layer, the epiblast, which is prospective ectoderm, mesoderm, and intraembryonic endoderm. Individual workers have provided maps of the epiblast that differ somewhat in details, yet the general layout is fairly well agreed upon (Figure 8-4). We center attention on the epiblast of the area

pellucida, for the epiblast of the entire area opaca is destined to provide extraembryonic ectoderm. The anterior two thirds of the area pellucida is prospective ectoderm. The greater part of the ectodermal area is prospective epidermis, and the bulk of this is extraembryoic. A crescentic area occupying the rear of the ectodermal area is prospective neural tissue. Immediately posterior to this is the prospective notochord. Just behind the notochord, the endoderm of the gut and part of the yolk sac occupy the center of the fate map, flanked by lateral and extraembryonic mesoderm. The posterior end of the map is completed by the area of extraembryonic mesoderm. It should be evident that despite the exaggeration of the horizontal dimension, the general arrangement of these areas, with respect to both each other and the future long axis, is comparable to that in Amphibia (Figure 7-8).

The morphogenetic movements that will segregate the prospective mesoderm from the ectoderm and bring the mesoderm to its ultimate position between the developing hypoblast below and the residual ectoderm above are fundamentally like those

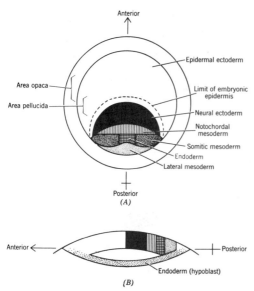

Anterior

Epidermal ectoderm

Area opaca

Limit of embryonic epidermis

Area pellucida

Neural ectoderm

Notochordal mesoderm

Somitic mesoderm

Endoderm

Lateral mesoderm

Posterior
(A)

Anterior ← → Posterior

Endoderm (hypoblast)

(B)

Figure 8-4 Prospective organ regions in the chick blastoderm immediately prior to incubation. (*A*) Surface view of epiblast. (*B*) Diagrammatic section along midline axis.

already seen in amphibians: involuting, convergent streaming, involution, and spreading or elongation. The one important difference is that whereas in Amphibia the total mass of the embryo remains fairly constant during gastrulation, the avian blastoderm increases in mass and diameter through steady growth. The delineation of definitive mesoderm is accompanied by supplemental contributions to the hypoblast.

With the resumption of development caused by the heat of incubation, the lateral mesoderm begins to converge on the posterior midline (Figure 8-5*A*). The resulting aggregation of cells begins to create a broad, thickened area in the epiblast known as the *primitive shield.* Within the next hours of development, the convergence of more epiblast cells toward the center of the shield causes it to become elongated along what will be the central axis of the embryo. This elongation results in a structure consisting of a longitudinal groove, the *primitive groove,* separating two slightly elevated ridges along its sides, the *primitive folds*

(Figure 8-5*E*). At its anterior end, the folds bend toward each other and converge on a slightly elevated area, the *primitive knot,* or *Hensen's node.* The entire structure is designated the *primitive streak,* and its completion is accompanied by a change in shape of the area pellucida from circular to pear-shaped in outline (Figure 8-5*B*).

A second major distinction between development in the chick and in the frog is now evident. While invagination in the frog involves folding and retraction of a sheet of cells into the blastocoel, forming an archenteron and a blastopore, the separation of these layers from the ectoderm in the chick involves the process of *ingression.* From the earliest stages of convergence of epiblast cells toward the midline of the primitive streak, epiblast cells begin to separate from each other as they move down through the primitive groove into the blastocoel. Their epithelial relationships are lost, and they contiue to migrate as individual cells, or mesenchyme (Figures 8-5 and 8-6). The first cells to undergo ingression are those destined to form the gut and the head mesoderm. The prospective gut endoderm cells insinuate themselves into the hypoblast endoderm already lying under the primitive streak, and the head mesoderm forms between them and the overlying epiblast. These cells are soon followed by epiblast that was lateral in position to them, the lateral mesoderm. The majority of ingressing cells move medially toward the streak, down through the groove, and then back laterally toward their final positions in the embryo. Epiblast cells directly in front of the primitive streak now move posteriorly toward it, progress down through the front end of the streak, and then move anteriorly again over the hypoblast. These cells will become the notochord, and are designated the *head* or *notochordal process.*

These processes of convergence and ingression not only initiate the arrangement of mesodermal and endodermal tissues between the epiblast and hypoblast, but also

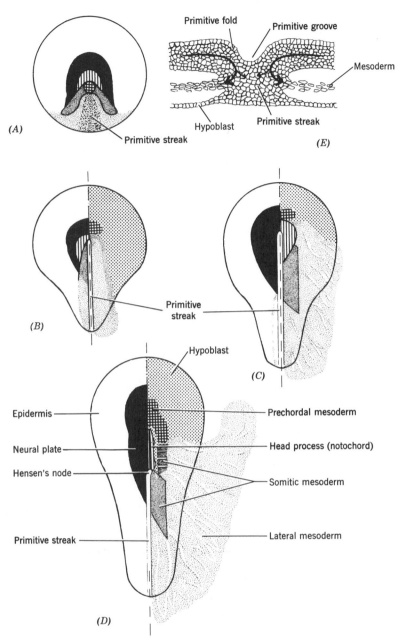

Figure 8-5   Primitive streak and germ layer formation in the chick embryo. (*A*) Surface view of the epiblast. (*B–D*) The left side pictures the surface and the right side the pattern of the mesodermal areas as they spread over the hypoblast below. (*E*) Diagrammatic cross section showing the involution of mesoderm through the primitive streak. See text for further description. White = epidermal ectoderm, black = neural ectoderm, lined = notochord, crosslined = prechordal mesoderm, coarse stipple = endoderm (hypoblast), dense fine stipple = somitic mesoderm, light fine stipple = lateral mesoderm.

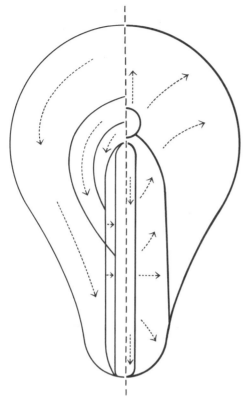

Figure 8-6 General pattern of morphogenetic movements during the formation of the primitive streak in a chick embryo. (Based on data from Rosenquist, *Contributions to Embryology,* Vol. 38.)

cause the primitive streak to be elongted to its full length (Figure 8-5*D*). The final stages of gastrulation in the avian embryo involve a continuation of convergence and ingression, but now accompanied by shortening of the streak as the elements of mesoderm that will lie closest to the axis of the embryo are laid down. These are the epimere, or somite, mesoderm, and the notochord.

The establishment of the head process marks the completion of the elongation of the primitive streak. All further lengthwise streaming movements within the streak are directed rearward. Hensen's node, in a sense, travels backward, and the streak gradually shortens. As it does so, those areas that originally flanked the streak come to lie in front of it (Figure 8-5*D*). The streak ultimately becomes reduced to a fragment in

the tail bud and finally disappears, signifying that the deployment of mesoderm is complete; thus the entire embryo develops anterior to the streak. The nature and fate of the streak suggest to us that, like the amphibian blastopore, the primitive streak should be looked upon simply as the morphological expression of a transitory developmental event rather than as an organ in the strict sense.

As the node and streak move posteriorly, the head process and prechordal mesoderm stretch out anteriorly in accompaniment to the forward elongation of the blastoderm as a whole. Closer scrutiny of the head process reveals that it consists of a thicker central mass of cells and more diffuse lateral wings. Initially it is also blended in the midline with the hypoblast. The thicker central portion alone provides the definitive notochord, whereas the lateral wings contribute to the epimere (somitic) mesoderm that comes to flank the notochord. With its differentiation, the notochord becomes detached from the hypoblast (endoderm) below, except for a more persistent fusion at the extreme anterior end.

At the start of the morphogenetic movements that culminate in the laying down of the various mesodermal components, the hypoblast was an incomplete layer that was spreading forward from the rear of the blastoderm. There is some evidence to indicate that the developing hypoblast may actually exercise a causal influence on the primitive streak and thus on the deployment of mesoderm. It has been shown, for example, that if the hypoblast is detached from the epiblast and reoriented, the long axis of the streak is correspondingly changed; or if the hypoblast is simply removed, the primitive streak fails to form. These observations suggest that the hypoblast provides a necessary substratum for the oriented deployment of prospective mesoderm. Reciprocally, the completion of the hypoblast depends in part on cellular contributions from the epiblast, which migrate through the streak. The definitive endoderm, therefore, has a dou-

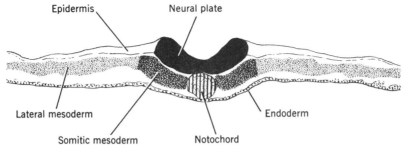

Figure 8-7 Diagrammatic cross section of chick embryo to showing early organogenesis.

ble origin: (1) in the multiplication and spreading of cells originally set off at the rear of the area pellucida (p. 133) and (2) in the involution of supplementary cells through the primitive streak.

The events that have been described serve to create an embryo consisting not only of three primary germ layers, but also of the beginnings of definitive parts and organs (Figure 8-7). The midline neural ec-

toderm exhibits the thickening and, anteriorly, the marginal elevations that form the neural tube; the notochord becomes separated from the adjoining sheets of mesoderm; the median portion of the endoderm represents the prospective digestive tract. This is the point at which study of the chick embryo is undertaken in the laboratory and at which organ systems are introduced in the chapters to come.

Chapter Nine
Early Human Development and Placentation

Although the various orders of mammals have much in common, the details of their early development are sufficiently different so that no one form can be selected as typical of the entire class. For this reason any example chosen for description must to a degree be considered a special case. In our choice of the human subject, therefore, we are deferring to the intrinsic interest that we all have in human development rather than the human's status as a representative mammal. At the same time, we shall not hesitate to introduce facts derived from other forms, even nonmammals, whenever these facts may help clarify the special human conditions that confront us—a reminder that the pattern of human morphogenesis is a reflection of the history of the human's vertebrate relatives. Moreover, no discussion of human development

or, for that matter, development of any mammal can proceed very far without involving a consideration of extraembryonic membranes and the relationship of the embryo to the uterus. *The title of this chapter, "Early Human Development and Placentation," thus indicates only the main thread of the discussion, a thread that is woven in a broader fabric of reproductive physiology and comparative morphogenesis of vertebrate embryos and their membranous envelopes.*

THE EXTRAEMBRYONIC MEMBRANES OF VERTEBRATES

In the development of every vertebrate there are produced certain tissues or structures that, temporarily or permanently, do not enter into the formation of the embryo itself, but are external to and devoted in one way or another to the care and maintenance of the embryo. Collectively these parts are spoken of as *extraembryonic membranes.* A completely logical treatment of these structures would begin with the cyclostomes and proceed methodically upward to mammals and humans. Instead of pursuing this course, let us begin with the chick embryo, in which all the basic membranes found in vertebrates are represented.

THE EXTRAEMBRYONIC MEMBRANES OF BIRDS

In the development of the avian blastoderm the somatopleure and splanchnopleure gradually spread peripherally over the yolk mass, far beyond the area where the body

of the embryo is taking form. Shortly, however, the embryo begins to be undercut by a series of body folds that serve to delimit the embryonic regions from the more peripheral extraembryonic somatopleure and splanchnopleure (Figure 9-1).

The extraembryonic splanchnopleure continues to spread over the yolk mass and, as a *yolk sac,* eventually encloses the yolk (Figures 9-2 to 9-4). Coincidently, the intraembryonic splanchnopleure is subjected to folds, analogous to the more superficial body folds, which serve to establish a walled sigestive tract, or gut, in the body of the embryo. The middle of the gut, though, remains open to the yolk beneath, and at this level the walls of the yolk sac are continuous with the walls of the gut as a constricted *yolk stalk.* Despite the fact that the yolk sac is connected to the digestive tract by the yolk stalk, the yolky food reserves are not transmitted to the embryo by this

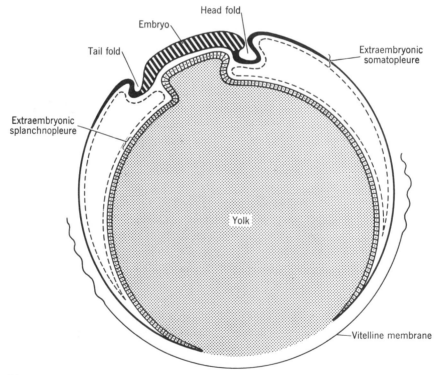

Figure 9-1 Schema of an early chick embryo showing body folds delimiting embryo from extraembryonic areas.

route. Rather, the endodermal surface of the sac is thrown into folds that penetrate the yolk mass and, through the mediation of appropriate enzymes, steadily digest the yolk, after which it is absorbed and transported to the embryo by the *vitelline veins* provided by and carried in the mesoderm of the sac.

The extraembryonic somatopleure is elevated over the embryo by a folding process consisting essentially of a doubling of the somatopleure upon itself. The initial elevation is over the head end of the embryo, producing a double somatopleuric hood. As the hood gradually works backward, its caudally extending side limbs arch over the embryo from each side, to be joined finally by a similar elevation over the tail. All these folds ultimately converge so as to encase the embryo in two sheets of somatopleure. The inner somatopleuric sheet becomes the *amnion,* and the outer, the *chorion* (Fig-

ures 9-2 to 9-4). The cavity between the amnion and the embryo is bounded by ectoderm and is termed the *amniotic cavity;* that between the amnion and chorion is the *chorionic cavity,* which, being bounded by mesoderm, is the *extraembryonic coelom.* Muscle fibers differentiate within the mesoderm of the amnion, and the amniotic cavity comes to be filled with fluid. The function of the amnion with its muscle fibers and contained fluid is to protect the embryo from adhesions and mechanical shock, as well as to buoy the embryo in a salty "pond" at a time when its own soft, collapsible tissues lack adequate skeletal support. The chorion joins with the allantois, as described later, to serve in a respiratory and nutritional capacity.

The *allantois* arises as a splanchnopleuric outgrowth from the rear of the gut, from where it extends into and eventually largely fills the extraembryonic coelom (Figures 9-

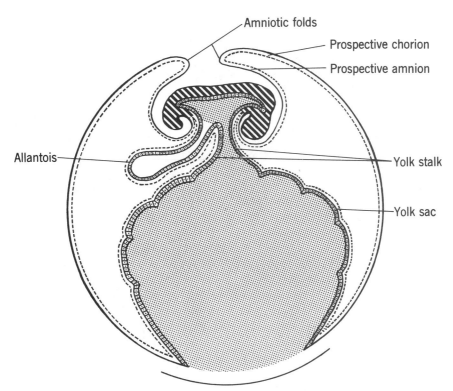

Figure 9-2   Early stage in the development of the extraembryonic membranes of the chick.

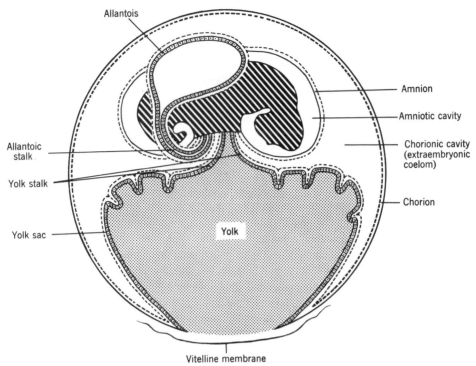

Figure 9-3   Later stage in the development of the extraembryonic membranes of the chick.

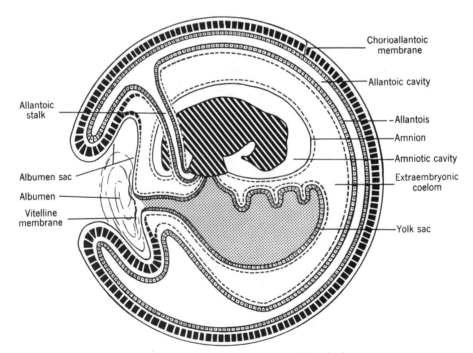

Figure 9-4   The fully matured extraembryonic membranes of the chick.

2 to 9-4). Its mesoderm fuses with that of the chorion and, to a considerable degree, also with that of the amnion and yolk sac. Together with the chorion, the allantois also surrounds the albumen to form the *albumen sac.* The conjoined mesoderm of the chorion and allantois becomes highly vascular; that is, an allantoic circulation develops. The close proximity of this vascular area to the inner surface of the porous shell allows the "chorioallantoic membrane" to function as an efficient organ of respiration. In addition, the cavity of the allantois serves as a receptacle for excretory waste, the mesoderm of its interior wall contributes to the muscle supply of the amnion, and as a part of the albumen sac it assists in the absorption of nutritional albumen (Figure 9-4).

These, then, are the four extraembryonic membranes that concern us: amnion, chorion, yolk sac, and allantois. And they concern us not just as pieces of anatomy to be described, but as devices utilized in providing for adequate nutrition, protection, and satisfaction of metabolic requirements of the developing embryo.

EVOLUTION OF EXTRAEMBRYONIC MEMBRANES

In the embryos of cyclostomes, fishes, and Amphibia the only extraembryonic membrane formed in some type of yolk sac. (Lacking the other membranes, including the amnion, these vertebrates are commonly referred to collectively as the *anamniota* to distinguish them from the amnion-possessing reptiles, birds, and mammals called *amniota.*) The yolk sac actually presents itself in two variations. In the Amphibia, whose eggs are moderately yolky, the yolk material is within the endodermal cells of the gut (Figure 9-5), whereas the arrangement deriving from the macrolecithal eggs of cyclostomes and teleost fishes is one in which the yolk comes to lie as a noncellular mass within the gut cavity (Figure 9-6). In either case the gut endoderm

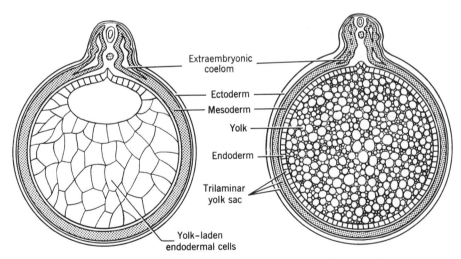

Figure 9-5 (*left*). Schematic cross section of an amphibian embryo showing yolk incorporation into cells of gut. (From Hamilton, Boyd, and Mossman, *Human Embryology,* courtesy W. Heffer and Sons Ltd., Cambridge.)

Figure 9-6 (*right*). Schematic cross section of a teleostean embryo showing yolk invested by a trilaminar yolk sac. (From Hamilton, Boyd, and Mossman, *Human Embryology,* courtesy W. Heffer and Sons Ltd., Cambridge.)

is bordered by mesoderm, and both are enclosed within the ectoderm of the body wall. Because in Amphibia the yolk lies within the endodermal cells, the yolk sac is really an integral part of the gut. But in the other anamniotes there exists a truly definitive sac composed of three layers—ectoderm, mesoderm, and endoderm. The mesodermal layer becomes vascularized, and the nutritive material digested and absorbed by the endodermal layer is transported to the embryo by vitelline blood vessels. As the embryo grows, the yolk sac shrinks and eventually is incorporated within the ventral body wall and gut.

The origin of the amnion, chorion, and allantois almost surely went hand in hand with the "invention" of the terrestrial egg, but we can only speculate on what actually occurred. Possibly something like the following took place (Figure 9-7). As the embryo grew and the yolk mass was used up, the embryo might have sunken into a collapsing yolk sac whose margins would concomitantly elevate over the body of the embryo. This very thing happens, in fact, in the developing bird embryo. But recall that the basic yolk sac is a trilaminar one. This means that originally the elevation of the yolk sac would bring all three layers over the embryo. In the meantime, the saccular allantois originated from the rear of the gut, providing a reservoir for nitrogenous waste, which in a terrestrial egg could no longer be disposed of in surrounding water. The wall of this allantois, consisting of splanchnopleure, was well vascularized. Its enlargement was facilitated by the initiation and expansion of a split in the mesoderm of the trilaminar yolk sac. In other words, an extraembryonic coelom come to lie between a now definitive splanchnopleuric yolk sac on the interior and an extraembryonic somatopleure on the exterior. The latter, folding upon itself, provided the amnion and chorion. The complete enclosure of the embryo by the amnion allowed for the enclosure of the embryo in a water-filled chamber; continued expansion of the allantois brought an increasing blood supply to the chorion, with resultant improvement in respiratory exchange.

The primitive pattern of reproductive behavior is one in which eggs, laden with yolk, are simply deposited in water with the inevitable high mortality rate compensated for by the deposition of a large number of eggs. The yolk provides the nutritional reserves; the aquatic environment furnishes oxygen and serves as a medium into which metabolic wastes may diffuse. When eggs came to be deposited on land, supplementary devices served the metabolic requirements of the embryo: yolk was still carried in a sac, but water resources were conserved by an investing shell and augmented by amniotic fluid, nitrogenous waste was poured into the allantois, and the chorion, vascularized by the allantois, took care of respiration.

Animals developing from externally deposited eggs, independent of the parents, are described as being *oviparous.* Most fishes, amphibians, and reptiles and all birds are oviparous.

But along the way (we shall not speculate on how it came about) some animals began to enhance the probability of survival of eggs and embryos by producing fewer of them and employing some variety of parental care. This care, of many forms and degrees, culminated in the retention of the eggs by a parent, either at some site within the female reproductive tract or in some specially prepared brood-pouch in either the male or female. The young were then "born alive," that is, they were *viviparous.* So we find some "live bearers" in every class of vertebrates, except the cyclostomes and birds. The mammals, expecially, have exploited this reproductive device, for all of them with the exception of the monotremes bear their young alive.

Notable examples of viviparity are found among the Chondrichthyes. In some sharks the well-vascularized yolk sac is directly op-

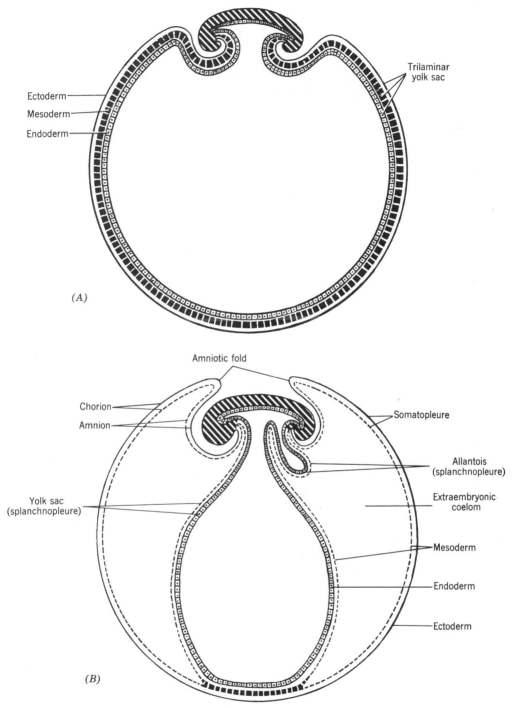

Figure 9-7 Schemata illustrating theoretical origin of extraembryonic membranes of amniotes from a trilaminar yolk sac. (*A*) Downward sinking of embryo into roof of trilaminar yolk sac whose margins are elevating over embryo. (*B*) Mesoderm split so as to separate splanchnopleuric yolk sac from external somatopleure. Allantois projects into extraembryonic coelom and somatopleure folds over embryo to provide amnion and chorion.

posed to the inner surface of the uterus, with the resultant transfer of metabolites; in other sharks embryos are nourished by swallowing surplus immature eggs. In some skates and rays the uterus provides a "milk" that is absorbed either through the yolk sac or the gills of the embryo. Certain teleost fish are also viviparous. These fish have no true oviducts, and embryos develop either within the ovaries themselves or in the "gonaducts," which are extensions of the ovarian walls. Embryos may draw support from ovarian secretions by way of the gills or elaborate modifications of the embryonic gut. Among amphibians, there are instances in which the larvae are carried in pockets of skin on the back of the female, using their well-vascularized tails for metabolic exchange with the cutaneous circulation of the mother.

Among the amniota, viviparity is the rule in all mammals except the monotremes. Conversely, it never occurs in birds. In reptiles, although egg laying is the common practice, some snakes and lizards are viviparous. Some of them use the yolk sac in association with the lining of the uterus, others use the chorion, and still others use both.

In retrospect, it is difficult to point to an orderly series of steps toward the evolution of true viviparity. Although it is obvious that viviparity has been derived from oviparity by the retention of eggs or embryos by the female, the devices and methods for their nurture are too wide ranging to permit absolute conclusions on which arrangement may have preceded another. This is especially true among the anamniotes, in which the wide range of devices suggests a randomness of invention rather than orderly succession. The anamniotes may have approached the achievement of viviparity, with varying degrees of success, by trying out many alternative methods within the limits of the structures available to them. The lack of an allantois precluded the kind of arrangement exploited by the reptiles and mammals. The best one can say is that among the anamniotes viviparity was achieved to a variable degree by exploiting the yolk sac and an assortment of other devices; with the amniotes there came a new and more successful evolution of viviparity in a declining use of the yolk sac and a greater exploitation of the allantois (in association with the chorion). Hand-in-hand there also came a still incompletely understood physiological evolution of the corpus luteum, described later. Although it plays a major role in the conditioning of the mammalian uterus for pregnancy, the corpus luteum in the lower vertebrates is more concerned with the regulation of breeding behavior; it may, in fact, have originated as a nutritive gland.

We can now look at human development in the light of the evolution of extraembryonic membranes.

THE OVUM AND OVARIAN FOLLICLE: OVULATION AND FERTILIZATION

The mature eggs of eutherian mammals are minute in size and of the microlecithal type. They range in diameter from 0.08 to 0.15 millimeter, with the human ovum (Figure 9-8) measuring on the average about 0.14 millimeter. Maturation of the ovum goes hand in hand with development of the *ovarian follicle* (Figure 9-9). A young oocyte is surrounded by a single layer of cuboidal ovarian cells, designated the *granulosa layer* (Figure 9-9A), constituting what is termed the primary follicle. The granulosa cells then rapidly multiply to produce an increasingly thick, multilayered covering about the oocyte (Figure 9-9B). Coincidentally, a transparent noncellular area, known as the *zona pellucida,* also arises around the oocyte. The structure of the zona permits passage of materials from the granulosa cells into the egg.

The appearance of the zona pellucida between the oocyte and the multilayered granulosa marks the secondary stage of follicle development. Clefts now begin to appear within the granulosa. These clefts ultimately coalesce to form a single *follicular cavity,* or *antrum* (Figure 9-9*C*), which from the very start is filled with fluid whose nature and function are discussed later. A follicle with an antrum is termed a *tertiary follicle.* Conversion of a tertiary follicle into a mature follicle is largely a matter of increase in size. As the follicle grows, the antrum continues to enlarge in such a way that the oocyte, still embedded in a portion of the granulosa, remains attached to the granulosa proper on one side only (Figure 9-9*D*). The granulosa investing the oocyte is now called the *cumulus.* While these

changes have been going on, the connective tissue framework of the ovary has laid down an additional sheath, the *theca,* outside the parietal granulosa.

The first of the two maturation divisions is now inaugurated by the oocyte. As this proceeds, the cumulus gradually becomes undercut so that first the oocyte is carried in a slender stalk of cells and then, still encased in the zona pellucida and remnants of the cumulus, it floats into the follicular fluid. The release of the cumulus cells from the other follicle cells causes them to extend out into the follicular fluid in a mass of projections called the *corona radiata.* Now the liquid-filled follicle gradually enlarges so as to balloon from the surface of the ovary and finally bursts, releasing the oocyte (Figure 9-10), still surrounded by

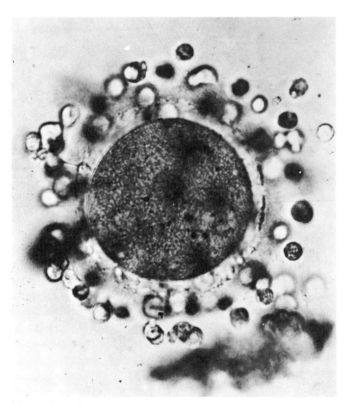

Figure 9-8 Photomicrograph of a living human ovum surrounded by the corona radiata. (Courtesy Professor W.J. Hamilton.)

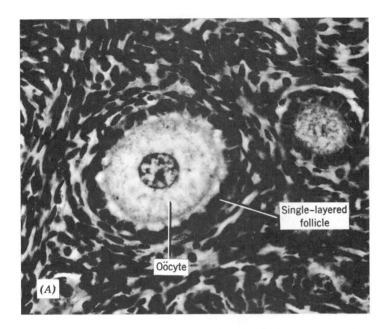

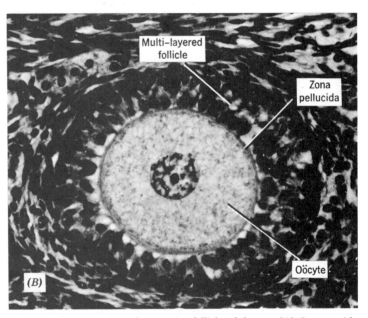

Figure 9-9   Photomicrographs of stages in the maturation of an ovarian follicle of the cat. (*A*) Oocyte with single-layered follicular capsule (granulosa), a primary follicle. (*B*) Oocyte with zona pellucida and multilayered granulosa, a secondary follicle. (*C*) Tertiary follicle showing beginning of follicular cavity. (*D*) Mature follicle. (*A* and *B*) ×215, (*C*) ×154, (*D*) ×53. (Photos by Torrey.)

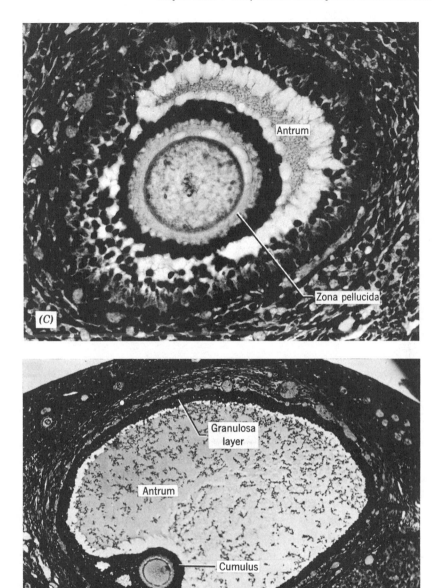

Figure 9-9 (*Continued*)

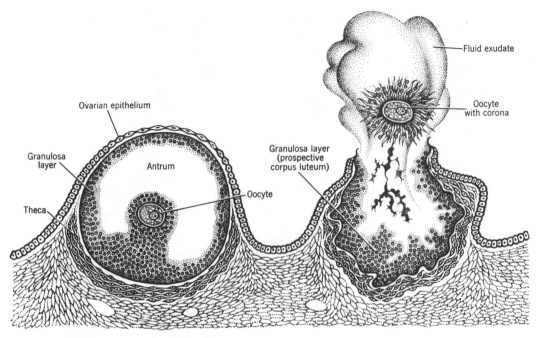

Figure 9-10   Diagram of mammalian ovulation.

the zona pellucida, and in many mammals by the corona as well, into the funnel of the uterine (Fallopian) tube. This shedding of the ovum is known as *ovulation*. In most mammals, ovulation is accompanied by completion of the first meiotic division and the appearance of the first polar body.

In the human female only a single follicle and ovum normally mature at a time; in mammals that produce litters, several ova are prepared and shed simultaneously. If spermatozoa have been deposited in the lower end of the female tract, these sperm cells are carried up the tract and fertilization occurs in the uterine tube. The second maturation division of the ovum is then completed and cleavage inaugurated.

THE ESTRUS CYCLE

The patterns of breeding behavior among vertebrates are remarkably varied but, in general, they fall into two categories: seasonal cycles and intraseasonal cycles. In the

first case, there is a single period during any given year when breeding occurs; in the second, breeding may occur several times a year. In either case, there has been an evolution of patterns that ensure optimal conditions for the development of young. That is, natural selection has sorted a host of internal and external control systems whereby both individuals and populations are prompted to initiate reproduction at times most favorable to the appearance and survival of offspring. The controlling factors are either inherent to the internal physiology of the breeding individuals or environmental in origin, or a combination of the two. The former refers to assorted endocrinological controls and the "biological clocks" that regulate the behavior of so many animals; the latter refers to such factors as duration and intensity of light, temperature, visual, olfactory, and auditory stimuli, rainfall, and lunar tides.

The Mammalia, one class of vertebrates alone, present many such variations. Some mammals mate periodically throughout the year; others mate only during a restricted breeding season. In the majority of wild mammals, both males and females are capable of mating only at specific, synchronized periods of the year. The males of domesticated forms, however, tend to be sexually potent at all times; this is especially true of the primates, including humans. However, all female mammals, with the exception of the higher primates, permit mating only at definite times, which are termed the periods of "heat," or *estrus*. In other words, at regular intervals the female comes into a state of physiological and psychological readiness for mating, and only in this state does she accept the male. Behind this state of readiness there lies a complex of structural and functional transformations that, in the absence of pregnancy, are regularly repeated.

This repeated series of events constitutes the so-called *estrus cycle*. The pattern of breeding behavior of a given mammal is thus an indication of the frequency with which the estrus cycle is repeated. In some mammals, such as the deer family and wild sheep, the cycle occupies the entire year; that is, estrus and mating occur only once a year. In others, such as the rat and mouse, estrus reoccurs every 4 or 5 days. Between these two extremes there are the familiar cases of the dog, with two periods of heat yearly, and the domestic cow and pig, which come into heat every 3 weeks. Sometimes there are special departures from the pattern of regularly recurring cycles, as in the rabbit in which the cycle is suppressed during the winter but is repeated periodically in the spring and summer. In some domestic animals, also, what has been termed *quiet heat* often occurs: all the morphological and physiological phenomena, including ovulation, appear in the usual fashion, but the behavioral manifestations are absent.

Whether long or short, occupying 1 year or 5 days, the estrus cycle consistently proceeds through a series of phases: (1) a short period during which sexual desire is at its peak, *estrus;* (2) *metestrus* in which prep-

arations for pregnancy are completed and, if pregnancy does not ensue, a regression of these preparations; (3) *diestrus,* a period of quiescence; and (4) *proestrus* in which preparatory changes lead up to the next estrus. Actually, the variations in length of cycle among mammals are largely reflections of variations in the length of the diestrus phase. Thus, in the case of a seasonal breeder, the female is in diestrus for the greater part of the year. (When so extended, this period is commonly referred to as *anestrus.*) Once the cycle resumes, it proceeds at a pace comparable to that in continuous breeders.

Ovarian Changes during the Estrus Cycle. The maturation and growth of a follicle proceed during that phase of the estrus cycle that constitutes proestrus. Growth of the follicle culminates in its rupture and the release of the ovum, ovulation. This event is timed to coincide with the onset of estrus. In other words, the ovum is made available for fertilization at the very time the female arrives at a state of readiness for mating and thus when spermatozoa are most likely to be present. Yet there may be considerable variability in the time of ovulation. Sometimes it occurs several hours before or after heat, and in the case of the human female records show that ovulation may occasionally occur almost any time in the estrus cycle. Also, ovulation is not a spontaneous event in all mammals. In some, notably the rabbit, the cat, and the mink, ovulation takes place only after nervous stimulation deriving from the act of copulation.

After the discharge of the ovum, the ruptured follicle collapses and its wall becomes folded. The cells of the granulosa multiply, increase in size, frequently develop a yellowish pigment, and become vascularized by blood vessels growing in from the theca. The result is termed the *corpus luteum.* Development of the corpus luteum is a feature of the early part of metestrus, which for convenience may be designated *metestrus I* (Figure 9-11). If the liberated ovum

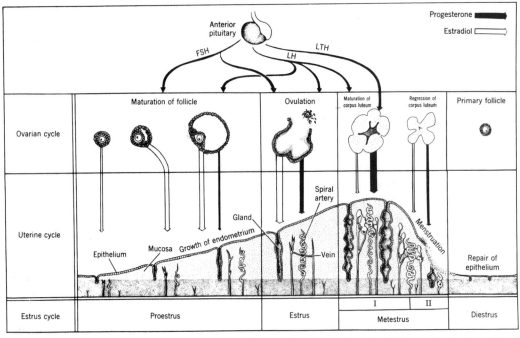

Figure 9-11 Hormonal relationships and changes in functional anatomy of ovarian follicle and uterine endometrium during the estrus cycle. Rise and decline of estradiol and progesterone indicated by width of arrows.

is now fertilized, growth (and function) of the corpus luteum is prolonged through the duration of pregnancy, toward the end of which the corpus degenerates. By the same token, metestrus I is maintained until the end of gestation. If, however, fertilization does not occur, not only does the ovum itself disintegrate, but the corpus leuteum promptly regresses. This regression proceeds during the latter part of metestrus, termed *metestrus II,* and consists of a breakdown of the cellular part of the corpus with a concomitant invasion by fibrous connective tissue from the theca. The result is a mass of whitish-appearing scar tissue called the *corpus albicans.*

The diestrus phase of the cycle represents for the ovary a period of relative inactivity, especially in those mammals in which diestrus is greatly prolonged. The oogonium(a) and neighboring ovarian cells remain quiescent until a new cycle of maturation is inaugurated in proestrus.

Uterine Changes during the Estrus Cycle. Although other portions of the female reproductive tract exhibit parallel changes, it is the uterus that undergoes the most profound modifications during the course of the estrus cycle. The part of the uterus primarily involved is the so-called *endometrium.* This consists of a thin uterine *epithelium* bordering the cavity of the uterus and the highly vascular *mucosa* beneath the epithelium. It also includes numerous *uterine glands* which, developmentally, are tubular invaginations of the uterine epithelium projected into the mucosa.

At the beginning of proestrus the endometrium is relatively thin and its glands are short and stubby. As an accompaniment of follicular maturation during proestrus, the mucosa then gradually thickens and the uterine glands deepen. These processes continue into estrus. After ovulation and with the growth of the corpus luteum, the endometrium builds up still more rapidly. The glands not only become even longer, but their deeper parts become coiled and markedly dilated and a considerable amount of secretion collects within them. There is also an increase in the size of the arteries supplying the endometrium, and they tend to follow a spiral course. Add to this an accumulation of fluid in the mucosa, and we see in metestrus I an endometrium that has thickened five- or sixfold.

The subsequent performance of the endometrium depends on whether fertilization occurs. If it does not, then during metestrus II regression of the endometrium follows. This regression for most mammals is essentially a reversal of the events of buildup: the blood supply declines, the glands straighten out and shorten, and the mucosa becomes dehydrated. There may be some sloughing of superficial epithelium, but by and large the endometrium simply returns to its original state in which it remains through diestrus and until a new cycle is inaugurated. In the primates, however, regression of the endometrium during metestrus II is considerably more dramatic. As a consequence of reduced blood flow, the tissues begin to deteriorate, blood vessels rupture, and fluid pours from the open mouths of the glands. Once started, the process of deterioration proceeds rapidly, with the result that the greater part of the endometrium together with extravasated blood is sloughed away. This denuding of the uterine lining constitutes *menstruation.* A period of repair follows. Regeneration of the uterine epithelium proceeds from the remnants of the glands; the open blood vessels are healed and new capillary networks are established. All this happens early in diestrus. The endometrium remains in a state of rest for the duration of diestrus, long or short as the case may be, after which the cycle of events is repeated.

The buildup of the endometrium that has been described occurs in anticipation of a possible association with an embryo. Thus in the event of fertilization, regression of the endometrium does not occur; on the contrary, the endometrium maintains itself and undergoes further modifications which will be described in due course. For the present it need only be emphasized once more that pregnancy means an extension of metestrus I. Termination of pregnancy and birth of the fetus involve a destruction and loss of endometrium, followed by a period of repair. In other words, the sloughing of endometrial tissues at birth may be equated with menstruation and, in terms of the estrus cycle, with metestrus II.

Ovarian Hormones and the Estrus Cycle. Figure 9-11 shows that the cyclical events in the ovary and endometrium are neatly coordinated. Each developmental circumstance in one is timed to correspond with a developmental circumstance in the other. It has long since been established that this timing is controlled by chemical agents circulating in the bloodstream, that is, hormones. The total picture of the kinds, activities, and chemistry of these hormones is exceedingly complex, and despite many years of study all the facts are still not in.

Recall that as an accompaniment of the growth and maturation of the follicle during proestrus, the enlarging follicular cavity, or antrum, becomes filled with a fluid. This fluid has been demonstrated to contain a hormone called *estradiol.* (Estradiol is only one of several closely related chemical substances that collectively are known as *estrogens.*) Estradiol is poured into the ovarian circulation and from there is distributed over the body, including the uterus and other regions of the female tract. Appropriate tests have shown that it is estradiol that initiates the histological differentiation of the endometrium. Estradiol also promotes the buildup of mating desire, which in nonprimates reaches its peak at estrus. It is during estrus, of course, that the follicle ordinarily ruptures, freeing the ovum and at the same time releasing a maximum amount of estradiol, which incites maximal sexual activity.

After ovulation, the collapsed follicle is converted into the corpus luteum. This structure produces a second hormone, *progesterone,* which elicits several reactions. Most importantly, it brings the endometrium into its final readiness to receive an embryo. Progesterone also inhibits the occurrence of additional ovulations by any other mature follicles that may exist, and stimulates the endometrial glands to begin secretion. If fertilization of the ovum does not occur, the corpus luteum regresses to a corpus albicans, and with the decline in concentration of progesterone the endometrium regresses or, in primates, engages in menstrual sloughing. But with pregnancy the corpus luteum, and thus the supply of progesterone, is maintained to support the endometrium. The corpus luteum is necessary in the early stages of pregnancy, after which time the placenta produces progesterone for the remainder of the pregnancy. The decline in luteal production coincides with the end of the third month.

The production and the effects of estradiol are allocated precisely to the periods of proestrus and estrus, and those of progesterone to the period of metestrus. In actuality, however, there is a considerable overlap in the presence and operation of these two hormones. The many refined experimental analyses of recent years have revealed a modest carryover of production and action of estradiol into metestrus, and the secretion of progesterone has been shown to begin as the follicle is completing its maturation. Growth of the endometrium is thus a consequence of the cooperative action of both substances. Figure 9-11 suggests the overlapping rise and decline of these ovarian hormones.

Pituitary Hormones and the Estrus Cycle. The cyclical events in the ovary are not automatic and self-contained, but are under the control of hormones produced by the *anterior* lobe of the pituitary gland

(see p. 318). These hormones, because they act through the intermediation of the ovaries rather than directly on the uterus, are termed *gonadotropic hormones.* This term distinguishes them from other anterior pituitary hormones that exert specific effects on other hormone-producing organs, such as the thyroid and adrenal glands. There are at least two gonadotropic hormones and probably also a third. One of these is the so-called *follicle-stimulating hormone* (*FSH*). As its name implies, FSH is responsible for directing the maturation of the follicle and the production of estradiol. The second is *luteinizing hormone* (*LH*), which joins with FSH in promoting follicular maturation and estrogen production and then controls the development of the corpus luteum and the production of progesterone. FSH and LH together in the proper proportion are responsible for ovulation, with LH playing the major role. The third hormone is *luteotropic hormone* (*LTH*), which in fact may be identical to the hormone *prolactin* that is known to regulate milk production by the mammary glands. LTH is believed to stimulate progesterone secretion by the corpus luteum in rodents.

All the factors responsible for the cyclical production of the gonadotropic hormones by the anterior pituitary are by no means fully understood, but a reasonable scheme of operation is one in which the ovarian hormones have a reciprocal effect on the pituitary. Increased production of estradiol under the impetus of FSH (and LH) may inhibit production of FSH. This leads to diminished secretion of estradiol which in turn, results in the diminution of its inhibitory action and a new rise in FSH activity and production of estradiol. A comparable relationship may prevail between LH and progesterone. Be this as it may, the hypothalamus of the brain (p. 433) is also involved.

The hypothalamus has the unique status of being an intermediary between the nervous system, of which it is a part, and the

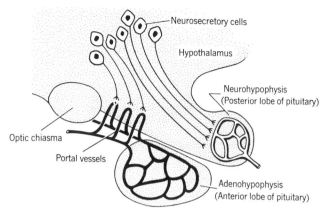

Figure 9-12 Schema of hypothalamic–pituitary neurosecretory system of mammals. (After Scharrer and Scharrer, *Recent Progress in Hormone Research,* Vol. 10.)

endocrine system, to which it sends chemical information. It operates through neural mechanisms, on the one hand, in the regulation of such things as body temperature and emotional behavior; on the other hand, it utilizes chemical means for the regulation of water balance and the release of hormones from the anterior pituitary.

Figure 9-12 presents the principal features of the hypothalamus–pituitary system. Certain *neurosecretory cells* within the hypothalamus, possessing the qualities of both nerve cells and gland cells, respond to stimulation from other regions in the brain by releasing chemical substances. Some of these secretions are delivered by the fibers of the neurosecretory cells directly to storage depots in the *posterior* lobe of the pituitary from where they are distributed via the bloodstream to distant target organs. However, the *anterior* lobe of the pituitary does not receive nerve fibers from the hypothalamus. Instead, neurosecretory cell fibers deliver their products to the blood vessels constituting the portal system with which the anterior pituitary is equipped. The portal vessels are locally distributed to different types of cells which themselves are grouped in specific regions of the anterior pituitary, so that different neurosecretory control centers in the hypothalamus have

specific vascular connections with and regulate different cell types within the pituitary. Of the numerous chemical regulators provided in this way, two are involved in the cyclical production of the gonadotropic hormones. One is the so-called *FSH-releasing factor;* the other is *LH-releasing factor.* As the terms suggest, their presence is required for the production and FSH and LH.

The Estrus Cycle in Synthesis. Figures 9-11 and 9-12 summarize the relationships of hypothalamic, pituitary, and gonadal roles in mammalian reproduction.

Under the stimulus of FSH (and LH) provided by the anterior pituitary, the ovarian follicle matures. In so doing it produces estradiol and a small amount of progesterone, which incite the endometrium to hypertrophy. Follicle maturation culminates in ovulation, an event that coincides with the height of the mating impulse. Thereupon, under the primary influence of LH, the follicle is converted to a corpus lutem, which secretes progesterone and some estradiol, bringing the differentiation of the endometrium to fullness. In the event of fertilization, the endometrium is maintained until pregnancy has run its course; without fertilization, the corpus luteum degenerates, the concentrations of progesterone and estradiol decline, and the endometrium either regresses or sloughs off. With or without pregnancy, ovary and endometrium ultimately return to a resting condition from which a new cycle takes off.

We are now prepared to resume the story of embryogenesis.

CLEAVAGE

Our account of eutherian embryology is derived mainly from studies of human, monkey, and mouse embryos. Cleavage of the eggs of eutherian mammals is holoblastic, but despite a microlecithal condition the blastomeres tend to show size differences from the start. Moreover, the numbers of blastomeres do not increase by a regular doubling sequence but tend to show arithmetical progression. The egg of the monkey conforms on both counts.

The initial divisions take place as the ovum makes its way through the uterine tube (Figures 9-13 and 9-14). For some time, too, the zona pellucida remains intact. The first division occurs within 24 hours of ovulation and results in two oval blastomeres, one of which is slightly larger than the other (Figure 9-15*A*). Between the 36th

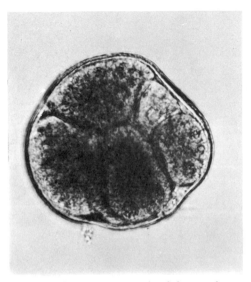

Figure 9-14 Photomicrograph of five-cell human ovum (slightly abnormal). × 500. Carnegie No. 8630. Intact zona pellucida closely applied to blastomeres. (From Hertig and Rock, courtesy Carnegie Institution of Washington.)

and 48th hours two divisions succeed each other. The first involves the larger of the first two blastomeres and results in a three-cell stage (Figure 9-15*B*); the smaller cell of the two-cell stage then divides. These two cleavage planes are approximately at right angles to each other so that the cells of the resulting four-cell stage, consisting of two larger and two smaller cells, lie crosswise (Figure 9-15*C*). In the interval between the 48th and 72nd hours, the larger cells of the four-cell stage divide before the smaller cells, so as to produce stages of five, six, seven, and eight blastomeres (Figures 9-15*D–F*). The planes of division are again at right angles, so that at the eight-cell stage the blastomeres are so arranged that the four derived from the larger cell of the original two-cell stage are at one pole and the four derived from the smaller cell are at the other pole. Soon after the third cleavage division, the process of *compaction* occurs, during which the blastomeres are pulled tightly together, obliterating the single cavity at the center of the embryo. The blastomeres continue to divide at different

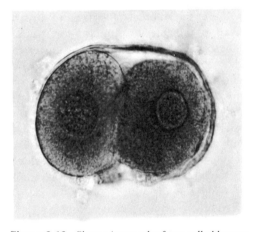

Figure 9-13 Photomicrograph of two-celled human ovum. × 420. Carnegie No. 8698. Zona pellucida intact; polar bodies faintly visible. (From Hertig and Rock, courtesy Carnegie Institution of Washington.)

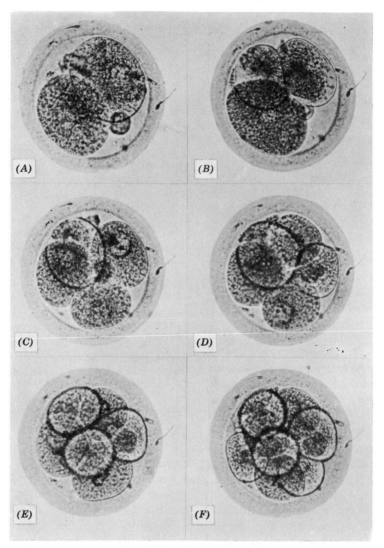

Figure 9-15 Photographs of living cleavage stages of the macaque monkey. × 200. Description in text. (From Lewis and Hartman, Courtesy Carnegie Institution of Washington.)

rates so as to produce stages with 9, 10, 12 and ultimately 16 blastomeres. The cavity opened up during the process has dubious affinities, but is probably safely designated the *blastocyst cavity* or *primary yolk sac*. The 16-cell stage is attained by the 96th hour after ovulation, by which time the cleaving ovum has neared the uterus. So far as our knowledge presently goes, a similar timetable obtains for man.

At this point more adequate human material is available; therefore our account from here on refers to human development. Henceforth, cell divisions proceed with much greater rapidity so that not only is the number of blastomeres increased quickly (Figure 9-16), but it becomes difficult to follow the maneuverings of individual cells. Thus by the end of the fifth day the total number of blastomeres has reached 100. In

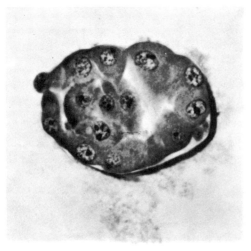

Figure 9-16   Photomicrograph of 58-cell human embryo. × 500. Carnegie No. 8794. Zona pellucida present only in part. (From Hertig and Rock, courtesy Carnegie Institution of Washington.)

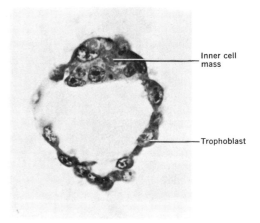

Inner cell mass

Trophoblast

Figure 9-17   Photomicrograph of section of 107-cell (approximately 5-day) human blastocyst. × 500. Carnegie No. 8663. (From Hertig and Rock, courtesy Carnegie Institution of Washington.)

the meantime, fluid passes from the uterine cavity through the zona pellucida and begins to accumulate between the cells. Gradually the fluid-filled intercellular spaces become confluent, forming a single cavity. There is also a sorting out and rearrangement of blastomeres in such a way that the outermost members of the cell cluster become flattened, and those on the inside become aggregated and attached eccentrically to the inner aspect of the outer flattened cells. The embryo, essentially a minute hollow sphere at this stage of development, is now referred to as a *blastocyst* (Figure 9-17). Its outer shell is termed the *trophoblast*; the inner cluster is called the *inner cell mass.* Although the blastocyst superficially appears comparable to a conventional blastula, it more nearly corresponds to the avian blastoderm prior to the formation of the blastocoel. In other words, the fluid-filled blastocyst cavity is equivalent to the subgerminal space beneath the avian blastoderm.

It is at about this time that the zona pellucida disappears, thus permitting the trophoblastic cells of the blastocyst to make

contact with the endometrium of the uterus.

PLACENTA FORMATION

Our next concern is with the areas external to the inner cell mass, and here again we discover a developmental pattern comparable to that in birds. In the pig (Figures 9-18A, B), for example, the extraembryonic somatopleure is observed to elevate itself over all sides of the embryo to provide a conventional amnion and chorion. Concomitantly, the allantois grows out of the hindgut, expands into the extraembryonic coelom, and fuses broadly with the chorion. Although no yolk is present, a large yolk sac is initially established in a manner comparable to that in birds. But with the enlargement of the allantois, the yolk sac rapidly declines and ultimately becomes reduced to a shriveled remnant.

Meanwhile, the endometrium of the uterus has been completing its preparations to receive the embryo. The association to come between the embryo, by way of its membranes, and the endometrium will provide an entity known as the *placenta.* In a

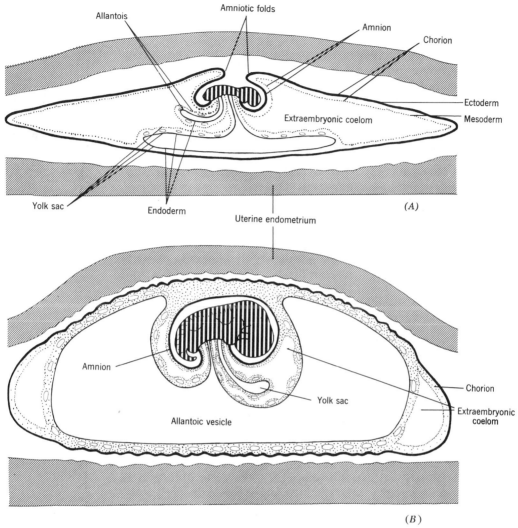

Figure 9-18   Early development of the extraembryonic membranes of the pig. (*A*) Amniotic folds starting; yolk sac relatively large; allantois beginning to project into extraembryonic coelom. (*B*) Amniotic folds complete; allantois broadly expanded in extraembryonic coelom; yolk sac reduced. (Based on figures by Mossman, courtesy Carnegie Institution of Washington.)

preliminary way a placenta may be defined as a structure produced by the apposition or fusion of the extraembryonic membranes with the endometrium for the purpose of physiological exchange. It therefore follows that the placenta, from the point of view of origin, consists of two parts: a *fetal placenta* furnished by the extraembryonic membranes and a *maternal placenta* furnished by the uterine endometrium.

Now it should be obvious that although on the maternal side a single component, the endometrium, is involved, on the fetal side we have to consider the prospective roles of four elements: amnion, chorion, yolk sac, and allantois. The first of these, the amnion, may be ruled out immediately as making no direct contribution to the placenta, although it does participate in an accessory way in a manner to be described later. This leaves the other three, of which the chorion, because of its most external position, is the membrane making immediate contact with the endometrium. But we have seen in the chick embryo that the chorion plays its role by way of a vascular supply, which it acquires from the allantois. In mammals there are two possible sources of chorionic vascularization: the vitelline circulation provided by the yolk sac and the allantoic circulation provided by the allantois.

It is not practicable in an account of this scope to describe all the variations in the chorion—yolk sac—allantois association exhibited by mammals. We can, however, point to some generalized patterns. In some mammals, notably the marsupials, the allantois remains relatively small and never makes contact with the chorion, whereas the yolk sac becomes very large and fuses broadly with the chorion (Figure 9-19). In these forms the chorion gains its blood supply from the yolk sac, and the fetal placenta is thus said to be of the *choriovitelline* type. This is true even though in marsupials only a portion of the yolk sac, and thus the chorion, is provided with vascular mesoderm. Among many carnivores, rodents, and insectivores a similar situation may prevail temporarily or permanently. In some the yolk sac provides the initial vascular supply, only to regress as the later-developing allantois reaches and vascularizes the chorion; in others the yolk sac shares with the allantois the task of vascularizing the chorion. In most eutherian mammals, however, the yolk sac remains rudimentary, and it is

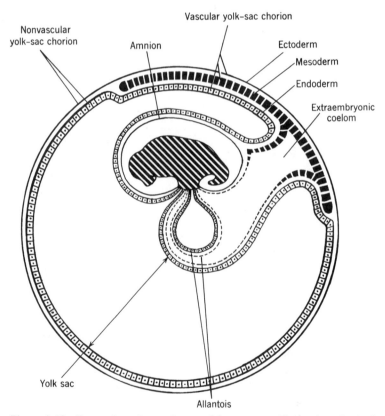

Figure 9-19 Extraembryonic membranes of the opossum. The moderately sized allantois is splanchnopleure, but the greatly expanded yolk sac bears mesoderm over a restricted area only. Chorion and amnion also largely devoid of mesoderm.

the allantois that furnishes the chorionic circulation. The fetal placenta is then said to be *chorioallantoic.* It is this type with which we shall be solely concerned from here on.

Because the pig provided our original illustration of mammalian extraembryonic membranes, we may continue to use it with profit. As noted earlier, the expansion of the allantois leads to its broad fusion with the chorion. There is thus established a vascular chorioallantoic membrane quite comparable to that in the bird embryo, except that in this instance, instead of being applied to the inner surface of an egg shell, it lies in apposition to the uterine endometrium (Figure 9-20). Moreover, as the enveloping chorioallantoic membrane continues to enlarge, the greater part of its surface is thrown into shallow folds, which fit into corresponding depressions in the endometrium, adding to the intimacy of contact between the two components of the placenta. As the fetal placenta (chorioallantoic membrane) has allantoic blood vessels running to and from the fetus, and maternal

vessels run to and from the maternal placenta (endometrium), the two circulations are brought very close together. However, there is *no fusion of these two blood systems.* Fetal blood does not circulate in the mother; maternal blood does not circulate in the fetus. All food, wastes, gases, and other materials pass from one circulation to the other through tissue barriers interposed between the two bloodstreams.

In the case of the pig, no fewer than six membranes lie between the fetal and maternal streams (Figure 9-21A). A molecule of oxygen, for instance, in going from mother to fetus would pass through in this order: endothelium of the maternal blood vessel, endometrial connective tissue, uterine epithelium, chorionic epithelium, chorionic connective tissue, and endothelium of fetal blood vessel. Because the immediate contact of the two halves of the placenta involves chorionic epithelium and uterine epithelium, this type of placenta is described by the term *epitheliochorial* (Figure 9-21A). The generally held view is that the epitheliochorial placenta, presenting

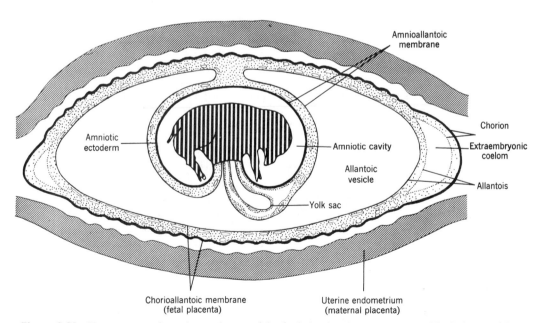

Figure 9-20 Mature extraembryonic membranes of the fetal pig, showing composition of fetal placenta (chorioallantoic membrane) and relationship to endometrium of uterus.

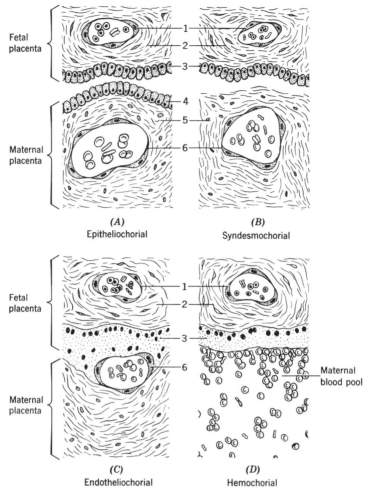

Fetal placenta

Maternal placenta

(A)
Epitheliochorial

(B)
Syndesmochorial

Fetal placenta

Maternal placenta

Maternal blood pool

(C)
Endotheliochorial

(D)
Hemochorial

Figure 9-21 Histological types of mammalian placentae. *1* = endothelium of fetal blood vessel, *2* = chorionic connective tissue, *3* = chorionic epithelium, *4* = uterine epithelium, *5* = endometrial connective tissue (mucosa), *6* = endothelium of maternal blood vessel. (Adapted from illustrations by Flexner and Gellhorn.)

apposition of fetal and maternal components and six tissue barriers between the two circulations, is the primitive type from which others have been derived. These derivations reflect a changeover from apposition to some degree of fusion of the two placental components, which in turn involves some measure of destruction of maternal tissues.

In the ruminants (cattle, sheep) the fetal and maternal components are fused so in-

timately as to result in a destruction of the uterine epithelium, thus bringing the chorion into contact with the connective tissue of the uterine mucosa. Only five barriers therefore lie between the two bloodstreams. This situation is termed *syndesmochorial* (Figure 9-21*B*). An *endotheliochorial* placenta (Figure 9-21*C*) is one in which the uterine mucosa is reduced and the chorionic epithelium comes in contact with the walls of the maternal blood vessels. A placenta of this type is characteristic of the carnivores. A reduction of the barriers to three is found in the primates and many rodents. This, the *hemochorial* type (Figure 9-21*D*), is one in which the maternal endothelium also disappears and the chorionic epithelium is bathed directly in maternal blood.

This scheme of classification of placentae is by its very nature somewhat artificial and should not be taken too literally. In the first place, there occur many gradations from one type to another and many variations around any one type. Furthermore, a classification is subject to revision as new knowledge is acquired. For instance, one category of placental types (the so-called hemoendothelial placenta) has recently been judged to be nonexistent, and some professional experts are eyeing a redefinition of the other categories. But even though the classification that we have presented may fall short of absolute reality, it does have descriptive usefulness, especially for the student receiving her or his first introduction to the form and function of the placenta. It is in a similarly cautious vein that the matter of placental function should be considered. Not only do placentas vary one from another in the number of tissue barriers interposed between the two circulations, but placentas tend to exhibit a thicker barrier at earlier stages of development than later. For example, the primate placenta is initially epitheliochorial and acquires its hemochorial status later. Therefore, conclusions on placental physiology reached from studies of one type do not necessarily apply to other types; and within any given type, functional operations vary from early to late.

Generally speaking, the placental barrier, whatever its status, appears to operate as an ultrafilter. Accordingly, there is a relationship between placental transmission and molecular size, smaller molecules passing more readily than larger ones. This means that water, oxygen, carbon dioxide, and such soluble inorganic materials as chlorides and phosphates pass by diffusion. As for carbohydrates, fats, and proteins, a great deal of variability in their transmission is shown by different placental types and at different ages of a given placental type. In any event, their passage is more than a matter of simple diffusion; biochemical work with expenditure of energy is almost surely involved. The situation is further complicated by the fact that transmission is only one manifestation of placental activity. The placenta also stores materials such as fat, glycogen, and iron; it participates in the metabolism of proteins; and it is an endocrine gland. The details on all these matters are beyond our province.

Because of its functional implications, the classification of placentas in terms of histological types, on which we have dwelt at some length, is a fundamental one. But there is also usefulness in a terminology based on shape. Placentas like that in the pig, where the greater part of the chorionic surface is associated with the endometrium, are said to be *diffuse* (Figure 9-22*A*). In the sheep and the cow the contact is restricted to numerous localized patches; here the term *cotyledonary* is applied (Figure 9-22*B*). A contact involving a girdlelike band encircling the blastocyst produces a *zonary* placenta, as in carnivores (Figure 9-22*C*); when restricted to a disc or plate, as in humans and some rodents, it is termed *discoidal* (Figure 9-22*D*).

The degree of union of the fetal and maternal components of the placenta varies, as we have seen, from mere apposition to intimate fusion. In the event of apposition only, separation of the components is easily effected at birth and there is no loss of maternal tissues. The placenta is then said to be *nondeciduous.* A *deciduous* placenta is one in which union of the components is so intimate that at birth a variable amount of maternal tissue is shed.

Table 9-1 provides a convenient summary of all the foregoing.

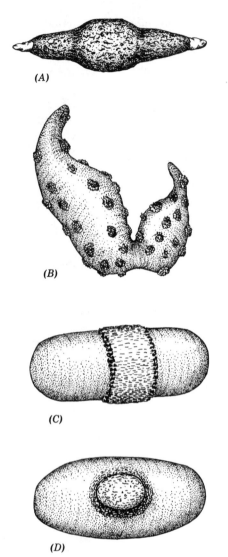

(A)

(B)

(C)

(D)

Figure 9-22 Diagrams of chorionic sacs of representative mammals showing gross forms of the placenta. (*A*) Diffuse placenta of the pig. (*B*) Cotyledonary placenta of the sheep. (*C*) Zonary placenta of the dog. (*D*) Discoidal placenta of the bear.

HUMAN DEVELOPMENT AND PLACENTATION

We are now prepared to resume the story of human development. We left off with a blastocyst approximately 5 days of age, consisting of a minute hollow sphere whose outer surface was designated trophoblast, to which was attached an inner cell mass.

By this time the blastocyst has, or shortly thereafter will have, completed its tubal journey and lies in the uterine cavity. The zona pellucida will also have disintegrated so that the surface of the trophoblastic wall is exposed.

That portion of the trophoblast overlying the inner cell mass, an area that for convenience may be termed *polar trophoblast,* now makes contact with the endometrium. These trophoblastic cells multiply rapidly and begin to insert themselves between the epithelial cells of the endometrium, cells that concomitantly show signs of degeneration. This represents the first step in a process of "digging in" by the blastocyst called *implantation.* As the polar trophoblastic cells continue to multiply, they spread into the endometrial mucosa; in so doing, those cells, which invade the mucosa most deeply, begin to lose their walls. In other words, the invading trophoblast differentiates into two layers, the original inner cellular *cytotrophoblast* and an outer syncytial *syntrophoblast* (Figure 9-23). It is the latter that has the most immediate association with the mucosa, and as an accompaniment of its rapid spreading, the syntrophoblast absorbs the fluids and extravasated blood resulting from the breakdown of mucosal tissues. This is not to suggest that the invasion of the mucosa is a one-way operation in which the mucosa is the passive victim of digestion by the trophoblast. There is evidence to show that the erosive activities of the trophoblast are supplemented by an autonomous disintegration of the mucosal tissues themselves. In addition, the syntrophoblast cells produce and release substances that block any rejection of the embryo by the mother. This, the syntrophoblast serves as both a nutritive and protective organ.

In the meantime, other events will have occurred. First, certain cells in the lowest level of the inner cell mass sort themselves out and become grouped as a ventral plate of *primary endoderm* (Figure 9-24), indeed in a manner quite reminiscent of the origin

TABLE 9-1

	Maternal Tissue (Endometrium)			Fetal Tissue (Chorioallantois)			Gross Shape	Fate of Endometrium	Familiar Example
	Endothelium	Connective Tissue	Epithelium	Epithelium	Connective Tissue	Endothelium			
Epitheliochorial	+	+	+	+	+	+	Diffuse	Nondecidous	Pig Horse
Syndesmochorial	+	+	0	0	+	+	Cotyledonary	Transitional	Sheep Cow
Endotheliochorial	+	0	0	0	+	+	Zonary to discoidal	Deciduous	Cat Dog
Hemochorial	0	0	0	0	+	+	Discoidal or zonary	Deciduous	Primates Rodents

+ = Tissue present; 0 = tissue absent.

167

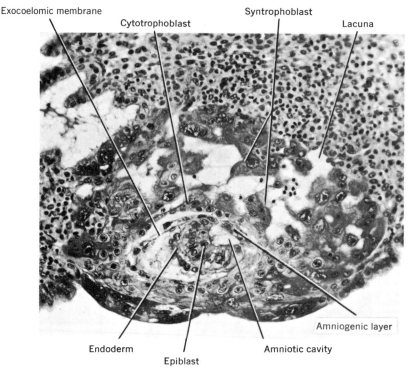

Exocoelomic membrane   Syntrophoblast

Cytotrophoblast   Lacuna

Amniogenic layer

Endoderm   Amniotic cavity

Epiblast

Figure 9-23   Photomicrograph of a section through a fully implanted 9-day human embryo. × 220. Carnegie No. 8171. (From Hertig and Rock, courtesy Carnegie Institution of Washington.)

of endoderm in the avian blastoderm. Coincidently, the remaining cells of the inner cell mass, often termed *formative cells,* arrange themselves into a flattened disc. Between this disc and a layer of cells derived from the covering polar trophoblast a narrow space appears. This space is the forerunner of the *amniotic cavity,* and the trophoblastic cells that bound it dorsally constitute the prospective amnion, or *amniogenic layer* (Figure 9-24). This direct and precocious method of establishing the amnion and its cavity, by "cavitation" rather than by the familiar folding of extraembryonic somatopleure, appears to be linked to the early acquired and intimate association of the blastocyst with the endometrium. When, as in humans and other primates, the early destiny of the blastocyst is to become deeply implanted in the endometrium, an accompanying accom-

modation is a direct amniogenesis by cavitation. A conventional folding of extraembryonic somatopleure occurs only in those mammals whose association with the endometrium is established relatively late and is to be more superficial. One of the earliest known human embryos in the first stages of implantation, an embryo approximately $7\frac{1}{2}$ days old (Figure 9-24), reveals an incipient amnion and amniotic cavity.

By the end of 9 days the blastocyst will be completely beneath the uterine epithelium and fully embedded in the mucosa (Figure 9-23). This comes about through the extended activity of the trophoblast, an activity that spreads beyond the confines of the polar area, so as now to involve the entire trophoblastic capsule, and thus progressively enlarges the scope of the invasion of the endometrium. The continued sprawling growth of the syntrophoblast results in

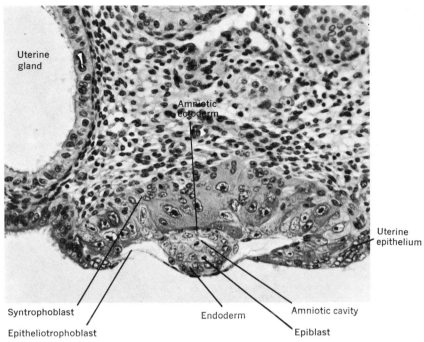

Uterine
gland

Amniotic
ectoderm

Uterine
epithelium

Syntrophoblast

Epitheliotrophoblast

Endoderm

Amniotic cavity

Epiblast

Figure 9-24 Photomicrograph of section through a $7\frac{1}{2}$-day embryo. $\times 260$. Carnegie No. 8020. Description in text. (From Hertig and Rock, courtesy Carnegie Institution of Washington.)

a maze of anastomosing strands, which come to enclose irregular spaces called *lacunae*. Because, from the very beginning, the invasion and breakdown of the endometrial tissues have resulted in the freeing of tissue fluids and blood, these liquids begin to accumulate in the trophoblastic lacunae.

The interior of the 9-day-old blastocyst (Figure 9-23) shows the amniotic cavity to have become larger and the amniogenic layer better organized. The plate of cells, which represents the ventral boundary of the amniotic cavity, may now be called, following the terminology established for the chick embryo, the *epiblast*. The endoderm beneath the epiblast is now an integral part of a more extensive layer lying immediately beneath the trophoblast and encompassing the entire cavity of the blastocyst. This layer is known variously as the *primary yolk sac* or *exocoelomic membrane*.

The precise nature and manner of origin of this structure are not clear. Some workers consider the cells constituting it to have

come from the adjacent trophoblast, hence the designation exocoelomic membrane; others believe that it results from a spreading of the original plate of primary endoderm, hence the designation primary yolk sac. In any event, what we now have is an embryonic region in the form of a two-layered disc consisting of epiblast above and endoderm below and associated extraembryonic parts. The embryonic endoderm is continuous with the primary yolk sac; the epiblast is continuous with the amniogenic layer. All are enclosed by the elaborating trophoblast (Figure 9-23).

In embryos 11 to 12 days of age (Figure 9-25), a still deeper implantation within the depths of the endometrial mucosa is seen to have occurred. Organization of the trophoblast has progressed so that the bounding wall of the blastocyst consists of a fairly regular layer of cytotrophoblast from which the branching system of irregular syntrophoblast extends. The lacunar spaces in the latter now intercommunicate and contain greater quantities of maternal

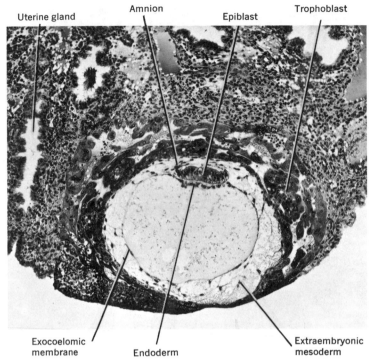

Uterine gland   Amnion   Epiblast   Trophoblast

Exocoelomic membrane   Endoderm   Extraembryonic mesoderm

Figure 9-25 Photomicrograph of a section thorough a 12-day human embryo. ×88. Carnegie No. 7700. (From Hertig and Rock, courtesy Carnegie Institution of Washington.)

blood. In the interior, mesodermal cells will have delaminated from the cytotrophoblast to form a loose meshwork between it and the exocoelomic membrane and the amniogenic layer. This is *extraembryonic mesoderm,* and it is important to note that its appearance antedates that of the embryonic mesoderm that is yet to form. That is, the prospective embryo is still a bilaminar disc.

Between the 13th and 19th days, the following are the essential developmental events (Figures 9-26 and 9-27).

1. Spaces arise within the extraembryonic mesoderm and coalesce to produce an *extraembryonic coelom.* The external mesoderm layer of the coelom lines the inner-surface of the ectodermal trophoblast whose sprawling syncytial strands concomitantly have become molded into branching structures known as *villi.* At

first purely ectodermal in composition, the villi progressively become invaded by cells from the mesoderm; that is, the primitive extodermal, or *primary,* villi acquire mesodermal cores within which blood vessels will ultimately emerge. (These vessels are derived from the allantoic system and foreshadow a villous circulation that will start operation by the 23rd day.) The mesodermalization of the trophoblast and its villi essentially means the creation of a somatopleure, so it is appropriate to abandon the term trophoblast and speak henceforth of *chorion* and *chorionic villi.* Vascularized villi are further designated *secondary* villi, in contrast to the original growths.

2. A small *secondary yolk sac* is established. The mechanism of its formation is not fully understood, but it appears to arise as a result of a constriction of the primary yolk sac (exocoelomic membrane)

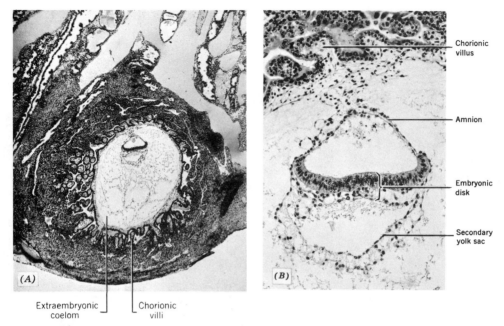

Chorionic villus

Amnion

Embryonic disk

Secondary yolk sac

(A)

(B)

Extraembryonic coelom

Chorionic villi

Figure 9-26 Photomicrographs of a section through a 16-day human embryo. Carnegie No. 7802. (*A*) Medium-power view, × 24. (*B*) Higher-power view of embryonic area, × 110. (From Heuser, Rock, and Hertig, courtesy Carnegie Institution of Washington.)

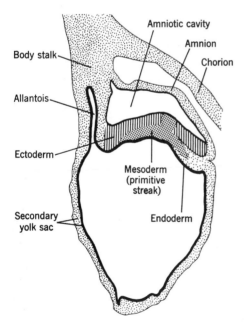

Amniotic cavity

Amnion

Chorion

Body stalk

Allantois

Ectoderm

Mesoderm (primitive streak)

Secondary yolk sac

Endoderm

Figure 9-27 Drawing of sagittal section through a 19-day human embryo and immediately adjacent parts, showing relationship of rudimentary allantois to body stalk and beginning of conversion of bilaminar embryonic disc to trilaminar one. (After Jones and Brewer, courtesy Carnegie Institution of Washington.)

whose distal portion disintegrates while the proximal portion remains intact and still continuous with the embryonic endoderm. Just as the external mesoderm bounding the extraembryonic coelom joins with and converts the trophoblast to a somatopleuric chorion, the internal mesoderm adjoins the secondary yolk sac to create a true splanchnopleure.

3. The rise and spread of the extraembryonic coelom also serve to leave a mesodermal covering over the original amniogenic layer, thereby creating a definitive somatopleuric *amnion*. However, at the prospective posterodorsal end of the embryo, the extraembryonic mesoderm does not become coelomated but remains as a continuous band, known as the *body stalk*, which connects the amnion to the chorion. A fingerlike projection of endoderm, a rudimentary *allantois*, ultimately grows into the matrix of the body stalk. It is important to note that, in contrast to the pig and many

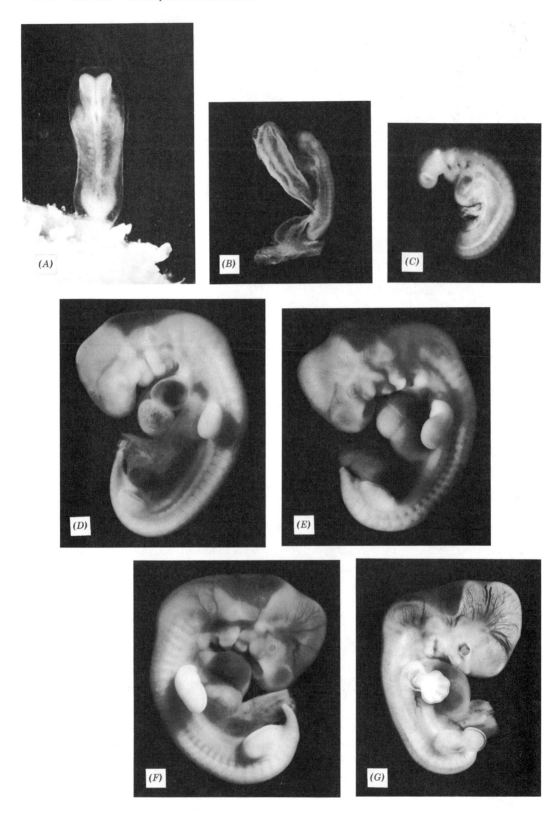

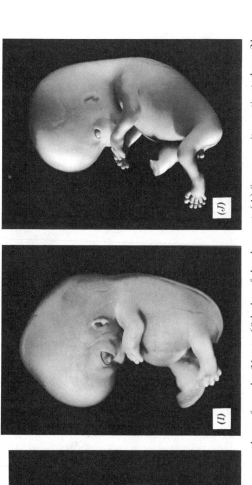

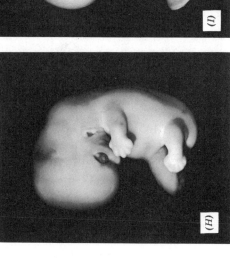

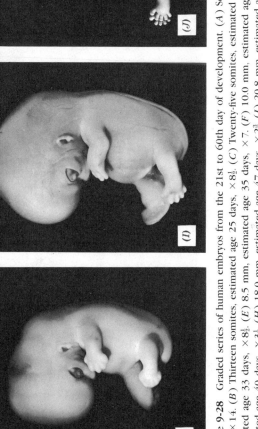

Figure 9-28 Graded series of human embryos from the 21st to 60th day of development. (*A*) Seven somites, estimated age 21 days, $\times 14$. (*B*) Thirteen somites, estimated age 25 days, $\times 8\frac{1}{2}$. (*C*) Twenty-five somites, estimated age 28 days, $\times 9$. (*D*) 7.3 mm, estimated age 33 days, $\times 8\frac{1}{2}$. (*E*) 8.5 mm, estimated age 35 days, $\times 7$. (*F*) 10.0 mm, estimated age 37 days, $\times 5\frac{1}{2}$. (*G*) 15.5 mm, estimated age 40 days, $\times 3\frac{1}{2}$. (*H*) 18.0 mm, estimated age 47 days, $\times 2\frac{3}{4}$. (*I*) 20.8 mm, estimated age 50 days, $\times 2\frac{3}{4}$. (*J*) 30.0 mm, estimated age 60 days, $\times 1\frac{3}{4}$. (All photographs courtesy Carnegie Institution of Washington.)

173

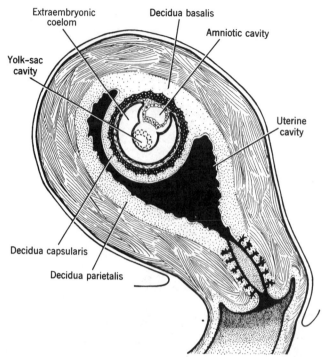

Figure 9-29 Early relationship of the implanted blastocyst to the endometrium. (From Hamilton, Boyd, and Mossman, *Human Embryology,* courtesy W. Heffer and Sons Ltd., Cambridge.)

other mammals, the human allantois does not expand into the extraembryonic coelom, but remains as a rudiment within the body stalk.

4. The bilaminar embryonic disc is converted to a trilaminar one (Figure 9-27). As far as can be determined, events are comparable to those in the chick embryo. That is, prospective mesoderm and notochord in the epiblast move into and through a primitive streak and Hensen's node, respectively, to assume their final positions between the residual ectoderm above and endoderm below. The essential departure from the pattern seen in the chick embryo is that in humans, as in mammals generally, the extraembryonic mesoderm is emancipated from these formative movements, appearing instead as a direct product of the trophoblast.

From this point on we take only incidental note of the embryo itself and concentrate our attention on the further history of the extraembryonic membranes. Head, lateral, and tail folds undercut the embryo in the same fashion as we have seen in the chick embryo. These processes move the heart primordia and the allantois into their respective final positions anterior and posterior to the yolk stalk, marking the full separation of the body from the extraembryonic membranes. Figure 9-28 pictures representative stages in the molding of the gross body form of the embryo; the morphogenesis of individual organ systems is given attention in later chapters.

We have seen that by the end of the 16th day the blastocyst is fully implanted in the endometrium and that chorionic villi are elaborated over its entire surface. Although the normal implantation site varies, most

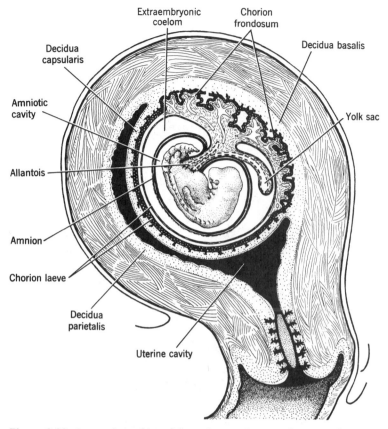

Figure 9-30 Later relationships of the embryo and its membranes to the endometrium. (From Hamilton, Boyd, and Mossman, *Human Embryology,* courtesy W. Heffer and Sons Ltd., Cambridge.)

sites are found on the posterior wall of the uterus. Once implantation occurs it provides a landmark for distinguishing three topographical areas in the endometrium. That area immediately beneath the blastocyst is termed the *decidua basalis;* that encapsulating the lumen-ward surface of the chorion is the *decidua capsularis;* that lining the remainder of the uterus is the *decidua parietalis* (Figure 9-29). As development advances, the chorionic villi on the side facing the decidua basalis continue to enlarge and branch, while those on the side facing the decidua capsularis gradually regress, and by the fourth month of gestation this portion of the chorion is left almost smooth. This villus-free chorion is termed the *chorion laeve;* the area bearing the persisting villi is termed the *chorion frondosum* (Figure 9-30). It follows, then, that the placenta is a product of the decidua basalis and chorion frondosum, the former representing the maternal component and the latter, the fetal component.

From the very beginning of implantation, of course, the villi have been associated with the excavation of the uterine endometrium and the resulting opening of spaces filled with tissue detritus and blood. Now confined to the chorion frondosum, the villi continue to branch and increase in size and length, while burrowing still more deeply into the decidua basalis. However, this does not occur uniformly over the

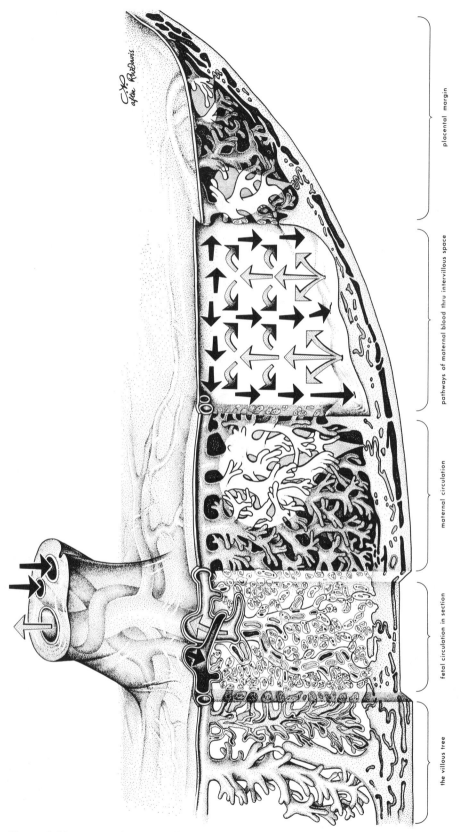

Figure 9-31 Functional anatomy of the human placenta. (From Ramsey and Harris, Courtesy Carnegie Institution of Washington.)

chorion frondosum. Scattered clumps of villi engage in this process, whereas intervening areas remain relatively stable. Consequently, clusters of villi come to project into endometrial crypts that are more or less walled off by septa remaining by virtue of the lesser erosion by the neighboring less active villi.

The tips of some of the villi fuse with the maternal tissues and serve to anchor the fetal to the maternal component of the placenta. The other villi branch freely in the crypts, or intervillous spaces, between the septa. Continued enlargement of these spaces serves to open more and more endometrial blood vessels so as to create pools of maternal blood within which the free villi are bathed. Physiological studies on mature placentas have revealed that the blood pools are by no means stagnant. Instead, there is a circulatory pattern within them. Under the head of arterial pressure, arterial blood pulses in fountainlike spurts. As the pressure is intermittently dissipated, there is a lateral flow and finally a return flow through the open ends of the uterine veins. Physiological exchange between mother and fetus is therefore mediated via two sep-

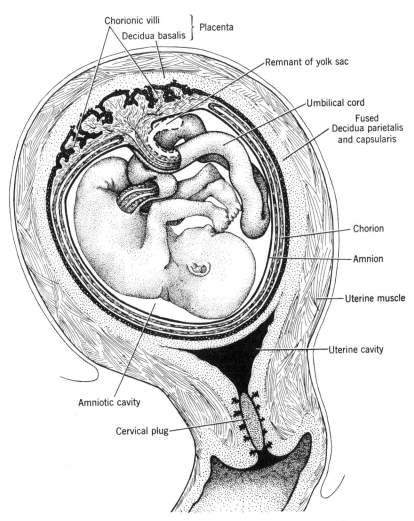

Figure 9-32 Relationships of the advanced fetus and its membranes to the uterus.

arate, though intimately associated, circulatory arcs: one fetal, through the chorionic villi, and the other maternal, within the blood pools of the intervillous spaces. All these structural and functional features are summarized pictorially in Figure 9-31.

As the embryo enlarges, the amnion expands rapidly within the extraembryonic coelom and progressively ensheathes the body stalk and proximal portion of the secondary yolk sac. Because the rudimentary allantois also lies within the body stalk, further expansion of the amnion serves to press the mesoderm of the body stalk into a cordlike structure within which the allantois and neck of the yolk sac are embedded. This is the *umbilical cord* (Figure 9-30), whose mesodermal core becomes converted to a gelatinous tissue through which umbilical (allantoic) arteries and veins course. The right umbilical vein eventually disappears, and only rudiments of the yolk sac and allantois persist.

With the advance of gestation the mesoderm of the external surface of the amnion fuses with the mesoderm of the inner aspect of the chorion, thus obliterating the extraembryonic coelom. The chorion keeps pace with the enlarging fetus and, by pushing out into the uterine cavity, causes the decidua capsularis to meet and fuse with the decidua parietalis. Because the expansion of the amnion obliterates the extraembryonic coelom and the chorion with its

encapsulating decidua fuses with the parietalis, the cavity of the uterus in the later stages of pregnancy is monopolized by the amnion with its contained fluid and fetus. Keeping pace, the originally short umbilical cord becomes relatively long. Figure 9-32 pictures these associations at about the fifth month of pregnancy.

To summarize, then, as in eutherian mammals generally, fertilization and embryogeny in humans are internal to the female reproductive tract with all preparations and accommodations therefore part of the estrus cycle. Early development proceeds in a direct and abbreviated fashion and provides for early association with the uterus. In this association there is a utilization of the basic extraembryonic membranes employed by amniotes generally for protection and maintenance of the embryo. These membranes combined with the endometrium constitute the placenta, which for the human embryo may be characterized as follows:

1. Discoidal (in reference to gross shape)

2. Hemochorial (chorionic villi bathed in maternal blood)

3. Deciduous (loss of endometrial tissues at birth)

4. Chorioallantoic (by reason of allantoic blood supply to chorion even through the allantois proper is rudimentary)

Part Three
Morphogenesis of Organ Systems

Chapter Ten
The Skin and Its Derivatives

It is appropriate to open the account of the morphogenesis of organ systems with the skin (integument) for, like the wrapping about a parcel, it is the object first encountered in the examination of any animal. But unlike a wrapping, which serves only to enclose the parts within, the skin in conjunction with its assorted derivatives exhibits a wide range of functional versatility.

The skin inevitably serves as a *protective organ,* shielding the body against mechanical injuries, the invasion of infective organisms, and excessive loss of moisture. It is protective too, in the sense that its many adaptations may contribute to an animal's survival in the face of enemies and predators. There come to mind in this connection the many instances of protective coloration afforded either by pigmentation of the skin proper or through hair and feather patterns,

odor- and poison-producing glands, and external armor featured by many forms. Physiological protection is also furnished by immunological responses of the skin to antigenic agents.

In birds and mammals, which maintain a fairly constant body temperature, the skin is involved in the *regulation of heat loss.* It plays this role in two ways, physiologically and physically. Physiologically, the skin operates as a radiator, dissipating heat brought to it by the bloodstream, the volume of which is regulated by the nervous system. Through the action of appropriate central nervous "thermostats" and peripheral nerves, the sizes of the arteries and capillaries are adjusted as required. Relaxation of the blood vessels brings an increase in blood volume and favors radiation of heat from the surface of the skin. Conversely, constriction of the vessels brings a decline in blood volume and reduces heat radiation. Physical regulation is accomplished in numerous ways. Evaporation of sweat by mammalian skin makes for cooling; heat conservation is furthered by such insulating materials as oil and fat, mammalian hair, and the feathers of birds.

In addition to evaporative cooling of the skin as a component of temperature regulation, the sweat glands supplement the kidneys as a mechanism for *excretion*. Although sweat is about 99% water, the sweat glands selectively dispose of a variety of solutes. The most concentrated solute in sweat is sodium chloride, but most of the water-soluble substances in blood plasma also appear in some degree. Varying concentrations of potassium, calcium, magnesium, and other mineral components of plasma have been recorded, as have lactic acid and such nitrogenous products as ammonia, urea, uric acid, and creatinine.

The integument is also an organ of *secretion* in that a variety of cutaneous glands provide useful products of one kind or another. Mucus-secreting glands of fishes and amphibians are illustrative, and so are the numerous oil-secreting glands found in mammalian skin. Especially important are the milk-secreting glands of mammals.

Cutaneous *respiration* plays a significant role in the respiratory economy of many vertebrates and probably occurs to some degree in all. In some fishes and amphibians, for instance, a large fraction of the oxygen demand (and carbon dioxide discharge) may be handled by the skin. Terrestrial vertebrates employ the skin in respiration to a much lesser extent, yet even in such representatives as the turtle, the pigeon, and the human, some gaseous exchange takes place through the skin.

One ordinarily thinks of *locomotion* in terms of appropriate skeletal parts and associated muscles, but the skin constitutes an important adjunct. It is the membrane of skin stretched over the skeletal supports that gives to the fin of a fish its operative properties. So it is, too, with the wing of a bat or the web-footed animals such as ducks or frogs. And the flight of birds obviously depends on the feathers of the wings and tail.

Finally, the skin serves as the vehicle for a great array of *cutaneous sense organs* (see p. 450) by which animals maintain contact with their external environment. Notable among these are the receptors for touch and the allied receptors concerned with the senses of pressure, temperature, and pain.

STRUCTURE AND ORIGIN OF THE SKIN

General Structure of the Skin. The skin is not a single structural entity; it consists of two parts: an outer *epidermis* of ectodermal origin, and an inner *dermis* of mesodermal origin (Figure 10-1).

The cells of the epidermis are arranged in stratified layers. The deepest layer is columnar in form and is called the *stratum germinativum* in reference to the fact that here, and only here, does cell multiplication occur. As new cells are formed in the germinative layer, they are gradually pushed peripherally, becoming more and more flattened in the process. Although the entire thickness of the epidermis in cyclostomes and fishes consists of live cells, in amphibians and amniotes the cells die as they approach the surface and their cytoplasm is converted to some variety of a tough, fibrous type of protein known as *keratin*. Keratin is the stuff of which hair, feathers, claws, and the like are made. It is also because of the presence of keratin that the outermost cells of the epidermis constitute an impermeable membrane designated the *stratum corneum,* or horny layer, of the skin; it is composed of flat, squamous cells.

The epidermis has no blood supply of its own. It is dependent on the rich vascularization of the dermis to which the stratum germinativum is immediately adjacent. This accounts for the limitation of cell division

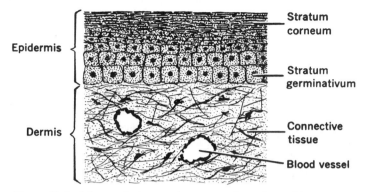

Figure 10-1 Schema of microscopic anatomy of vertebrate skin.

to the germinative layer. The steady reproduction of cells here would create a steadily thickening epidermis were it not for the fact that the cells of the stratum corneum are constantly sloughed or worn from the surface. The germinative layer has the ability to accommodate, however, to special circumstances, as in the production of a callus in response to excessive friction. It is also the principal agent in wound healing and skin regeneration, and, as we shall see, it produces all the glands of the skin. Shedding of the corneal cells may be constant and piecemeal as in mammals, or the stratum corneum may be sloughed in whole or in part at regular intervals. Thus, in amphibians and many lizards large sheets of corneum are shed at intervals of a few days, and snakes periodically slough the outermost layers of the stratum corneum as a whole.

The dermis, in contrast to the epidermis, consists more of cell products than of cells. Its bulk is composed of a feltwork of connective tissue fibers derived from the original segmented dermatome, and produced by a relatively small population of cells. This feltwork serves as the vehicle for an abundance of blood and lymph vessels, nerve endings, sense organs, deposits of fat, smooth muscle, and an assortment of epidermal and dermal derivatives. It is the "tanned" dermis from which the epidermis has been removed that provides our familiar leather.

Origin of the Skin. We have seen that the skin is composed of two parts, epidermis and dermis. By the same token, the skin of all vertebrates traces its origin to two embryonic sources, ectoderm and mesoderm.

The entire surface ectoderm exclusive of that dedicated to the neural tube may be regarded roughly as prospective epidermis. There are, however, some special instances of the employment of the embryonic skin ectoderm for the creation of things other than epidermis. The inner ear and the lens and conjunctiva of the eye (see Chapter 17) are cases in point, as are the epithelial lining of the mouth and parts of the teeth (see Chapter 13). Yet it is interesting to note that the embryonic ectoderm in such loci is capable of forming conventional epidermis when grafted to atypical positions. For example, in urodeles, mouth ectoderm grafted to the trunk forms epidermis.

It is customary to trace the dermis to that subdivision of the epimere known as the dermatome. The conventional view is that the dermatomal components of the somites resolve themselves into mesenchymal cells, which multiply and deploy to all regions of the embryo. But certain experimental analyses suggest that the mesodermal source of the dermis is by no means restricted to dermatome. For example, pieces of somatopleuric body wall (consisting of somatic mesoderm covered by skin ectoderm) of chick and mouse embryos, when grown in isolation, produce structurally normal skin and skin derivatives, feathers and hair, respectively. In birds and mammals, therefore, it appears that while the dermatome furnishes the dermis of the dorsal and dorsolateral portions of the body, the dermis of the flank and ventral surface of the body is a derivative of the mesoderm of the somatopleure. Whether the same conclusion may be applied to other vertebrates is not entirely clear, although there is some evidence from experiments on amphibian embryos pointing to a similar situation.

Whatever the source of the dermis, it is perfectly clear that from the beginning of embryogeny the dermal and epidermal components of the integument are very intimately associated. So as we turn to the morphogenesis of the skin and its derivatives, we shall give attention not only to the developmental anatomy of parts, but to the causal relations and interactions of the dermal and epidermal constituents.

GLANDS

Many of the functional operations of the skin are carried out through the medium of glands, which in one form or another are found in every vertebrate class. With reference to structure, the glands of the skin are either *unicellular* or *multicellular.* The former are modified single cells scattered among the other cells constituting the epidermis. The latter are formed as ingrowths from the stratum germinativum into the dermis, sometimes hollow from the beginning of development, sometimes solid at the start but ultimately hollowing out. In any event, they discharge their products through ducts leading to the surface. Multicellular glands are either *tubular* or *alveolar* (saclike). Tubular glands may be straight, coiled, or branched; alveolar glands may be simple sacs or moderately or elaborately branched (Figure 10-2).

Cyclostomes. The only skin glands found in cyclostomes are of the unicellular type (Figure 10-3). Scattered throughout the ep-

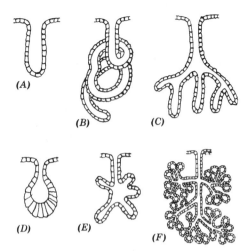

Figure 10-2 Diagrams of various types of epidermal integumentary glands. (*A*) Straight tubular. (*B*) Coiled tubular. (*C*) Branched tubular. (*D*) Simple alveolar. (*E*) Simple branched alveolar. (*F*) Compound branched alveolar.

idermis and lying near the surface, they secrete relatively large amounts of mucus, which renders the surface of the body extremely slippery. In the hagfishes, the more generalized slime glands are supplemented by others, which discharge long mucous threads that, added to the other mucous secretion, constitute a thick, slimy protective coat over the body.

Fishes. The skin of fishes contains both unicellular and multicellular glands (Figure 10-4). The unicellular glands are like those of cyclostomes; the multicellular glands are of the simple alveolar type. Most of these glands, like the unicellular glands of cyclostomes, are concerned with secretion of mucus. In several species of sharks, chimaeras, and teleosts, however, some of the multicellular glands associated with spines on the fins, tail, and gill covers produce an irritating poison. Another interesting adaptation of multicellular glands is found in many deep-sea elasmobranchs and teleosts. These forms, living in total darkness, feature luminous (light-producing) organs, judged to be modified mucous glands. In some teleosts these luminous organs display com-

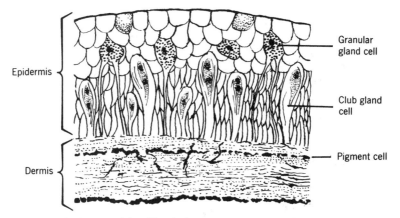

Figure 10-3 Section of the skin of a lamprey.

plicated accessory parts, including a lens in front and a pigmented reflector behind.

Amphibia. Because the skin of many amphibians plays an important role in respiration, the need for keeping the skin moist is paramount. A moist or aquatic habitat takes care of this need in large part, but numerous *mucous glands* also contribute. These glands are usually of the simple alveolar type embedded in the dermis (Figure 10-5). Besides the many mucous glands,

many amphibians possess *granular glands*, so called because of the granular appearance of the protoplasm in the secreting cells. The secretions of these granular glands are milky in character and of a poisonous nature, ranging from mildly irritating to highly toxic. Frequently the secretions are odorous as well. Granular glands are particularly abundant in toads, where they are aggregated on the dorsal side of the body and legs in localized thickenings of the epidermis known as "warts." Al-

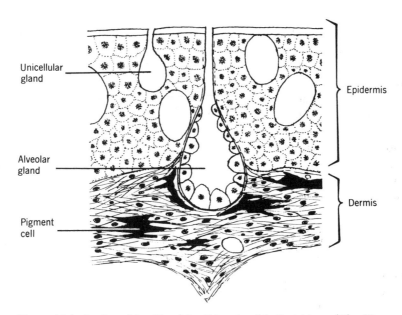

Figure 10-4 Section of the skin of the African lungfish, *Protopterus*. (After Kingsley.)

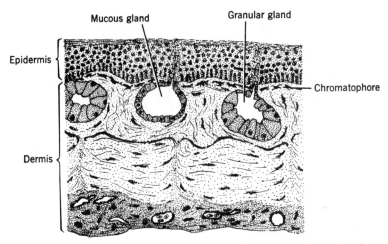

Figure 10-5 Section through the skin of the frog, *Rana pipiens.* (From Weichert, Anatomy of the Chordates, 3rd ed., courtesy McGraw-Hill Book Company, New York).

though the alveolar mucous and granular glands predominate in amphibians, there are instances of very specialized tubular glands as well. They include the suctorial discs on the feet of certain tree-dwelling frogs and the thumb pads of male frogs and toads, which enlarge during the breeding season and aid in the clasping of the females during mating. Unicellular glands also occur in amphibians, being found on the snouts of larval frogs and salamanders.

Reptiles and Birds. The skins of reptiles and birds are virtually devoid of glands. Among reptiles, such glands as do occur secrete products that appear to be related to breeding behavior or defense. In crocodiles and turtles, for instance, there are glands beneath the lower jaw (in crocodiles within the cloacal aperture as well), which during the breeding season secrete an odorous substance known as "musk." There are no skin glands in lizards, but some snakes have glands in the cloacal region, which produce a milky fluid with a nauseating odor. As for birds, the only conspicuous gland is the *uropygial gland,* a branched alveolar gland on the dorsal side of the body at the base of the tail. It supplies oil, which

the bird collects on its beak and uses in the preening of its plumage.

Mammals. Despite their reptilian origins, mammals are equipped with an abundance and great variety of integumentary glands. All these variants have probably been derived from two essential types: *sebaceous glands* and *sweat glands.*

Sebaceous glands are of the alveolar type, secrete oil, and are ordinarily associated with hairs (Figure 10-17). They may, however, be present in regions where hair is absent. Their oily secretions serve as a protection and lubricant for the hair and skin and also impart to the animal an individual scent or odor. Special sebaceous glands in the eyelids help to lubricate the eyeball and prevent the overflow of lacrimal fluid (tears). Others, located near the anus in certain mammals, serve as scent glands.

Sweat glands are simple coiled tubular glands (Figure 10-7). It is by way of their watery secretion that the skin plays its role in temperature regulation and excretion mentioned earlier. In many mammals, including the human, the horse, and the bear, sweat glands are widespread over the body. Other mammals, by contrast, have no sweat

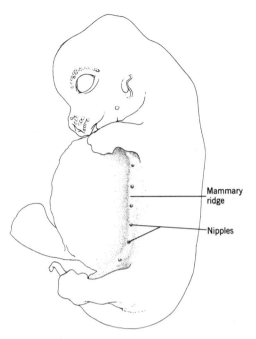

Figure 10-6   Mammary ridge of a 20-mm pig embryo. (After Minot.)

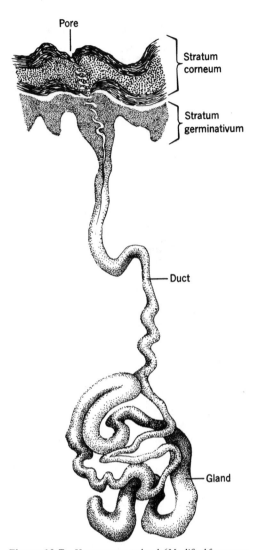

Figure 10-7   Human sweat gland. (Modified from von Brunn.)

glands at all. Among these are moles, whales, sea cows, and the scaly anteater. Still others have a limited number of sweat glands confined to specific localities: in mice, rats, and cats they are restricted to the undersides of the paws; in deer they are found about the base of the tail; rabbits have them about the lips; and in sheep, cattle, and pigs they occur on the muzzle and snout.

The most noteworthy of the specially modified sweat glands are the *mammary,* or milk-producing, *glands* that are present in all mammals and from which this class of vertebrates derives its name. One can only speculate as to the evolutionary origin of mammary glands and the lacteal feeding of young. There are no obviously related precedents in other classes of vertebrates, and no clues are provided by the embryogeny of mammary glands other than the obvious resemblance to developing sweat glands.

In eutherian mammals, development of the mammary glands is inaugurated by the appearance of a pair of bandlike thickenings of epidermal ectoderm along the ventrolateral body wall from the level of the forelimbs to the groin. Each of these thickenings is known as a *mammary ridge* or *milk line* (Figure 10-6). It is from the epithelium along the course of the milk line that the mammary glands arise. The number and position of the mammary glands that are to form vary with the species of mammal. In

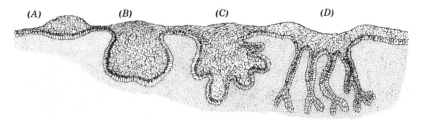

Figure 10-8   Stages in the histogenesis of a mammary gland. Description in text.

humans, sea cows, elephants, and bats a single pair of pectoral mammae ordinarily develop, whereas in cattle and horses one or two pairs of inguinal mammae appear. Three pairs of thoracic and three pairs of abdominal glands are usually formed in rats and mice, whereas pigs and dogs have five or six pairs arranged in a continuous series from chest to groin. Occasionally, as an abnormality, more than the number characteristic of the species may develop, as witnessed by the not infrequent instances of extra mammae in humans, a condition known as hyperthelia.

Development of a given mammary gland proceeds from continued local thickening of the epithelium of the milk line so as to produce a roughly spherical mass projecting into the subjacent dermis (Figures 10-8*A, B*). Cords of cells then gradually ramify within the surrounding dermis (Figure 10-8*C*), so as finally to produce a branchwork (Figure 10-8*D*). In due time the cords hollow out to create a duct system converging at the surface, where an epidermal nipple is elevated. At birth and for some time thereafter the duct system shows only a moderate degree of secondary branching and the glands are about equally developed in both sexes. No further development occurs in males, but elaboration of the mammae resumes in females with the advent of sexual maturity.

The suctorial projections (nipples) on mammary glands are of two types. One is a *true nipple* through which a single mammary duct (e.g., in rodents) or several ducts (e.g., in carnivores and humans) pass to open directly to the outside (Figure 10-9*A*).

The other is the *false nipple,* or *teat,* found in horses and cattle, in which the ducts open into a chamber at the base of the projection, from where a secondary tube leads to the exterior (Figure 10-9*B*).

SCALES

The history of the skin of vertebrates is distinguished by the elaboration of a variety of derivatives that constitute a protective armor. When we use the term in its broadest sense, the units of which this armor is composed are called scales. Some are epidermal in origin; others are dermal. Because the dermal scales play the more conspicuous part in vertebrate history, let us open the discussion with them.

Dermal Scales. The fossil record clearly shows that the most ancient of vertebrates, the jawless ostracoderms, were completely ensheathed in a bony armor. There has been much speculation about the selective factors that may have been responsible for this situation. Some workers have suggested that it may reflect the evolutionary benefit of a

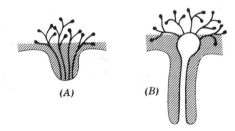

Figure 10-9   Mammary nipples. (*A*) True nipple. (*B*) False nipple, or teat.

seal against excessive absorption of water, because the ostracoderms presumably lived in fresh water and osmotic forces would have tended to drive water into the body. Others have attempted to "explain" it in terms of defenses against large invertebrate predators. Be that as it may, the ostracoderms were completely covered by large bony plates presumably underlying the epidermis. These plates were particularly prominent over the head and gill region and down over the "shoulders." Placoderms, the first jawed vertebrates, retained extensive armor except that the large plates tended to become subdivided into smaller units.

Structurally, the bony scales of placoderms consisted of three layers (Figure 10-10). On the bottom (inside) there was a layer of very compact bone. This was followed by a spongy layer with numerous spaces, which presumably were filled with vascular tissue. The top (outside) was again composed of compact bone or material very much like it. In this arrangement, that is, spongy bone sandwiched between two layers of compact bone, one sees the very architecture that is featured by the dermal bones of the skull. Commonly, too, the top-

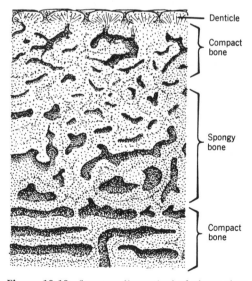

Figure 10-10 Structure (in section) of a bony plate of a Devonian placoderm.

side was ornamented by tiny projections or ridges called *dermal denticles.* Opinions vary as to whether the dermal denticles were integral parts of the scales or constituted an independent and separately formed integumentary system. In either case, each consisted of one or two layers of very hard material and lay over a cavity, which in the living animal was undoubtedly filled with a pulp of vascular tissue. Over the margins of the jaws the dermal denticles were especially enlarged and served as *teeth.*

In the Chondrichthyes the skin is devoid of armor except for the diffusely distributed dermal denticles, termed *placoid scales* (Figure 10-11). Again, as in placoderms, those denticles on the jaws and for some distance within the mouth cavity are enlarged as teeth. The gradation of teeth and placoid scales about the mouth (Figure 10-11) and the similarity of their development and structure establish a homology between teeth and placoid scales. Modified placoid scales take the form of the sting of the stingrays and the claspers of male sharks and skates, as well as the spine of the familiar spiny dogfish, to mention only a few. The ratfish or chimaera is totally devoid of placoid scales.

In the line that led toward the Sarcopterygii, there was retention of the three-layered bony scales. Termed *cosmoid scales,* they were characteristic of the primitive sarcopterygians, in which for the first time they assumed a regularity of position, shape, and size presaging their incorporation into the skeleton. Thus the story of the evolution of the tetrapod skeleton from its beginnings in crossopterygian ancestors involves the exploitation of cosmoid scales in the skull (dermatocranium) and shoulder girdle. In later sarcopterygians the scales tended toward a simpler architecture, and true cosmoid scales are nonexistent in living representatives of these fishes. Modern Dipnoi, for instance, have extremely thin scales of a fibrous, bony texture.

In the Actinopterygii the teeth and der-

Denticle

Compact bone

Spongy bone

Compact bone

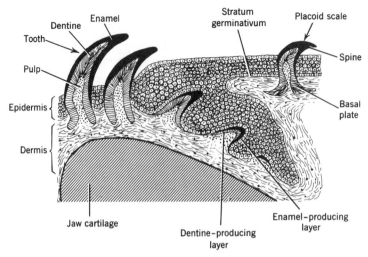

Figure 10-11   Placoid scales and successive generations of teeth in a shark. (Designed after several sources.)

mal components of the skull and shoulder girdle, all originating in the primitive armor of placoderms, were retained, and the bony scales evolved independently. The cosmoid scale was replaced by the *ganoid scale,* so called because of the presence in the upper surface of layer upon layer of a hard, enamellike substance known as ganoin. Representative ganoid scales are found presently in the Chondrostei and Holostei. But just as modern lungfishes have only the vestiges of the cosmoid scales of their ancestors, the Teleostei have been left with only thinned out and ganoin-free derivatives of onetime ganoid scales. Teleost scales are either *ctenoid* (toothed) or *cycloid* (circular) (Figure 10-12) and are arranged so as to overlap like shingles on a roof. In some instances (e.g., eels and catfishes) even these reduced scales may be absent.

Four categories of hard parts originating in the dermal armor of the first fishes carry into the tetrapod line of evolution via the route from Crossopterygii to Amphibia: (1) the dermal components of the skull, (2) the dermal components of the shoulder girdle, (3) the teeth, and (4) body scales. The further history of the skeletal parts will be con-

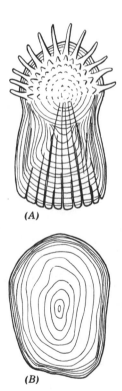

(A)

(B)

Figure 10-12   Ctenoid scale (*A*) and cycloid scale (*B*) of teleosts. (From Portmann, *Vergleichende Morphologie der Wirbeltiere,* courtesy Benno Schwabe and Company, Basel.)

sidered in Chapter 11; that of the teeth in Chapter 13. The present account, therefore, concerns itself only with the permutations of the scales in the tetrapods.

The general trend in tetrapods as a whole has been toward steady reduction and loss of dermal armor. As a heritage from their fish ancestors, the older amphibians retained rows of scales on the flanks and underside of the body. Modern amphibians, however, are devoid of scales except for vestiges buried in the skin of the Apoda. The ancient reptiles, too, featured many elaborations of dermal scales, but the only representatives of fish scales found in present-day reptiles are the rows of riblike structures, termed *gastralia,* found in the ventrolateral body wall of crocodilians, *Sphenodon,* and some lizards (Figure 11-35). In birds and mammals there is no carryover whatsoever of dermal scale equipment.

Although, as noted, the general trend in tetrapods has been toward the elimination of dermal scales, the dermis appears to have retained a potential for bone formation that has manifested itself in some notable instances. The stout bony shell that encases most turtles is illustrative, as is the carapace of the armadillo. But these are surely instances of secondary bone formation, and the individual plates of bone cannot be homologized with ancient bony scales. The same is probably true of the bony scales found in some lizards and the scattered nodules of bone that reinforce the skin of certain edentate mammals.

Epidermal Scales. Epidermal scales are derivatives of the stratum corneum, the keratinized surface component of the epidermis. Because the stratum corneum is best expressed in terrestrial vertebrates, epidermal scales are lacking in fishes and are found only rarely in amphibians. It is possible that the ancient armored amphibians were equipped with epidermal scales, but in modern forms they are virtually absent. The nearest approach to them is the localized thickenings of the stratum corneum, the "warts" of toads.

In contrast to amphibians, epidermal scales are extremely well developed in reptiles and are found to be of two general types. One variety occurs in snakes and lizards, the other in crocodiles and turtles. In the former type a scale is initiated embryonically by an elevation of epidermis into which the subjacent dermis projects (Figure 10-13*A*). The dermis and stratum germinativum subsequently retract, leaving the flat, cornified scale at the surface (Figure 10-13*B*). Although the final scale itself is thus purely epidermal, the dermis appears to be involved in its initiation. Generally speaking, the scales of snakes and lizards are arranged in longitudinal rows, each scale being directed backward and overlapping the one behind like a roof shingle. In snakes, however, those scales on the ventral side

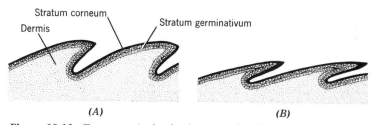

Stratum corneum

Dermis

Stratum germinativum

(A) *(B)*

Figure 10-13 Two stages in the development of epidermal scales in a lizard. (From Weichert, *Anatomy of the Chordates,* 3rd ed., courtesy McGraw-Hill Book Company, New York.)

of the body are arranged transversely and are utilized in locomotion (p. 242). In certain snakes and lizards too, special modifications of epidermal scales are found. The so-called horns of the horned lizards are specialized epidermal scales, as are the rattles of the rattlesnake. As noted earlier, there is in terrestrial vertebrates generally a steady shedding of the stratum corneum and its constant replacement from below. In snakes and many lizards the entire "skin" is periodically shed as a whole. Through the activity of the thyroid gland and anterior lobe of the pituitary gland, and also depending on the amount of food eaten, a new crop of epidermal scales is produced at intervals, following which the animal literally crawls out of its old sheathing.

The epidermal scales of turtles and crocodiles do not overlap as do those of lizards and snakes, and each is an exclusive product of the stratum germinativum without participation of the subjacent dermis. In turtles the epidermal scales arise on top the bony armor beneath, but their pattern does not correspond to the bony plates. (Scales also cover the neck and legs.) Each scale is produced independently through the activity of the stratum germinativum. During the life of a turtle, the germinative layer under each scale periodically creates a new scale of a diameter slightly larger than the preceding one. The older scale is then pushed away from the surface. The older scale may be worn or peeled off, but commonly the successive generations of scales simply pile up one upon the other. Because each new scale is larger than its predecessor, there is created an impression of "growth rings" indicative of the number of scales produced. The entire body of crocodilians is also covered by epidermal scales which, wearing away steadily, are replaced as needed.

Birds and mammals have a limited heritage of the epidermal scales possessed by their reptilian ancestors. Such as they have are of the overlapping type and are formed in the same manner as those of snakes and lizards. In birds they are confined to the legs

and feet. When they do occur in mammals, it is usually on the tail, as in the beaver, rat, mouse, and opossum. The pangolin, or scaly anteater, of Asia and Africa, however, has a complete body covering of horny scales, and epidermal scales are superimposed on the dermal bony plates investing the armadillo.

FEATHERS

The distinguishing mark of birds is the covering of feathers, the uses of which range from heat insulation and flight to protective coloration and sexual display. Feathers are believed to have been derived from reptilian scales, but the similarities between them are evident only during early embryogeny. We shall examine feather development in some detail later, but first let us consider certain matters of their architecture.

There are three general types of feathers (Figure 10-14): the *hair feather,* the *down feather,* and the *contour feather.* The hair feather is the smallest variety and the simplest in structure, consisting of no more than a long, slender *shaft* from whose distal end a few hairlike *barbs* flare out. Hair feathers are usually scattered rather widely over the body beneath the contour feathers.

The down feather is somewhat more complex. It is composed of a short, hollow basal *quill,* buried in the skin, from which there spray a number of filamentous barbs bearing tiny *barbules* along their margins. The first plumage of young nestling birds consists of down feathers. This first crop, however, is shed and replaced by a later generation that accompanies the appearance of contour feathers. Underlying the contour feathers over much of the body, the down feathers provide effective insulation for the adult bird.

Contour feathers are the large feathers that ensheathe the body and provide its characteristic configuration. Special varieties of contour feathers are also present in

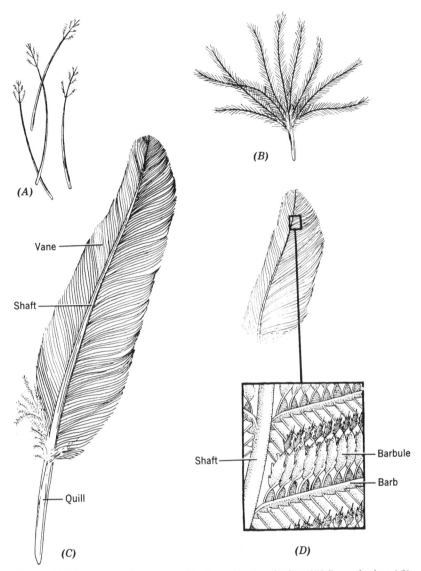

Figure 10-14   Types and structure of feathers. (*A*) Hair feather. (*B*) Down feather. (*C*) Contour feather. (*D*) Detail of structure of the vane of a contour feather.

the wings and tail. Unlike hair and down feathers, which have a fairly general distribution, contour feathers tend to be restricted to definite tracts, yet because of their manner of overlapping, the body is completely ensheathed by them. A typical contour feather consists of a lengthwise *axis* and a flat *vane.* The axis begins in a hollow quill embedded in the skin and continues as a semisolid shaft whose interior is filled with a spongework of horny, air-filled cells. The vane is composed of numerous barbs that project obliquely from either side of the shaft. Each barb in turn bears rows of barbules on its distal and proximal sides. The barbules on adjacent barbs are interlocked by hooklets, in consequence of which the barbs are linked together and the vane presents a flat, firm, unbroken surface.

The remarkably varied colors and color patterns of birds are due either to chemical pigments or to the physical structure of the feathers. The reds, yellows, and blacks are due to pigments whose deposition is known to be regulated by the hormones of the gonads and the thyroid gland. The blues and the iridescent colors result from refraction and interference of light rays through structural irregularities of the feathers. The greens result from structural blue combined with pigment yellow, whereas white results if no pigment is present and the cells of the barbs and barbules break up light and reflect all wavelengths equally.

Replacement of feathers occurs throughout the life of a bird. In some birds the replacement may be seasonal, that is, molting occurs; in others there is gradual replacement throughout the year. In each case the replacement process is in essence a repetition of embryonic development, to which we may now logically turn.

The first indication of a feather, like the scale of a snake or lizard, is found in an aggregation of dermal cells immediately beneath the epidermis. Continued proliferation of the dermal aggregation results in a conical elevation, the *dermal papilla*, ensheathed by epidermis (Figure 10-15A). Thus formed, the feather germ grows rapidly outward as a tapering epidermal cylinder filled with a vascular mesodermal core (Figure 10-15B). Meanwhile, the epidermis surrounding the base of the papilla sinks inward so that the feather germ comes to project from a pit, or pocket, termed a *follicle*. From this point onward, down and contour feathers show developmental differences.

HAIR

Hair is as characteristic of mammals as are feathers of a bird. Hair, like feathers, is also

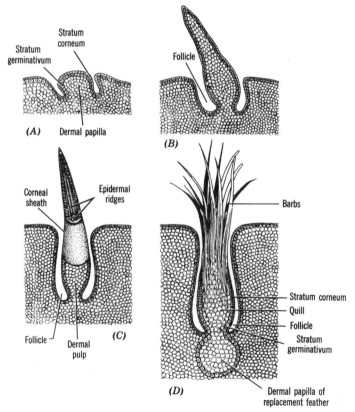

Figure 10-15 Four stages in the development of a down feather.

entirely epidermal in origin and composition. But unlike feathers whose source in repitilian scales seems well established, the phylogenetic origin of this distinctive mammalian integumentary derivative remains uncertain.

A typical hair (Figure 10-16) includes two portions, a projecting *shaft* and a *root,* the latter of which lies in a pit, termed the *hair follicle,* sunken slantwise into the dermis. At its base in the bottom of the follicle, the root is expanded into a hollow bulb into which a small *dermal papilla* projects, carrying blood vessels and nerves. It is only in the bulbous end of the root that living cells are found. Here the cells constantly multiply and, as they are produced, steadily push distally and add to the length of the root and shaft. In the process, the cells gradually die so that the hair proper becomes composed of dead, cornified cells. These cells are found to be arranged in either two or three layers. In coarser hairs (it is not demonstrable in fine ones) there is a narrow central core, or *medulla,* consisting of shrunken cells and large air spaces. About the medulla is the *cortex,* which provides the bulk of the hair and is composed of cells modified so as to provide a rather dense, horny material. The cortex carries most of the pigment of the hair. The exterior is clothed by a thin *cuticle* made up of a single layer of scaly cells which often overlap like shingles. Within the confines of the follicle, the root and shaft are also invested by an additional sheath produced by the follicular wall. This sheath is often of considerable complexity, showing a number of layers, each with a distinctive structure and name—all are structural details beyond our concern. Adjacent to the follicle, and a derivative of its epithelial lining, is a sebaceous gland that pours its oily secretion into the follicular cavity. Each hair is also equipped with a small, smooth muscle that, arising superficially, passes obliquely downward to insert into the side of the follicle. Its contraction, under the control of the autonomic nervous system, serves to move the hair into a more nearly vertical position and elevates the skin immediately surrounding the hair. This is the action that makes the hair "stand on end" in an angry dog and creates "gooseflesh" in humans.

Although hairs are constantly cast off and replaced, some mammals may produce a thick, dense coat during the winter months and then shed a good part of it in the spring. And in addition to its primary function as an insulating and heat-regulating agency, hair may be adapted for many special uses. Long tail hairs, eyelashes, eyebrows, and the dust-catching hairs in the nose are examples, as are the tactile whiskers and vibrissae of cats and dogs, and the defensive quills of the porcupine.

Unlike feathers, whose development is initiated by a preliminary aggregation of dermal cells, the embryogeny of a hair begins with the epidermis alone. At the site of a future hair an epidermal nodule is formed by a local proliferation of the stratum germinativum (Figure 10-17A). The nodule then extends downward into the

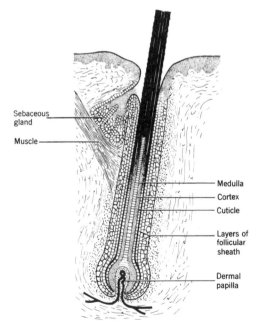

Figure 10-16 Diagram of structure of a hair. (After Portmann, *Vergleichende Morphologie der Wirbeltiere,* courtesy Benno Schwabe and Company, Basel.)

Sebaceous gland

Muscle

Medulla

Cortex

Cuticle

Layers of follicular sheath

Dermal papilla

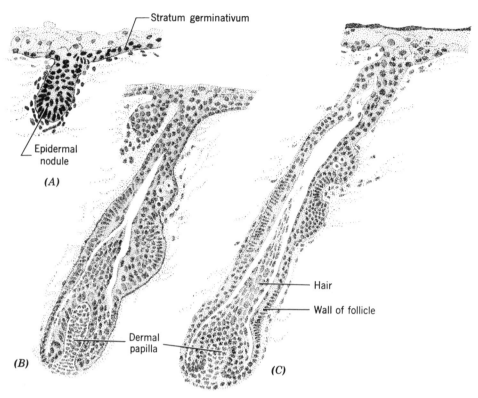

Figure 10-17 Three stages in the development of a human hair. (From *Textbook of Histology*, 5th ed., by Bremer. © 1936. Used with permission of Blakiston's, McGraw-Hill Book Company, New York.)

dermis as a tongue of tissue whose deeper end becomes enlarged to form a bulb (Figures 10-17*B, C*). This bulb is soon molded into an inverted cup into which the vascular dermis pushes and aggregates to form a dermal papilla. Meanwhile, the deep-lying portion of the epidermal tongue splits so as to separate a central strand from a surrounding epithelium. The central strand becomes the root and shaft of the hair; the surrounding tissue provides the wall of the follicle. The follicular wall at the same time exhibits a local proliferation destined to provide a sebaceous gland, and a tiny smooth muscle arises in the dermis and attaches to the follicular wall. As the embryonic hair grows, its shaft pushes toward the surface by making an opening for itself through the center of the still solid epithelial cylinder, the cells of which thus come to surround the hair

root and shaft and constitute its epithelial sheath.

EPIDERMAL RIDGES

In all of the primates, the skin on the undersurfaces of the tips of the digits and at restricted sites on the palms and soles is characteristically marked by narrow, parallel epidermal ridges. The patterns of these ridges—arches, loops, and whorls—are so distinctive to each individual and so fixed for life that they provide a basis for personal identification.

The embryogeny of epidermal ridges begins in folds of the stratum germinativum, which grow downward into the dermis. Coincidently, the dermis sends alternate elevations upward into the epidermis. In due time the previously smooth surface of the epidermis becomes converted to alternating ridges and valleys, the ridges matching the dermal elevations and the valleys matching the germinal downgrowths. That the mesodermal substratum plays an essential part in the determination of the epidermal ridge pattern is suggested by known facts relating to regeneration. If, for instance, the ridge patterns of the fingertips, that is, fingerprints, are damaged by friction or burning, the identical prints are restored provided there has been no damage to the underlying pattern of dermal ridges. But damage to the dermis precludes normal regeneration of characteristic epidermal ridges.

CHROMATOPHORES

All vertebrates possess pigment-containing cells known as *chromatophores*. They are usually stellate cells, with long, slender processes, within which the pigment may either be concentrated as a central mass or dispersed throughout, even into the slender processes. Although chromatophores may be found in the connective tissue investments of nerves and blood vessels and in the mesenteries and peritoneal linings of the coelom, they occur most conspicuously in the skin where their presence and activities play a large part in determining the characteristic colors of vertebrates.

A convenient classification of chromatophores is one based on the color and biochemical nature of the pigments they contain. The most common variety is the so-called *melanophore,* containing a type of pigment known as *melanin.* Melanin pigments are believed to be polymerized indole quinones that result from the oxidative metabolism of the amino acid tyrosine, and they show a color range from pale yellow through various shades of red-brown to black. A second type features red pigment and is called an *erythrophore.* A third one is the *xanthophore,* containing yellow pig-

ment. The pigments in erythrophores and xanthophores are, with some exceptions, *carotenoids* and *pterins.* These are fat-soluble pigments, that animals obtain directly or indirectly from plants, for no animal is known to be able to synthesize them. A fourth type is the *iridiophore,* which contains a white pigment composed of *guanine,* a nucleotide of ubiquitous occurrence in plants and animals. Sometimes the guanine occurs as fine granules that can disperse and concentrate as do the other pigments. Commonly, however, it is present as nonmigrating crystals that reflect and disperse light.

Of all the vertebrates, the bony fishes show the most brilliant and varied color patterns. Melanophores, erythrophores, xanthophores, and iridiophores are abundant in the dermis (Figure 10-18) and in various combinations produce a wide range of colors. Not only are characteristic colors provided, but many fishes have considerable ability to change their colors and color patterns in response to environmental background changes and variable light intensities. Particularly adept in this respect are the flounders and their relatives, which have remarkable capacities for matching background patterns. These changes are probably mediated by stimuli routed through the eyes and brain and thence by either hormonal or nervous agencies. The chromatophores of many teleosts are equipped with nerve fibers that belong to the autonomic system and control the dispersion and concentration of pigment via neurosecretions elaborated at the nerve fiber terminals. Although the evidence is not so strong for teleosts as for certain other vertebrates, dispersion of the pigment in the chromatophores also appears to be incited by secretions from the intermediate lobe of the pituitary gland, whereas adrenalin from the adrenal glands incites concentration.

Color is also well displayed by Amphibia and again results from varying combinations of pigment colors and light-dispersing iri-

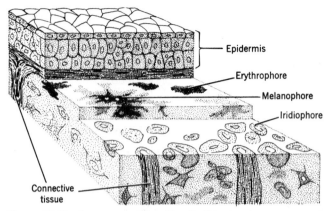

Figure 10-18   Stereogram of the skin of a fish, showing arrangement of chromatophores. (Based on a figure by Andrew, *Textbook of Comparative Histology,* courtesy Oxford University Press.)

diophores that lie for the most part between the epidermis and dermis (Figure 10-5). The general color tone of amphibians is green, resulting from the fact that the pigment cells are arranged in three layers. The melanophores, with dark pigment, lie deepest and are overlain by iridiophores, which by diffraction produce a blue-green color. Xanthophores overlie these and filter out the blue, so that the overall effect is green. Other colors are sometimes created, notably orange and red produced by assorted shades of erythrophores and xanthophores. In contrast to the bony fishes, there is no peripheral nervous control over the chromatophores. The pigment cells respond either directly to light or temperature changes, or indirectly through the eyes and thence via secretions from the adrenal and pituitary glands. It is also interesting to note that the embryogenesis of amphibian chromatophores is dependent on secretion by the developing pituitary.

Although some reptiles are rather drab in appearance, many snakes and lizards have conspicuous and elaborate patterns of coloration. The ability to change color, however, is much less common in reptiles than in fishes and amphibians, except in the instance of numerous lizards. Some lizards use color change as a means of temperature regulation, becoming pale in the sunshine,

hence reflecting light and retarding temperature increase, and becoming darker in the shade. Others change color, often rapidly, after any sensory stimulation. The physiological control of color change also shows some variations. In some lizards the control is purely hormonal: secretion by the intermediate pituitary causes expansion, and secretion by the adrenal medulla causes contraction of the pigment. In others, the well-known chameleons, for example, the slower-acting hormonal mechanism is supplemented by autonomic nerve fibers, which produce rapid contraction of the melanophoric pigments.

The colors of the feathers of birds are produced partly by pigments and partly by physical phenomena of diffraction and reflection. As in other vertebrates, the pigments involved are the melanins, carotenoids, and pterins. The melanins show a range of color from yellow through yellow-orange and reddish brown to dark brown and black, and are the products of definitive melanophores. In contrast to fishes, amphibians, and reptiles, however, the yellow and red pigments of birds are not produced in specialized chromatophores, but when present are manufactured by the cells of the stratum germinativum itself. These pigments are involved primarily in the coloration of the skin and of such special parts as beaks and bills, although in many birds they may supplement the melanin colors of the feathers. It is to melanophores, then, that the principal coloration of feathers is due.

Coloration of the skin and hair of mammals derives from various mixtures and hues of melanin. No carotenoids and pterins are found in mammalian skin. As with birds, the melanins are manufactured by melanophores, which are a constant component of the stratum germinativum of the epidermis. Present in large numbers and linked together by their branching processes, they actually constitute a reticular system within the epidermis. Their secretory product, melanin granules, is deposited directly into the cells of the germinative layer. Thus, as these cells in turn proliferate to build up the superficial epidermis, or multiply in the bulbs of growing hair, epidermis and hair acquire their characteristic pigmentation. Because both epidermis and hairs are expendable, that is, they are constantly being replaced, there is a concomitant loss of cells from the melanophore system.

So far we have considered chromatophores only in terms of their association with the integument. We should now inquire into their embryonic source. It is now firmly established that vertebrate pigment cells are products of the neural crest that arises on the occasion of the fusion of the embryonic neural folds to form the neural tube (see p. 415). It therefore appears that a fundamental homology exists among the pigment cells of all the vertebrates.

From their site of origin in the neural crest, prospective pigment cells migrate along prescribed pathways until they reach their final destinations. The factors regulating this migration are by no means fully understood. It seems clear that the cells themselves possess inherent powers of movement, but it is also clear that these movements are guided and channeled by influences emanating from other cells and tissues contacted along the way. Having reached their destinations, the embryonic pigment cells proceed to create given pigment patterns. This creation is guided, on the one hand, by the genetic equipment of the cells themselves, equipment that apparently operates through enzyme systems that control the amount and kind of pigment to be produced, and, on the other hand, by the tissue environment in which the pigment cells are located. The tissue environment in turn is conditioned by a host of factors, such as hormones, vitamins, light, and temperature.

MISCELLANY

Horny epidermal sheaths enclosing the bones of the upper and lower jaws constitute the *beaks,* or *bills,* of birds. Similar cornifications of epidermis on the edges of the digits serve as *claws* (in reptiles and birds, some amphibians, and many mammals), *hoofs* (horses, cattle, deer, pigs, and so on), and *nails* (humans and other primates). The *antlers* of deer, elk, and moose are given sufferance as skin derivatives only because they are temporarily covered by skin. Antlers themselves are composed of bone and connected with the frontal areas of the skull. The *horns* of cattle, goats, antelopes, and the like have a bony core but, unlike antlers, they are not shed and the bony core is encased in a keratin shell of epidermal origin. As can be seen in an antique "powder horn," the keratinaceous shell is hollow and fits like a shoe on the bony core that projects from the frontal bone of the skull. The so-called hair horns of the rhinoceroses are no more than tight bundles of epidermal keratin fibers and are devoid of a bony core.

Chapter Eleven

The Skeleton

DEFINITIONS AND GENERAL CONSIDERATIONS

Every animal, invertebrate or vertebrate from coelenterate to human, employs certain tissues for protection and support of the soft, pliable organs of the body and in the mediation of many functional operations of these organs. Wide though they are in variety, all may be grouped in a single category termed the *connective tissues*. This is done in recognition of one feature that all possess in common: the constituent cells lie in a matrix of extracellular material containing fibers. There are several varieties of fibrous components, but the most abundant type is that composed of a protein known as *collagen*. Collagenous fibers are glistening white in the living state and are characterized by great tensile strength. The kind and amount of nonfibrous matrix vary widely from one connective tissue to another. Mucopolysaccharides and proteins in varying quantities and combinations are the most common materials. Both matrix and fibers are manufactured by the cells, and thus the differences between the many kinds of connective tissues are manifestations of the differences in manufacturing prowess of the cells. All this variety notwithstanding, we may distinguish two broadly embracing classes of functional connective tissues: binding and supportive.

Binding connective tissues include such entities as *tendons, ligaments,* and *fasciae.* Tendons and ligaments feature an intercellular matrix packed with collagenous fibers oriented in parallel rows, the tendons serving to bind muscles to bone or cartilage and the ligaments uniting skeletal parts. Fasciae are sheets of connective tissue serving to bind constituent cells together into a definitive organ, such as muscle cells into the mass of an individual muscle or nerve fibers into a nerve. There is also much generalized, loose connective tissue both between and within organs, which provides an elastic highway for the transport of nerves and blood vessels.

Supportive connective tissues provide the *skeleton* around or within which an animal is built. The skeleton also serves as a mechanical framework for locomotion. When the skeleton is on the outside, it is known as an *exoskeleton;* when on the inside, it is known as an *endoskeleton.* The vertebrates have both an exoskeleton and an endoskeleton. Such derivatives of the skin as plates and scales constitute an exoskeleton; furthermore, many bones, especially of the head, are believed to have originated evolutionarily as plates in the skin. Thus, in the final analysis they belong to the exoskeleton. But because such bones come to assume an internal position and are intimately associated with the true endoskeleton, they are customarily considered together with it. For practical purposes, therefore, we may limit the exoskeleton to the parts associated with the skin and assign everything else to the endoskeleton. Of what kinds of materials is the endoskeleton composed?

SKELETAL MATERIALS AND THEIR DEVELOPMENT

Three types of materials are utilized in the endoskeleton of vertebrates: *notochordal tissue, cartilage,* and *bone.*

The first of these obviously is found in that ancient, primitive structure, the *notochord,* still present in the lower chordates and in the embryos of all vertebrates. But with the exception of a few of the lower categories, the notochord is replaced in adult vertebrates by the vertebral column and the base of the skull. The manner of

development of the notochord has already been dealt with in conjunction with earlier discussions of general embryogeny and need not be repeated. Suffice it to say that in those adult vertebrates in which it is retained, for example, the cyclostomes and elasmobranchs, it appears as a soft, flexible rod lying immediately beneath the neural tube. It is composed of a distinctive vesicular connective tissue and surrounded by a cylindrical sheath and external membrane (Figure 11-1). Although in the lower vertebrates it runs as a continuous, unsegmented rod from the infundibulum of the brain back to the terminus of the fleshy part of the tail, it is observed to be progressively restricted as it is replaced by the vertebrae in the higher vertebrates. In adult elasmobranchs, for instance, it shows alternate constrictions and expansions within and between the vertebrae, respectively; in most tetrapods it is represented in adults only as components of the discs that lie between successive vertebral centra.

Cartilage is a form of connective tissue whose intercellular matrix is composed of a complex protein of a rubbery texture

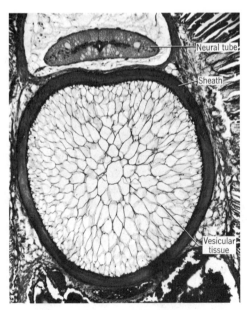

Figure 11-1 Notochord of a larval lamprey as seen in cross section. × 57. (Photograph by Torrey.)

within which there is spread a network of collagenous fibers. It exists in several variations: the matrix may be homogeneous and translucent with relatively few fibers (hyaline cartilage), it may be very fibrous and elastic (elastic cartilage), or it may contain deposits of calcium salts giving it a hard but brittle texture (calcified cartilage). Whatever its form, the matrix is a product of cartilage cells, or *chondrocytes*, which lie in small spaces (*lacunae*) within the matrix; and any given mass of cartilage is surrounded by a layer of dense connective tissue called the *perichondrium*. It is to the perichondrium that tendons and ligaments become attached.

The connective tissue character of cartilage is especially revealed by the manner of its development. At the site where cartilage is to form in an embryo, embryonic connective tissue cells (mesenchyme) begin to aggregate. The branching mesenchy-

mal cells then gradually round up and begin to secrete matrix. As more and more material is laid down, the cells (chondrocytes) are forced farther and farther apart, so that they become isolated in lacunae within the matrix they have secreted (Figure 11-2). For a time the cells may continue to divide; thus frequently a pair or quartet of chondrocytes may be seen grouped together within a single lacuna. The continued laying down of matrix by these internal chondrocytes is supplemented by depositions from cells at the surface provided by the perichondrium to be. This ability of cartilage to grow both by internal expansion and by external deposition stands in strong contrast to the growth of bone that we shall examine shortly.

Cartilage is the prominent skeletal material of the embryos of all vertebrates. But only in the cyclostomes, Chondrichthyes, and a few degenerate groups of Ostei-

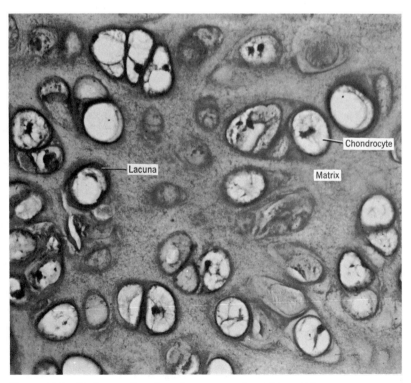

Figure 11-2   Hyaline cartilage from the sternum of a rabbit. × 215.

chthyes is cartilage a major skeletal material of adults; the adult skeleton of the higher vertebrates is predominantly bony.

Bone, like cartilage, is a derivative of mesenchyme and, except for some acellular bone in certain agnathans, consists of a matrix within which cells are buried. This is about as far as the resemblance goes, however. In the first place, bone matrix, in addition to being fibrillar, contains depositions of a hard, complex mineral substance composed chiefly of crystals of a calcium phosphate–calcium hydroxide combination known to mineralogists as an *apatite.* Moreover, the bone cells that secrete the matrix, unlike cartilage cells, do not become isolated from one another. The spaces (lacunae) within which they lie communicate by branching systems of minute canals into which project the radiating processes of the cells themselves. During embryonic life the cytoplasmic processes of neighboring bone cells interconnect, but whether these connections are maintained in adults is uncertain. Finally, again in contrast to cartilage with its generally uniform makeup, bone has a complex architecture. This architecture will be better understood in terms of its developmental history.

Embryonically, bone is always preceded by some other member of the connective tissue family. A preliminary aggregation of mesenchyme invariably anticipates bone formation. In many instances the mesenchymal precursor is succeeded directly by bone. Bone arising in this fashion is said to be *intramembranous* in origin. Speaking of definitive elements, bones so formed are known variously as *membrane,* or *dermal, bones.* In other instances the original mesenchyme first provides a cartilaginous model, which is then destroyed and replaced by bone. Bone development is then said to be *intracartilaginous,* and the resulting bone per se is known as *cartilage,* or *replacement, bone.* It should be clearly understood that these terms refer only to the sequence of events in bone develop-

ment, for when fully formed, bones of both types are fundamentally alike in structure.

Because the histogenesis of membrane bone is not only more direct but to a certain extent simpler, the story of bone development may be introduced with it. As a first step the mesenchymal cells congregated at the site of a future bone tend to cluster and arrange themselves in an interlacing network of strands. Each strand becomes the site of secretion of collagen. The result is a preliminary framework of collagenous fibers, with each fiber being invested by the connective tissue cells that produced it. Deposition of calcium salts then begins. The mesenchymal cells, which now may appropriately be called *osteoblasts,* or boneforming cells, extract the necessary raw materials from the blood supply pervading the area and lay down bone salts around the fibers. This serves gradually to create a scattering of bars and plates (*trabeculae*) of bone. As successive layers of bone are added to the trabeculae, some of the osteoblasts are caught up in their own deposits and there, in their individual lacunae, cease to function as active bone formers. Rather, as definitive bone cells, or *osteocytes,* they begin to play a role in the maintenance of the bone already formed. Neighboring osteocytes interconnect by delicate cytoplasmic processes running through the interlacunar canaliculi, and materials thus picked up by those osteocytes adjacent to blood vessels are distributed throughout the bone matrix (Figure 11-3). In the meantime, the remaining osteoblasts continue to add to the trabeculae which, as they enlarge, come together and fuse into a continuous latticework (Figure 11-4). By reason of its architecture, bone in this condition is termed *spongy,* or *cancellous, bone.* The areas between the trabeculae are occupied by connective tissue, which rapidly becomes highly vascular and as such constitutes the *bone marrow* (Figures 11-3 and 11-4). The bone marrow represents a primary source of new blood cells

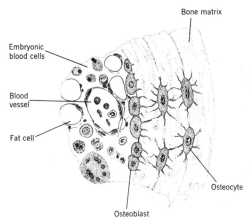

Figure 11-3 Area of bone tissue and adjacent marrow. (From Patten and Carlson, *Foundations of Embryology*, courtesy McGraw-Hill Book Company, New York.)

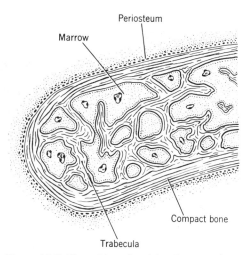

Figure 11-5 Deposition of peripheral compact bone about an area of cancellous bone. (Redrawn from Patten and Carlson, *Foundations of Embryology*, by permission of McGraw-Hill Book Company, New York.)

throughout life and with progressing age becomes a site of fat storage as well.

With the laying down of the primary mass of cancellous bone, there appears about this mass a peripheral concentration of mesenchyme. This membranous sheath is termed the *periosteum,* and within it certain cells again take on the qualities of osteoblasts and resume the deposition of bone. Now, however, the bone is laid down in dense, parallel sheets. The result is an external layer of *compact bone* investing the cancellous bone within (Figure 11-5). Sometimes cancellous bone may be supple-

mented or replaced in some measure by depositions of compact bone.

It is important to appreciate that there is really no fundamental difference between compact and cancellous bone, other than the degree of density and certain details of architecture. Both are products of osteoblasts; both are vital tissues maintained by entrapped but intercommunicating osteocytes. In the case of compact bone, physiological maintenance is facilitated by an array of canals constituting the so-called

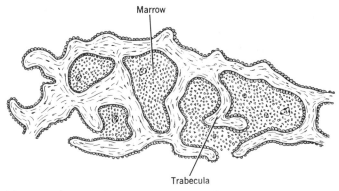

Figure 11-4 Area of early cancellous bone. (Redrawn from Patten and Carlson, *Foundations of Embryology*, by permission of McGraw-Hill Book Company, New York.)

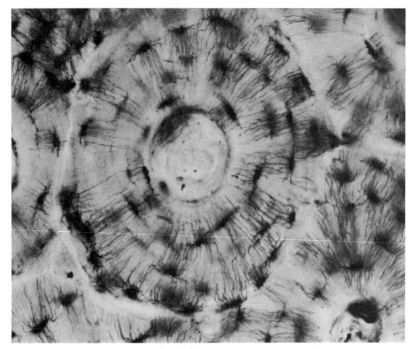

Figure 11-6   Ground section of human compact bone showing Haversian system. Concentric lamellae of bone surround a central Haversian canal; osteocytes lie in lacunae between the lamellae. ×150. (Photograph by Torrey.)

Haversian system (Figure 11-6). The origin of this system is a complex one and will not be discussed. Suffice it to say that compact bone becomes perforated by long, branching, and anastomosing *Haversian canals,* which communicate externally with the periosteum and internally with the marrow cavities. These canals provide routes for the distribution of blood vessels throughout the bone. The osseous tissue itself is found to be arranged in concentric layers (lamellae) about each canal. Between adjacent lamellae are the lacunae within which the osteocytes are situated. The lacunae communicate with each other and a given Haversian canal by canaliculi, thus enabling the widely scattered osteocytes to collaborate in the physiological maintenance of the tissue. It has also been suggested that the Haversian system constitutes a structural arrangement facilitating tissue replacement without disturbance to the gross form of a bone and its muscular attachments.

To keep the record straight, it should be pointed out that Haversian systems are not of universal occurrence among vertebrates. Most Amphibia, some reptiles, and many smaller mammals do not have them. In this event, the physiological economy of compact bone is maintained by widely distributed vascular channels.

The intramembranous formation of bone just described is largely confined to parts of the skull and the shoulder girdle. Elsewhere bone is preceded by cartilage. However, the cartilage does not become converted to bone; rather, the cartilage is destroyed and its place is taken by bone. The events of intracartilaginous bone formation are essentially the same as those in intramembranous bone development. The only difference is the added complication of the rise and decline of the cartilaginous precursor (Figure 11-7).

When an area of cartilage is about to be replaced by bone, it begins to undergo pro-

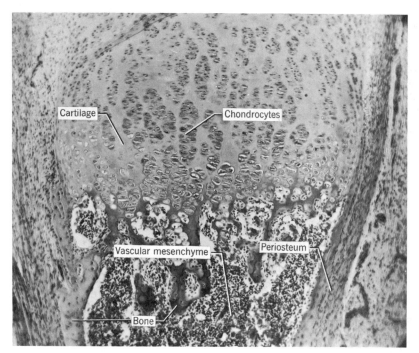

Figure 11-7 Endochondral bone formation in a human fetus. ×60. (Photograph by Torrey.)

found changes. In one or more regions of the cartilaginous model the cartilage cells enlarge and arrange themselves in longitudinal columns. Coincidently, strands of vascular mesenchyme deriving from the perichondrium begin to penetrate the cartilaginous matrix between the columns of chondrocytes. This serves to bring both a blood supply and a stock of bone-forming cells (osteoblasts) into the matrix. Because it now inaugurates a role in bone formation, the perichondrium may henceforth be identified as the periosteum. With the invasion by periosteal tissue, the cartilage gradually becomes eroded away. This erosion may be a direct consequence of the invasion; then again, many of the cartilage cells themselves may acquire destructive powers and tear down the material that they originally deposited. In any event, the cartilage becomes extensively honeycombed. Shortly, then, osseous tissue is deposited by apposition upon the fragments of cartilage. Some of the bone is a product of the osteoblasts brought

in by the invading mesenchyme; some is a product of surviving cartilage cells that transform to osteoblasts. The result is the creation of cancellous bone consisting of the familiar latticework of trabeculae, within the interstices of which the invasive mesenchyme differentiates as marrow. This phase of intracartilaginous bone formation, involving as it does the interior of the cartilaginous precursor, is described by the term *endochondral.*

It should be self-evident that this preliminary endochondral formation of cancellous bone is essentially like its formation intramembranously. The only real difference is that, whereas in membrane bone the trabeculae are laid down around fibers of collagen, in endochondral bone the fragments of cartilaginous matrix serve as axes for bone deposition. In due time this cartilage is also replaced and collagen is added supplementally.

Continuation of the story calls for attention to several concomitant events. We have noted the start of ossification in the creation of one or more areas of cancellous bone within the original cartilaginous model. These areas expand steadily, and a total bony element of small size would obviously result in short order if something else did not occur. Two events create increasing dimensions. Increase in length is provided by the formation of new cartilage at each end of the skeletal part in question. Thus, as rapidly as the older cartilage is replaced by cancellous bone, new cartilage is added. In fact, ossification does not catch up with the cartilage, so to speak, until adult size is reached. Increase in diameter is provided by the activity of the periosteum investing the growing bone. Such peripheral deposition of bone is described by the term *periosteal.* (Because the periosteum corresponds to the perichondrium, the term *perichondral* is sometimes used synonymously with periosteal.) But hand in hand with this comes a reworking of the cancellous interior. Many of the osteoblasts that deposited the original trabeculae now become bone destroyers, *osteoclasts,* and proceed to break down the cancellous area to some degree. This results not only in the element becoming more or less hollow, but in a blending of the originally separate islands of marrow into one. In other words, replacement bones tend to feature a maximum of compact bone and a hollow interior filled with marrow.

In summary, then, the development of replacement bone in general involves these steps. A miniature of the part in question is first laid down in cartilage. This becomes eroded as an accompaniment of periosteal invasion, and a spongework of bone is deposited in its place. Continued addition of cartilage at the ends of the original and its steady replacement by cancellous bone provide for elongation. Increase in diameter is provided by the periosteal deposition of compact bone around the cancellous core. Eventually the cancellous bone is destroyed to a considerable degree, so that the final product is composed largely of compact bone investing a marrow-filled cavity.

BONE AS A TISSUE

Histophysiology. We have seen that in embryonic life bone is formed either by direct ossification within a mesenchymal precursor (intramembranous ossification) or by the replacement of a preexisting cartilaginous structure (intracartilaginous ossification). Although these two methods of bone formation provide prime criteria for the characterization of bones as skeletal units, the origin of bone as a tissue is essentially the same in both instances and, therefore, there is no distinction to be made in tissue structure between membrane and cartilage bone. It has surely been self-evident from the foregoing that the embryonic creation of osseous tissue involves a dynamism of events. What may be less evident is that even after it is fully differentiated, bone is not a static tissue. Both during and after its formation, bone is one of the most dynamic tissues of the vertebrate body and as such is influenced by and, in turn, participates in a variety of physiological circumstances.

With respect to its embryogeny, it is clear that a variety of complex biochemical transformations take place. A collection of the necessary raw materials and their synthesis into the appropriate organic compounds and inorganic salts, all under the direction of the osteoblasts, require the operation of an intricate biochemical factory. Like such biochemical transformations in general, these are under the mediation of certain organic catalysts, enzymes. For example, the enzyme alkaline phosphatase is believed to play an important part in the manufacture of fibrous protein. It is also thought to be involved in the deposition of calcium salts. In addition, a number of vitamins are known to be concerned: vitamin A is employed in the endochondral growth of bony tissue; vitamin C regulates the deposition of collagen; and vitamin D promotes the acquisition of calcium and phosphate and regulates their supply. Among the several hormones that exert an influence on bone histogenesis, the growth hormone of the pituitary gland and thyroxine from the thyroid are the most notable. Elongative growth via the continued proliferation of cartilage and its progressive ossification are under the control of these hormones.

A laboratory student examining a dried and cleaned skeleton understandably may gain an impression of bone as an inert substance that plays only a passive role as a mechanical framework. We should remember, however, that living bony tissue is just as vital a substance as heart, kidney, or any other tissue. First, the quantities of marrow housed within much bone and its importance as a source of new generations of blood cells should be recalled. Less obvious is the fact that throughout the life of an individual, bone is in a constant state of flux: its internal architecture is constantly being worked over by the steady destruction of old bone and the addition of new.

Breakdown and resorption of bone appear to result from the behavior of certain osteocytes that operate as bone destroyers (osteoclasts). It is not at all clear how the osteoclasts accomplish the task. However, this much is certain. The process is not a random one, but is under hormonal control and is directly associated with certain mineral requirements of the body. Two materials are absolutely essential to the proper operation of many physiological mechanisms. One of these is calcium, whose presence in appropriate concentration in the blood is necessary to the blood coagulation mechanism, proper heart action, nerve metabolism, and other functions. The other is inorganic phosphate, which plays a manifold role in the enzyme-controlled pathways pertaining to the metabolism of glycogen that are common to all tissues and cells of the vertebrate body. Bone is the

reservoir for both these materials, which are added to and subtracted from the bloodstream as required through the turnover mechanism in bone. As for calcium, the entire operation appears to be largely under control of *parathyroid hormone*. The parathyroid glands are sensitive to the concentration of calcium ions in the bloodstream. A deficiency of calcium leads to increased activity of the glands with a resulting increase in hormone secretion, which in turn incites the osteoclasts to break down the bone salts and release calcium. A rise in calcium reverses the process, that is, leads to increased osteoblastic manufacture of bone salts. Opposing the action of parathyroid hormone is *calcitonin* (*thyrocalcitonin*), a substance that interferes with bone resorption and tends to lower blood calcium. Although believed to be a hormone produced by the thyroid gland, calcitonin may be a product of the ultimobranchial bodies (p. 303), which in mammals become incorporated within the thyroid.

In addition to serving as the source of supply of calcium and phosphate required for the maintenance of these subtle physiological balances in the blood, bone is also the reservoir that is drawn upon to satisfy certain structural requirements. One case is found in birds where calcium for the egg shell is drawn from the bone bank. Another is found in deer where the periodic formation of antlers is accompanied by skeletal destruction and its attendant release of mineral resources.

A final illustration of bone as a dynamic system comes as a by-product of the atomic age. Strontium-90 (chemically allied to calcium) has a special affinity for osseous tissue and when, as a consequence of "fallout" from nuclear weapons or other sources, it enters the body via the digestive or respiratory tracts, it fixes itself on bone crystals and builds up to relatively great concentrations. Strontium may not only do direct damage to the bony tissue itself, but through internal radiation of the blood-forming tissues in the marrow and through its wide distribution as an accompaniment of the activity of bone as a purveyor of ions for other functions, it may produce malignancies and other destructive effects.

BONES AS ORGANS

Now that we have become familiar with the development of bone as a tissue and certain of its functional activities, we can turn attention to some aspects of the development of definitive skeletal parts. There are special problems to consider with respect to the attainment of characteristic types of bones, as, for example, flat bones in the skull, long bones in the appendages, and irregularly shaped vertebrae.

Whether intramembranous or intracartilaginous, ossification of flat bones ordinarily is inaugurated in the center of the mesenchymal membrane or cartilage, as the case may be, and then proceeds radially from this center, primarily in one plane, until the final area is attained. Such thickness as is to be created is provided by periosteal deposition of compact bone upon the can-

cellous interior. In due time some of the original cancellous bone is dissolved so as to bring a blending of many of the initially separate areas of marrow, and some of it is transformed to compact bone. The final form of a typical flat bone is that of a reduced cancellous area, containing masses of marrow, sandwiched between upper and lower layers of compact bone.

Prospective long bones are anticipated by a cartilaginous model, which replicates in miniature the bone to come. In this model one can distinguish a main shaft, or *diaphysis*, with an *epiphysis* at each end (Figure 11-8*A*). Ossification begins in the center of the diaphysis and steadily progresses toward each end. But, as mentioned earlier, this process by itself would result only in a bone of miniature size. Elongation is provided by the steady creation of new cartilage at the epiphyseal ends, and ossification follows in its wake until adult length is attained. In lower vertebrates the epiphyses may actually remain cartilaginous throughout adult life. In mammals and some reptiles, however, accessory ossifications sooner or later appear in the epiphyses, ossifications that fuse with the diaphysis only after full adult size is reached (Figure 11-

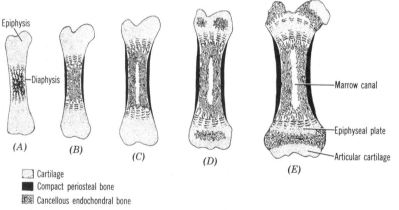

Figure 11-8   Progressive ossification in a long bone. (*A*) Primary ossification center in diaphysis. (*B*) Beginning of shell of periosteal bone. (*C*) Diaphysis largely ossified; beginning dissolution of cancellous interior to form marrow canal. (*D*) Secondary ossification centers in the epiphyses. (*E*) Ossification complete except for epiphyseal plates and articular surfaces.

8). The occurrence of the epiphyseal centers varies with the long bones. In the larger ones, such as those of the arm or leg, there is always at least one epiphyseal ossification center at each end, and sometimes two or more; in the smaller ones, such as those of the hands, an epiphyseal ossification occurs only at one end; and in the very small bones of the ankle and wrist, ossification proceeds from a single endochondral center.

Meanwhile, as a long bone increases in length, there is a proportional increase in its diameter and a reworking of its cancellous interior. During the reworking, many of the osteoblasts that laid down the original trabeculae become bone destroyers (osteoclasts) and proceed to dissolve the cancellous area. Speaking more precisely, it is the interior of the diaphysis that is so destroyed; the spongy bone of the epiphyses is left intact. Destruction of its cancellous interior serves to convert the diaphysis to a hollow cylinder. By the same token, the originally separate islands of marrow are brought together. In other words, this central resorption produces a marrow-filled *marrow canal* within the diaphysis (Figure 11-8). While all this is going on, the diameter of the bone is increased by the periosteal deposition of layers of compact tissue. These layers are thickest in the middle of the shaft and taper off toward the epiphyses, which remain essentially cancellous.

In the case of irregular bones such as the vertebrae and those of the limb girdles, several primary endochondral centers occur, and these are supplemented by secondary centers providing for various margins and processes. Again, the original cancellous bone is replaced in part and supplemented by periosteal deposition of compact bone.

Numerous tests of the developmental powers of skeletogenous tissues in cultures and grafts have revealed a striking ability of early primordia to attain characteristic form independently of any environmental influences that may exist within the embryo. For example, if the mesodermal aggregate foreshadowing the skeleton is excised from the hindlimb bud of a 4-day-old chick embryo and cultivated in a culture medium, cartilaginous skeletal parts of characteristic shape will be differentiated. Or if the cartilaginous rudiment of a long bone of a chick or mammalian embryo is cultured, it will continue to grow for a time, will acquire normal form, and may show a degree of ossification. These results signify that early primordia possess inherent morphogenetic properties that can and do express themselves up to a point. Sooner or later, however, extrinsic factors come into play in the molding of architectural details and in the attainment of appropriate size and proportions.

Some of these factors are mechanical in nature, as shown by such observations as the following. Orientation of the trabeculae in the spongy areas of developing long bones is subject to ensuing pressures and tensions imposed by weight, thrust, and muscle attachments; the contours of an entire bone may be modified if the muscles that normally pull upon it are detached or become nonfunctional because of injury or disease. Even localized growth of bone will cease in the absence of a proper muscle attachment, as shown by an experiment on newborn dogs. If the temporal muscle serving the lower jaw is detached from the side of the cranium, the ridge of bone that normally arises at the point of attachment fails to develop. A striking illustration is also provided by the skull in conjunction with the brain and sense organs. The size and shape of the bones of the skull accommodate to the structures they invest. Oversized eyes bring an enlargement of the orbital areas of the skull; the genetically induced anomaly of a reduced or virtually absent brain is accompanied by skull bones of greatly reduced size and abnormal shape; and abnormally large and thin skull bones accompany an anomalous brain of excessive size.

Nutritional and hormonal factors also play a major role in bone development and growth. Most prominent in this category is one of the many products of the anterior lobe of the pituitary gland (adenohypophysis), *somatotropin* (*STH*), commonly known as "growth hormone." Inadequate producton of STH leads to dwarfism of the entire body, including the skeleton; excessive secretion of this hormone leads to a disproportionate overgrowth of the skeleton. The dwarf and the giant of the circus are personifications of abnormal production of STH. Other hormones involved are those provided by the gonads, which govern the differences in form and proportion of the skeletal parts in males and females. Important nutritional factors are the vitamins. Vitamins C and D are essential for proper ossification; conversely, an excess of vitamin A inhibits bone formation.

THE SKULL

The morphogenesis of no part of the vertebrate body has been more thoroughly explored and documented than that of the *skull*, the skeletal framework of the head. The paleontological record is a full one, and alongside it lies a large reservoir of comparative anatomical and embryological data

that allow us to trace the changing form of the skull from fish to human.

From the developmental point of view, the skull may be divided into three parts: (1) the *neurocranium* (sometimes called the *endocranium*), which includes the capsules bounding the olfactory organs, eyes, and internal ears, and the box enclosing the brain, composed of cartilage, replacement bone, or both; (2) the *splanchnocranium,* or *visceral skeleton,* which is the skeleton of the jaws and gill arches or their derivatives (Figure 11-9); and (3) the *dermatocranium,* the so-called surface skull whose parts consist exclusively of dermal bone and are believed to have evolved from the external armor (exoskeleton) of primitive fishes (Figure 11-9). As long as the endocranium is composed entirely of cartilage, it may be spoken of as a *chondrocranium.*

Looking now to general embryonic origins, the material source of the neurocranium is found in the connective tissue membranes that initially envelop the embryonic brain and associated sense organs. There may be several such connective tissue layers, depending on the vertebrate, but it is the outermost one alone that is concerned with the formation of the neuro-

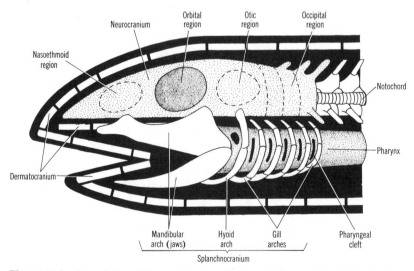

Figure 11-9   Ground plan of the vertebrate skull. (From Portmann, *Vergleichende Morphologie der Wirbeltiere,* courtesy Benno Schwabe and Company, Basel.)

cranium. At the site of the prospective endocranium a number of local aggregations of mesenchyme (see later) make their appearance, aggregations that then chondrify. If the vertebrate in question is one destined to have a bony skeleton in whole or in part, the preliminary cartilage will be replaced by bone in some measure. Coincidentally, additional aggregations of mesenchyme ossify directly to provide the overlying dermatocranium.

The origin of the tissue that ultimately gives rise to the splanchnocranium is a little uncertain. The conventional view is that the condensations of mesenchyme from which the visceral arches arise are derived from the splanchnic mesoderm investing the pharyngeal gut. However, there is considerable evidence indicating that the endocranial components of the arches are formed from cells that are encompassed within the neural crest (see p. 415 and Figure 17-6*D*) and that migrate from the crest to their appropriate pharyngeal positions. It is, in fact, possible that the dermal components of the visceral arches and even parts of the neu-

rocranium, especially the trabeculae, or prechordal plates, are also products of neural crest.

Let us move now from the general matters of the origin of the skull-forming materials to more particular embryonic events, but for the present still speaking in terms of vertebrates as a group. Cartilage formation is initiated in a number of isolated centers (Figure 11-10). With respect to the neurocranium, there are six pairs of such centers. Two of these lie ventral to the brain: *parachordal cartilages* form beneath the posterior part of the brain and flank the notochord; *prechordal cartilages* (*trabeculae*) form beneath the anterior part of the brain. The parachordals tend to be fused almost from the start with *occipital cartilages,* which appear at the level of transition from brain to spinal cord and resemble the primordial cervical vertebrae forming immediately behind. The remaining centers of chondrification are associated with the developing sense organs. A cartilaginous shell, the *auditory* (*otic*) *capsule,* forms around each primordial inner ear; medial to each

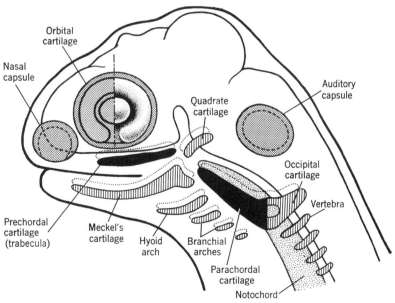

Figure 11-10 Embryonic primordia of the skull. (Adapted from a figure by Portmann, *Vergleichende Morphologie der Wirbeltiere,* courtesy Benno Schwabe and Company, Basel.)

eye an *orbital cartilage* provides a preliminary sidewall to the neurocranium; a *nasal capsule* forms around each olfactory sac.

Separate paired cartilages also appear within the mesodermal cores of the embryonic visceral arches and, collectively, constitute the splanchnocranium (Figure 11-10). For the present it will be sufficient to identify them in only a general way. The *first* or *mandibular arch* is represented by a pair of *quadrate cartilages* signifying the upper jaw and a pair of *mandibular (Meckel's) cartilages* in the lower jaw. Other cartilages arise in the prospective *second* or *hyoid arch* and the subsequent *branchial arches.*

As we continue with the events of cranial embryogeny, it is desirable to shift from the general to the particular. That is, we shall take the six pairs of neurocranial elements and the parts of the splanchnocranium that appear in vertebrates generally and pursue their histories in that one vertebrate that holds our prime interest, the human. In so doing, we shall confine ourselves to major developmental events and in no way consider the origin and structure of numerous anatomical supplements.

The dominating transformation in the neurocranium is one of a gradual spreading and fusion of the original paired cartilages. Accordingly, the parachordals, united from the start with the occipitals, fuse with one another to form the *basal plate.* Toward the front, the basal plate unites with the fused trabeculae (prechordals), which in turn blend with the nasal capsules to form the *ethmoid region.* Dorsolaterally, the basal plate and occipital area blend on each side with the auditory and orbital cartilages. The end result is the creation of an irregular, one-piece cartilaginous neurocranium (chondrocranium) supporting the brain (Figure 11-11). Nevertheless, the neurocranium may still be marked off descriptively into a number of regions which, on the one hand, reflect their separate origins and, on the other, will pursue somewhat different histories. Reading from back to

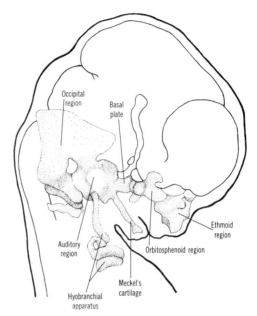

Figure 11-11   Chondrocranium of an 8-week human fetus. (After Lewis, courtesy Carnegie Institution of Washington.)

front these regions are (1) the occipital region, (2) the auditory region, (3) the basal plate, (4) the orbitosphenoid region, and (5) the ethmoid region (Figure 11-11).

BASIC COMPONENTS OF THE VERTEBRATE SKULL

We have seen that the skull is composed of dermal bones and cartilages or their bony replacements of both somatic and visceral nature. It is now necessary to identify these separate elements in their primitive number and arrangement, for they constitute the "stockpile," so to speak, that has been selectively utilized in the evolving skull. To this end it will be helpful to employ a schematic diagram (Figure 11-12) and a summarizing table (Table 11-1) to supplement word description. Our approach will be to begin with the elements that appear in each of the subdivisions of the endocranium and then follow up with a listing of the dermal components that are roughly associated with these endocranial regions.

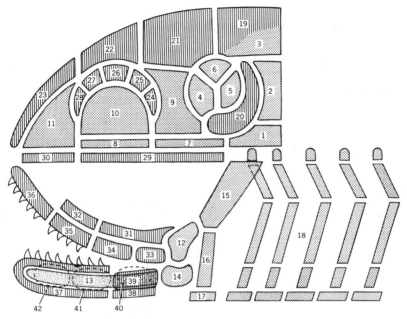

Figure 11-12 General plan of the vertebrate skull. Dermal bones, lined; replacement bones or cartilage, stippled. The numbers correspond to those in Table 11-1. Except for the omission of the opercular series, the relationships shown are such as prevail in a bony fish. In tetrapods, particularly reptiles and mammals, the bones of the upper jaw are affixed to the floor and sides of the neurocranium, more or less obscuring the primary cranial bones. (Based on a similar figure by Rand.)

We shall consider the neurocranium first. In the occipital region four elements, usually ossified, ring the *foramen magnum* through which the spinal cord emerges: a median ventral *basioccipital,* a median dorsal *supraoccipital,* and a pair of lateral *exoccipitals.* Immediately in front of the occipital ring, three bones arise in each auditory capsule: a *prootic* in front, an *opisthotic* behind, and an *epiotic* on top. Commonly, the epiotics seem either to be fused with or even replaced by extensions of the supraoccipital. Forward from the auditory region the original basal plate area provides two elements serving as the floor of the braincase. The major one is the *basisphenoid,* anterior to which is the narrow *presphenoid,* which extends forward into the ethmoid region. Providing sidewalls to the braincase are the *alisphenoid* and *orbitosphenoid,* which form on each side in the orbitosphenoid area. The status of the

ethmoid region varies widely from vertebrate to vertebrate, but commonly ossifies as an *ethmoid complex* associated with the nasal passageways, including, in humans, the distinctive *cribriform plate,* perforated by the olfactory formanina, and the thin, scroll-like *turbinate bones.*

In the splanchnocranium, the first visceral arch is composed of two pairs of components, either cartilage or bone, which constitute the upper and lower jaws; this is the *mandibular arch.* The upper jaw is represented by the *quadrate cartilage* (or bone), and the lower jaw by *Meckel's cartilage.* The proximal end of the latter is identified as the *articular cartilage* (or bone). The second or hyoid arch consists of a median ventral *basihyal* and on each side a *ceratohyal* topped by a *hyomandibula.* The remaining visceral arches, III–VII, are branchial or true gill arches. Each is composed of several parts, but we shall

TABLE 11-1

BASIC COMPONENTS OF THE VERTEBRATE SKULL

	Replacement Bones and/or Cartilages			Dermal Bones		
	Region	**Components**	**Region**	**Components**		
Neurocranium	Occipital	1. Basioccipital 2. Exoccipital 3. Supraoccipital	Rear roof	19. Postparietal (interparietal) (supraoccipital)		
	Auditory	4. Prootic 5. Opisthotic 6. Epiotic	Cheek area	20. Squamosal		
	Basal plate	7. Basisphenoid 8. Presphenoid	Midroof	21. Parietal 22. Frontal		
	Orbitosphenoid	9. Alisphenoid 10. Orbitosphenoid	Circumorbital series	24. Postorbital 25. Postfrontal 26. Supraorbital 27. Prefrontal 28. Lacrimal	Palatal Series	29. Parasphenoid 30. Vomer 31. Pterygoid 32. Palatine
	Ethmoid	11. Ethmoid complex	Foreroof	23. Nasal		
Splanchnocranium	Visceral arch 1 (mandibular arch)	**Upper Jaw** 12. Quadrate	Maxillary series	33. Quadratojugal 34. Jugal 35. Maxilla 36. Premaxilla		
		Lower Jaw 13. Meckel's cartilage 14. Articular	Mandibular series	37. Dentary 38. Angular 39. Surangular 40. Splenial 41. Prearticular 42. Coronoid		
	Visceral arch II (hyoid arch)	15. Hymoandibula 16. Ceratohyal 17. Basihyal				
	Visceral arches III–VII (branchial arches	18. Branchials				

make no attempt to homologize these individually with particular derivatives in the higher tetrapods.

The dermatocranium consists first of a longitudinal series of paired bones that provide a roof for the neurocranium. The rearmost are associated with the occipital region and are known variously as *postparietals* or *interparietals.* Next in order come

the *parietals,* behind the orbits, and the *frontals,* between the orbits. The region of these bones coincides with the basal plate area of the endocranium. The forepart of the cranial roof, coinciding with the ethmoid region, is provided by the *nasals.*

The endocranial elements of the sidewalls of the neurocranium are supplemented by dermal bones in the cheek and orbital areas. One bone alone occupies each cheek area behind and below the auditory region. This is the relatively broad, platelike *squamosal.* But each orbit is ringed by a series of five elements, which, from rear to front, are the *postorbital, postfrontal, supraorbital, prefrontal,* and *lacrimal.*

On the ventral side of the neurocranium, a palatal series of dermal bones develops in the skin lining the roof of the mouth, and this comes to supplement the endocranial floor of the braincase. Considerable confusion attends the identification of these elements in various vertebrates and we shall deal here with only the most important. One of these is single; the other three are paired. The single bone is the median *parasphenoid.* In front of and overlapping the parasphenoid is a pair of *vomers,* followed by the *palatines* and *pterygoids.* The pterygoids are commonly differentiated into *ectopterygoids* and *entopterygoids.* Parasphenoid and vomers tend to be closely applied to the floor of the braincase, but the palatines and pterygoids are more often linked to the upper jaw.

Finally, mention should be made of two other series of bones that are not shown in Figure 11-12, nor listed in Table 11-1. One of these is a set of *operculars* that provide support for the flaplike operculum covering the gills of bony fishes. The other is the set of so-called *temporal bones,* a series of three small bones on each side behind the orbit and interposed between the parietals and squamosals. Because the operculars are absent in all tetrapods, and because the temporals are reduced or wanting in all but the earlier tetrapods, we are not greatly concerned with them.

Associated with each half of the upper jaw, in addition to the palatine and pterygoid just mentioned, is a set of four dermal bones constituting the maxillary series: *quadratojugal, jugal, maxilla,* and *premaxilla.* The maxillae and premaxillae are the tooth-bearing upper jaw bones (as may also obtain for any or all of the palatal series). In the tetrapods, where the upper jaw is fused to the neurocranium, the jugal bounds the ventral margin of the orbit and therefore is often included in the circumorbital series, whereas the quadratojugal is more closely linked to the squamosal in the cheek area.

Meckel's cartilage and the articular cartilage or bone of each half of the lower jaw become encased in a number of separate dermal bones constituting the mandibular series. On the outer surface the major element is the tooth-bearing *dentary.* Below and behind the dentary and extending around to the inner surface are the *splenial(s), angular,* and *surangular.* The inner surface, however, is largely covered by the *prearticular* and one or more *coronoids.*

No dermal elements are associated with the branchial arches.

PHYLOGENY OF THE SKULL

We are now prepared to trace the principal modifications exhibited by the cranial components in the gradual transition from fish to mammal. The main thread of morphogenesis of the skull is found in the succession Crossopterygii → labyrinthodont amphibians → cotyolsaurs and therapsid reptiles → mammals.

There are two possible ways in which to approach the story. One is to treat the skull pattern as a whole in each of the vertebrate groups under consideration. The other is to take each category or group of skull components separately and follow their histories through the series of representative vertebrates. We shall do both, first surveying the total skull pattern vertebrate by vertebrate, and then returning to summarize the major transformations in the several categories of bones.

Crossopterygii. In contrast to present-day Dipnoi, Holostei, and Chondrostei in which the endocranial neurocranium shows a retention of considerable amounts of embryonic cartilage, this portion of the neurocranium of ancient crossopterygians was thoroughly ossified. Unfortunately, the fossil record gives little information as to the individual bones present. Although the five basic subdivisions (see Table 11-1) appear to have been represented, this unit of the neurocranium was built in two sections movably articulated with one another. The anterior section, consisting of the ethmoid and orbitosphenoid regions, was firmly linked to the upper jaw and palatal complex; the rear section included the basal plate, auditory region, and occipital region. The notochord ran forward without interruption to terminate at the rear of the anterior section. (Interestingly enough, such an intracranial joint is still found in the coelacanth *Latimeria,* the surviving specialized "cousin" of the ancient crossopterygian fishes.) The origin of the two sections of the endocranium may have reflected the basic architecture of the brain, which also appears to have originated in two parts under the separate inductive influences of the anterior prechordal mesoderm and the posterior notochord.

It is probable that the decking of dermal bones laid down about the endocranium originated in a mosaic of plates deriving from the skin. By the time the Crossopterygii arrived on the evolutionary scene, there had been a reduction in the total number of such plates as a consequence of the elimination of some and the fusion of others. But they still remain numerous and the variations from one crossopterygian fish to another are so many that there is considerable uncertainty in establishing homologies. One can never be quite sure, in comparing two skulls, that bones occupying corresponding positions are actually equivalent. Nevertheless, there is fairly general agreement on a pattern, which is exemplified by the upper Devonian crossopterygian *Eusthenopteron* (Figure 11-13). This pattern will be seen to approximate the idealized schema suggested in Figure 11-12.

On each side of the dorsal midline we see a roof series of paired bones reading, from front to back, as nasals, frontals, parietals (bounding the pineal opening), and postparietals. On each side there is a circumorbital series consisting of a prefrontal, postfrontal, postorbital, jugal, and lacrimal. (Note the inclusion of the jugal in the circumorbital series rather than the maxillary series.) The cheek area is occupied by the conspicuous squamosal. The dermal investment of each half of the upper jaw includes the tooth-bearing premaxilla and maxilla and a posterior quadratojugal. The dermal elements of the lower jaw consist of a large external dentary bounded below by splenials, angular, and surangular, and internally by a prearticular and several coronoids. In addition to these basic components, there is a series of operculars involved in the covering of the gills, a set of temporals constituting the rear margin of the neurocranium, and an array of rostrals supplementing the nasals. The rostrals are probably reminiscent of the primitive mosaic work of plates that featured the primordial skull and are not readily homologized with any components in tetrapod skulls.

The dermal bones constituting the palatal complex on the undersurface of the skull are almost schematically arranged. A narrow, median parasphenoid lies beneath the anterior part of the braincase. It does not, however, extend backward to the basioccipital region. There is a considerable gap between the rear of the parasphenoid and the front of the occipital, a gap whose presence appears to be associated with the earlier-mentioned hinge in the endocranium. The parasphenoid is flanked by a pair of conspicuous pterygoids. Each pterygoid is in turn bordered by a lateral series consisting of a vomer, palatine, and ectopterygoid, reading from front to back. The internal nares (nostrils) lie between the vomers and palatines.

There remain to be added some points regarding replacement bones in the jaws and the relationship of the jaws to the neurocranium. That the dermal components of

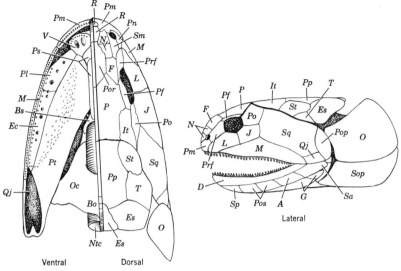

Figure 11-13 Skull of the crossopterygian fish, *Eusthenopteron* (with supplementary data derived from *Osteolepis*). (Based on illustrations by Jarvik, Romer, and Stensiö.)

A—Angular	Ntc—Notochord	Prf—Prefrontal
Bo—Basioccipital	O—Opercular	Ps—Parasphenoid
Bs—Basisphenoid	Oc—Otic capsule	Pt—Pterygoid
D—Dentary	P—Parietal	Qj—Quadratojugal
Ec—Ectopterygoid	Pf—Postfrontal	R—Rostral
Es—Extrascapular	Pl—Palatine	Sa—Surangular
F—Frontal	Pm—Premaxilla	Sm—Septomaxilla
G—Gular	Pn—Postnasal	Sop—Subopercular
It—Intertemporal	Po—Postorbital	Sp—Splenial
J—Jugal	Pop—Preopercular	Sq—Squamosal
L—Lacrimal	Por—Postrostral	St—Supratemporal
M—Maxilla	Pos—Postsplenial	T—Tabular
N—Nasal	Pp—Postparietal	V—Vomer

the upper jaw are fused to the neurocranium is self-evident. But in all bony vertebrates whose embryogeny is known this dermal armor is laid down around a cartilaginous embryonic jaw, and there is no reason to doubt that the same occurred in the Crossopterygii. The fossil record does not supply full understanding of all the upper jaw replacement bones, but a definitive rearmost quadrate bone is present. Similarly, the dermal bones of the lower jaw develop as an investment of the mandibular (Meckel's) cartilage, whose basal end is ossified as an articular bone. The hinge between lower and upper jaw on each side is therefore one involving two replacement bones, articular and quadrate, with the for-

mer abutting on the latter. The quadrate, however, does not articulate directly with the neurocranium. Rather, there is interposed between it and the auditory region of the neurocranium the dorsalmost unit of the hyoid arch, the hyomandibula. This arrangement by which the jaws are braced against the braincase by the hyomandibula provides a type of jaw suspension termed *hyostylic* (Figure 11-14). A peculiar feature of the crossopterygian hyomandibula is that it articulates with the sidewall of the neurocranium by a double head, and in addition to its union with the quadrate of the upper jaw and the ventrolateral components of the hyoid arch, it has a connection with the operculum (Figure 11-14).

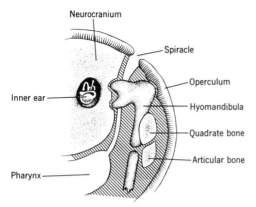

Figure 11-14 Jaw suspension in a crossopterygian fish.

The branchial arches are so variably ossified in crossopterygians that no clear fossil record of their architecture is provided.

Amphibia. The skull of primitive labyrinthodonts, as exemplified by *Palaeogyrinus* (Figure 11-15), shows a remarkable resemblance to that of the Crossopterygii. There is, however, a marked change in proportions. Whereas in the crossopterygian skull the "table" behind the orbits is long and the snout relatively short, in the labyrinthodont Amphibia, as with tetrapods generally, the skull table is much shortened and the snouth greatly lengthened. Hand in hand with this change goes a modification of proportions and contours of the bones concerned. There has also been a stabilization of the dermal bones in the sense that such highly variable elements as the rostrals and others that occur in fishes have been eliminated so as to leave a fixed stockpile of elements whose history can be traced without very much difficulty throughout the tetrapods. And with the loss of gills, the elements involved in the opercular cover are removed from the morphogenetic scene. If we keep these general modifications in mind, the labyrinthodont skull presents the following pattern.

With respect to the dermal roof and sides, elimination of the rostrals and the overall lengthening of the snout cause the nasals and frontals to become more prom-inent and considerably elongated. The frontals are succeeded by broad parietals and these in turn by postparietals. Flanking these, on each side, are a pair of temporals and a tabular. Each circumorbital series includes the same five bones seen in Crossopterygii: prefrontal, postfrontal, postorbital, jugal, and lacrimal. The cheek areas are largely occupied by the conspicuous squamosals, supplemented by the quadratojugals from the upper jaw series. Tooth-bearing premaxillae and maxillae complete the dermal upper jaw series. The primary upper jaw is represented by a quadrate bone at the rear and a small epipterygoid, a new element, farther forward and dorsal to the pterygoid of the palatal series. Each external nostril (naris) is bounded by premaxilla, maxilla, lacrimal, and nasal.

On the palatal side the pattern again agrees in most essentials with that of the Crossopterygii, except that now the parasphenoid extends all the way back to the basioccipital, a situation correlated with the elimination of the crossopterygian hinge in the endocranium. The median parasphenoid itself is quite narrow, and a space intervenes on each side between it and the long and broad pterygoids that reach back to join the quadrates. As noted earlier, a new element, an epipterygoid, lies above each pterygoid. This bone joins with the sphenoid region of the endocranium and may also extend back to contact the quadrate. A pair of vomers again lies between the pterygoids and the ventral margins of the premaxillae. An anterior palatine and posterior ectopterygoid lie between each pterygoid and maxilla. Each internal naris is bounded by premaxilla, maxilla, vomer, and palatine.

The endocranium is well ossified and is a single structure, that is, it lacks the movable joint that distinguishes the crossopterygian skull. At the rear are the four elements of the occipital ring: median ventral basioccipital, median dorsal supraoccipital, and a pair of exoccipitals. The basioccipital bears a single condyle for articulation with the first vertebra. The auditory region is occu-

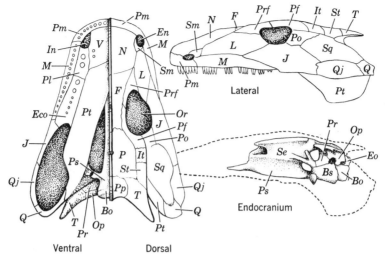

Figure 11-15 Skull of the labyrinthodont amphibian, *Palaeogyrinus.* (Based on figures by Romer.)

Bo—Basioccipital	*N*—Nasal	*Ps*—Parasphenoid
Bs—Basisphenoid	*Op*—Opisthotic	*Pt*—Pterygoid
Eco—Ectopterygoid	*Or*—Orbit	*Q*—Quadrate
En—External naris	*P*—Parietal	*Qj*—Quadratojugal
Eo—Exoccipital	*Pf*—Postfrontal	*Se*—Sphenethmoid
F—Frontal	*Pl*—Palatine	*Sm*—Septomaxilla
In—Internal naris	*Pm*—Premaxilla	*Sq*—Squamosal
It—Intertemporal	*Po*—Postorbital	*St*—Supratemporal
J—Jugal	*Pp*—Postparietal	*T*—Tabular
L—Lacrimal	*Pr*—Prootic	*V*—Vomer
M—Maxilla	*Prf*—Prefrontal	

pied by two bones, opisthotic and prootic. Ventrally, in front of the basioccipital, lies a well-developed basisphenoid, which articulates on each side with a pterygoid and is covered below by the parasphenoid. The anterior part of the braincase is a voluminous tubular ossification representing a sphenethmoid composite. Unlike in the crossopterygians, the nasal capsules are not ossified.

The lower jaw of the labyrinthodonts closely resembles that of the crossopterygians. The dermal elements are the same and need not be relisted (see Figure 11-13). The only manifestation of the primary jaw is the small articular bone at the rear. This element abuts on the quadrate, which, in contrast to the crossopterygian situation, is now an intergral part of the skull and, as was noted earlier, lies at the rear of the

pterygoid. With this arrangement has come an alteration of the relationships of the jaws to the neurocranium and the status of the hyomandibula of the hyoid arch (Figure 11-16). With the quadrate of the upper jaw now in the sidewall of the cranium itself and the articular of the lower jaw abutting it, it follows that the jaws are now suspended directly on the neurocranium, a type of suspension termed *autostylic.* The hyomandibula, which in the fishes was interposed between the quadrate and the auditory region of the skull, now assumes new relationships (Figure 11-16). The one-time spiracular opening is closed over externally by a thin *tympanic membrane* and attached to its inner surface is the *stapes* (*columella*), the reduced and modified homologue of the hyomandibula. If one can conceive of the tympanic membrane as

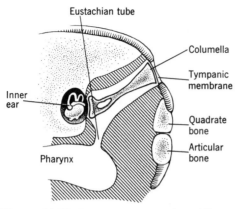

Figure 11-16 Jaw suspension and middle ear of amphibians.

being in a measure equivalent to the fish operculum, then the stapes has the same relationship to it as did the hyomandibula to the operculum. By the same token, the inner end of the stapes abuts the auditory region of the skull within which the inner ear lies, just as did the hyomandibula. In this fashion the stapes comes to serve as a device transmitting sound vibrations from the tympanum across the one-time spiracular (middle ear) cavity to the inner ear.

With the loss of gill breathing in adults, the branchial arches are greatly reduced. Knowledge of them, however, derives al-

most entirely from modern Amphibia, for they are seldom found in fossils.

Present-day Amphibia are extremely small as compared with their ancestors, and many elements of the dermal bone decking have been lost, others have been reduced in size, and much of the endocranium remains cartilaginous (Figure 11-17). In larval amphibians the hyoid and branchial arches are rather well formed (Figure 11-18A), but with the loss of gills in most adult forms, the hyobranchial skeleton shows a concomitant reduction (Figure 11-18B).

Reptilia. The skulls of reptiles usually are higher and narrower than those of Amphibia. In fact, this difference in proportion is about the only thing that distinguished the skulls of the earliest stem reptiles (cotylosaurs) from those of their amphibian relatives. With the later reptiles, however, there came distinctive modifications.

One change was that of the elimination of certain elements of the dermatocranium. The notable deletions in the roof and sides of the neurocranium were the postparietal at the rear margin and components of the temporal region on either side. The elements of the circumorbital series also showed a measure of instability as indicated

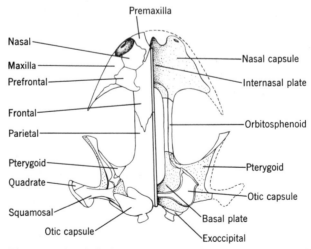

Figure 11-17 Skull of a salamander, dorsal view. Investing dermal bones removed on right side. Cartilage stippled.

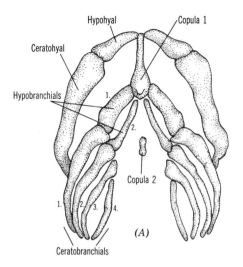

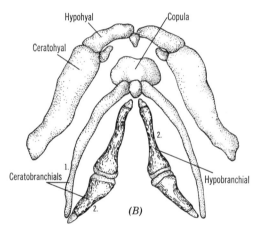

Figure 11-18 Hyobranchial skeleton of the salamander, *Cryptobranchus.* (*A*) Larva, (*B*) adult. All parts cartilaginous. (From deBeer, *The Development of the Vertebrate Skull,* courtesy Clarendon Press, Oxford, England.)

by the variability of their occurrence from one group to another. Likewise, the splenials and coronoids were reduced to a single one of each for each half of the lower jaw.

But a more important modification had to do with the rise of certain openings in the temple and cheek areas. Whereas these regions in the stem reptiles, and even today in the modern chelonians, were solidly roofed over, in all the other reptiles one or two openings make their appearance. In fact, the number and position of these open-

ings provide the primary basis for the classification of the reptiles (see Chapter 2 and Figure 2-26). Accordingly, one may distinguish three groups of reptiles featuring one opening on each side, with a differentiation based on the elements bounding it, and a fourth group featuring two openings on each side. The fourth group, the Diapsida, encompasses the major categories of ancient reptiles, and birds and all the modern reptiles, except the Chelonia, derive from it. It was from one of the single-opening groups, the Synapsida, that the mammals stemmed.

Generally speaking, the makeup of the palatal complex conforms to that of the ancient amphibians. However, two modifications may be noted in modern reptiles. In the Chelonia, the palatal elements are broadly fused with the floor of the braincase, thus increasing the solidity of the total skull structure. This is also true in the Crocodilia, with the added feature of the development of secondary shelves of bone from the maxillae, palatines, and pterygoids, which fuse to form a *secondary palate* (Figure 11-19). This structure, about which

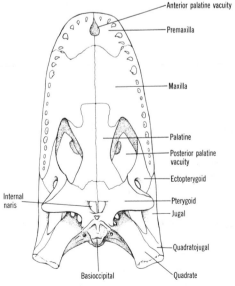

Figure 11-19 Ventral view of the skull of the alligator.

more will be said in conjunction with the mammals, serves to separate the food and respiratory passageways, and in so doing brings the internal nostrils far rearward to open just above the inlet (glottis) to the windpipe (trachea).

There remain to be added a few points regarding the endocranium, both neural and visceral. With respect to the braincase, the ossifications present are generally those seen in the early Amphibia. But unlike the later and modern Amphibia in which, as we have seen, there is an increasing tendency toward retention of embryonic cartilage, the reptilian braincase remains largely os-sified. Fusion of the palatal bones with the floor of the braincase in modern turtles and crocodiles obviously brings certain modi-fications in bony detail, but the basic struc-ture with which we are already familiar remains essentially unchanged. Except in the line leading to the mammals, there is a single occipital condyle.

The articulation of the jaws is again bas-ically like that of Amphibia (Figure 11-16). That is to say, the articular bone at the rear of the lower jaw abuts against the quadrate lying in the rear sidewall of the neurocran-ium. In most reptiles the quadrate is solidly anchored (Figure 11-19), but an interesting variation occurs in the Squamata. Although fundamentally diapsids, the bar of dermal bone bounding the lower temporal opening ventrally is missing in lizards, and in snakes the upper bar is absent as well. This clears the entire cheek area of dermal bone and leaves the quadrate to hinge loosely against the squamosal, thus providing useful motil-ity in the swallowing of large prey (Figure 11-20).

Exclusive of the hyomandibula, which again is represented by the stapes (colu-mella) in the middle ear, the hyobranchial apparatus appears as elements supporting the tongue. These may be fairly well de-veloped in lizards, but in crocodiles and turtles they are reduced to short processes and in snakes to very slender remnants. The homologies of these parts with the basic

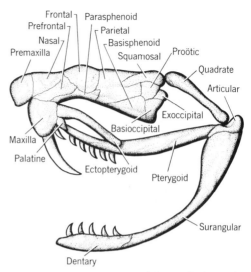

Figure 11-20 Lateral view of the skull of a snake, showing loosely hinged jaw suspension.

elements in the piscine hyobranchial skel-eton are extremely obscure.

To make a passing reference to birds, we should note that their skulls are constructed essentially as those of their reptilian ances-tors. The principal variations on this basic theme are expansion of the cranial roof to accommodate the enlarged brain; a thinning of bones to provide lightness; considerable fusion of individual elements; a movable quadrate akin to that in snakes and lizards; elongation of the premaxillae and nasals; and loss of teeth and the maxilla that housed them.

Mammals. The synapsid reptiles occupy the key position in the history of mammals for, on the one hand, the earliest members of this group are linked to the stem reptiles (cotylosaurs) and, on the other, the later ones (Therapsida) lead to the mammals. In fact, the gradation between the therapsid reptiles and the first mammals is so gradual that often it is purely a matter of opinion whether a given form is a mammallike rep-tile or a reptilelike mammal. Thus, within the Therapsida we see the following fea-tures of cranial form and change, which reach their final expression in the definitive mammals: a pair of occipital condyles; a single temporal opening on each side; a secondary palate in various stages of completion; reduction in the number of individual elements; fusion of elements to create new bone complexes; translocation of parts with resulting functional changes and associations; alteration of proportions of elements.

When translating these generalities of morphogenesis into specificities in the mammalian skull, it is convenient to utilize a representative type such as the dog, and at the same time to point up comparable conditions in the human skull. This will serve to emphasize that, although there are many variations in proportions of skull parts among different groups of mammals, especially in the dermatocranium, there is a consistency of skull pattern within mammals as a whole.

We have emphasized that the mammals came from a synapsid stock and that the synapsids were those reptiles that had a lateral temporal opening behind the orbit (Figure 11-21A). In the beginning the opening was separated from the orbit by the postorbital. As the synapsids began to grade into the mammals, the temporal opening enlarged to such an extent that the postorbital bar separating the orbit from the temporal opening disappeared. Thus in primitive mammals, as exemplified today in the skulls of marsupials, the orbit and temporal space became confluent (Figure 11-21B). Subsequently, however, the bar was rebuilt in part or in whole by extensions from the frontal and jugal bones (Figure 11-21C). The disappearance of the postorbital was accompanied by the reduction and eventual loss of all the other elements in the circumorbital series except the lacrimal, and by the fusion of the postparietal with the occipital complex. The cranial roof and sidewalls were consequently provided by the nasals, frontals, and parietals (Figure 11-21C). In the human and other primates, the frontals and parietals become tremendously expanded to accommodate to the size of the brain (Figure 11-22). The remaining dermal contributions to the roof and sides of the braincase are the squamosals in the cheek areas and the reduced lacrimals in the front of the orbits.

Before we turn to the palatal side of the skull, it will be helpful to consider the internal braincase. As a preliminary generalization, it may be pointed out that with the enlargement of the mammalian brain and the concomitant expansion of the covering dermal vault, the endocranium provides little more than the floor and rear wall of the

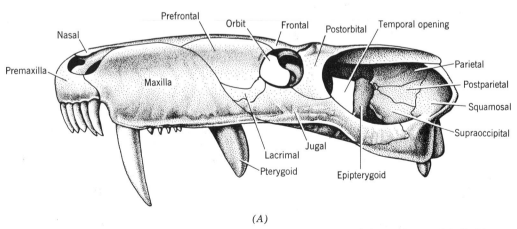

(A)

Figure 11-21 Evolutionary relationships of the orbit and temporal opening. (A) Lateral view of skull of *Scymnognathus,* a Permian therapsid. (After Watson.) (B) Lateral view of the skull of the opossum. (C) Lateral view of the skull of the cat.

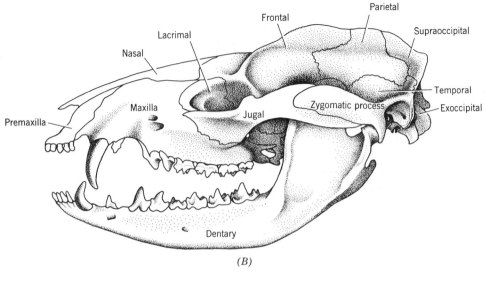

(B)

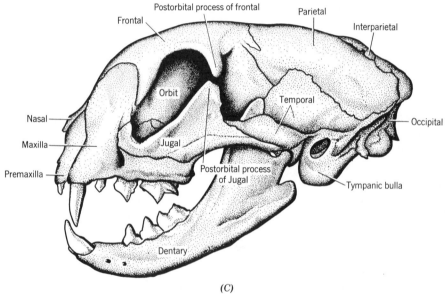

(C)

Figure 11-21 (*Continued*)

cranial cavity (Figure 11-23). Starting at the rear, the four conventional occipital elements arise separately in the embryo, but in the adult mammal these are usually fused into a single *occipital bone* (Figure 11-21C). In the same fashion, the original separate auditory elements unite to form a single *periotic bone.* To this is added a new structure, the *tympanic bulla.* The periotic

and bulla commonly unite with each other and with the adjacent squamosal to form a compound *temporal bone* (Figure 11-21C). Extending forward from the occipital in the ventral midline are the basisphenoid and presphenoid, plus in many mammals (for example, primates, carnivores, and rodents) a new element termed the *mesethmoid,* which extends forward as a median

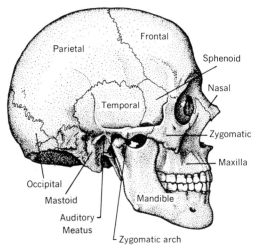

Figure 11-22 Lateral view of human skull.

partition between the nasal passages (Figure 11-23). These midline parts are supplemented by bilateral orbitosphenoids and alisphenoids, the latter being considered the equivalent of the amphibian and reptilian epipterygoids. All these sphenoidal elements are frequently fused in a single complex termed simply the *sphenoid bone* (Figure 11-22).

Turning to the palatal side, the parasphenoid, which sheathed the braincase floor in lower tetrapods, is absent in mammals, and the pterygoids are reduced to

small processes on the alisphenoids. The palatines and vomers remain as before, except that the latter are united in one. More importantly, all the mammals feature a secondary palate similar to that seen in crocodilian reptiles. The progressive formation of this structure is clearly revealed by the mammallike therapids (Figure 11-24). In brief, the secondary palate is established by the approximation of flanges from the premaxillae, maxillae, and palatines—flanges that first extend ventrad and then send horizontal shelves to meet in the midline. It has been noted previously that the palate serves to separate the food and air passages. Actually, the bony secondary palate of mammals does not extend as far backward as it does in crocodilians; rather, it is supplemented by a membranous "soft palate."

As we turn to the first visceral arch, it may be well to recall the basic conditions found in amphibians and reptiles (Figure 11-25). The upper jaw is composed of a series of dermal bones and one replacement bone, the quadrate, all united to the neurocranium; the lower jaw is similarly composed of several dermal parts and a basal replacement bone, the articular. Junction of lower and upper jaw is by way of the articular and quadrate. The history of these elements through the Therapsida and into the

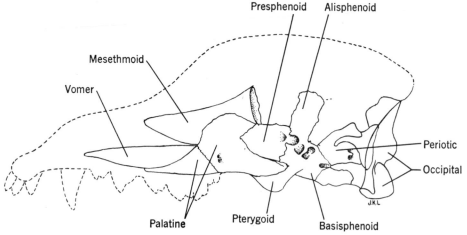

Figure 11-23 Lateral view of the endocranium of a dog. All dermal bones of the skull roof removed. (After Romer and Parsons, *The Vertebrate Body,* 5th ed., courtesy W.B. Saunders Co., Philadelphia.)

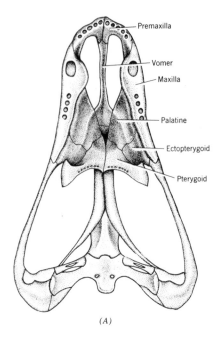

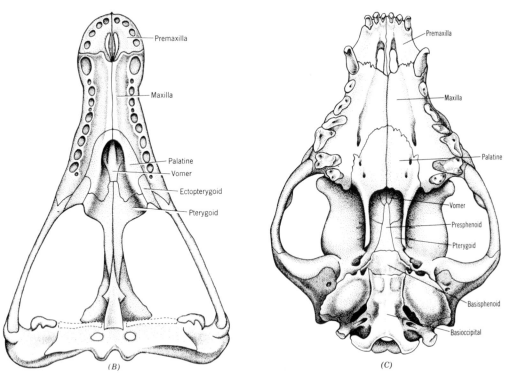

Figure 11-24 Ventral views of a series of skulls to show evolution of the mammalian palate. (*A*) Primitive mammallike reptile. (*B*) Advanced mammallike reptile. (*C*) Dog.

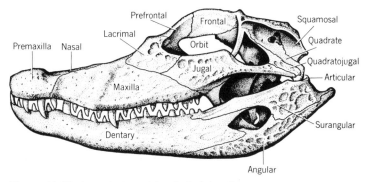

Figure 11-25 Lateral view of the skull of the alligator.

mammals is one of loss of some elements and translocation of others. On the loss side we may note the elimination of the quadratojugal from the upper jaw series and all of the dermal elements of the lower jaw series except the dentary. That this loss of postdentary dermal bones was gradual rather than abrupt is revealed by the later therapsids, in which there was a steady reduction in the size of postdentary parts and concomitant enlargement of the dentary itself. This tended to bring the rear of the dentary closer and closer to the squamosal, resulting finally in a transitional arrangement, exemplified by a late mammallike therapsid appropriately dubbed *Diarthrognathus* ("double-jointed jaw"), in which there was a simultaneous articulation of articular with quadrate and dentary with squamosal. Though a structural intermediate, *Diarthrognathus* probably should not be considered a true reptile–mammal transition form first because it coexisted with true mammals and, second, because its immediate affiliates lay off the main line of mammalian descent. Much more illuminating is the middle Triassic therapsid *Probainognathus,* in which again there is a coincidental articulation of articular with quadrate and dentary with squamosal. *Probainognathus* is otherwise reptilian in architecture and its immediate relatives are actually on the pathway of mammalian ascent.

With the rise of the definitive mammals, the quadrate (as did the hyomandibula

in the transition from fish to amphibian) moves into the middle ear, becomes connected with the stapes, and as the *incus* serves as a sound-transmitting device (Figure 11-26). The articular follows suit and, keeping its contact with the quadrate (incus), makes a third sound-transmitting bone, the *malleus* (Figure 11-26). As a consequence of this translocation and functional alteration of the articular and quadrate, the articulation of mammalian jaws with the neurocranium comes to involve two dermal bones exclusively, dentary and squamosal (Figure 11-27).

The mammalian hyobranchial skeleton shows considerable variation. Basically it

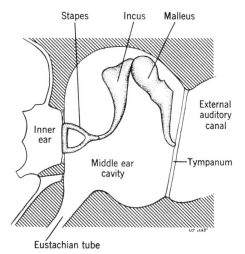

Figure 11-26 Arrangement of the hyomandibula (stapes), quadrate (incus), and articular (malleus) in the middle ear of mammals. Compare with Figures 11-13 and 11-15.

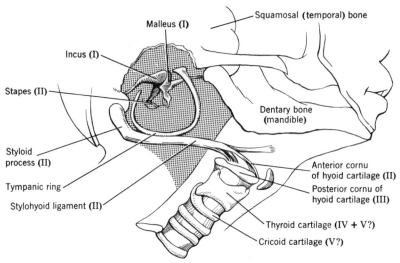

Figure 11-27   Middle ear ossicles, hyobranchial apparatus, and jaw suspension of a human. (After Kollman.) Numbers in parentheses indicate visceral arch origins.

consists of a *hyoid apparatus,* associated with the rear of the tongue, and a *larynx,* comprising the upper end of the trachea (windpipe). The hyoid apparatus ordinarily consists of a body, or *basihyal,* and two pairs of horns, or *cornua.* The anterior cornua are attached to the auditory region of the skull, sometimes being made up of a chain of bones or, as in the case of humans, being reduced to fibrous ligaments (Figure 11-27). The posterior cornua are attached to the *thyroid cartilage* of the larynx. The basihyal and anterior horns are derivatives of the hyoid arch whose hyomandibulae, as we have seen, provide the stapes of the middle ears. The posterior horns are believed to be derivatives of the third visceral (first branchial) arches, and the remaining branchial arches provide the thyroid and other cartilages of the larynx (Figure 11-27). However, the precise homologies involved are debatable.

Retrospect. Our story of the morphogenesis of the skull may profitably be brought to a close by a brief survey of the principal alterations introduced in the broad interval between fish and mammal. Starting with the "stockpile" of components provided by the Crossopterygii and considering these in terms of the subdivisions and groupings laid out in Table 11-1, the following are the most notable circumstances.

1. The history of the replacement bones of the braincase has been on the conservative side. The elements making up the occipital group are ordinarily present in all the vertebrates, although in the adults of modern mammals the four occipital bones tend to become fused into one. The one variation of note has to do with the number and arrangement of the condyles, which articulate with the first vertebra. In the ancient Amphibia there was a single condyle associated with the basioccipital. The later ones and all the modern Amphibia, however, have a pair of exoccipital condyles. In the reptiles the single basioccipital condyle was the rule, as it is in modern forms and also the birds. But the line of reptiles leading to the mammals featured two condyles, and all mammals are similarly equipped. The homologies of these condyles are extremely obscure, and it is not at all certain to what extent one type of condyle may be equivalent to another.

In the auditory region the three otic bones tend consistently to be present, except that again in mammals they fuse into one and this unitary element, in turn, unites with the newly formed bulla and the adjacent squamosal.

The floor of the braincase is regularly composed of a basisphenoid and presphenoid in the midline and flanked by alisphenoids (epipterygoids) and orbitosphenoids. Many mammals add an anterior mesethmoid which, joining with the ethmoid complex, provides a median partition between the nasal passages. Again in mammals, all these sphenoidal parts tend to fuse together.

2. In contrast to the conservative history of the internal braincase, the dermal elements of the neurocranium are considerably altered. It is an alteration involving the steady loss of parts and a concomitant change in proportions of those that remain. In the roof, the frontals and parietals become the dominant bones; of the circumorbital series, a reduction that began with the therapsid reptiles culminates in the retention of the lacrimal alone in mammals.

3. On the palatal side of the braincase, the conspicuous parasphenoid of Amphibia and most reptiles is eliminated in mammals, and the mammalian pterygoids are reduced to small processes on the alisphenoids. The vomers are maintained throughout, except that they are fused in mammals. Beginning with the primitive therapsids and continuing through the advanced mammallike therapsids to the mammals themselves, horizontal shelves from the premaxillae and maxillae grow toward and eventually fuse in the midline. Coincidently, the palatines, originally widely separated, also send shelves medially to meet the maxillae. This is the source of the secondary palate bridging the roof of the mouth and separating the food and air passages. The presence of a similar palate in modern Crocodilia stands as a case of independent, parallel evolution.

4. The dermal bones constituting the upper jaw series are consistently maintained, except that in mammals the quadratojugal is absent. Although the jugal is described with this group, its history is more closely linked to the borders of the orbit. The premaxillae and maxillae represent the essential tooth-bearing bones of the upper jaw and, as noted, they play a major role in the establishment of the secondary palate. Otherwise, the most notable change has to do with the proportions of the maxillae. Starting out as long, narrow elements, they become steadily broader and higher, and in primates their vertical axes become longer than their longitudinal axes.

In contrast to the upper jaw series, the history of the dermal bones of the lower jaw is one of a steady reduction in number until in mammals only one, the dentary, is retained.

5. The two cartilages or replacement bones of concern in the first visceral arch are the articular of the lower jaw and the quadrate of the upper jaw. In the Crossopterygii, the hyomandibula of the second visceral arch is interposed between the quadrate and the otic region of the skull, providing a so-called hyostylic jaw suspension. In Amphibia and reptiles, however, the hyomandibula moves into the middle ear and, as the stapes (columella), assumes sound-transmitting functions. Coincidently, the quadrate is incorporated into the auditory wall and the articular–quadrate articulation becomes autostylic. With the steady reduction of the postdentary elements in the lower jaw of mammallike therapsids, the dentary gradually approaches the squamosal until, finally, there is attained a transitional double jaw suspension involving articular–quadrate and dentary–squamosal. In mammals the articular and quadrate move into the middle ear and,

as malleus and incus, respectively, join the stapes in the sound transmission system. This leaves the distinctively mammalian type of jaw suspension involving two dermal bones, dentary and squamosal.

6. The parts of the hyoid arch, exclusive of the hyomandibula, and the branchial arches provide in tetrapods the variably formed hyoid apparatus and laryngotracheal cartilages.

VERTEBRAE, RIBS, AND STERNUM

Although the notochord appears early in the ontogeny of vertebrates, it is ultimately replaced wholly or in part by cartilaginous or bony *vertebrae.* Collectively the vertebrae constitute the *vertebral column* with which are associated, when present, the *ribs* and *sternum.*

VERTEBRAE

The variations in vertebral structure are so many that there really is no such entity as a "typical" vertebra. However, a convenient way of presenting the parts that will be of concern is by the device of a fanciful composite (Figure 11-28). The main body of a vertebra, and the part that functionally replaces the notochord as the axial support, is the *centrum.* Above the centrum a *neural arch* surrounds and protects the spinal cord; similarly, a *hemal arch* projects ventrally and invests associated blood vessels. Both neural and hemal arches are extended as *neural* and *hemal spines,* respectively. Each vertebra is locked to those directly in front and behind by projections from the neural arch known as *zygapophyses.* Those projecting forward are termed *prezygapophyses* and those projecting backward, *postzygapophyses.* Any given prezygapophysis links with a postzygapophysis of the vertebra in front through a contact of *ar-*

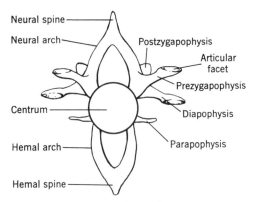

Figure 11-28   A "typical" vertebra.

ticular facets, that facet on the prezygapophysis being on its upper surface and that on the postzygapophysis being on its lower surface. Each vertebra also possesses a variety of laterally extending transverse processes. Those most consistently present are the *diapophyses* arising from the bases of the neural arch, and the *parapophyses* arising from the centrum. So much for anatomical fundamentals.

These parts, and others not mentioned, exhibit the widest variations in form and degree of expression from one vertebrate group to another, and our ultimate aim will be to assess their evolutionary course and present status in modern vertebrates. However, attention must first be given to certain embryological preliminaries.

Embryogeny of Vertebrae. The vertebrae trace their origin to embryonic connective tissue, mesenchyme, which aggregates on each side of the notochord and neural tube. This mesenchyme is derived from the lower edge of the inner wall of the epimere. Because the epimere consists of a series of paired blocks (somites), the vertebral mesenchyme derived from it is also segmented. Any one such block, or segment, of vertebral mesenchyme is known as a *sclerotome* (Figure 11-29), and initially the paired sclerotomes correspond to the flanking paired muscle segments (mytomes). From this point on, the variations among vertebrates are many. In some, there follows a

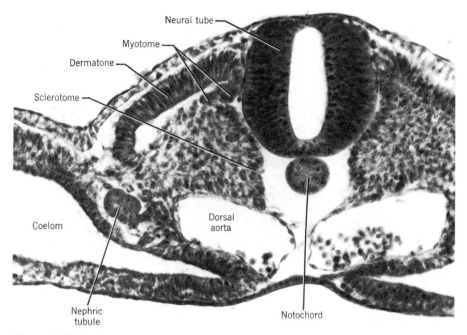

Figure 11-29 Architecture of a somite of a 48-hour chick embryo. (Photo by Torrey.)

complicated maneuver by which the cells that constitute the sclerotomes are regrouped so as to produce new sclerotomes lying *intersegmentally,* that is, alternating with rather than corresponding to the myotomes. However, the mesenchyme of the sclerotomes commonly spreads out as a continuous sheath enveloping the notochord and neural tube with only the vaguest suggestion of segmentation. In any event, nodules of cartilage, known as *arcualia,* shortly appear within the sclerotomal mesenchyme.

In many groups of vertebrates, two pairs of arcualia appear dorsolateral to the notochord and two pairs ventrolateral. In other cases, only one pair arises above and below the notochord, and these are situated intersegmentally with reference to the myotomes. In the latter situation, the dorsolateral arcualia grow upward and unite above the neural tube to form the *neural arch* (Figure 11-30). The ventrolateral arcualia grow downward and unite beneath the caudal blood vessels in the tail region to form the *hemal arch,* whereas in the neck

and trunk regions they extend laterally to provide the rudiments of ribs (Figure 11-30). At the same time, the primordial cartilaginous *centrum* is laid down. It may differentiate directly from the original mesenchymal sheath investing the notochord, or the arcualia above and below the notochord may extend around it to form the centrum. Combinations of both methods are known to occur. Whatever the method, the centrum is ordinarily deposited outside the notochordal sheath, but sometimes, notably in the elasmobranch fishes, sclerotomal cells may penetrate the sheath of the notochord and there produce a portion of the centrum. The zygapophyses and diapophyses originate from elaborations of the cells of the primordial neural arch; the parapophyses originate from the centrum.

With the completion of chondrification, the basic architecture of the prospective adult vertebra is fully established. The remainder of development is given over to ossification (except for the cartilaginous fishes), that is, to the replacement of the

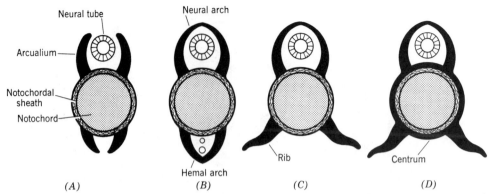

Figure 11-30 Development of vertebrae. Dorsal and ventral pairs of arcualia (*A*) give rise to the neural arch and the hemal arch in the tail (*B*), the neural arch and ribs in the trunk (*C*), and the centrum (*D*).

cartilaginous model by bone and the attainment of adult size and proportions.

The fact that from the beginning of development the vertebrae are intimately associated with the notochord and neural tube suggests the possibility of some developmental relationship other than mere anatomical alliance. The suggestion is borne out by experiments on amphibian embryos, in which the notochord is extirpated. The result is a very distorted vertebral column. But this could be a secondary consequence of the fact that the neural tube and somites are also distorted. More significantly, the experimental excision of the neural tube brings either a total absence of vertebral rudiments or their reduction to insignificant vestiges. Conversely, vertebral cartilages will appear in somites into which neural tube is grafted or in cultures of neural and somitic material. It is therefore concluded that differentiation of vertebral cartilage depends on an inductive stimulus provided by the neural tube.

It has already been pointed out that the arcualia (and subsequently each vertebra) develop intersegmentally with respect to the myotomes. Now a pair of spinal nerves with spinal ganglia arises in correspondence with each muscle segment. The fundamental arrangement, then, is that of each vertebra being connected to two consecutive innervated muscle segments (Figure 11-

31). Experimental analyses reveal that this system of alternating parts is dependent on the segmentation of the muscle primordia. If, for instance, extra somites are grafted into the sidewall of an amphibian embryo, the number of spinal nerves and ganglia is increased, although not necessarily in strict correspondence to the number of muscle segments. The number of dorsal arcualia will also be increased, but it *is* in accord with the number of spinal ganglia. The sequence of events thus appears to be the following: the myotomes roughly control the number of spinal ganglia to be created; sclerotomal mesenchyme aggregates between the ganglia; and the ganglia thus control both the position and number of vertebral rudiments.

Phylogeny of Vertebrae. Interpretations of vertebral evolution have long been influenced by the theoretical proposition that there are two vertebrae per two consecutive muscle segments—that is, each vertebra is basically double—and that the general trend has been toward the establishment of one vertebra between consecutive myotomes through the elaboration of certain elementary parts and the elimination of others. Support for this proposition is derived from several sources. There is, for instance, the circumstance of the occasional appearance of two pairs of em-

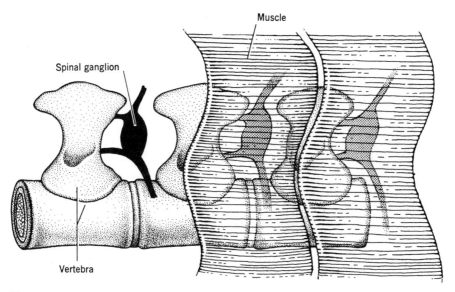

Figure 11-31 Relationship between the muscle segments, spinal ganglia, and the vertebrae in a larval salamander. (After Goodrich.)

bryonic arcualia above and below the notochord. The preliminary formation of a pair of centra in some amniote embryos and the actual presence of functional double centra in primitive tetrapods (see Figure 11-32 and accompanying discussion) and in the tails of some holostean fishes (Figure 11-32*A*) are also evidence in point. The presence of paired elements at the site of the neural arches in lampreys (Figure 11-32*B*), in the neural and hemal arches of chondrosteans (Figure 11-32*C*), and in the neural arches of sharks (Figure 11-32*D*) is likewise suggestive. Doubling of the neural arches in lizards is also common.

There is, however, another side to the picture. With respect to the vertebral elements in cyclostomes and fishes in general, about the only thing on which the experts agree is that the issue of homologies is in a state of utter confusion. This comes partly from honest disagreements on interpretation and partly from lack of sound embryological data. Any position that we take, therefore, will find its dissenters. But the following conclusions appear to be reasonable. The duplicated parts that we see in the neural and hemal arches of various

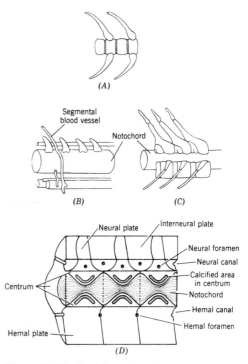

Figure 11-32 Doubling of vertebral parts. (*A*) Double centra in the caudal vertebrae of the holostean fish, *Amia.* (*B*) Paired elements constituting the neural arch in a cyclostome. (*C*) Double elements in the neural and hemal arches of the sturgeon. (*D*) Sagittal section through several consecutive caudal vertebrae of the shark.

fishes are more likely manifestations of secondary specializations than of primitiveness and are not to be homologized with each other. The presence of double centra in the tails of some fishes is also probably a secondary adaptation in the interest of flexibility rather than a carryover of a primitive condition. Thus not very much can be made of piscine vertebrae as an argument for the fundamental doubleness of vertebrae. But what does the fossil record tell us?

In the piscine ancestors of the tetrapods, as exemplified by the crossopterygian fish *Eusthenopteron,* there is a dorsal neural arch and two sets of ossifications associated with the still conspicuous notochord (Figure 11-33*A*). With regard to the latter, one set of bones consists of a pair of small plates lying dorsolateral on each side of the notochord. These constitute what has been termed the *pleurocentrum.* The other is a single element curved around the base and up each side of the notochord. This is the so-called *hypocentrum.*

This primitive pattern of two centra to a neural arch is duplicated in the oldest known tetrapods, the Ichthyostegalia of the upper Devonian, a group that truly bridges the gap between fishes and land forms (Figure 11-33*B*). From this starting point, then, several lines of evolution take off. In all, there is a trend toward a more solid, com-

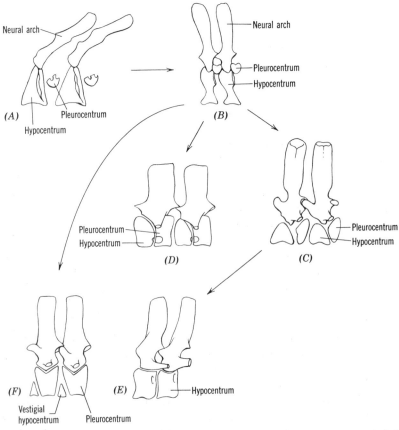

Figure 11-33   Phylogeny of vertebral types. (*A*) *Eusthenopteron.* (*B*) *Ichthyostega.* (*C*) Rachitomous type in *Eryops.* (*D*) Embolomerous type in *Archeria.* (*E*) Labyrinthodont amphibian. (*F*) Primitive amniote type in *Sphenodon.* Arrows indicate probable lines of succession. (Adapted from several sources.)

pact central area with concomitant reduction and/or elimination of the notochord. But in some lines there are "experiments" with two centra of approximately equal size; in others, one or the other of the centra is reduced or eliminated, leaving one functional centrum to a neural arch.

The double-centrum experiment took two forms: one, the *rhachitomous* type with a double centrum of interlocking wedges (Figure 11-33*C*); the other, the *embolomerous* type with a double centrum of successive rings (Figure 11-33*D*). Neither of these experiments was a success as measured by the persistence of the forms having double centra, for none lasted beyond the Jurassic period. It follows, then, that the vertebral history of the tetrapods is largely one of the elaboration of a single centrum. This history also pursued two lines. In one, the hypocentrum was the surviving element and was featured by the labyrinthodont amphibians (Figure 11-33*E*). However, in the amphibians that led to the reptiles, and they in turn to mammals and birds, the pleurocentrum was the surviving element, with the hypocentrum being reduced to intervertebral vestiges. So it is with all modern amniotes; the centrum is a pleurocentrum (Figure 11-33*F*).

Unfortunately, no satisfactory answer to the proper place of the modern Amphibia in this framework has been reached. One view considers the Urodela and Apoda to have a hypocentrum and the Anura, a pleurocentrum. Another more recent view is that all the modern Amphibia are alike and, as do the amniotes, feature a pleurocentrum. But still other students of the problem feel that there is yet insufficient evidence for correct interpretation of modern amphibian vertebral types.

Comparative Anatomy of Vertebrae. In most vertebrates, the vertebrae vary in some degree in different regions of the body. No attempt will be made to portray all the variations that occur, but some of the more outstanding features of the vertebral column in the principal classes of vertebrates should be of interest. As we remarked earlier, it is so difficult to equate the assorted vertebral components in the several categories of fishes with one another and with tetrapod vertebrae that it would seem wiser to restrict this survey of comparative morphology to the tetrapods.

Basically the column is divided into two regions, or perhaps we should say there are always two general types of vertebrae that may be distinguished: *caudal vertebrae* in the tail featuring hemal arches and *trunk vertebrae* lacking hemal arches.

The Amphibia exhibit the beginning of special deviations within the two general categories. Specifically, the last vertebra of the trunk is set off as a *sacral* vertebra, which, by the way of its transverse processes and ribs, articulates with and supports the pelvic girdle (see p. 249). Also, the first trunk vertebra is modified for articulation with the skull. This element is designated the *atlas*, which is the first and, in the Amphibia, the only *cervical* vertebra. Modern Amphibia also show great variation in the total number of vertebrae. The Apoda (which obviously lack a recognizable sacral vertebra) may have as many as 200. At the opposite extreme, the Anura have only 9 plus an elongated, rodlike *urostyle* (Figure 11-34). The urostyle is often interpreted as representing a number of fused caudal vertebrae, but there is no embryological evidence that this is truly the case.

The reptiles provide the first manifestation of the five subdivisions of the vertebral column that are henceforth featured. Using the sacral region as a landmark, we can say that the column consists of three primary regions: presacral, sacral, and postsacral (caudal). The presacral region is, in turn, subdivided into three regions: cervical, thoracic, and lumbar. From anterior to posterior, then, the amniote vertebral column consists of these five regions: *cervical, thoracic, lumbar, sacral,* and *caudal.* These subdivisions are clearly expressed by legged lizards and crocodiles (Figure

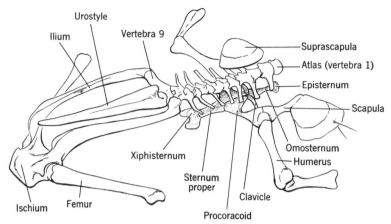

Figure 11-34 Axial skeleton and girdles of the frog. (Suprascapula and scapula are shown turned back on right side.)

11-35). The first two cervical vertebrae, termed the *atlas* and *axis,* differ markedly from the others in this region. Most notably, during embryogeny the centrum of the atlas, the first cervical vertebra, remains independent and associates with the centrum of the axis, the second vertebra. There, as the so-called *odontoid process,* it serves as a pivot upon which the head and atlas turn. The remaining cervical vertebrae are essentially alike, and all bear short ribs. But these ribs do not meet the sternum (p. 242) as

do those of the immediately following thoracic vertebrae. The succeeding lumbar vertebrae lack ribs. Next in order comes the *sacrum,* consisting of two sacral vertebrae fused together, to which the pelvic girdle is joined. Caudal vertebrae follow the sacrum. They are devoid of true hemal arches, but these are probably represented by the rudimentary *chevron bones* on the ventral side of the vertebrae, and thus the criterion for caudal vertebrae is technically satisfied.

The other groups of reptiles display nu-

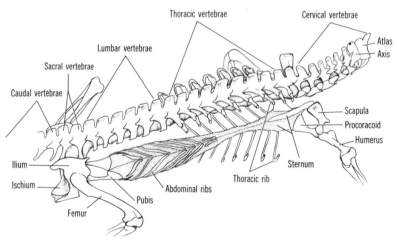

Figure 11-35 Axial skeleton and girdles of the alligator. (Ribs are cut away on right side and only the first two of the numerous caudal vertebrae are pictured.)

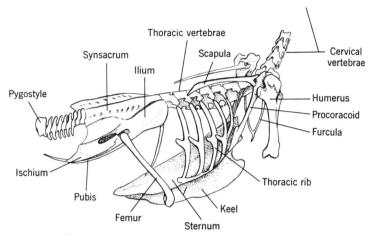

Figure 11-36 Skeleton of the trunk of the pigeon. (Most of the cervical vertebrae are omitted.)

merous departures from the basic pattern shown so well by crocodiles and legged lizards. Snakes and legless lizards exhibit, for instance, only the two general categories of vertebrae, trunk and caudal. The turtles also show some deviations. They lack true lumbar vertebrae, and their thoracic vertebrae, together with the sacrum and first caudal vertebra, are fused to the carapace of the shell.

The five subdivisions of the vertebral column are always present in birds, although obscured by a considerable measure of fusion (Figure 11-36). It is this fusion, involving all regions except the cervical, which remains highly flexible, that gives great rigidity to the column. The number of cervical vertebrae varies from 8 to 25, and all bear extremely short ribs. The thoracic vertebrae, 6 to 10 in number and bearing ribs joined ventrally with the sternum, are usually united. The last thoracic, the lumbar, two sacral, and the first few caudal vertebrae are fused into an elongated *synsacrum,* serving to support the pelvic girdle. The remaining few caudal vertebrae, ordinarily 6 to 10, are usually fused to form a short *pygostyle.*

The mammalian vertebral column (Fig-

ure 11-37), like that of most reptiles and of birds, is again divided into cervical, thoracic, lumbar, sacral, and caudal regions. The number of cervical vertebrae is remarkably consistent; with few exceptions, there are 7 of them, including the atlas and axis. Usually they all bear short ribs, but these are fused to the transverse processes and never reach the sternum. The number of thoracic vertebrae, by contrast, ranges from 9 to 25. They articulate with ribs, which usually extend ventrally to connect with the sternum. The lumbar vertebrae also vary in number, usually inversely with the thoracic vertebrae. That is, the greater the number of thoracics, the lesser the number of lumbars, and vice versa. Thus the total number of thoracic and lumbar vertebrae tends to be rather constant; it is the proportion of thoracics to lumbars that is the highly variable factor. There are usually 3 to 5 sacral vertebrae, firmly fused for support of the pelvic girdle. As for the caudal vertebrae, they may range all the way from 3 to 4, as in humans, to 40 or more. Although the more anterior caudal vertebrae possess the usual complement of parts, the more posterior ones tend to consist of little more than centra. Chevron bones, probably

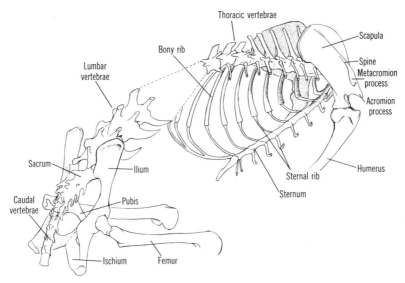

Figure 11-37 Skeleton of the trunk of the cat. (All cervical and some of the lumbar and caudal vertebrae are omitted. Ribs on right side are cut away.)

the equivalent of hemal arches, are commonly present on the ventral sides of the more anterior caudal vertebrae.

RIBS

Analysis of the development and structure of ribs will be greatly facilitated by some preliminary consideration of the fundamental layout of vertebrate musculature (Figure 11-38). The parietal, or body wall, muscles of a vertebrate are in part products of a linear series of paired embryonic myotomes. This means, on the one hand, that the muscles are laid down as segmentally arranged blocks and, on the other hand, that these blocks are paired. Any two successive muscle segments on one side are, therefore, separated by a sheet of connective tissue; likewise, the pairs of segments on each side are separated by dorsal and ventral midline

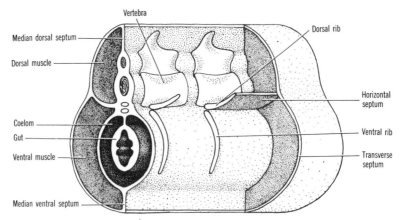

Figure 11-38 Section of the trunk of a vertebrate showing the septal system and relationship of ribs to it. (After Goodrich.)

connective tissue partitions. Throughout the trunk level the ventral partition is reflected about the coelom as the parietal peritoneum. In addition, the muscle blocks on each side are subdivided into dorsal and ventral portions with intervening partitions. These connective tissue partitions are termed *myosepta,* and it follows from the above that there are three sets of these: median dorsal and ventral myosepta separate the muscles into right and left sets, transverse myosepta separate successive muscle segments, and horizontal myosepta separate the dorsal and ventral muscles (Figure 11-38).

It is in association with the lines of intersection of the myosepta that the ribs are laid down. Ideally, there are two sets of ribs for each vertebra. *Dorsal ribs* develop where the transverse myosepta intersect the horizontal septa; *ventral ribs* develop where the transverse myosepta meet the ventral myoseptum (parietal peritoneum) (Figure 11-38). The material source of the ribs is aggregated mesenchyme derived from the sclerotomes and lateral mesoderm. The ribs are first preformed in cartilage and then may become ossified in whole or in part.

Among most fishes, notably in numerous teleosts, both types of ribs are present together. Many fishes, however, exhibit only the ventral series. So it is in the sharks, chondrosteans, and holosteans. But our concern is less with the vagaries of piscine ribs and more with the issue of tetrapod ribs, their origin and structure.

In all the tetrapods, only one type of rib is found. The upshot of long controversy, much of it only of historical interest nowadays, is that tetrapod ribs are the homologs of the dorsal ribs of fishes. A troublesome feature of the tetrapod rib is that typically it has two heads, a dorsal *tuberculum,* which articulates with the diapophysis of the associated vertebra, and a ventral *capitulum,* which articulates with the parapophysis—troublesome because the origin

of this double-headed condition has never been satisfactorily solved. Some investigators have attempted to account for the situation by proposing that tetrapod ribs actually represent coalesced dorsal and ventral ribs and that the two heads are manifestations of this, but no good embryological support for the idea has ever been forthcoming. But this much does seem to be true: however they may have originated, separate capitular and tubercular heads were present on the ribs of the first tetrapods. In the primitive Amphibia and Reptilia, the capitulum articulated with the hypocentrum and the tuberculum was anchored on the neural arch. With the decline of the hypocentrum and the rise of the pleurocentrum as the functional centrum in advanced and modern tetrapods, the capitulum either shifted to the pleurocentrum or, as it does in many mammals, came to articulate intercentrally on facets belonging to two consecutive centra.

To make a brief survey of the ribs of modern tetrapods, it may be noted first that the Amphibia, as in so many of their structural features, have departed widely from primitive conditions. Ribs are always absent on the atlas, and they are either reduced or lacking on the other vertebrae as well. In the urodeles, for instance, they are so short that they never reach the sternum, and their double heads are attached to transverse processes of a peculiar type that has no counterpart in any other vertebrate group. As for the Anura, ribs are usually absent entirely except for the sacrum (Figure 11-34).

The ribs of reptiles show a number of variations. One pertains to the head, which may be single or double. In the Squamata and Chelonia, the ribs are single-headed and articulate either with the vertebral centra (Squamata) or intersegmentally on adjacent vertebral centra (Chelonia). Crocodilian ribs, however, are double-headed, some attaching to the vertebral centra and others to transverse processes. Some of the other more conspicuous variations include the following. Turtles have no ribs whatsoever in the cervical region. Conversely, the ten trunk vertebrae bear long, flat ribs which, like the vertebrae themselves, are fused to the carapace. The ribs of the sacral vertebrae provide union with the pelvic girdle. In contrast to the turtles, the cervical vertebrae (except for the atlas and axis) of snakes and lizards bear ribs. The thoracic ribs of most lizards curve ventrad to meet the sternum, whereas the free ends of the trunk ribs in snakes, which have no sternum, have muscular connections with scales on the ventral side of the body and thereby assist in locomotion. All five regions of the crocodilian vertebral column (Figure 11-35) bear ribs, even the atlas and axis. Eight or nine of the thoracic ribs join the sternum. Each of these ribs usually consists of two sections, a proximal bony rib proper and a cartilaginous sternal rib. The joint between the two parts provides for flexibility essential to respiratory movements. The two sacral ribs are well developed for support of the pelvic girdle.

All five groups of vertebrae in birds bear ribs (Figure 11-36). Those in the cervical region are usually fused to the vertebrae. The anterior thoracic ribs unite with the sternum. Like those of crocodiles, they consist of two sections, except that both are bony. The remaining thoracic and all the lumbar, sacral, and anterior caudal ribs are fused to the synsacrum.

Mammalian ribs (Figure 11-37) are usually two-headed. Although at first sight the cervical vertebrae appear to lack ribs, the embryogeny of the transverse processes reveals they are composed in part of ribs. All the thoracic vertebrae bear ribs. Again, as in crocodiles, each consists of a bony section and a cartilaginous sternal section. Most of the ribs join the sternum directly, but the more posterior ones are bound to the sternum only indirectly by ligamentous connections with their anterior neighbors. Mammals lack ribs in the lumbar and caudal regions.

STERNUM

The sternum, or "breastbone," is found only in tetrapods. This is not to say that all tetrapods have one. It is absent, for instance, in the Apoda and in some salamanders among the Amphibia, and in snakes, legless lizards, and turtles among the reptiles. The sternum consists of one or more ventral plates attached anteriorly to the pectoral girdle and posterolaterally with the ventral ends of the thoracic ribs.

The evolutionary origin of the sternum is obscure. Several suggestions have been made to account for its origin, but the lack of good paleontological evidence makes a sound decision impossible. One view holds it to be a derivative of the middle portion of the pectoral girdle; another view maintains that it results from the fusion of the ventral ends of the thoracic ribs; still another calls for its origin independently of either girdle or ribs, that is, these parts become secondarily associated with the sternum. Support for the last is provided by sternal embryogeny in reptiles and mammals, where the sternum clearly develops independently of the ribs. It may even be possible that amphibian and amniote sterna have evolved independently and are not homologous structures.

Be that as it may, the sternum exhibits some interesting variations from one tetrapod group to another. In Amphibia, its association is exclusively with the pectoral girdle; thoracic ribs make no contact with it. It may be a simple median, triangular plate as in salamanders. In frogs, it may consist of four elements, two in front of the shoulder girdle and two behind (Figure 11-34). Among reptiles, the sternum of lizards consists of a cartilaginous plate to which the sternal sections of some of the thoracic ribs attach, whereas in crocodiles the sternal plate is single anteriorly and split into two arms posteriorly, with the sternal sections of the thoracic ribs attached on each side (Figure 11-35). The avian sternum is a very large, ossified structure and, except in flightless birds such as the ostrich, has its ventral portion expanded as a prominent bladelike keel, which provides extensive surface for the attachment of the muscles employed in flight (Figure 11-36). The mammalian sternum is typically made up of a series of separate bones, a kind of jointed rod, with the sternal sections of the thoracic ribs attached at the joints between the segments (Figure 11-37). In some cases, however, the separate segments are fused to some degree. So it is in humans, whose sternum consists of a three-part plate, the middle section of which represents several sections united in one.

THE APPENDICULAR SKELETON

The appendicular skeleton, in its broadest sense, includes the supports of all the projections from the body that are concerned with locomotion and steering. It therefore encompasses the *median fins* of fishes and the *paired appendages* of fishes and land animals. The former include the *caudal fin* embracing the tail end of the body, one or two *dorsal fins* along the dorsal midline, and a ventral *anal fin* posterior to the anus. The latter include the *pectoral appendages* anteriorly and the *pelvic appendages* posteriorly. We shall restrict our discussion to the paired appendages.

Any given one of the paired appendages includes the *free appendage* per se and an associated *girdle* lying within the trunk and serving as a support for and an area of anchorage of muscles for the free appendage. Thus, in speaking of the pectoral appendages we have in mind the pectoral fins or limbs and the pectoral girdles to which they are attached; the pelvic appendages refer to the pelvic fins or limbs and their girdles.

If the story of the morphogenesis of the paired appendages is to be complete, it must encompass answers to a number of critical questions. First, paired appendages were not present in the original vertebrates; they arose during the course of early evo-

lution of the fishes. Therefore, what was their source and manner of origin? Because the first free appendages, whatever their origin, were in the fishes and thus were fins, and because the fishes preceded the first vertebrates with limbs (the Amphibia), there is posed a second question of how fins were transformed into limbs. And, finally, what has been the general evolutionary history of limbs and girdles in the tetrapods? Let us consider each of these questions separately.

THE ORIGIN OF PAIRED APPENDAGES

The issue of the origin of the paired appendages has long been debated and even today remains unresolved. One early, ingenious idea was that the appendages derived their skeleton from rearmost gill arches. The gill bars themselves were presumed to provide the girdles, and the fins were envisaged as coming from flaps of skin at the edges of the gills, flaps whose internal stiffenings became fin supports. Quite aside from the fact that this fanciful notion had no embryological or morphological support whatsoever, it left the dermal elements in the pectoral girdle unaccounted for and was no answer at all to the source of the pelvic appendages unless a rearward migration of branchial parts could be imagined.

A second explanation derives from the fact that certain ostracoderms featured a series of paired spinous projections, which apparently were attached to the muscles of the flanks. It is suggested that as the pectoral and pelvic levels these spines aggregated to provide a skeleton for the membranous flap of skin that invested them. It is then speculated that some of the basal spinous elements may have worked inward to provide the endoskeletal components of the girdles. In addition, certain dermal plates of the carapace at the rear of the skull contributed the dermal components of the pectoral girdle. In this last we have an explanation of the fact that in bony fishes the pectoral girdle is attached to the rear of the skull, a point about which more will be made later.

A third scheme traces the paired appendages to continuous folds on each side of the body. The thought is that these folds became interrupted at intervals, providing a series of paired appendages; then, through the dropping out of the intermediate ones, pectoral and pelvic pairs alone were left fore and aft. As an accompaniment, radial skeletal pieces projecting parallel to one another and perpendicular to the body were presumed to be concentrated in the persisting appendages and worked over into fin and girdle skeleton. In support of this idea, attention has been drawn to some of the Devonian acanthodians in which there were numerous accessory fins between the pectoral and pelvic pairs, which consisted of little more than batteries of spines tied together by a web of skin (Figure 11-39). Another argument derives from the fact that in the embryogeny of some modern sharks there is, in early stages of development, a continuous series of muscle buds given off from all the segments of the trunk, but ultimately all regress except those buds supplying the pectoral and pelvic fins. However, recent experimental analyses of amphibian and chick embryos throw doubt on the epimeric origin of limb musculature; rather, the material source of the skeleton and musculature of these tetrapod appendages appears to be the more ventral somatopleuric wall.

Ingenious though these ideas are, especially the latter two, let us hasten to reiterate that they remain largely speculative; the origin of the fins and girdles is really unknown. For now, therefore, we can only accept their existence and move on to the second issue, the origin of the tetrapod limb.

THE ORIGIN OF THE TETRAPOD LIMB

As a first step it is necessary to become acquainted with the basic skeletal anatomy of a tetrapod limb, so that we may know what requisites a fin would have to possess if it were to be potentially convertible to a limb. Despite the many variations in functional anatomy among the limbs of tetrapods, these prove to be variations on one structural theme that prevails in both fore- and hindlimbs and is consistently maintained from amphibians to mammals (Figure 11-40). Reduced to its essentials, a limb is composed of three major divisions—proximal, middle, and distal. The proximal segment consists of a single bone, which

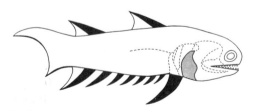

Figure 11-39   Fins of the Devonian acanthodian *Euthacanthus.* (After Watson.)

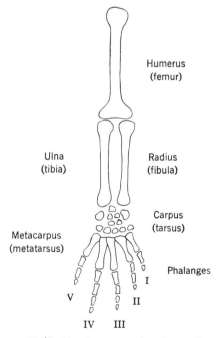

Figure 11-40   Vertebrate pentadactyl appendage.

articulates with the girdle and projects laterally from the body. In the forelimb this is termed the *humerus;* its counterpart in the hindlimb is the *femur.* At the elbow or knee, as the case may be, the single proximal bone articulates with the middle segment composed of two elements—*radius* and *ulna* in the forelimb, *tibia* and *fibula* in the hindlimb. These bones articulate in turn with the third and most distal division which descriptively can be broken down into three subdivisons: a *carpus* (wrist) or *tarsus* (ankle) composed of numerous *carpals* or *tarsals; metacarpals* or *metatarsals* composing the palm of the hand or sole of the foot; and *phalanges* making up the fingers or toes (digits). The phalanges are basically arranged in five rows, which is to say that the primitive number of fingers and toes is five. As we shall see, this number may be reduced, but only in one or two exceptional instances is it ever exceeded. Thus, the tetrapod limb is often referred to as a pentadactyl (five-fingered) appendage.

In seeking a piscine fin whose skeletal architecture would be amenable to conversion to a tetrapod limb, our thoughts naturally turn to the Crossopterygii whose other anatomical qualifications place them in the key position as progenitors of the Amphibia. Or let us put it this way. An exploration and evaluation of the many varieties of fin skeletons exhibited by cartilaginous and bony fishes reveal that none is a likely candidate for "limb making" except the type found in the rhipidistian Crossopterygii. Accordingly, the fin of the upper Devonian crossopterygian fish *Eusthenopteron* is submitted in evidence. (*Eusthenopteron* itself is not considered to be the immediate ancestor of the tetrapods, but some close rhipidistian relative probably was.)

The pectoral and pelvic fins are essentially the same, except that the pelvic ones are approximately one fourth smaller. Each fin contains a chain of bones arranged as shown in Figure 11-41. In terms of the orientation of the fin with respect to the body,

Figure 11-41   Skeleton of the pectoral fin of *Eusthenopteron.* (After Gregory and Raven.)

there may be recognized a series of so-called postaxial (dorsomedial) elements identified for conveinence in Figure 11-41 as *A, B, C, D, E,* and an accompanying series of radials along the preaxial (anteroventral) border identified as *a, b, c, d.* It is believed that the basal element *A* is equivalent to the humerus (femur) and the second element *B* is equivalent to the ulna (fibula). This brings the first radial *a* into correspondence with the radius (tibia). The remaining parts are thought to have provided the carpal (tarsal) bones. Because these parts are not sufficient in number to account for the metacarpals (metatarsals) and digits as well, there seems to be no alternative to the assumption that these arose independently as new ossifications.

Although paleontological evidence is slowly accumulating, there is still no complete record of the transition from crossopterygian fin to tetrapod limb. Nor is the gap filled by the facts of embryogeny of fins and limbs in modern species. Nevertheless, it is important that some attention be given to *embryological* events.

Fins and limbs have a similar origin in aggregations of mesenchyme derived from the somatic layer of the hypomere covered by somatic ectoderm (Figure 11-42*A*), but there the similarity ends. In fishes, the ectoderm projects as a disc, or flange, at whose base the mesenchyme remains packed. Skeletal parts (and muscles) arise within the mesenchymal core; horny fibers and bone-like rays differentiate within the ectodermal flange. In tetrapods, however, the mesen-

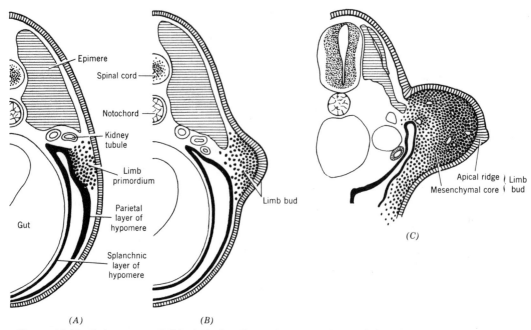

Figure 11-42 Embryonic origin of a limb. (*A, B*) Based on an amphibian. (*C*) Limb bud of a chick embryo. (From Balinsky, *An introduction to Embryology,* 4th ed., courtesy W.B. Saunders Company, Philadelphia.)

chymal aggregate quickly increases in mass and with its ectodermal covering bulges outward to produce a *limb bud* (Figure 11-42*B*). At first a rounded protuberance, the limb bud soon flattens to form a paddle, and in amniotes, but not amphibians, the epidermis along the edge of the flattened bud becomes thickened to form an *apical ridge* (Figure 11-42*C*). It has been demonstrated in avian embryos that further differentiation of the limb bud is dependent on the apical ridge of ectoderm. If the ridge is removed, all distal parts of the limb fail to form, not because their material source has been removed, but because the apical ridge normally provides some kind of inductive stimulus to the mesenchyme. Other experiments with avian embryos show that there is a two-way interaction between the mesenchymal and ectodermal components of the limb bud. For example, a reciprocal combination of apical ridge of a leg bud with the mesenchyme of a wing bud results in the formation of *wing* structures. Thus, although the apical ridge furnishes the in-

ductive cue, the mesenchyme itself furnishes the specific properties of the parts to be formed.

After the limb bud has grown so far that its length exceeds its breadth, it becomes marked off into distinct regions (Figure 11-43). The distal portion of the bud becomes even more flattened and distinctly broader. The flattened and broadened distal part is the hand (or foot) plate. The edge of the plate is initially circular, but soon becomes pentagonal, with the projecting points indicating the rudiments of the digits. The tips of the rudimentary digits then elongate while the intervening sections regress, so that the digits become separated by clefts. Meanwhile, the elongating limb primordium becomes subdivided by two constrictions that delineate middle and proximal parts.

The gross modeling just described is accompanied by internal differentiation. The mesenchyme of the early limb bud is relatively compact and shows no overt regional organization. However, histochemical stud-

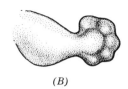
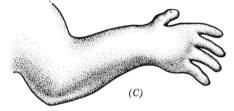

(A)   (B)   (C)

Figure 11-43 Stages in the development of the human forelimb. (*A*) 9-mm embryo. (*B*) 12-mm embryo. (*C*) 25-mm embryo.

ies of later stages reveal that the prospective skeleton is derived from the center of this mesenchymal mass, whereas the more external materials will produce the muscles, tendons, and generalized connective tissues. As the limb bud elongates proximo-distally, there is first a gradual condensation of the central mesenchyme at the proximal level to create the primordial humerus (femur), followed by a mesenchymal condensation foreshadowing the radius and ulna (tibia and fibula). The latter is initially a common primordium, but shortly segregates into the two forearm (foreleg) components. Flattening of the distal hand (foot) plate is accompanied first by the appearance of condensations that will provide the carpals (tarsals) and metacarpals (metatarsals), followed by the primordial digits. The digital mesenchyme first consists of five relatively broad condensation centers that subsequently become set off more sharply as a consequence of the death and destruction of cells in the interdigital areas. As illustrated by the forelimb of a 16-mm (7-week) human embryo (Figure 11-44), the mesenchymal primordia of the principal skeletal components have emerged in concert with the early molding of gross form. These primordia very shortly become converted to cartilaginous miniatures of the bones to come and subsequently become ossified in the manner already described (p. 207).

It is self-evident from a cursory examination that a fully formed limb is asymmetrical: it has proximal and distal ends, dorsal and ventral surfaces, and anterior and posterior sides, which are different from each other. Experiments on amphibian embryos have demonstrated that the anteroposterior axis is established first, the dorsoventral axis later, and the proximodistal axis still later. To illustrate, if a very young limb rudiment is inverted, the anterior and posterior sides of the resulting limb become reversed, whereas the other axes adjust to the new orientation and are normal. Inversion of an older primordium brings a reversal of top and bottom as well, although the parts are still normal from base to tip. Only at a later stage does reversal of a primordium bring a distortion of the parts along the proximodistal axis. Among amniotes the apical ridge appears to play some role in axial determination, for if the ridge on the wing bud of a chick embryo is rotated 180°, the symmetry of the wing from the elbow out is reversed.

One can only speculate about how these embryological events became imprinted on

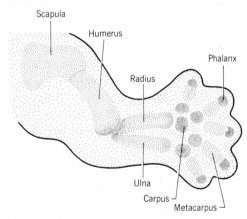

Figure 11-44 Primordia of limb skeleton in a 16-mm human embryo.

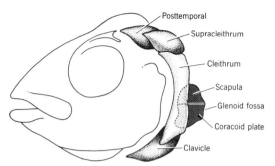

Figure 11-45 Piscine pectoral girdle.

the evolutionary history of the tetrapod limb. But it must be surmised that the transition from crossopterygian fin to tetrapod limb and the subsequent evolution of the limb itself involved the perfecting of these developmental operations.

THE HISTORY OF TETRAPOD GIRDLES AND LIMBS

The Pectoral Girdle. The history of the pectoral girdle must begin with its status in bony fishes. Whatever its evolutionary source, it consists of two general subdivisions: (1) an endoskeletal component composed of cartilage and/or bone, and (2) an exoskeletal component of dermal bone. With regard to the endoskeletal girdle (Figure 11-45), the *glenoid fossa* serving as the site of articulation of the fin provides a useful landmark. Above the fossa there extends a blade of bone or cartilage termed the *scapula;* below the fossa is the *coracoid plate.* The associated and investing dermal bones are a ventral *clavicle,* a long dorsal *cleithrum,* and still farther dorsally one or more additional elements (for example, supracleithrum and posttemporal) that connect to the posterior margin of the skull. Because the endoskeletal and exoskeletal components have different histories in the vertebrates following the fishes, it is convenient to consider them separately.

With respect to the dermal components, the most striking departure found in the first land vertebrates (Figure 11-46) is the loss

of the dorsal elements and with it the connection to the rear of the skull. However, the clavicle and cleithrum are retained, but as relatively narrow bones rather than broad plates. There is also introduced a new element, the *interclavicle,* a median plate, to which the clavicle on either side attaches. This pattern carries over into the early reptiles, but from here on the dermal parts have a varied history, although on the whole it is one of reduction and loss. The cleithrum is the first to go, and it is never found among modern tetrapods, except in anuran amphibians where it occurs as a small splint in the shoulder girdle. The interclavicle does persist in many reptiles and

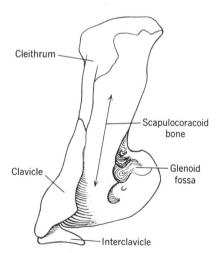

Figure 11-46 Pectoral girdle of the Paleozoic amphibian *Eryops.* (From Romer and Parsons, *The Vertebrate Body,* 5th ed., courtesy W.B. Saunders Co., Philadelphia.)

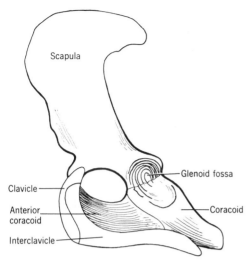

Scapula

Clavicle

Anterior
coracoid

Interclavicle

Glenoid fossa

Coracoid

Figure **11-47** Pectoral girdle of the prototherian
mammal *Ornithorbynchus.* (From Romer and Parsons
[after a figure by W.K. Parker], *The Vertebrate Body,*
5th ed., courtesy W.B. Saunders Co., Philadelphia.)

in prototherian mammals (Figure 11-47),
but is otherwise absent. Although absent in
urodele amphibians, some lizards, and in
snakes that have lost the entire girdle and
appendages, the clavicle tends to be more
persistent. It is found in anurans (Figure 11-
34), many lizards, and all birds except some
flightless forms. In the last-named, the two
clavicles are fused to form the *furcula,* or
"wishbone" (Figure 11-36). Some mammals
have no clavicle at all, and where it does
occur, it ranges in size from a small vestige,
for example, the cat, to a strong "collarbone" arching from the scapula to the sternum, as in humans.

In contrast to the exoskeletal parts, the
endoskeleton is retained throughout the
tetrapods. But considerable confusion attends the identification of the parts appearing in various forms, and it is quite unlikely
that the corresponding names given to parts
by various investigators are always indicative of true homologies. To cut through the
confusion, we shall take a rather arbitrary
position on nomenclature that seems to fit
the facts best.

The endoskeletal girdle of the first Amphibia (Figure 12-46) is much like that of

their piscine ancestors, except that it is ossified as one piece, which may be termed a
scapulocoracoid bone. It is only with the
later and modern Amphibia that a definitive
dorsal *scapula,* often topped by a cartilaginous *suprascapula,* may be distinguished
from a separate ossification in the coracoid
plate termed the *procoracoid* (Figure 11-
34). These two parts, scapula and procoracoid, are consistently present in ancient
and modern reptiles (Figure 11-35) and in
birds (Figure 11-36). But with the therapsid
line leading to the mammals, a new ossification, the definitive *coracoid,* appears in
the rear of the coracoid plate. This reptilian
condition is exhibited by prototherian
mammals (Figure 11-47). In the placental
mammals, however, the entire coracoid
plate largely disappears, leaving the girdle
to consist solely of a scapular blade to which
there is attached a small acromion process
(Figure 11-37). (The spine that runs down
the middle of the mammalian scapula may
represent the anterior margin of the ancestral scapula, and thus the scapular area in
front of the spine may be a new mammalian
creation.)

The Pelvic Girdle. The pelvic girdle is
composed of endoskeletal elements alone.
It has its beginning in fishes as a simple bar
of either cartilage or bone embedded in the
tissues of the ventral body and having no
connection with the vertebral column.
Commonly, the pelvic bars on either side
may be fused in the ventral midline.

It is with the earliest tetrapods that the
girdle exhibits three major transformations
(Figure 11-48). First, its ventral portion
expands greatly and provides two ossifications, an anterior *pubis* and posterior
ischium. Second, it sends a process dorsally,
which ossifies as the *ilium.* Third, the ilium
becomes anchored to the vertebral column,
either directly or through the mediation of
sacral ribs. At the site where the three pelvic
bones converge, a socketlike *acetabulum*
provides for articulation of the head of the
femur.

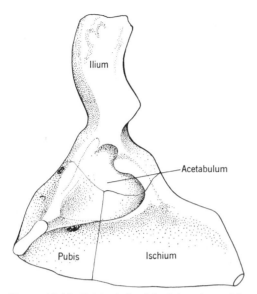

Figure 11-48 Pelvic girdle of the Paleozoic amphibian *Eryops.* (From Romer and Parsons, *The Vertebrate Body,* 5th ed., courtesy W.B. Saunders Co., Philadelphia.)

Except in those forms in which the paired appendages are reduced or lost, this tripartite organization of the pelvic girdle is maintained throughout the tetrapods (compare Figures 11-34—37). There are, of course, many variations on this structural theme. In addition to differences in proportions of the three elements from group to group and the retention of considerable cartilage in the girdles of modern Amphibia, one may note such variables as the degree of breadth of fusion of the ilia with the sacrum, the two distinctive types of orientation of pubes and ischia in reptiles (Figure 2-27), and the extent of ventral midline fusion of the pubes and ischia. But except for the occurrence in monotreme and marsupial mammals of an extra pair of forward-projecting *epipubic bones* of uncertain homology, the architecture of the pelvic girdle remains remarkably stable.

The Limbs. The principal circumstances in the history of the limbs are changes in orientation with respect to the body trunk and a great variety of functional adaptations.

In the first tetrapods, the limbs tended simply to project at right angles to the sides of the body. They are essentially that way still in present-day urodeles, chelonians, and many lizards, except that the body is moderately elevated. To accomplish this, there is a moderate bend at elbow and knee, so that the forearm and shank are directed somewhat vertically. But mechanically this provides for a most laborious kind of elevation, somewhat akin to a person lying on the ground doing "push-ups." Complete and more efficient elevation comes only from the arrangement found in mammals (but not confined to mammals, as witness birds and many of the dinosaurian reptiles). This is an arrangement accomplished by bringing the limbs to a vertical position and at the same time rotating them. The hindlimbs are rotated forward, so that the knees and feet point to the front. By contrast, the forelimbs are rotated backward, so that the elbows point to the rear. But this process alone would result in the forefeet pointing backward. To bring a foot forward, there is an additional rotation of the wrist through 180° about the vertical axis of the forearm. In consequence, the forearm is also twisted with a resulting crossing of radius and ulna. With the limbs so oriented vertically, a free fore-and-aft swing is facilitated—a far more effective mode of locomotion than that provided by the paddlelike or crawling movements permitted by limbs oriented horizontally.

In the scope of our treatment, it is impossible to go into all the functional adaptations of limb structure, but some of the more notable ones associated with locomotion on land, in the air, and in water may be cited (Figure 1-3).

Among land travelers, there may be noted those, such as bears and humans, who apply the whole surface of the foot to the ground; dogs and cats, which rest only the undersurfaces of the fingers and toes on the ground; and horses and cattle, which walk on the tips of their digits (the third digit only in horses in which the others are reduced to vestiges; the third and fourth in cattle.) In addition, there are numerous variations of limb structure operating in the interest of digging, climbing, grasping, and so on.

Three separate and independent groups of vertebrates have accommodated to life in the air by the modification of the forelimbs into wings—the Pterosauria ("flying reptiles"), birds, and bats—each in a different way. In the pterosaurs, the fourth digit of the hand was enormously elongated, and a web of skin was stretched between it and the side of the body. A bird's wing is built quite differently. The wing surface is provided by feathers inserted on the forearm and hand, the skeleton of the latter being a very specialized assembly of the first, second, and third metacarpals and phalanges. The wing of a bat is different again, for in it the second, third, fourth, and fifth digits are greatly elongated and support a web of skin that extends from the side of the body.

The accommodation to water takes the form of a modification of the limbs to provide paddles or flippers. Superficially, these resemble the fins of fishes, but their skeletons quickly reveal their status as pentadactyl appendages. The whales, sea cows, and seals are the possessors among mammals; among birds, the penguins have modified the wings into paddles; and in reptiles, paddles were featured by several ancient groups (for example, Ichthyosaurs and Plesiosaurs) and are presently found in certain turtles.

Finally, mention should be made of some forms that lack appendages. Among Amphibia, certain urodeles lack pelvic appendages, and the Apoda have lost both front and rear girdles and limbs. Snakes, among the reptiles, lack pectoral and pelvic appendages, although vestiges of the latter may remain. There are also a number of lizards that have become superficially snakelike through loss of limbs. Among mammals, the pelvic girdle and limbs vanish almost completely in whales and sea cows.

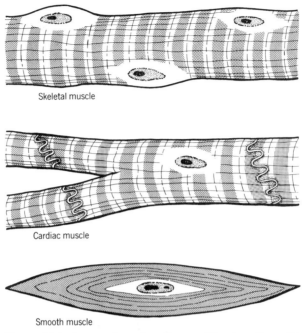

Skeletal muscle

Cardiac muscle

Smooth muscle

Figure 12-1 Survey of muscle cell types. (Based on Figure 5.1 [p. 85] in *Comparative Vertebrate Histology* by Donald I. Patt and Gail R. Patt. Harper & Row, 1969, by permission of the publisher.)

smooth musculature of the intestine or urinary bladder separately from these organs. As one writer put it, "As the heart goes, so goes cardiac muscle tissue," and the same might be said of any organ containing smooth muscle tissue. Skeletal muscle tissue, however, is assembled in definitive organs. These are the muscles themselves that are as independent and self-contained as any other organs of the body. The bulk of our attention is directed to the skeletal muscles, but smooth and cardiac muscle deserve some prior consideration.

SMOOTH MUSCLE

The cells (fibers) of smooth muscle tissue are spindle-shaped and contain a single nucleus (Figure 12-1), but the length and broadest diameter of the cells vary widely with animal and organ. In the pregnant human uterus, for instance, the fibers

(cells) may reach a half millimeter in length, whereas in the walls of small blood vessels they may be as short as 20 micrometers. The broadest are found in fishes and amphibians, the shortest in amniotes. The associations of smooth muscle cells also vary. They may occur singly, as in the smallest blood vessels, they may be assembled in interlacing bundles oriented in different directions, as in the urinary bladder; they may be arranged in well-defined longitudinal and circumferential layers, as in the intestine (Figure 12-2).

The contractions of smooth muscle cells are much slower than those of the other two types, and the cells are able to maintain forceful contraction for relatively long periods. Depending upon the site, contraction may be initiated by nerve impulses or local changes arising within the cells themselves, or through hormonal stimulation. Since nervous stimulation is by way of the autonomic nervous system, smooth muscle is said to

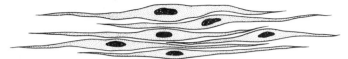

Figure 12-2   Isolated cells of smooth muscle from the stomach of a cat. (After Bloom and Fawcett, *A Textbook of History,* courtesy of W.B. Saunders Company.)

be "involuntary." The most important kind of local stimulation initiating contraction is that of stretching of the cells such as occurs in a filled urinary bladder or gastrointestinal tract. As for hormonal stimulation, uterine smooth muscle is responsive to oxytocin elaborated by the neurohypophysis, and smooth muscle in other parts of the body is generally susceptible to stimulation by adrenaline.

CARDIAC MUSCLE

Cardiac muscle, the contractile tissue of the heart, shares one distinctive feature with skeletal muscle, namely, its cells are characterized by longitudinal striations, *myofibrils,* that show a pattern of registered cross-banding (Figure 12-3). Although there are differences in detail, the nature and significance of the cross-striated (banded) myofibrils are essentially the same in cardiac and skeletal muscle. Accordingly, a detailed analysis of this distinctive architecture will be dealt with in conjunction with skeletal muscle and only those matters that are unique to cardiac muscle are considered now.

Cardiac muscle tissue is composed of individual mononucleated cellular units joined end to end by surface specializations termed *intercalated disks* (Figure 12-3). Some cells are simple cylinders; others branch and connect with similar branches from adjacent cells so as to create a complex three-dimensional network. The spaces within the branchwork are occupied by connective tissue serving to carry blood vessels, lymphatic vessels, and nerves. Electron micrographic studies of the intercalated disks reveal that they are areas where

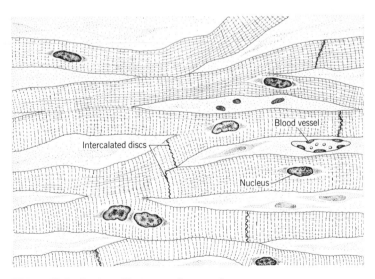

Figure 12-3 Section of human cardiac muscle.

the opposing ends of the cells are sculptured in a complex pattern of ridges and projections arranged so that the projections of one cell fit into corresponding pits and depressions in the other cell. In this fashion the cells are bound together in a structural and functional whole: firm connection of the cells under conditions of rigorous contraction is ensured; rapid spread of excitation from cell to cell promotes coordinated pulsation of the total heart muscle.

Cardiac muscle, like smooth muscle, is controlled by the autonomic nervous system and therefore is also involuntary. However, the control is only that of regulating the rate of pulsations of the heart, for contraction is innate to heart muscle, as evidenced by the fact that the embryonic heart pulsates prior to its acquisition of innervation and an adult heart continues to beat even if its nerve supply is severed. The subsequent mediation of the rate of pulsation by the autonomic nervous system exploits the remarkable capacity of cardiac tissue itself to conduct electrical impulses. In the embryonic heart of all vertebrates, the original rate of pulsation is established by the sinus venosus. Originating here, the impulses spread over the muscle cells and provide for sequential pulsations of the other heart chambers. The autonomic fibers that ultimately innervate the heart associate themselves primarily with the sinus venosus and, in cold-blooded vertebrates, it is the sinus venosus, with its rhythm regulated by the electrical firing of the autonomic system, that sets the pace of pulsations of the other chambers. In birds and mammals, however, the sinus venosus becomes reduced to a nodule of modified muscle and connective tissue, the *sinoatrial node,* in the wall of the right atrium. Impulses originating in the sinoatrial node and regulated by nerve discharge travel over the muscle cells of the atria and initiate their pulsations. Some impulses also reach another specialized nodule, the *atrioventricular node,* lying near the tricuspid valve, guarding the orifice between the right atrium and right ventricle. Bundles of modified cardiac muscle constituting a special conducting system then radiate from the atrioventricular node and distribute coordinating impulses throughout the ventricles.

THE ARCHITECTURE AND ACTIONS OF SKELETAL MUSCLES

Every skeletal muscle consists of a large number of cylindrical *muscle fibers* grouped in bundles of assorted sizes. The historical issue of whether these fibers are elongated multinucleated cells or syncytial arrangements of nuclei and cytoplasm has been resolved by exquisite studies on the behavior of individual embryonic muscle cells grown in culture. Embryonic muscle cells (myoblasts) have a spindlelike configuration and contain a single nucleus. Each myoblast divides and subdivides so that in a remarkably short time a colony of cells results from the original. Members of the colony then fuse to produce a multinucleated fiber. The number of nuclei fluctuates within wide limits, depending on the length of the fiber. The position of the nuclei also shows considerable variability: in muscles of lower vertebrates nuclei may be scattered throughout the fiber, but in mammals they tend to lie immediately under the sarcolemma (see later).

The properties and behavior of the muscle fiber, in concert with all the other fibers, confer on the muscle itself its distinctive operational qualities. In the final analysis, then, to find out how the muscle machine is constructed and how it functions we must turn to the muscle fiber. But rather than look immediately to this unit, let us begin with the general architecture of a muscle and then work downward through the diminishing magnitudes of its organization (Figure 12-4).

Any given muscle is invested by a relatively coarse connective tissue sheath, called the *epimysium,* from the undersurface of which septa pass into the interior of

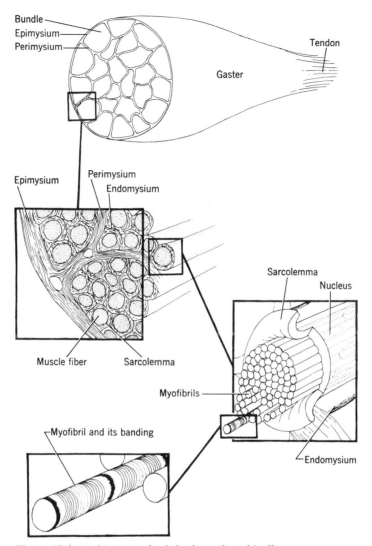

Figure 12-4 Architecture of a skeletal muscle and its fibers.

the muscle. These septa, collectively constituting the *perimysium,* serve to enclose variably sized bundles of muscle fibers. There extend, in turn, from the perimyseal septa delicate sheets of connective tissue, designated the *endomysium,* which provide an encasement for each muscle fiber. Thus, were it possible to remove all the pure muscle tissue, the form of the muscle down to the outlines of its fibers would be reproduced by this connective tissue framework. It is this scaffolding that provides for the distribution of nerves and blood vessels throughout the muscle. As we shall see, this scaffolding is also linked to the anchors that are required for a muscle to produce traction on bones and other movable parts.

Each muscle fiber is composed of numerous threadlike *myofibrils* running lengthwise through a cytoplasmic matrix of semifluid *sarcoplasm,* all encased by a thin membranous *sarcolemma* (Figure 12-4). Electron microscopic studies show that the sarcolemma is actually a compound structure consisting of a layer that may be equated with a conventional cell limiting membrane plus a protein–polysaccharide coating and a network of connective tissue. The sarcoplasm is essentially the same kind of ground substance that is found in all kinds of cells and includes such commonplace organelles as mitochondria. A unique organelle, however, is the *sarcoplasmic reticulum,* a network of minute canals that surrounds each myofibril. The sarcoplasmic reticulum is believed to further the electrical integration of the total muscle fiber, and it may be involved in the energy turnover indigenous to fiber contraction. Another important component of the sarcoplasm is *myoglobin,* an oxygen-binding pigment similar in structure and function to hemoglobin of the blood. Although the muscle fibers of vertebrates are otherwise consistent in structure, they do differ with respect to the content of myoglobin. *White fibers* that are relatively devoid of myoglobin may be distinguished from *red fibers* containing abundant myoglobin. The former operate anaerobically, contract rapidly, and are easily fatigued; the latter utilize myoglobin as an oxygen carrier, work more slowly, and can maintain contraction for longer periods. Generally speaking, the fast moving fishes have muscles containing a higher proportion of white fibers than do the slower moving tetrapods; and in any given class of vertebrates, there are "sprinters" whose muscles contain many white fibers and "stayers" whose muscles have a preponderance of red fibers.

It is to the myofibrils that muscle owes its contractile qualities. Myofibrils range from 1 to 3 micrometers in diameter and along their lengths exhibit a repeating variation in the density of their substance. This manifests itself in alternating light and dark bands (Figure 12-5). The characteristic striation of skeletal muscle results from the fact that the alternating bands of all the myofibrils within a fiber are in register. Studies with the electron microscope have revealed many details of fine structure of the myofibrils and their striations (Figure 12-6).

The pattern of striations is that of a regular alternation of dense areas known as *A-bands* and lighter areas known as *I-bands.* The central region of an A-band is often less dense than the rest of the band and is called the *H-zone.* Each I-band is bisected by a narrow, unusually dense band known as the *Z-line.* Thus from one Z-line to the next the repeating structural units of the myofibril read as follows: Z-line, I-band, A-band (interrupted by the H-zone), I-band, and Z-line. Any one such succession of bands constitutes what is referred to as a *sarcomere,* and the Z-lines may therefore be regarded as septa separating successive sarcomeres.

Careful studies of the ultrastructure of the myofibrils have revealed the presence of at least two kinds of filaments within

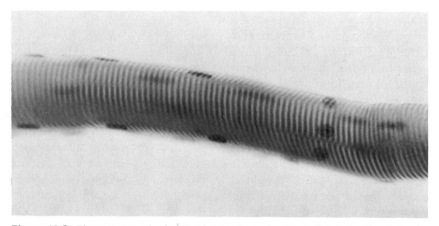

Figure 12-5 Photomicrograph of a slip of striated muscle. ×215. (Photo by Torrey.)

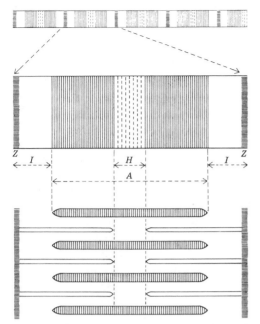

Figure 12-6 Ultramicroscopic structure of a myofibril. Explanation in text. (From H.E. Huxley, The Contraction of Muscle, *Scientific American*, Vol. 199, No. 5. © by Scientific American, Inc., November, 1958. All rights reserved. Courtesy Scientific American, Inc.)

because of the relatively enormous amount of energy made available whenever a phosphate group is detached from it. The steps in the degradation of glucose to the ultimate waste products CO_2 and H_2O do not, then, directly provide the energy required for muscle contraction; much of the chemical energy originally in the glucose molecule is used to build up the supply of ATP with its high-energy phosphate bonds, and it is from ATP that the energy for muscle contraction immediately derives.

Although a full understanding of the circumstances leading to muscle contraction has not been reached and there are honest differences of opinion regarding the significance of certain observations, the picture presently shapes up as follows. From the chemical point of view, the contractile property of muscle is associated with certain proteins, which constitute about 80% by dry weight of a muscle fiber. Specifically, the two principal proteins are substances known as *actin* and *myosin*. Myosin is believed to be the material that constitutes the thick filaments traversing the A-bands of the myofibrils; actin is believed to be the material constituting the intermingled thin filaments. In the resting muscle fiber the filaments of actin and myosin are not associated. Moreover, neither of these proteins by itself has the ability to contract. Somehow they must come together in an electrostatic association termed *actomyosin*, which does have contractile powers. The association is brought about by interaction of cross-bridges on the myosin that come into close contact with the actin. The precise chemical nature of the interaction is beyond the scope of this text; however, in any event, actomyosin utilizes energy-rich ATP to form an actomyosin–ATP complex.

It is at this point that the sarcolemma comes into play. In the resting fiber the sarcolemma is electrically polarized, its inside being about -0.1 volt with respect to the outside. An impulse arriving by a motor nerve serves to depolarize the sarcolemma.

them. (1) Relatively thick filaments extend across the length of the A-band. (2) Thinner filaments intermingle in an ordered pattern with the thick filaments and in each sarcomere extend from the Z-line to the edge of the H-band. The significance of these filaments with respect to the mechanism of muscle contraction is revealed by related biochemical and physiological studies.

The source of energy for muscle contraction is carbohydrate, which is brought to the muscle through its blood supply as glucose and then stored as glycogen. The glycogen is reconverted to glucose as it is needed. The derivation of energy from glucose involves a complicated routine of biochemical reactions. It is beyond our province to consider the multiple steps in this routine, but it is important to note that the ultimate source of energy lies in the energy-rich compound known as *adenosine triphosphate (ATP)*; it is energy-rich

This is believed to activate the myosin fraction of the actomyosin–ATP complex, which proceeds to operate as an enzyme serving to split a phosphate group from the ATP. The ATP is thereby converted to *adenosine diphosphate* (*ADP*), with a concomitant release of energy. Under this energy thrust the filaments constituting the actomyosin complex respond in such a way as to cause a shortening of the myofibrils. The observation that fibril contraction is accompanied by the disappearance of the I- and H-bands suggests that the actin filaments glide between the myosin filaments, and thus the name "sliding filament theory" has been applied to this phenomenon. In recharging a muscle, excess phosphate interacts with contracted actomyosin–ADP, energy is transferred, and actomyosin–ATP is reformed. The actomyosin and ATP separate, the actin and myosin filaments disengage and glide apart (the fibrils, that is, relax), and the stage is set for the process to be repeated.

We have seen that a skeletal muscle is composed of bundles of fibers with each fiber surrounded by a fine sheath (endomysium), each bundle clothed in a sheath of perimysium, and all the bundles of the total muscle invested by the epimysium. It is through this connective tissue system of sheathing that the power generated by the contracting muscle fibers is transmitted to the skeleton and other body parts. Thus, on the one hand, the endomysium is tied to the sarcolemmas of the muscle fibers and, on the other hand, the epimysium leads without interruption into strands or bundles of fibrous tissue which, in turn, are tied to the parts upon which traction is to be exerted. Although a muscle and its connections constitute a functional unit, it is customary to distinguish between the *belly*, or *gaster*, of a muscle to which the contractile fibers and the sheathing system are confined, and the *tendons*, whose fibers run from the epimysium of the gaster and blend with the periosteum and/or perichondrium of the skeleton or join the connective tissues of other organs such as the skin.

The bellies of muscles differ greatly in gross form. Some are spindle-shaped; others are broad and sheetlike. They vary, too, in complexity: many muscles have only a single gaster, others have two or more. The tendinous connections of muscles also exhibit a variety of form, ranging from cord- or ribbonlike tendons per se to flat sheets (aponeuroses). Occasionally, in fact, tendons may actually be lacking, as in the instance of the epimysium of one muscle being attached to that of another, or the direct union of epimysium and periosteum.

Muscles perform their tasks through their ability to exert force by contraction: they pull, never push. This means they must be anchored at two points or areas, one designated the *origin* and the other the *insertion*. The term *origin* is usually applied to the proximal and less mobile point of attachment, the term *insertion* to the distal and more mobile point. However, these are only relative terms, for all parts of the body are capable of some movement and the difference in degree of movement at the sites of origin and insertion may sometimes make for arbitrariness of definition.

With respect to the actions of muscles in general, almost without exception they operate in teams rather than individually. Because movement is the principal manifestation of muscle contraction, we may say that any given movement is effected by what may be termed a *prime mover*. A prime mover cannot operate, however, without the coincidental relaxation of its opposite member, the *antagonist*. These principals in the play of movement are in turn aided by supporting actors termed *fixation muscles* and *synergists*. Fixation muscles steady otherwise movable areas so that they may serve as the origins for prime movers and antagonists. For example, fixation muscles stabilize the scapula to enable the appropriate prime movers to act on the humerus. Synergistic muscles are called into play to prevent a waste of power by prime movers as they pass over intermediate joints. In the clenching of a fist, for instance, the muscles to the digits are prevented from bending the wrist by synergists in the wrist.

The patterns of movements created by muscle teams may be classified in terms of pairs of opposing actions, and by the same token the muscles concerned can be categorized in pairs of opposites. Accordingly, *sphincters* constrict openings and *dilators* enlarge them; *levators* raise parts in reference to other parts and *depressors* lower them; *retractors* withdraw parts and *protractors* extend them; *abductors* move parts away from the central axis of the body and *adductors* move them toward the central axis; *flexors* bend parts over an angle and *extensors* straighten them; and *pronators* rotate parts downward or backward and *supinators* turn parts upward or forward.

The names of individual muscles derive largely from those applied by generations of anatomists to the muscles of the human body. An unfortunate consequence of this is the extension of these appellations to many muscles whose homologues in the lower vertebrates are either unknown or of doubtful validity. Moreover, the roster of names is something of a hodgepodge, in which the only element of consistency is a description of some obvious structural feature. Thus, muscles may be named in terms of origin and insertion (sternohyoid), position (external intercostal), shape (triangularis), direction of fibers (oblique), size (vastus), number of bellies (digastric), function (levator), or some combination of these terms.

EMBRYOGENY OF SKELETAL MUSCLES: GENERAL CONSIDERATIONS

The embryonic history of the skeletal muscles of vertebrates begins with the early de-velopment of the mesoderm (see Chapter 3). The original mesodermal sheet, of course, is interposed between the ectoderm and endoderm. As the neural tube and notochord take form, the mesoderm on each side of the notochord thickens to produce a longitudinal band of *paraxial mesoderm* (epimere). This band leads to a narrow zone of *intermediate mesoderm* (mesomere) which, in turn, leads peripherally into *lateral plate mesoderm* (hypomere). In due time the paraxial mesoderm becomes segmented to form paired cubical masses called *somites*. The lateral plate remains unsegmented, but does split into two sheets— an outer *somatic mesoderm* associated with the ectoderm (somatic mesoderm and ectoderm constituting the *somatopleure*) and an inner *splanchnic mesoderm* associated with the endoderm (splanchnic mesoderm and endoderm constituting the *splanchnopleure*). (Refer to Figure 3-3.) The intervening space between the somatic and splanchnic sheets is the *coelom,* the forerunner of the peritoneal and other body cavities. The intermediate mesoderm between the somites and lateral plate provides the excretory system and is not of present concern. It is to the paraxial (somitic) mesoderm and lateral plate mesodern that the skeletal muscles trace their origin.

With respect to the paraxial mesoderm, the last statement requires some refinement, for the somites in their entirety are not involved in the formation of skeletal muscles. When first established, the somites are cuboidal masses, the cells of any one of which are arranged radially about a small central cavity. Subsequently each somite becomes extended dorsoventrally and flattened mediolaterally. As a consequence, the central cavity becomes a narrow slit lying between outer and inner walls of the somite (Figure 12-7). The outer wall will contribute to the formation of the dermis of the skin, and is therefore termed the *dermatome.* A considerable part of the inner wall resolves itself into a loose mesenchyme

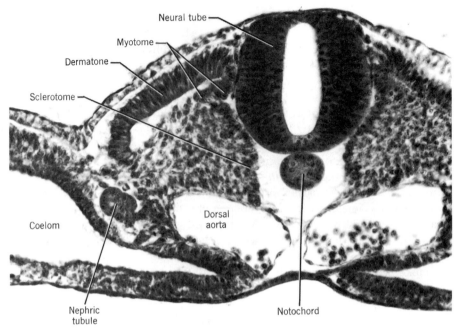

Figure 12-7   Architecture of a somite of a 48-hour chick embryo. (Photo by Torrey.)

constituting the *sclerotome* from which the vertebrae derive. That portion of the inner wall remaining after the sclerotomic cells have dissociated themselves is called the *myotome*. It is this myotomic fraction of the somites that is concerned with muscle formation. The general statement of the origin of skeletal muscles should therefore be revised to read: skeletal muscles are products of the myotomes and lateral plate mesoderm.

For the purpose of this preliminary survey of the embryogenesis of muscles, it is convenient to consider the head and neck regions separately from the trunk and tail. With regard to the head and neck, there are two sources of muscles: myotomes and the hypomeric mesoderm associated with the pharyngeal portion of the gut. The myotomes of the head and neck may be described in terms of their position relative to the developing inner ear. Those in front of the ear are designated as preotic myotomes, and in vertebrates generally there are three of these (speaking of each side of the body). Behind the ear, depending on

the vertebrate, there are a variable number of occipital and cervical myotomes (postotic myotomes), which lead without break into the myotomes of the trunk. Although the three preotic myotomes are consistently exhibited and it is generally agreed that these represent the first three segments of the vertebrate body, an unresolved problem attends the myotomes that immediately follow. Apparently the evolution of the vertebrate head has witnessed the elimination of one or more of the postotic myotomes, producing a gap in the series between the third preotic segment and the next segment to follow. So not only does the number of myotomes at the occipital and cervical levels vary from vertebrate to vertebrate, but in any particular case there is uncertainty as to their numerical identity.

In due course we shall see that the preotic myotomes are concerned with the formation of the extrinsic muscles of the eye, whereas the occipital and cervical myotomes, whatever their number and identity, provide the musculature of the tongue and neck. But in addition to these muscles of myotomic origin, there is in the head and neck an extensive system of voluntary muscles associated with the visceral arches and their derivatives. This musculature is a product of the lateral plate mesoderm associated with the pharynx. It is, in essence, a portion of that musculature which is associated with the entire alimentary canal. But whereas in the digestive tube proper this musculature is of the smooth variety (except in the esophagus of fishes and mammals where striated muscle develops) and is an integral part of the tube itself, in the pharyngeal region the muscles are striated and have abandoned the gut, so to speak, to serve the visceral skeleton and even the shoulder region in part. There is also this further point of developmental difference. The smooth muscle of the gut is provided by the splanchnic layer of the hypomere; but no coelom extends into the pharyngeal region, so that it is the thickened, unsplit hypomere as a whole that is involved there. Moreover, in contrast to the hypomere elsewhere, that of the pharyngeal region becomes segmented as a consequence of the projection of the endodermal pharyngeal pouches toward the surface. It should be emphasized, however, that the pharyngeal segments of hypomere do not correspond to the myotomic segments of the paraxial mesoderm. Accordingly, it is customary to employ the following distinguishing terms: hypomeric pharyngeal segmentation is called *branchiomerism;* myotomic segmentation is called *myomerism.*

In the trunk, the issue is the source of the muscles of the paired appendages and the body wall. As for the appendages, two situations appear to exist. The muscles of the fins of fishes are known to arise from ventrolateral extensions of the adjacent trunk myotomes into the mesenchyme of the embryonic fins. In tetrapods, however, there is no convincing evidence of similar myotomic outgrowths. Rather, the bulk of observational and experimental evidence (for example, the cultivation of isolated limb primordia) favors the view that the

limb muscles differentiate *in situ* from the mesenchyme of the limb bud. The source of this mesenchyme is the somatic layer of the lateral plate (Figure 11-42). In other words, the primordial limb bud represents a localized extension of the somatopleure, whose ectodermal component provides the epidermis and whose mesodermal component provides the dermis, musculature, and skeleton. It is not impossible, of course, that the mesenchymal cells destined to furnish the musculature may originate in the myotomes. Indirect evidence favoring this possibility is found in the fact that the nerve supply to the limbs derives from branches of the nerves that serve the adjacent myotomes; but no direct demonstration of a myotomic source of limb bud mesenchyme has yet been forthcoming.

As for the muscles of the pectoral and pelvic girdle, the situation is somewhat obscure, but it is probable that in such transitional areas between limbs and trunk, muscles might be formed in part from myotomic ingrowths and in part by mesenchymal differentiation *in situ.*

Conventional accounts of the development of the trunk muscles assign their origin to the myotomes. The thought is that in addition to providing the dorsolateral musculature directly, individual myotomes send budlike outgrowths ventrally into the somatic layer of the lateral plate and it is from these buds that the ventrolateral musculature derives. By this view, the lateral plate participates in muscle formation only in a passive way; that is, it serves only as a matrix or vehicle for the myotomic buds. Fairly recently, however, experimental analyses conducted on the chick embryo have opened other possibilities. "Tagging" of the myotomes with carbon particles so that their movements may be followed and tests of the developmental capacities of isolated pieces of somatopleure reveal that the musculature of only the dorsal third of the body forms from the myotomes. The muscles of the ventral half of the body come from the somatic layer of the lateral plate, whereas the muscles of the intervening area appear to derive from both lateral plate and myotomes. Yet, in experiments involving the excision of somites in urodele embryos, defects in the ventrolateral musculature materialize, which suggests that in these forms all the trunk muscles trace to a myotomic source. For the present it is difficult to reconcile these results. It is possible of course, that different developmental operations prevail in different kinds of animals, but one might hope that an extension of observational and experimental methods on a broadly comparative basis would one day clarify the roles played by myotomes and lateral mesoderm in the formation of trunk musculature. There is no question of the derivation of some muscles from the myotomes. The issue to be resolved is whether the myotomes are the *sole* source of the trunk musculature, or does the somatic layer of the lateral plate participate in some measure. As for the muscles of the tail, there is no evidence to contradict the conclusion that they derive exclusively from the myotomes.

So that we may have the essential facts clearly before us, let us, with the aid of Figure 12-8, make a summarizing rundown of the embryonic materials involved in the fabrication of the skeletal musculature. Two categories of mesoderm are concerned: myotomes and lateral plate. With respect to the former, the basic condition is probably that of an uninterrupted series of myotomes extending from a level immediately in front of the first visceral arch back to the end of the tail. This condition actually prevails in cyclostome embryos, but in all jawed vertebrates (gnathostomes) one or more myotomes just behind the otocyst are suppressed, creating a gap in the series. Thus, whereas three preotic myotomes are consistently exhibited, the number and numerical identity of the myotomes at the immediately following occipital and cervical levels cannot be stated categorically. From the cervical level the series runs without break to the end of the tail. Speaking generally, this axial series of myotomes is dedicated as follows: the preotic myotomes provide the extrinsic muscles of the eye; the occipital and cervical myotomes contribute to the musculature of the head and neck; the trunk and caudal myotomes participate in varying degrees, depending on the vertebrate, in the formation of the musculature of the trunk, tail, and paired appendages.

With respect to the lateral mesoderm, the pharyngeal and trunk regions of the embryo must be considered separately. In the trunk, where a coelom exists, it is the somatic layer that is concerned. Except in fishes, the muscles of the appendages derive from this layer. There is also some evidence

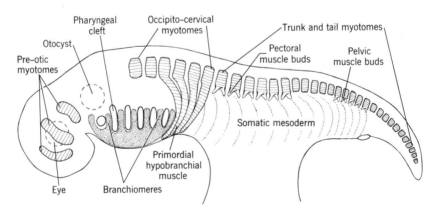

Figure 12-8   Vertebrate embryo, with special reference to a shark, showing the embryonic primordia of muscles.

that the ventrolateral muscles of the body wall itself come from this source as well, but this may not be true in all vertebrates. In the pharyngeal region, where no coelom exists, the lateral plate is interrupted by the pharyngeal clefts and is therefore subdivided into masses that will provide the musculature of the visceral skeleton.

With these embryological fundamentals before us, we may now turn to the evolutionary history of the skeletal musculature. To this end, it will be helpful first to lay out the architectural pattern found in a representative fish such as a shark (Figure 12-9). The morphogenesis of each of the separate groups of muscles constituting this pattern may then be traced separately through the tetrapods.

THE MUSCLES OF ELASMOBRANCHS

The bulk of the muscle in fishes is laid down in segmental masses along the length of the body and is termed *axial muscle.* The characteristic segmentation of the axial musculature reflects its origin from the embryonic myotomes. Ignoring for the moment such special locales as the appendages and head and concentrating instead on the trunk and tail, it is found that the dorsally placed myotomes grow down on each side so as to create a succession of muscle segments known as *myomeres.* Not only do the myomeres arise separately from individual myotomes, but this separation is preserved in the adult by sheets of connective tissue, *myocommata (transverse myosepta),* that are interposed between successive myomeres. The myocommata serve both to tie the myomeres to the vertebrae and to anchor the short, longitudinally oriented muscle fibers composing the myomeres.

When first formed, the myomeres are straight vertical bands, but eventually they become folded in such a manner that they take on a zig-zag form. It will be recalled (p. 241 and Figure 11-38) that because of the presence of a *horizontal septum* each myomere is divided into dorsal and ventral portions. Consequently, the axial musculature on each side is divided into two major units: the *epaxial musculature* above the

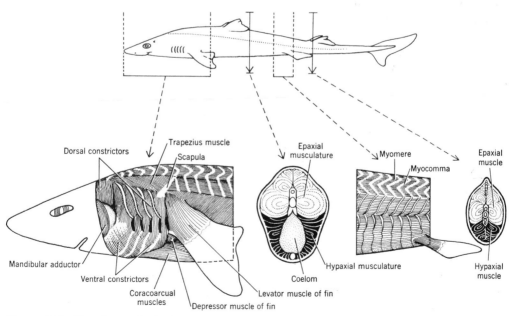

Figure 12-9 Musculature of a shark.

septum and the *hypaxial musculature* below the septum. Throughout the trunk and tail the epaxial muscles appear as bulky masses above and to the sides of the vertebrae, filling in the total area between the vertebrae and the skin. In the tail, the hypaxial muscles are similarly elaborated on the sides and below the vertebrae, but because the coelom and viscera occupy much of the trunk, the hypaxial musculature here appears as a relatively thin enveloping sheath.

The axial muscles of the trunk are interrupted only to a minor degree at the levels of the pectoral and pelvic appendages. In the shark embryo, myotomes at these levels give off buds, which grow into the somatopleuric projections destined to become the paired fins. Each of the muscle buds concerned divides into dorsal and ventral halves; thus each fin comes to be equipped with dorsal and ventral masses of muscle. The dorsal muscle is a *levator,* serving to elevate and extend the fin; the ventral muscle is a *depressor,* serving to lower and adduct the fin. These muscles have peripheral insertions within the fin itself; centrally, they fan out and are anchored, that is, have origins, on the adjacent girdle. The levators originate on the iliac processes of the pelvic girdle or scapular processes of the pectoral girdle, as the case may be, whereas the depressors originate on the puboischiac bar and coracoid bar, respectively.

In front of the pectoral girdle, the succession of axial myomeres is modified by the presence of the pharyngeal system. The modification consists essentially of a wide separation of the epaxial and hypaxial portions of the myomeres because of the interposition of the pharyngeal apparatus. The epaxial myomeres, which at this level constitute the *epibranchial musculature,* are essentially unchanged over those in the trunk. In other words, the epaxial myomeres of the trunk are simply continued without break until they terminate at the rear of the neurocranium. By contrast, the hypaxial myomeres provide a *hypobran-*

chial musculature specialized to expedite movements of the floor of the mouth and, in association with the muscles of the gill arches, to assist in the inspiration of water. As a consequence of the presence of the pharynx, the hypobranchial musculature is not established by direct downward extensions from the dorsally located embryonic myotomes. Rather, there is a ventroanterior growth of the lower ends of the occipitocervical myotomes immediately behind the pharynx. It is this embryological circumstance that accounts for the fact that the hypobranchial muscles of the shark, and of vertebrates generally, have a nerve supply originating at the occipital and cervical levels of the central nervous system. The hypobranchial musculature of the shark consists of a complex of *coracoarcual muscles* originating from the coracoid bar of the pectoral girdle and inserting within the connective tissue constituting the floor of the pericardial cavity. From these muscles there extend lesser muscle slips, each bearing a specific name, to the lower segments of the adjacent visceral arches.

The one remaining component of the axial musculature is that represented by the six muscles that move the eyeball (Figure 12-10). Four of them, the *rectus muscles,* originate in the posterior wall of the orbit and insert at such positions on the eyeball as to be identified individually as follows: *superior rectus,* dorsal; *inferior rectus,* ventral; *internal rectus,* anterior; and *external rectus,* posterior. The other two are the *oblique muscles,* which originate in the anterior wall of the orbit. The *superior oblique* inserts on the dorsal side of the eyeball; the *inferior oblique* inserts on the ventral side.

The three preotic myotomes represent the embryonic source of these six eye muscles. The first myotome forms four of them (superior rectus, internal rectus, inferior rectus, and inferior oblique, in that order from top to bottom; Figure 12-10). The dorsal half of the second myotome provides the superior oblique muscle. The external rec-

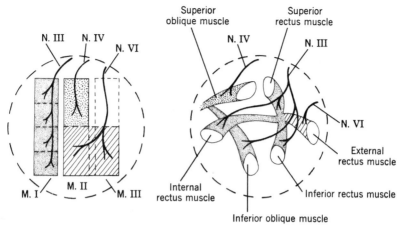

Figure 12-10 Origin of the extrinsic muscles of the eyeball from the three preotic myotomes.

tus is a product of the ventral half of the second myotome and the ventral half of the third myotome. More will be made of this point later (Chapter 17), but it is appropriate to point out that this pattern of development—four muscles from one myotome and two muscles from parts of two myotomes—provides an explanation of the fact that one cranial nerve (N. III) innervates four muscles, whereas each of the remaining two muscles is supplied by a separate nerve (N. IV and N. VI).

The last category of muscles in the shark to be considered is the *branchiomeric system* whose origin traces to the blocks of hypomere between the pharyngeal clefts. As previously noted, this is the one area of skeletal musculature that in fishes is not derived from the axial myotomes. Associated as it is with the elaborate visceral skeleton and related respiratory operations, the branchiomeric musculature presents a complicated architecture. It is convenient to describe the system in terms of four series of muscles (Figure 12-9 and 12-11).

Most superficial is the *constrictor musculature* consisting of six *dorsal constrictors* above and six *ventral constrictors* below the external gill slits. With respect to the dorsal constrictors, the first one is a relatively small muscle lying immediately in front of the first gill slit (spiracle). It orig-

inates on the auditory capsule and inserts on the upper jaw. The second one, a much larger muscle, also originates on the auditory capsule, but inserts on the hyomandibula of the second arch. The remaining four are similar to each other and are anchored to fasciae of the gill chambers and adjacent epibranchial muscles. With respect to the ventral constrictors, the first one is a broad muscle originating in the fascia of the ventral midline and inserting on the lower jaw. The second one is attached to the hyoid arch, and the remaining four are linked to the gill chambers and fasciae of the hypobranchial region.

The other three series are largely buried beneath the superficial constrictors. One consists of the *levators* (Figure 12-11) which are attached to the dorsal ends of the skeletal arches. The first levator serves to elevate the upper jaw. The second one, however, is fused with the overlying dorsal constrictor; and the remaining ones are, in most sharks, fused together to form a single muscle (the homologue of the tetrapod trapezius muscle? see Figure 12-9), which slants back to attach to the scapula of the pectoral girdle. The branchial levators, operating in conjunction with the dorsal and ventral constrictors, elevate the gills and effect a compression of the pharynx so that water is expelled. The remaining two series

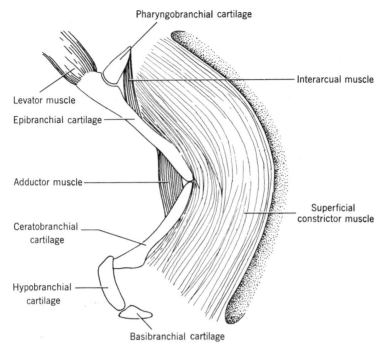

Figure 12-11 Musculature of a single gill arch of a shark. (From Romer and Parsons, *The Vertebrate Body,* 5th ed., by permission of W.B. Saunders Company.)

of deep muscles are the *adductors,* which bend the arches at the middle so that their dorsal and ventral halves are brought toward each other, and the *interarcuals,* which bend the upper ends of the arches backward.

To remake our earlier point, the foregoing analysis of the origin and general form of the musculature of the shark serves to bring before us the principal groups of muscles whose morphogenetic histories we wish to trace in the tetrapods. In summary, then, we have to deal with six categories: (1) epaxial muscles, (2) hypaxial muscles, (3) hypobranchial muscles, (4) eye muscles, (5) appendicular muscles, and (6) branchiomeric muscles. In addition to these six groups, we also give attention to (7) a special set of integumentary muscles.

MORPHOGENESIS OF THE MUSCLES IN TETRAPODS

Epaxial Muscles. The epaxial musculature, which in fishes consists of massive seg-

mented columns running all the way from the rear of the skull to the end of the tail, exhibits two general kinds of modifications. First, there is a steady reduction in volume—a reduction correlated with the transfer of propulsive duties from the trunk and tail to the appendages—so that these muscles become concerned essentially with the support and movements of the vertebral column. Second, there is an increasing tendency toward loss of the original segmentation and the creation of new architectural complications. An accompaniment of this is a considerable lengthening of individual fibers, so that instead of extending just the distance of one segment, they cover an expanse corresponding to many original segments. Let us translate these generalities into specificities.

Although the epaxial muscles of the trunk and tail of caudate amphibians (urodeles) continue to show the primitive segmental arrangement (Figure 12-12), the myomeres are less bulky and are strictly vertical rather than showing the zig-zag pat-

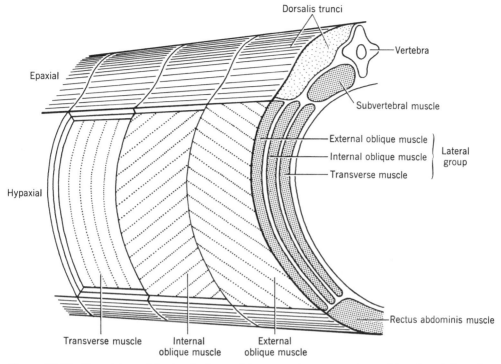

Figure 12-12 Musculature of the trunk of a urodele as seen in cross section and in a lateral dissection. Epaxial musculature, fine stipple; hypaxial musculature, coarse stipple.

tern found in the faster swimming fishes. In anurans the epaxial muscles are still further reduced. The tail, of course, is lacking in these forms, and in the trunk the muscles are now largely concerned with the support of the vertebral column and its dorsal rather than lateral bending as in fishes and urodeles. The one site of any conspicuous regional differentiation is in the "neck," where in both anurans and urodeles the main epaxial mass is split into several muscles, which are attached to the skull and effect a turning of the head.

Beginning with the reptiles, the segmental composition of the epaxial musculature tends to become obscured and secondary complications of build become manifest. The latter expresses itself in the appearance of two major groups of muscles. One is the *longissimus dorsi system*. Originally (in amphibians) a single segmented column (*dorsalis trunci*), it is now subdivided into several muscles (Figure 12-13)—a

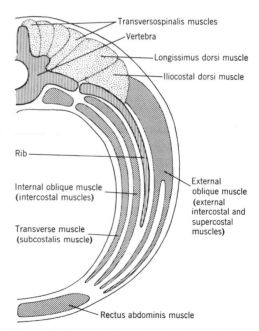

Figure 12-13 Musculature of the trunk of an alligator. Epaxial musculature, fine stipple; hypaxial musculature, coarse stipple.

longissimus dorsi proper lying above the transverse processes of the vertebrae, a *longissimus capitis* (Figure 12-14) along the sides of the neck and extending to the temporal region of the skull, and an *iliocostal dorsi* inserting on the proximal parts of the ribs. The other is the *transversospinalis system.* Here is included a crisscross of muscles between the neural spines and successive vertebrae, between spines and transverse processes, and between transverse processes. These epaxial muscles find

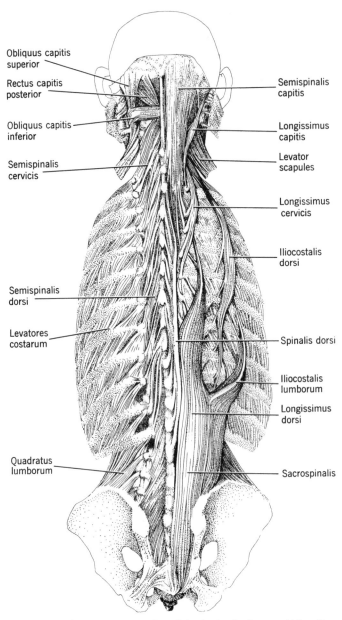

Obliquus capitis superior
Rectus capitis posterior
Obliquus capitis inferior
Semispinalis cervicis
Semispinalis dorsi
Levatores costarum
Quadratus lumborum

Semispinalis capitis
Longissimus capitis
Levator scapules
Longissimus cervicis
Iliocostalis dorsi
Spinalis dorsi
Iliocostalis lumborum
Longissimus dorsi
Sacrospinalis

Figure 12-14 The deep muscles of the back of a human. (After Cunningham.)

their most complex expression in snakes, where locomotion depends on the ribs and lateral movements of the trunk and tail rather than on limbs. In turtles, by contrast, the epaxial system is greatly reduced, and the same may be said of birds where extensive fusion of the trunk vertebrae and virtual loss of the tail preclude well-developed epaxial muscles.

The epaxial system in mammals (Figure 12-14) is basically similar to that of reptiles, except that little remains of the original segmentation. The two major categories of the system are represented by a maze of relatively small interlacing muscles that tie the vertebrae together as well as to the ribs and the rear of the skull. In the lumbar region the latissimus dorsi and iliocostal dorsi are commonly united in a strong *sacrospinalis muscle,* which supports the arch of the vertebral column.

Hypaxial Muscles. Unlike the epaxial system in which the muscles of the trunk continue into the tail without break, the hypaxial muscles are subdivided into two groups, trunk and caudal. This subdivision is brought about by two circumstances: interruption by the pelvic girdle and the limitation of the coelom to the trunk. Other than to mention that the hypaxial tail musculature consists of longitudinal columns running forward beneath the transverse processes of the vertebrae to insert upon the pelvic girdle, we shall confine our attention to the trunk.

In fishes, and probably also in Amphibia, the hypaxial musculature of the trunk derives from ventrolateral downgrowths from the dorsal myotomes, but in amniotes the somatic layer of the embryonic lateral plate may provide this system in part. The hypaxial muscles arise within the thin somatopleuric wall enveloping the coelom and thus, unlike their epaxial neighbors, are relatively thin and sheetlike. The history of the hypaxial musculature is, then, linked to the role it plays in the support of the viscera of the trunk, and complicating this history is the elaboration of ribs in all tetrapods, except modern amphibians.

Caudate amphibians present the basic pattern of the hypaxial musculature (Figure 12-12). Here we find a subdivision of the system on each side into three groups of muscles, reading as follows from top to bottom: (1) a *subvertebral group* beneath the transverse processes of the vertebrae; (2) a *lateral group* on the sides; and (3) a *rectus abdominis* group ventrally.

The developmental anatomy of the subvertebral group is quite conservative. Throughout the tetrapods it consists of a longitudinal band running the length of the vertebral column beneath the transverse processes. In this position it serves as the antagonist for the epaxial muscles powering the dorsoventral movements of the column.

Beginning with the urodeles, the lateral group comes to consist of three superimposed sheets of muscle. The most external sheet is the *external oblique muscle,* whose fibers course in a posteroventral direction. The middle sheet is the *internal oblique muscle,* whose fibers course in a posterodorsal direction. The inner sheet, abutting the coelomic peritoneum, is the *transverse muscle,* whose fibers run almost vertically. In the abdomen, these three sheets of muscle are maintained with little modification in all tetrapods, including mammals (Figure 12-15). But the presence of ribs in the thorax of amniotes creates complications. The external oblique becomes split into two layers and the transverse layer as well may be split (Figure 12-13). The layers also tend to be broken up into discontinuous units, with the result that the thoracic basket presents a maze of interlacing and overlapping muscles whose individual relationships and names are far beyond our concern.

The rectus abdominis group includes only the muscle of the same name. Ideally it runs on each side of a midventral tendinous area, the *linea alba,* from the pectoral

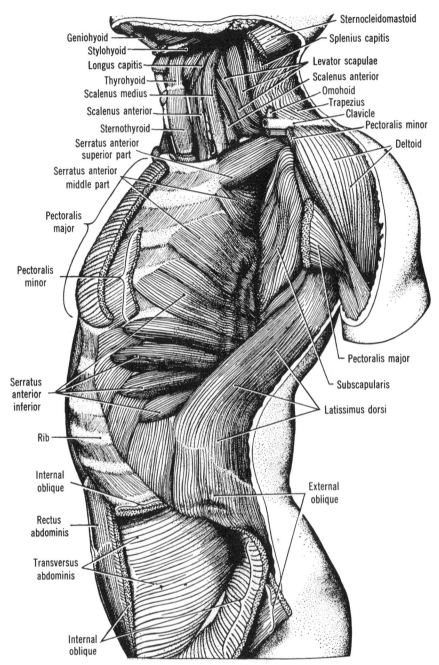

Figure 12-15   Deeper trunk and neck muscles of a human. (From Neal and Rand, *Comparative Anatomy*, by permission of Blakiston's, McGraw-Hill Book Company.)

girdle to the pelvic girdle, but with the elaboration of the sternum it becomes restricted to the abdomen (Figure 12-15).

Hypobrancial Muscles. In elasmobranchs the hypaxial portions of the occipitocervical myotomes send buds ventrally and anteriorly to form a muscular system running from the pectoral girdle to the underside of the pharyngeal skeleton. This basic situation prevails in all tetrapods. But with the working over of the hyoid and branchial arches, so that they ultimately (in mammals) become identified in the components of the larynx and hyoid apparatus, there is a concomitant differentiation of an assortment of hypobranchial muscles running from the sternum and shoulder girdle to the hyoid and to the thyroid cartilage of the larynx, and from these parts in turn to the lower jaw. Moreover, the developing tongue, which originates from the floor of the mouth and pharynx, also acquires its musculature from the hypobranchial system. (Figure 12-15). As a consequence of the source of this musculature in the occipitocervical myotomes, its innervation also traces to this source—an innervation that in amniotes is represented by the hypoglossal nerve (N. XII) and cervical plexus.

Eye Muscles. The six muscles that move the eyeball show a remarkable consistency of embryogeny and final form throughout the vertebrates. True, the three preotic myotomic primordia are often vaguely defined, yet there is no good reason to believe that there is in any tetrapod any consequential deviation from the pattern of embryogeny already described. And except for those few vertebrates with degenerate eyes, the basic adult pattern of six muscles innervated by three cranial nerves is constantly exhibited. Occasionally these six muscles may be supplemented by a retractor muscle that serves to pull the eyeball deeper into its socket, and lesser slips from the eyeball muscles may be associated with the upper eyelid or the nictiating membrane of the eye itself.

Appendicular Muscles. The history of the appendicular musculature is part and parcel of the complete changeover from the relatively voluminous axial muscles and modestly developed appendicular muscles of fishes to the reduced axial musculature and bulky, complex limb musculature of tetrapods.

The appendicular musculature includes all those muscles that (1) extend from the body wall to the paired appendages and/or girdles and (2) extend from one part of the appendicular skeleton (including girdles) to another. With a few exceptions, those muscles of the first group are derived from some division of the trunk musculature. Specific illustrations are the following: The *rhomboideus muscle,* anchoring the upper end of the mammalian scapula, comes from epaxial myotomes. Other muscles that in tetrapods serve as supporting slings for the pectoral girdle (which unlike the pelvic girdle is not articulated with the vertebral column) and the *pectoralis muscles* (Figure 12-15), spreading from the sternum and ribs to insert on the proximal end of the humerus, are special modifications of the external oblique musculature whose origin is in hypaxial myotomes. The *trapezius complex,* running from the rear of the neurocranium and the spines of the cervical and thoracic vertebrae to the front of the shoulder girdle and (in many mammals) to the humerus, is a product of the gill arch levators of fishes and thus traces to the branchiomeric system.

The basic architecture of the muscles of the second group is presented in fishes where, as we have seen, each of the paired fins is equipped with a dorsal levator and ventral depressor. Although in fishes these dorsal and ventral muscle masses derive from dorsal and ventral slips from the adjacent axial myotomes, it has already been noted that in tetrapods the limb muscles differentiate *in situ* from the mesenchyme of the primordial limb bud. Nevertheless, the same basic pattern is evident. That is, as the bones of the limb skeleton differentiate, the mesoderm from which the muscles will take form aggregates into masses above and below the skeleton. These masses are surely the equivalents of the opposed dorsal and ventral muscles in fishes, and by their subdivision and differentiation all the complicated musculature of the tetrapod limb will arise. In general, the muscles originating in the dorsal mass become the extensors and those originating in the ventral mass become the flexors. Likewise, outgrowths from the dorsal mass toward the trunk furnish the abductor muscles, and these from the ventral mass, the adductors.

Branchiomeric Muscles. Consideration of the history of the branchiomeric muscles calls for a recollection of their layout in fishes (Figures 12-9 and 12-11). Remember that they are categorized in four series: (1) *constrictors* above and below the gill slits, (2) *levators* attached to the dorsal ends of the visceral arches, (3) *adductors* pulling the dorsal and ventral halves of the gill arches together, and (4) *interarcuals* bending the upper ends of the gill arches backward. The story, then, of the branchiomeric musculature is that of its modification in company with the profound working over of the visceral skeleton. In part, as gill breathers give way to lung breathers, it is a story of the elimination of muscles; in part, too, it is a story of the reduction of some muscles, the elaboration of others, and the conversion of still others to new uses as the visceral skeleton pursues its evolutionary course.

With respect to the first visceral arch, the three principal muscles of concern are the levator, adductor, and ventral constrictor. In fishes the levator serves to support and raise the upper jaw, but with the fusion of the upper jaw to the braincase in most tetrapods, this muscle becomes superfluous and is either abolished or exists only as a relic. The adductor, by contrast, is a prominent and important muscle in all gnathostomes, serving to pull together the upper and lower jaws, that is, the two halves of the first arch (which serve either as jaws or as the framework about which the jaws are formed) for biting or grasping purposes. In tetrapods, the adductor musculature is divided into two components: temporal and pterygoid. The former finally becomes subdivided in mammals into *temporalis* and *masseter muscles* which, originating on the side of the neurocranium and inserting on the lower jaw (mandible), serve to pull the jaw upward and forward (Figure 12-16). The *pterygoideus muscles* (Figure 12-17) represent smaller, deeper divisions of the original adductor and extend from the pterygoid region of the palate to the medial surface of the jaw.

As the foregoing indicates, the support and elevation of the lower jaw are implemented in tetrapods by adaptation of the adductor system. The lowering of the jaw, however, is accomplished in a variety of ways. In fishes, as observed earlier, hypobranchial muscles run from the pectoral girdle to the underside of the jaw and mouth, and a backward pull of these muscles depresses the jaw. But in most tetrapods other than mammals an anterior extension of the dorsal constrictor of the hyoid arch does this job. A new and unique complication is introduced in mammals, where the lowering of the jaw is accomplished by a compound muscle with two bellies termed the *diagastric* (Figure 12-18). The digastric muscle features an anterior belly homologous to the ventral constrictor of the first arch, and a posterior belly derived from the ventral constrictor of the second arch.

We have just noted the employment of the dorsal and ventral constrictors of the hyoid arch in the musculature serving to depress the lower jaw. Keeping in mind, now, the conversion of the hyomandibular to one of the bones of the middle ear (stapes) and the incorporation of other components of the second arch into the

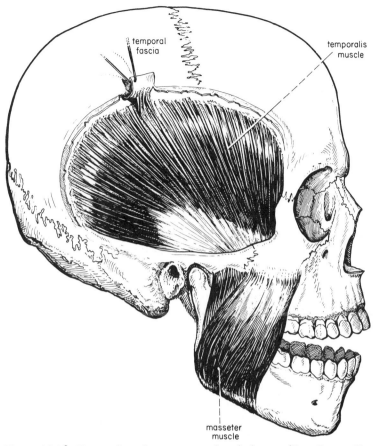

Figure 12-16 Temporalis and masseter muscles of a human. (From Berger, *Elementary Human Anatomy*. John Wiley and Sons.)

hyoid apparatus, we may observe these two other derivatives of the hyoid branchiomere. The levator is represented in mammals by the *stapedius muscle* and the ventral constrictor becomes the *stylohyoid muscle* (Figure 12-18). A third important derivative, this one from the dorsal constrictor, makes its first appearance in amphibians. This is the so-called *sphincter colli,* circling the neck ventrolaterally. Beginning with the reptiles, the sphincter colli associates itself with the skin where, in mammals, it has a spectacular history—a history that will be treated further in the coming consideration of the integumentary musculature.

The morphogenesis of the muscles of the gill arches may be disposed of briefly. At-

tention has already been drawn to the origin of the trapezius complex, that is, the trapezius proper and lesser components, from the gill arch levator. Other derivatives of the gill arch branchiomeres ultimately (in mammals) provide the musculature of the larynx and associated parts.

Integumentary Muscles. The history of the muscles of the skin really begins with the reptiles, for the skin of fishes is closely tied to the underlying skeletal muscles, and that of amphibians, although it lies relatively loosely over the underlying musculature, is virtually devoid of muscle in its own right.

The muscles of the skin fall into two groups: (1) intrinsic muscles, whose origins and insertions lie within the skin, and (2)

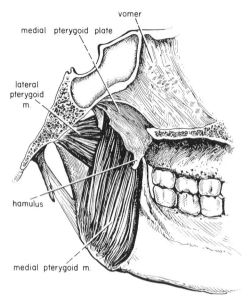

Figure 12-17 Medial and lateral pterygoid muscles of a human. (From Berger, *Elementary Human Anatomy*, John Wiley and Sons.)

extrinsic muscles, whose insertions are within the dermis but whose origins are on the subjacent axial or appendicular musculature. All the intrinsic muscles are of the smooth variety and are products of the embryonic dermatomes. Moreover, they are found only in birds and mammals: in the former they are the small muscles serving to erect the feathers, in the latter they are the small muscles serving the analogous function of erecting the hairs.

Except in snakes where special ventral slips of skeletal muscle are attached to and erect the scales so as to expedite movement, the integumentary muscle of living reptiles is confined to the earlier-mentioned sphinchter colli that encircles the neck region. It is important to recall that this muscle is branchiomeric in origin, indeed a

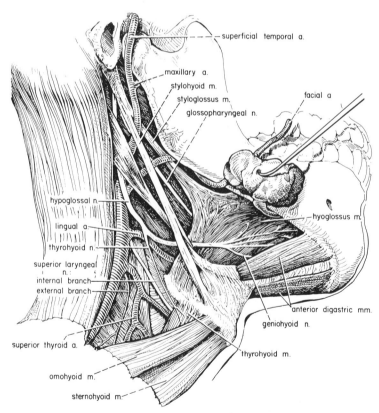

Figure 12-18 Deep muscles of the lower jaw and upper neck of a human. (From Berger, *Elementary Human Anatomy*, John Wiley and Sons.)

curious origin for a muscle having the remarkable history it does in mammals. For in mammals, where it is known as the *platysma,* it expands to cover much of the head and there comes to be subdivided into numerous components concerned with "facial expression." In humans, in whom facial expression (frowning, smiling, and so on) reaches its peak, the original single platysma is represented by more than 20 separate muscles.

In addition to the platysma, there is uniquely present in mammals a second integumentary muscle known as the *panniculus carnosus.* This muscle, which in primitive mammals ensheathes almost the entire neck and trunk, is myotomic rather than branchiomeric in origin. It is only slightly less extensive in eutherian mammals, except in primates where it has become reduced to vestiges in the armpit, groin, and the ventral surface of the chest.

Chapter Thirteen

The Alimentary Canal and Its Derivatives

Used in its broadest sense, the term *alimentary canal* refers to that internal tube which in all vertebrates runs from an anterior mouth opening to a posterior anal aperture at the base of the tail. The term *gut* also identifies this tube. Although the canal proper will call for some attention, we shall, in this chapter, give principal consideration to the various outgrowths and accessory structures derived from and associated with it.

EMBRYOLOGICAL PRELIMINARIES

The embryonic establishment of the gut is accomplished in one or the other of two ways. In the Cyclostomata, Holostei, Chondrostei, and Amphibia, vertebrates whose eggs feature a moderate amount of yolk ma-

terial, the primitive gut (*archenteron*) is one of the morphological consequences of gastrulation. Thus, as we have seen in Amphibia (Chapter 8), at the close of gastrulation the embryo consists essentially of two concentric tubes, an ectodermal one on the outside and an endodermal one on the inside, between which mesodermal constituents are interposed. But with the exception of teleosts, in which rather special events take place, all the other vertebrates initially exhibit a flat sheet of embryonic endoderm resting on a yolk mass and continuous peripherally with the endoderm of the prospective yolk sac. Even in eutherian mammals, despite a great reduction in the amount of yolk, this basic condition is found. It follows, then, that the rolling of the original plate of endoderm into a tubular gut calls for supplementary operations. These are operations linked to the general delimitation of the embryo from the adjacent extraembryonic areas. The avian embryo provides a good illustration of these events.

It will be recalled (Chapter 9) that the boundaries of the embryo are established through the intervention of body folds that progressively constrict the embryo from its extraembryonic annexments. The first of these is the *head fold* which, inaugurated as a transverse crescentic depression in the ectoderm immediately in front of the anterior terminus of the neural tube, steadily works downward and backward so that the future head becomes completely undercut (Figure 9-11). Meanwhile, the posterolateral margins of the head fold project themselves along the sides of the embryo as the *lateral body folds,* and shortly thereafter the posterior end of the body also becomes undercut by a *tail fold.* But our present concern is with the counterparts of these body folds that affect the embryonic endoderm.

Coincidentally with the inauguration of the head fold, the embryonic endoderm of the head region folds upon itself so as to produce an endodermal bay ventral to the neural tube. This is the *foregut* (Figure 9-11). Anteriorly the foregut ends blindly, but posteriorly it opens into the yolk-filled area beneath the yet unfloored gut by way of the *anterior intestinal portal* (Figure 13-1). Later, with the establishment of the tail fold, a concomitant endodermal folding produces the *hindgut,* which terminates blindly but opens forward by way of the *posterior intestinal portal.* The region of the prospective gut, which as yet has no floor and lies between the foregut and the hindgut, is designated the *midgut.* However, the midgut is destined to be eliminated because, with the progressive constriction of the embryo from its extraembryonic accessories, the foregut and hindgut become steadily longer (Figure 13-1). In effect, the foregut and hindgut develop at the expense of the midgut, and with the ultimate shriveling of the yolk sac and the closure of the yolk stalk canal, the midgut as such disappears.

As noted, when first set off from the yolk sac, the embryonic gut ends blindly both

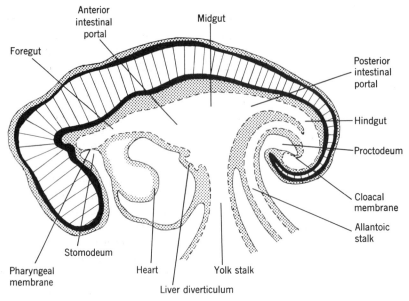

Figure 13-1 Longitudinal section of an amniote embryo showing the architecture of the embryonic gut.

fore and aft. Shortly, however, the somatic ectoderm on the underside of the body projects upward to contact the floor of the foregut near its terminus and likewise to contact the floor of the hindgut at the rear. (Experiments on amphibian embryos, in which foregut endoderm is excised, indicate that the ectodermal invagination fails to take place in the absence of the adjacent endoderm. That is, the endoderm normally provides some kind of stimulus essential for the inpocketing of the anterior ectoderm.) Initially rather shallow, these ectodermal indentations become deepened as a consequence of the flexure of the head and tail and thereby become conspicuous pockets. The one in front is termed the *stomodeum* or *oral cavity;* the one behind is the *proctodeum* or *anal cavity* (Figure 13-1). The plate of tissue created by the apposition of stomodeal ectoderm and foregut endoderm is known as the *oral,* or *stomodeal, plate* (*pharyngeal membrane*), and its counterpart at the rear created by the apposition of proctodeal ectoderm and hindgut endoderm is the *anal,* or *proctodeal, plate* (*cloacal membrane*) (Figure 13-1). Both plates rupture in due course, with the result that the primitive gut acquires anterior and posterior communications with the exterior. Stated differently, with the rupture of the pharyngeal and cloacal membranes, the stomodeum and proctodeum become integral parts of the alimentary canal; that is, short ectodermal units are added at each end of a canal whose epithelium is otherwise endodermal.

If we look ahead to the subsequent differentiation of this canal, the following regions along its length may be identified (Figure 13-2). At the front end is the *oral cavity,* or *mouth.* The epithelial lining of the mouth derives in large part from the original stomodeal ectoderm, but because, upon the disappearance of the oral plate, the boundary between the stomodeal ectoderm and gut endoderm becomes indistinguishable, it is difficult to determine where one leaves off and the other begins. Most investigators believe that the lining of the mouth includes an endodermal component, but it is not easy to decide on its limits and, in fact, there may be considerable variation from one animal category to another. Be that as it may, the mouth leads into the endodermal *pharynx* whose limits are defined by the paired pharyngeal clefts. The pharynx is continuous posteriorly with the relatively narrow *esophagus,* which in turn leads to the expanded *stomach.* An *intestine* follows the stomach and empties into the *cloaca,* a chamber also receiving the excretory and reproductive ducts. The *anal canal* is the ectodermally lined outlet to the rear. With reference to the original subdivisions of the embryonic gut, the foregut may be considered to be represented by those regions back to a level of the intestine where the liver and pancreas are destined to take form; the remaining units are assigned to the hindgut.

As we turn to the morphogenesis of these several components of the alimentary canal, it will be convenient to organize the discussion under three headings: (1) the mouth, (2) the pharynx, and (3) the digestive tube proper. The first two are considered separately, because their relationship

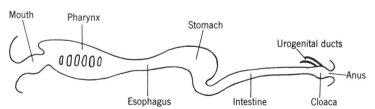

Figure 13-2 Representation of the subdivisions of the alimentary canal. (All accessory parts and derivatives omitted.)

to food processing is only ancillary; theirs is a complicated history involving respiratory structures and a variety of glands and other accessory parts. Conversely, all the components of the canal from the pharynx back to the cloaca are related to alimentation and thus may appropriately be treated as a unit. The cloaca and anal canal are omitted from the present account, because their history belongs primarily to that of the urogenital system (Chapter 15).

THE MOUTH

BOUNDARIES OF THE MOUTH

The first problem confronting us in beginning a study of the mouth is to define its limits. It would be convenient, if it were true, to equate it with the stomodeal invagination of somatic ectoderm. But we have already noted that posteriorly there is an indistinguishable gradation of stomodeal ectoderm into pharyngeal endoderm. This fact and the additional fact that endodermal contributions are made to certain oral accessories create difficulty in defining the rear limit of the mouth cavity. A comparable difficulty attends the setting of the anterior boundary, but the most useful landmark, or reference point, is the pair of nasal sacs that originate as invaginations of head ectoderm in vertebrate embryos.

The walls of the nasal sacs are the site for the olfactory organs, the structures concerned with the sense of smell. In elasmobranchs (Figure 13-3A) and in most other fishes, where the sacs end blindly, the external opening of each sac is ordinarily sub-

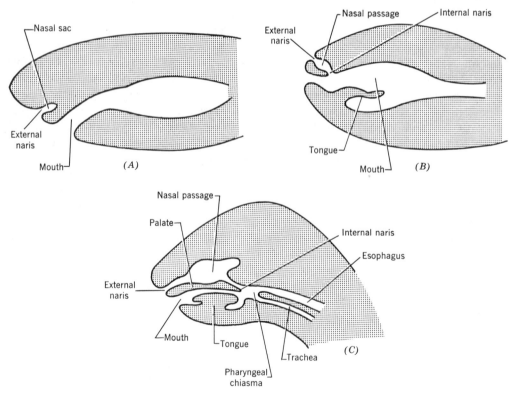

Figure 13-3 Relationships of the nasal passages to the mouth in vertebrates. (*A*) Shark, (*B*) amphibian, (*C*) mammal.

divided, so that olfaction is facilitated by a flow of water inward through one opening and outward through the other. The creation of interior openings (internal nares) to the mouth in air-breathing fishes presumably made olfaction more efficient by permitting a direct flow of water past the olfactory organs. So far as is known, no living lungfish employs these nasal passages to breathe air, but their presence in the immediate ancestors of the first amphibians made it possible for these amphibians and all other land forms since to utilize them for aerial respiration.

In amphibians, the internal nares open at the front of the mouth (Figure 13-3*B*), so that the mouth becomes a common passageway for air and food. All amniotes, however, feature some measure of subdivision of this common passageway: partial subdivision in birds and most reptiles, complete subdivision in crocodiles and mammals. In mammals the subdivision is provided by the secondary palate. Because of the palate, part of the original oral cavity becomes an air passage set off dorsally from the ventral mouth cavity proper (Figures 13-3*C* and 13-15), with resulting separation of breathing and food-manipulating operations. In effect, the internal nares are shifted far posterior to open into a very restricted area where food and air passageways remain confluent (Figure 14-3*C* and 13-15). The retention of even this limited region where food and air come together creates some mechanical hazards, as anyone who has choked on a bit of food can testify, yet it does provide the advantage of permitting an intake of air through the mouth in the event of obstruction of the nasal passages.

Except in cyclostomes, the mouths of vertebrates are, of course, bounded by the jaws, external to which are folds of skin constituting the *lips*. In fishes, amphibians, and most reptiles the lips are not very prominent; in birds, turtles, and some primitive mammals the lips are replaced by a horny beak or bill. In eutherian mammals, on the other hand, the lips are conspicuously developed, are separated from the jaw margins by deep clefts, and are rendered mobile by complements of the facial musculature.

DERIVATIVES AND ACCESSORIES OF THE MOUTH

The morphogenetic history of the mouth relates not only to it as a passageway, but involves an assortment of structures that either derive directly from its epithelial lining or become associated with it.

Anterior Lobe of the Pituitary Gland. Of all the endocrine organs of the vertebrate body, none is more important than the *pituitary gland (hypophysis)*. Anatomically, the pituitary consists of two lobes: a posterior lobe, or *neurohypophysis,* derived embryonically from the floor of the brain, and an anterior lobe, or *adenohypophysis,* derived from stomodeal ectoderm. The adenohypophysis is often described as consisting of three subdivisions, but some confusion attends the terminology applied to these, so we shall avoid trouble by not naming them. Suffice it to say that most of the hormones known to be produced by the pituitary gland are products of the most distal portion of the anterior lobe. These are the substances that exercise effects on the gonads and mammary glands, stimulate body growth, and influence other endocrine glands, such as the thyroid and cortical portions of the adrenals. Effects on the chromatophores of the skin are produced by substances coming from another region of the adenohypophysis. The somewhat lesser role played by the neurohypophysis will be considered in a later discussion of the brain (Chapter 17).

The anterior pituitary arises either as a solid bud or as a small pocket, *Rathke's pouch,* from the roof of the stomodeum just in front of the oral plate (Figure 13-4). This primordium pushes upward and comes to

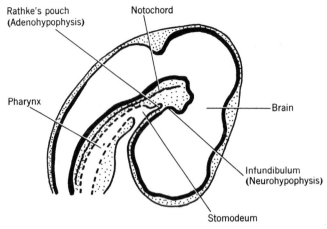

Figure 13-4 Rathke's pouch in an 80-hour chick embryo.

rest beneath and to the front of a downward-projecting pocket from the floor of the brain (diencephalon). In due course, Rathke's pouch loses its connection with the oral epithelium and proliferates the tissue constituting the anterior pituitary; concomitantly, the lobe from the brain differentiates as the posterior pituitary (Figure 17-18).

Oral Glands. Vertebrates exhibit a great variety of glands opening into the mouth. In fishes these are usually mucussecreting cells lying within the oral epithelium, but in tetrapods they develop as multicellular proliferations from the epithelium and present many forms and functions. The homologies among these glands are very uncertain, so that they are better considered in terms of their products. Within this reference, they fall roughly into two categories, *mucous glands* and *serous glands,* secreting lubricating fluids, enzymes, and poisons, either individually or in some mixture.

Amphibia possess only a few mucous glands, but reptiles are well equipped with both mucous and serous glands. The latter provide fluids, and occasionally enzymes, facilitating the processing and digestion of foods, and in some snakes and one lizard (the Gila monster) certain serous glands are poison producers. Instead of emptying directly into the mouth, the venom from the poison glands is poured out at the bases of modified teeth, from where it is transported by grooves or canals in the teeth. Most birds are also well supplied with both mucous and serous glands, but the oral glands find their greatest expression in mammals.

The most conspicuous of the oral glands in mammals are the *salivary glands* located in the cheeks and above and below the jaws (Figure 13-5). The salivary secretions, which are conducted to the mouth by long ducts, consist of mucus and watery fluid that moisten and lubricate food, and in a few mammals (including humans) contain a starch-splitting enzyme, *ptyalin.*

The Tongue. The tongue is more an adjunct to the oral cavity than a derivative, for, although it projects into the mouth and joins in the manipulation of food, its material source is largely external to the mouth.

Strictly speaking, the fishes have no tongue; it is represented at best by a fold of epithelium and connective tissue lying in the floor of the pharynx over the basal component (basihyal cartilage or bone) of the hyoid arch. But tongue or no, it is this tissue associated with the hyoid arch that represents one of the sources of the tongue in tetrapods. Thus, in amphibians this self-

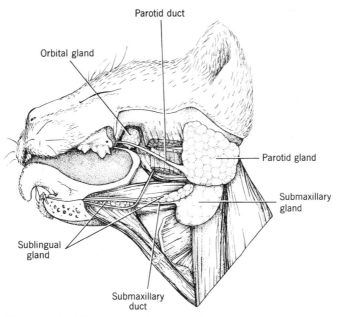

Figure 13-5 Salivary glands of the cat. (After Davison-Stromsten, *Mammalian Anatomy,* © 1947. Used with permission of McGraw-Hill Book Company.)

same tissue joins with a more anterior swelling in the pharyngeal floor between the embryonic mandibles to form a definitive tongue, which then, as a relatively bulky mass, projects into the mouth. The fully differentiated amphibian tongue becomes equipped with a modest muscular supply, but in this class of vertebrates extension of the tongue may be accomplished in part by the filling of blood and/or lymph sinuses within it.

In all amniotes the tongue is again inaugurated embryonically by two median swellings in the floor of the pharynx, one, now called the *copula,* between the hyoid arches and the following first branchial arches, and the other, the *tuberculum impar,* between the mandibular arches. In addition, a pair of *lateral lingual swellings* on the internal faces of the mandibles flank the tuberculum impar. So the definitive amniote tongue is compounded of four swellings, two median and two lateral (Figure 13-6A).

These primordia, relatively distinct in young amniote embryos, soon merge so completely that it is difficult to decide how much is contributed by each to the final adult tongue. In fact, the contributions of the primordia vary from one amniote class to another and possibly, too, among members of a class. In mammals, where the developmental anatomy of the tongue is best known, the greater part of the "body" of the tongue is a product of the lateral lingual swellings, whereas the base, or "root," is provided by the copula. The tuberculum impar also contributes to the body, but in what proportion remains a moot point. It is also difficult to settle on the embryonic source of the epithelium covering the various regions of the tongue. As the growing tongue pushes forward into the mouth cavity, it carries the stomodeal ectoderm with it, so that the greater part of the body becomes clothed with ectodermal epithelium. But it is not possible to distinguish where this ectodermal epithelium grades into the epithelium of endodermal origin that clothes the root of the tongue.

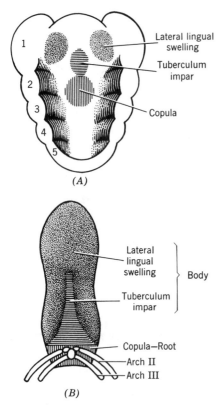

(A)

(B)

Figure 13-6   Development of the tongue of amniotes. (*A*) Embryonic primordia. (*B*) Adult tongue of a mammal.

Taken at face value, the foregoing account would identify the source of the entire bulk of the amniote tongue with the four embryonic swellings. In reality, however, only its epithelial and mucosal tissues derive from these primordia; the mass of the tongue comes from other sources. In the so-called "hard-type" tongue of birds, the hyobranchial skeleton extends throughout the length of the organ. Movement is provided by the muscles associated with this modified visceral skeleton, and these muscles are served primarily by motor fibers of the ninth (glossopharyngeal) nerve. Mammals and some reptiles, however, have a "soft-type" tongue, whose bulk consists of skeletal muscle tracing to the hypobranchial musculature, with a subordinate skeleton confined to the base. Because the musculature is a product of the occipito-

cervical myotomes, there is provided an explanation of the fact that the muscles receive their nerve supply from this area, that is, are innervated by the twelfth (hypoglossal) nerve. The origin of the root of the tongue from a level involving the hyoid and first branchial arch explains the employment of the hyoid apparatus as a basal skeletal support.

In addition to its motor innervation via the twelfth cranial nerve, the mammalian tongue is also generously supplied with sensory nerves serving the numerous tactile organs and taste organs (see Chapter 17) that lie in its epithelium. The source of these nerves once more reflects the origin of the tongue itself, namely, in conjunction with the first three visceral arches. As we shall see in Chapter 17, the fifth cranial nerve is associated with the first arch, the seventh nerve with the second arch, and the ninth nerve with the third arch. So it is, then, that the tongue receives its sensory innervation from branches of these three cranial nerves.

The principal role played by the tongue is in the manipulation of food. To this end it may, in many vertebrates, be elongated and greatly extensible and may function in the actual collection of food. But it also serves other ends. As already noted, it is the site for organs of touch and taste, and in humans it is an aid to speech.

Teeth. Although teeth are believed to have evolved from the tubercles on the dermal plates of armored ancestral vertebrates and thus in the final analysis belong to the integumentary system, they are inhabitants of the mouth and may appropriately be treated as oral adjuncts. All vertebrates, except the Agnatha, either possess teeth or have descended from toothed ancestors, so that the evolutionary history of teeth has been a long one and their architectural permutations have been extensive.

Variations and complexities of gross shape to the contrary, with few exceptions all teeth are built to a common plan (Figure 13-7). That portion of a tooth that is

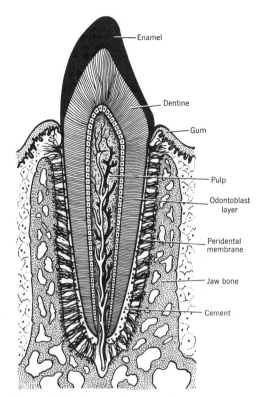

Enamel

Dentine

Gum

Pulp

Odontoblast
layer

Peridental
membrane

Jaw bone

Cement

Figure 13-7 Section of a mammalian canine tooth. (After Berger, *Elementary Human Anatomy,* John Wiley and Sons.)

exposed to the mouth cavity is sheathed externally by a layer of *enamel,* an exceedingly hard substance composed of elongate crystals of inorganic salts (primarily calcium phosphate) supplemented by an organic framework. Beneath the enamel is a layer of *dentine,* constituting the bulk of a tooth. Dentine is composed of collagenous connective tissue fibers embedded in a calcified ground substance called *hydroxyapatite.* Dentine is therefore chemically similar to bone, but it has a different structural configuration. Whereas in bony tissue bone cells are interspersed throughout its substance, the cells (*odontoblasts*) serving the dentine of a tooth lie in a layer immediately beneath the dentine and send fine protoplasmic extensions into the dentine via dentinal canaliculi. The interior of a tooth is occupied by *pulp* consisting of connective tissue, blood vessels, and nerves. In

the event that a tooth is anchored to or implanted within a jaw bone, as most teeth are, a bonelike *cement* clothes the dentine externally.

The embryogeny of a tooth shows a remarkable likeness to that of the skin and many of its derivatives in that the two principal tooth components, enamel and dentine, derive from separate embryonic sources. The enamel is a product of oral epithelium; the dentine is a product of embryonic connective tissue, mesenchyme. Reduced to its essentials, a prospective tooth is first represented by an epithelial cap overlying an aggregation of mesenchyme. The epithelial cap will secrete enamel and is therefore termed the *enamel organ;* the mesenchymal aggregation is termed the *dental papilla* which, after laying down the dentine, is retained as the pulp of the tooth.

The enamel organs of teeth are customarily said to originate in oral ectoderm; in fact, it is usually stated categorically that tooth enamel is an ectodermal product. However, in experiments on amphibian embryos involving the elimination of the stomodeum, teeth have been found developing in endodermal epithelium, as well, and a dual embryonic origin may extend to other vertebrates. As for the dental papilla, its mesenchymal cells appear to be part of the neural crest cells that produce much of the facial cartilage and musculature.

In the anamniotes, the growing dental papilla simply pushes into the oral epithelium, whose germinal layer concomitantly encases the papilla as the enamel organ (Figure 13-8A). The tooth is then formed in direct fashion: the enamel organ lays down enamel at its undersurface; the papilla deposits dentine about itself. In mammals, however, the fashioning of a tooth is considerably more involved. To start with, the oral epithelium over the embryonic jaws produces thickened ridges, which then push relatively deeply into the underlying mesenchyme. At the site of each prospective tooth, an epithelial bud is formed in the

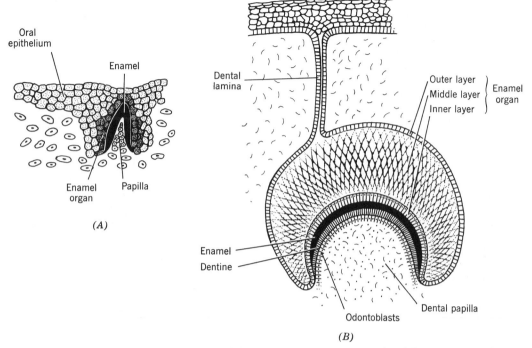

Figure 13-8   Embryonic primordia of a tooth. (*A*) Embryonic tooth in a salamander. (*B*) Embryonic tooth in a mammal.

ridge. Each such bud is an enamel organ. The enamel organ continues to push into the mesenchyme and, in so doing, takes on the contours of an inverted goblet enclosing a mass of mesenchyme, the papilla. At this time the enamel organ differentiates into three layers: an inner enamel layer surrounding the dental papilla, an outer epithelial layer, and a middle reticular layer (Figure 13-8*B*). It is the inner enamel layer alone that will produce the enamel of the definitive tooth. The cells of this layer become columnar and deposit enamel in the form of parallel prisms facing the dental papilla. The outer epithelial layer and middle reticular layer participate only as vehicles through which raw materials may pass to the inner layer, and with the completion of the tooth the entire enamel organ disappears.

Concomitantly with the laying down of enamel, the superficial cells of the dental papilla arrange themselves in a definite layer. The cells of this layer are the odontoblasts from the neural crest, and they proceed to deposit dentine against the inner face of the enamel. As the dentine increases in mass, the layer of odontoblasts retreats before it, but in the process the odontoblasts spin out threadlike fibers, which remain behind in small canals within the dentine. The entire odontoblast layer persists throughout the life of the tooth and may intermittently continue to deposit dentine. As for the more central mesenchyme of the dental papilla, it is retained as the pulpy framework for the support of nerves and blood vessels. The cement, if developed, derives from the general superficial mesenchyme investing the entire tooth germ. Originating as they do deep in the tissue of the jaws, mammalian teeth, in contrast to the more superficially forming teeth of other vertebrates, erupt to the surface only after they have almost reached full development.

Because of their hard mineral composition there is probably no chapter of vertebrate evolution more thoroughly documented than that pertaining to the teeth.

Whereas in fishes teeth are commonly found in the roof of the mouth as well as along the jaw margins, and even as far rearward as the branchial arches, there is a progressive limitation of the dispersal of teeth in the tetrapods. Thus, in amphibians and most reptiles the teeth are confined to the jaw bones and appear variably in the roof of the mouth (palate). Among the therapsid reptiles leading to mammals, palatal teeth are always missing; and in modern mammals, as an accompaniment of the reduction in the number of bones comprising the jaws, teeth are restricted to the dentary, premaxilla, and maxilla.

There has also been a general trend toward a reduction in the total number of teeth. Starting with the several scores present in fishes, the number declines steadily until in primitive mammals there are only 11 teeth on each side of each jaw (for a total of 44), and this number may be reduced even further in many mammals. The average human, for example, has only 8 in each jaw half. There are special instances, too, where the trend toward reduction reaches its extreme: complete elimination. Modern birds are totally devoid of teeth, as are turtles and some anurans. And there are many special cases, particularly among mammals, of the absence of certain categories of teeth.

In the anamniotes, new teeth are constantly produced during the life of the individual; as the earlier-formed teeth are worn away or lost, they are replaced by later generations. The tendency among the reptiles, however, is for regular replacement of teeth to occur only during the juvenile period of life, and at maturity a final set of permanent teeth is retained. With the mammals, the number of sets of teeth is reduced still further. The original mammals probably had two complete sets of teeth, but most modern mammals have only a partial replacement set, for the later-developing rear (molar) teeth have no successors, and in a few mammals only one set is ever formed. In still other mammals, for example, armadillos, anteaters, and baleen whales, embryonic teeth arise but fail to erupt.

The trend toward reduction in number of generations of teeth appears to be correlated with the nature and firmness of the anchorage of the teeth to the jaws. Where regular replacement prevails, the teeth are ordinarily attached relatively loosely to the inner surface of the jaw. But with the evolution of permanent teeth there came a firmer anchorage either on the crest of the jaw or, more commonly, through the implantation of the bases (roots) of the teeth in sockets. Socketing of the teeth is always characteristic of mammals and also is present in some reptiles, for example, crocodiles.

It is believed that the primitive tooth form is a conical one, and the vast majority of vertebrates have had and still retain conical teeth or some simple variation of the conical form. Yet the history of teeth is also marked by many adaptations of form and function. Some of the more notable ones are the triangular cutting teeth of some sharks, rounded or flattened crushing types in other sharks and some reptiles, grinding teeth in mammals, and an assortment of slashing or shearing forms in reptiles and mammals. Ordinarily, the teeth in any given animal are alike in form and the dentition is said to be *homodont.* When the teeth vary in form in various regions of the mouth, the term *heterodont* is applied. Heterodonty is particularly characteristic of mammals, for with few exceptions (for example, the toothed whales) mammals possess, in order from the front of the jaws to the rear, chisel-like *incisors,* pointed, conical *canines,* and *premolars* and *molars* with grinding surfaces of variable patterns. Most toothed nonmammalian vertebrates have a homodont dentition, although there are instances (for example, snakes and some lizards and fishes) of partial heterodonty.

THE PHARYNX

Perhaps the most characteristic feature of vertebrate development is the formation of the pharynx—the region of the gut tube between the mouth and the esophagus.

In embryonic stages of all vertebrates, the endodermal wall of the pharynx exhibits a series of paired outgrowths that project laterally to meet the ectoderm of the body wall (Figure 13-9). Each such outgrowth is termed a *pharyngeal,* or *visceral, pouch.* Where a pouch approaches the surface, there is a corresponding indentation of the somatic ectoderm to form a *pharyngeal,* or *visceral, groove* (*furrow*). This is more than a developmental coincidence because if, in some experimental fashion, the pouches are eliminated, the grooves fail to form; that is, the epidermal grooves are induced to develop by the endodermal

pouches. An important morphological consequence of the projection of the pouches toward the surface, and to a lesser extent of the ectodermal ingrowths, is a subdivision of the lateral plate mesoderm, so that it comes to consist of separate dorsoventral columns alternating with the pouches and grooves. These mesodermal columns, bounded externally by ectoderm and laterally and internally by endoderm, constitute the embryonic *pharyngeal,* or *visceral, arches.* The longitudinal series of paired pouches and the alternating arches give a segmental character to the pharynx.

It cannot be said with certainty how many pharyngeal pouches were present in the first vertebrates, that is, what the basic, primitive number was. That it may have been on the order of 10 to 14 is suggested by the fact that in modern myxinoids (hagfishes) the number ranges in different species from 6 to 14, and an interpretation of the fossil record of certain Paleozoic ostracoderms suggests 6 to 10. However, except in cyclostomes, where the lampreys customarily exhibit 7 pouches and the hagfishes may have as many as 14 most modern vertebrates have 6 or less. The customary number in the Chondrichthyes and Osteichthyes is 6; but in most amphibians there are only 5 and in amniotes only 4, the last one of which may actually represent a composite of two or three pouches.

The number of visceral arches varies with the number of pouches. Thus, in the sharks one sees 7 visceral arches alternating with the 6 pouches: the first, or mandibular, arch lying in front of the first pouch; the second, or hyoid, arch lying in front of the second pouch; and five succeeding gill, or branchial, arches. The hypomeric mesoderm (plus supplemental contributions from the neural crests), constituting the primordial arches in the embryo, will provide the skeletal arches and the associated branchiomeric musculature. Interestingly enough, the pharyngeal pouches have been shown to play an inductive role in the histogenesis of the skeletal arches: in the ab-

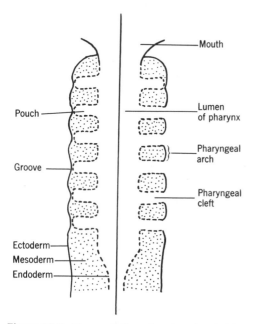

Figure 13-9 Pharynx in frontal section, showing the pattern of pharyngeal pouches, grooves, and arches.

sence of pouches, the arch mesenchyme fails to chondrify. Each differentiating arch is also provided with an aortic arch, an arterial vessel running from the ventral to the dorsal aorta, and one or more cranial nerve branches.

In all fishes and amphibians the endoderm–ectoderm contact areas (closing plates) of some or all of the pouches and grooves perforate in early embryonic life, so that the pharyngeal pouches open to the exterior. The cavity of each original pouch thus constitutes a passageway from the pharynx to the outside and is known as a *pharyngeal cleft,* or simply a *gill slit* (Figure 13-9). With the exception of the first one, which is commonly closed, and the last one, which shows some variability, the pharyngeal clefts of fishes are retained permanently. But in Amphibia the first pouch and usually the last (the fifth) do not become perforated, so that only three clefts are actually created in these forms. In most amphibians even these become closed once more at the time of metamorphosis, although in certain urodeles one or more of them may persist throughout adult life. As for reptiles, birds, and mammals, the embryonic pouches either fail to open externally or do so only temporarily.

RESPIRATION

The energy required to power the myriad of activities of an animal derives from the chemical breakdown of carbohydrates, fats, and proteins contained in cells. This breakdown calls for a consumption of oxygen in the mitochondria and ultimately results in the production of carbon dioxide. An animal's physiological economy is therefore dependent on arrangements that ensure a constant supply of oxygen and disposal of carbon dioxide.

The term *respiration* refers to the totality of events and conditions under which cells are supplied with oxygen, convert chemically bound energy into available

form, and release carbon dioxide. It is a term, however, that actually covers three different kinds of phenomena. It may refer to the mechanical operations that circulate and renew the two gas-carrying media, air and water, with which animals are associated. This is commonly spoken of as *ventilation.* Then again the term may refer to the oxygen-using biochemical transformations within cells, *cellular respiration.* Finally, it may signify the anatomical arrangements and processes that expedite the necessary exchanges of oxygen and carbon dioxide between the animal and its external environment. It is respiration in this context, *environmental exchange,* that is our principal concern.

There are two requisites that must be satisfied in a vertebrate if the exchanges between its cells and the environment are to be maintained. One is a transportation system that carries oxygen to and carbon dioxide away from the cells. The other is some kind of structural arrangement that will allow for the feeding of fresh oxygen supplies into the transportation system while carbon dioxide is disposed of at the same time. The transportation requirement is met by the blood-vascular system; the structural requirement is met by some form of extensive epithelial surface. So it is, then, that all the entities that in vertebrates are identified as respiratory organs possess two features in common: extensive surface and a voluminous blood supply.

Whatever the form of the particular respiratory organ, the operation of the transport-exchange system is based on the physical principle that a gas will tend to move from a region of higher concentration (tension) to one of lower concentration. The metabolism of a cell causes a depletion of oxygen and the production of carbon dioxide; but as blood possessing a relatively high oxygen tension and low carbon dioxide tension passes by the cell, the conditions for an exchange are created: oxygen will move out of the blood and into the cell, carbon dioxide will move from the cell to the blood. In the respiratory organ, conditions are reversed. The blood arrives with an oxygen deficit and a surplus of carbon dioxide, to be confronted across the respiratory epithelium with high oxygen and low carbon dioxide tensions. So once again diffusion exchange takes place.

The kinds of organs utilized by vertebrates to accomplish diffusion exchange are numerous indeed. The most primitive respiratory membrane was probably the skin, for it is the principal device utilized by cephalochordates and urochordates and is commonly employed by embryos and larvae of fishes and amphibians. We have noted that among adult amphibians the skin is still a major respiratory organ and may be employed to a slight degree in all vertebrates. Other less prominent sites are the epithelium of the mouth and pharynx, utilized by amphibians, and special pouches from the cloaca employed by freshwater turtles. In amniote embryos, the yolk sac and chorioallantoic membrane (or their placental counterparts) play a respiratory role. The two principal respiratory organs of vertebrates, however, are *gills* and *lungs,* the former designed to exploit water as a gas-carrying medium and the latter designed to exploit air. It is on gills and lungs, then, that we shall center attention; and because both ontogenetically and phylogenetically these organs are part and parcel of the pharynx, their story represents a chapter in the total history of the pharynx. But before turning to the specifics of these respiratory devices, two other matters deserve attention: (1) differences in the physical properties of air and water as respiratory media and (2) the general physiology of the oxygen-carrying pigment *hemoglobin* within the red blood cells.

A unit volume of air contains on the average about 20 times more oxygen than does a corresponding unit of water. That is, a liter of air contains sufficient oxygen to support an animal at least 20 times longer than a liter of water. A fish, then, must pass a far greater amount of water across its respiratory surface than does a terrestrial vertebrate breathing air in order to obtain 1 cubic centimeter of oxygen. Fishes are also at a disadvantage with respect to the effect of temperature on the oxygen content of water. Unlike air, the oxygen-carrying capacity of water declines rapidly with a rise in temperature. Thus, as its metabolic rate and need for oxygen go up with increasing temperature, a fish faces a declining oxygen supply. Water is also more dense and viscous than air and more energy is therefore required to move it. It has been estimated, for instance, that humans utilize about 1% to 2% of their oxygen intake for the work of pumping air, whereas a fish expends up to 25% of its oxygen intake to pump water.

With very few exceptions (some antarctic fishes, for example), oxygen is transported in a chemical combination with hemoglobin termed *oxyhemoglobin.* The advantage provided by this arrangement is revealed by calculations showing that if oxygen were transported in solution in the blood plasma rather than as oxyhemoglobin, humans would need either 30 times more blood or would have to circulate their supply 30 times faster for the needs of cells and tissues to be satisfied.

In keeping with the principle of diffusion, hemoglobin functions by giving up oxygen where the oxygen tension is low and absorbing it where the tension is high; in this respect, hemoglobins also vary as to the tensions in the tissues at which they *unload* oxygen. Another factor influencing the unloading of oxygen is the concentration of carbon dioxide (present in the form of sodium bicarbonate in the blood plasma or in unstable combination with hemoglobin), and hemoglobins also vary in the degree of their sensitivity to carbon dioxide. For example, fishes living in fresh water of high oxygen content bear hemoglobin that is highly sensitive to carbon dioxide and thus can survive only because ample oxygen resources counterbalance the tendency toward unloading, even with a modest rise in carbon dioxide concentration. Swamp-living fishes, on the contrary, have relatively insensitive hemoglobin and are therefore able to maintain adequate oxygen saturation in the face of relatively high concentrations of carbon dioxide. An important evolutionary implication derives from this circumstance. Although normal air contains 21% oxygen and only 0.03% carbon dioxide, the lungs of air-breathing vertebrates are never fully ventilated and thus contain considerably higher quantities of carbon dioxide, often as much as 5%. The point is that the evolutionary transition from water to land—fish to amphibian—must have been predicated on the mutational origin of a hemoglobin relatively insensitive to carbon dioxide. Fishes with sensitive hemoglobin cannot invade swamps and pools where carbon dioxide tension is high; a less sensitive type of hemoglobin must have evolved before terrestrial life and respiration by lungs were possible.

GILLS

We have already observed the manner in which pharyngeal pouches project laterally to meet ingrowths of somatic ectoderm and, in fishes and amphibians, break through to the exterior. Concomitantly, the lateral plate mesoderm becomes segmented and disposed as columns, the embryonic arches, between the pouches. In effect, then, each embryonic pharyngeal arch consists of a mesodermal core sheathed externally by an ectodermal epithelium, internally by an endodermal epithelium, and laterally by an epithelium deriving partly from the endoderm of the flanking pouches and partly from the ectoderm of the grooves (Figure 13-9). The sidewalls of the embryonic arches are, in other words, bounded by the epithelium lining the clefts, and it is difficult to say how much of this is ectodermal and how much endodermal. The situation is reminiscent of the developing mouth where, with the perforation of the oral plate, the limits of ectoderm and endoderm become obscure: the

rupture of each pouch-groove closing plate brings similar blending of ectoderm and endoderm.

Characteristic parts are now acquired by each embryonic arch. Internally (nearest the pharyngeal cavity), skeletal and muscular components differentiate and an aortic arch and cranial nerve branch are added. The more peripheral portion of the embryonic arch is converted to a sheet of connective tissue, which, as a *gill septum* containing intrinsic muscles and nerve fibers, extends outward. The epithelial sheathing of the sidewalls of the developing arch meanwhile becomes elaborately folded to form a stack of plates known as *filaments,* or *primary lamellae,* set at right angles to the long axis of the arch. The battery of filaments on one side of an arch constitutes a *hemibranch;* the two hemibranchs on the front and back side of an arch constitute a *holobranch,* or *gill* (Figure 13-10). Considered in another way, the hemibranchs developing from the epithelial walls bounding a given pharyngeal cleft are halves of separate holobranchs.

The basic structural theme just de-

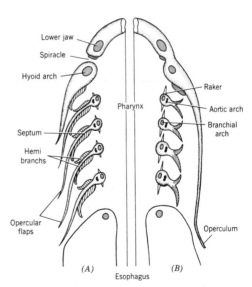

Figure 13-10 Frontal sections illustrating the gill arrangement in an elasmobranch (*left*) and a teleost (*right*).

scribed shows important differences in elasmobranchs and teleosts (Figure 13-10). Whereas in elasmobranchs the gill septum extends all the way to the skin and carries within it fine supporting rays of cartilage, the teleost septum is greatly reduced and the cartilaginous rays lie within the filaments. Attached as they are to the septa, the lamellae of shark gills project into individual branchial chambers (pharyngeal clefts), each of which leads independently to the outside via an opening covered by a skin flap. In teleosts, however, the filaments have basal attachments to the reduced septum, and all project into a common chamber covered externally by an *operculum* of membrane and bone.

Because water is the oxygen-carrying medium utilized for gill breathers, a mechanism providing for its circulation over the lamellae is called for. Some fast swimming teleosts (mackerel, for example) irrigate their gills simply by swimming continuously with their mouths open. But this is exceptional; branchial irrigation in teleosts usually depends on a complex of movements of skeleton and muscle. The mechanism is essentially that of the coordination of two pumps: a *pressure pump* in the mouth pushes water backward to the gills, and an opercular *suction pump* draws the water over the gills and sends it outward. Specifically, as the mouth opens in the breathing rhythm, the opercula close, and water enters the oral cavity. The mouth then closes, and the pressure rises in the constricted oral cavity, forcing the water into the pharyngeal cavity. The opercula and associated membranes open at the same time, so that the water is drawn over the gills and exits posteriorly. The alternation of interaction of these two pumps provides a fairly continuous flow of water.

The mechanism in sharks is basically the same, but it does differ in two ways from that of teleosts. The pressure pump again involves changes in the volume of the oral cavity (and pharynx) through closure of the mouth and constriction of the gill arches.

But instead of a pair of opercular suction pumps, there are five pairs governed by the skin flaps covering the separately opening branchial chambers. Another difference is the pressence in sharks of the *spiracles,* the reduced gill clefts between the mandibular and hyoid arches (Figure 13-10). Water reaches the pharynx by this route as well as through the mouth. It has been demonstrated that the water entering through the spiracles is pumped out mainly through the three anterior definitive gill chambers, whereas that entering by the mouth leaves by the three posterior gill chambers.

As water passes over the gills, gaseous exchange is facilitated by the operation of what is known as the *countercurrent principle.* The mechanism of this operation has its basis in the structure of the gills and in the relationship of the blood supply to it. The upper and lower surfaces of the aforementioned gill filaments are folded into a large number of *secondary lamellae* (Figure 13-11A). The afferent halves of the aortic arches send branches to the filaments, and by further branching of these vessels each of the secondary lamellae receives a capillary supply. These capillaries then lead to tributaries of the efferent halves of the aortic arches, which carry the now oxygenated blood to the dorsal aortae for distribution over the body. The functional arrangement is such that the flow of water past the secondary lamellae is in a direction opposite to that of the blood (Figure 13-11B). This pattern ensures that blood, which has already become partly oxygenated, meets water that is still loaded with oxygen. Thus it is possible for the blood to attain an oxygen saturation almost equal to that of the inhalant water.

The principle of counterflow has long been known and has been exploited by heating engineers, but its use by animals preceded human technology. Arctic animals employ it for temperature control, and it is known to operate in the placental circulation of mammals, in the kidney, and in the swim bladder of fishes.

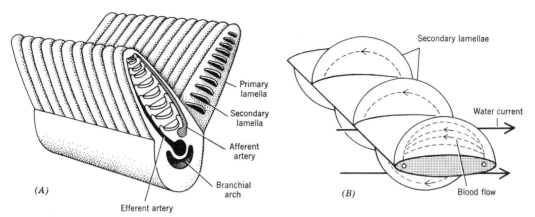

Figure 13-11 Structure and function of a teleost gill. (*A*) Primary and secondary lamellae. (After Hoar, *General and comparative physiology,* Prentice-Hall.) (*B*) Countercurrent system. (After Hughes, *Vertebrate respiration,* Harvard University Press.)

The gills just described, whose lamellae are derived from the epithelium (ectodermal and/or endodermal) bounding the pharyngeal clefts, are termed *internal gills,* and they are by far the predominant variety. However, *external gills,* formed as branching outgrowths from the external epithelium of the gill arches, also occur. These are found in larval Dipnoi and Amphibia and occasionally are retained permanently (for example, the urodele, *Necturus*). Thoroughly vascularized by the aortic arches and provided with muscles, external gills are simply waved in the water and no water passes through the gill clefts.

LUNGS AND SWIM BLADDERS

The first hint of lunglike structures is provided by an exceptionally favorable fossil specimen of a Devonian placoderm, a representative of the first group of jawed vertebrates. This primitive fish apparently possessed a rearmost pair of extended pharyngeal pouches, which failed to open to the exterior and were therefore capable of retaining air gulped in through the mouth. With the evolution of the higher categories of fishes, these pharyngeal air pouches then pursued three different historical pathways: (1) they were retained as devices for aerial

respiration, *lungs,* in the Sarcopterygii and a few actinopterygians; (2) they were specialized as *swim bladders* in most of the Osteichthyes; (3) they were lost in the Chondrichthyes and a few teleosts. Thus, despite the fact that lungs are most characteristically developed in tetrapods and tetrapods are younger than fishes, lungs did not appear on the scene with the tetrapods; they are phylogenetically much older and were undoubtedly present in the crossopterygian ancestors of the first land vertebrates.

Lungs. We have noted that in the one placoderm that has been amenable to analysis the last pair of gill pouches apparently served as primitive lungs. Add to this the fact that in some amphibians the lungs have a similar origin, and it becomes probable that primitively the lungs had a bilateral origin. In most air-breathing vertebrates, however, the lungs are initiated embryonically as a single median diverticulum from the gut floor just posterior to the branchial region, a diverticulum that then ordinarily becomes bilobed.

In the primitive chondrostean fish *Polypterus* (Figure 13-12*A*) the lung is a bilobed sac, with a short left and long right lobe, opening into the floor of the gut by a single duct. The Dipnoi, or lungfishes, pre-

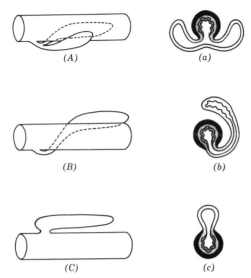

Figure 13-12 Lungs and swim bladders of fishes. (*A, a*) *Polypterus.* (*B, b*) *Ceratodus.* (*C, c*) A teleost.

sent a similar arrangement, especially the African lungfish, *Protopterus*, whose lung is essentially like that of *Polypterus*. In *Neoceratodus*, the Australian lungfish, the lung is a single sac lying dorsal to the gut, but still connected ventrally by a narrow duct passing down and around the right side of the gut (Figure 13-12*B*). Although the lungfishes possess internal nostrils, the nasal passages are exclusively subservient to olfaction; when branchial breathing is supplemented or superseded by lung breathing, atmospheric air is gulped through the mouth and taken into the highly vascularized lungs.

Except as noted earlier for some amphibians, the lungs of tetrapods arise embryonically as a pocketlike evagination from the floor of the gut (Figures 13-13*A, a*). As the pocket deepens, it gradually pinches away from the gut tube, except most anteriorly where it remains connected to and

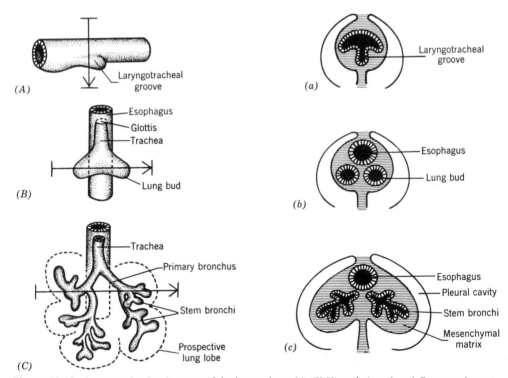

Figure 13-13 Stages in the development of the human lung. (*A–C*) Ventral view. (*a–c*) Cross sections at levels indicated.

in open communication with the rear of the pharynx. Meanwhile, the posterior end of the primordium bifurcates, and the two branches grow to the sides and rear (Figures 13-13*B, b*). The unpaired middle portion of the primordium becomes the *trachea,* whose opening to the pharynx is termed the *glottis.* The lateral branches (lung buds) give rise to the *bronchi* and *lungs* (Figures 13-13*C, c*).

Because the endodermal gut is embedded in a layer of mesodermal mesenchyme covered by visceral peritoneum, it follows that the developing lungs will also be invested by these materials (Figures 13-13*a–c*). That is, the endoderm produces only the inner epithelial lining; the mesoderm provides supporting connective tissue and smooth muscle and serves as a vehicle for blood vessels and nerves. When the mesenchyme from the early embryonic lungs

of the chick, the mouse, or the rabbit is removed, the branching of the epithelial component that normally ensues (Figure 13-13*C*) is disrupted. When the epithelium is recombined with mesenchyme, orderly branching is resumed. It is therefore concluded that the mesenchyme exerts an inductive influence on the endodermal epithelium.

The lungs of amphibians are essentially hollow sacs, sometimes smooth-walled, sometimes with an added respiratory surface provided by the development of supplementary ridges and septa from the internal epithelial lining (Figure 13-14*A*). The connective tissue covering is sparse and thin. In most reptiles, however, the internal septation becomes more prominent and involves the inclusion of greater and greater amounts of highly vascularized connective tissue in the partitions, so as to cre-

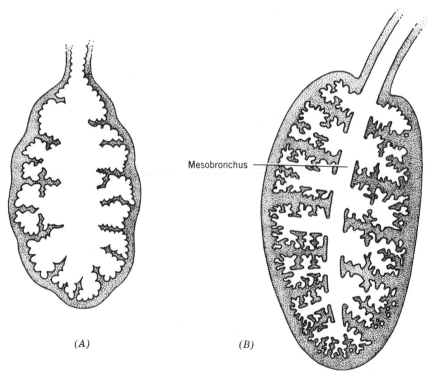

Mesobronchus

(A)   (B)

Figure 13-14 Lung types. (*A*) Frog. (*B*) Turtle. (Adapted from William F. Ballard, *Comparative Anatomy and Embryology,* © 1964 by the Ronald Press Co., New York.)

ate lungs of a decidedly spongy character (Figure 13-14*B*). A common arrangement in the reptilian lung is for the bronchus to lead into a roomy, nonrespiratory central chamber (*mesobronchus*), from which branches lead to finer passages and chambers where gaseous exchange takes place. In snakes and many lizards the lungs may also be extended posteriorly as thin-walled, bloodless air sacs that are disposed among the abdominal viscera. And snakes usually have a vestigial left lung and a very much elongated right lung.

The lungs of amphibians are force-filled by the swallowing of air, but in reptiles (with the exception of turtles) the lungs are alternately filled and emptied, as the volume of the body cavity is increased and decreased by the movements of the ribs by the trunk muscles. In turtles, whose ribs are

fused to the carapace, specialized abdominal muscles act to enlarge and contract the lungs.

The original embryonic lung buds of mammals divide again and again (Figure 13-13*C*), so that the definitive lungs come to consist of an elaborately branched "repiratory tree," whose fine terminal branches lead into countless tiny pouches (*alveoli*) where gas transfer with the equally elaborate blood supply takes place (Figure 13-15). The exchange of oxygen and carbon dioxide is facilitated by the enormous number of alveoli and the extreme thinness of their walls. It has been estimated that there may be as many as 350 million alveoli in each human lung, and the tissue that separates the blood in a capillary from the air in an alveolus is no more than half a micrometer in thickness. These structural

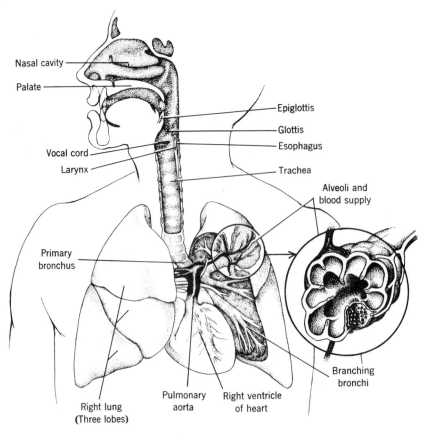

Figure 13-15 The human respiratory system.

features also expedite a number of non-respiratory functions. The mammalian lung acts as a filter that sieves organic and inorganic particulate matter from both air and blood; it contains a variety of vasoactive substances that regulate blood flow, and, in addition to the exchange of respiratory gases, it eliminates numerous volatile substances that are characteristically produced as metabolic by-products elsewhere in the body. The lung also has a remarkable capacity to cope with infection and disease. Inhaled particles and microorganisms are engulfed by several varieties of scavenger cells (macrophages), and antibodies conferring immunological protection are produced by lymphocytes within the lung and surrounding tissues. Finally, it has been demonstrated that the mammalian lung actively participates in both the synthesis and the breakdown of various organic substances, notably lipids.

The movement of air in and out of mammalian lungs is promoted in part by the movement of the ribs, as in reptiles, but is accomplished in greater measure by the movements of the diaphragm (Chapter 14). We have already had occasion to see how the channel for the intake of air is set off from the definitive mouth cavity by the palate. But immediately behind the palate there remains a limited region, where food and air utilize a common passageway. Here, the routes for air and food actually cross; food moves from the ventral mouth into the dorsal esophagus; air moves from the dorsal nasal chambers into the ventral trachea. The upper end of the trachea is surrounded by a complex of cartilages of hyobranchial origin, constituting the *larynx,* within which the vocal cords lie. The opening to it, the *glottis,* is guarded by a special muscularly controlled flap, the *epiglottis.* The remainder of the trachea leading to the bronchi and lungs is stiffened by cartilaginous rings in its walls. Figure 13-15 provides a convenient pictorial summary of the mammalian respiratory system as exemplified by the human.

Swim Bladders. The air bladders of fishes probably originated as modified pharyngeal pouches and were ventral in position. In the choanichthyan-tetrapod line they retained their ventral position and were exploited as lungs. But they disappeared in the Chondrichthyes, and in the Actinopterygii, notably in many teleosts, the paired bladder became single and moved to the dorsal side of the gut. By "moved" we mean to say that, so far as the scanty evidence goes, there was probably a gradual shift of origin from and connection with the gut from the ventral to the dorsal side, so that in all the teleosts in which it is present (with one exception) the air bladder arises as a dorsal sacculation. Unfortunately, there is only a single known case of a structural intermediate that directly supports this view of the migration of the air bladder: in the teleost *Erythrinus* the air bladder has a lateral attachment to the gut.

Although there are instances of the dorsal air bladder functioning as an accessory air-breathing organ (for example, *Amia,* a holostean fish), its principal role in teleosts is that of a hydrostatic organ. The quantity of gas in the bladder may be regulated so that the fish is able to adjust its specific gravity and maintain an appropriate depth in the water in which it is swimming; hence the appellation *swim bladder.* The only teleost fishes lacking a swim bladder are those that habitually live on the sea bottom (flounders and sole, for example) and a few rapid-swimming surface rovers.

In the more generalized groups of teleosts, the swim bladder retains connection with the gut via a duct. The duct is under the control of the autonomic nervous system and a well-defined reflex system opens and closes it for regulation of the volume of gas in the bladder. Gas may be expelled under water, but ordinarily the bladder can be filled only by the fish coming to the surface and gulping air. In the more specialized teleosts, the duct is lost during ontogeny, and the gas supply depends on the blood transport system.

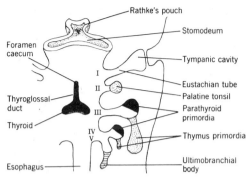

Figure 13-16 Ventral view of the embryogeny of nonrespiratory derivatives of the pharynx in mammals.

NONRESPIRATORY PHARYNGEAL DERIVATIVES

Although the pharynx is primitively subservient to respiration, even in fishes it is responsible for the provision of certain nonrespiratory parts, and with the relegation of respiration to the lungs in tetrapods, the pharynx becomes principally identified with the origination of these nonrespiratory accountrements (Figure 13-16).

Thyroid Gland. The first nonrespiratory pharyngeal derivative to be considered is the thyroid gland, a major endocrine gland that is "standard equipment" in all vertebrates. It originates embryonically as a midventral outpocketing from the floor of the pharynx at about the level of the first pair of pharyngeal pouches (Figure 13-17). In teleost fishes, clumps of thyroid tissue disengage from the original primordium and become dispersed along the ventral aorta, but, in the tetrapods, the primordium either remains single or, more commonly, is lobed. In the human embryo, for example, the initial median evagination expands and becomes bilobed. It gradually migrates back to the level of the developing larynx where it assumes its final position (Figure 13-18) and undergoes its characteristic histogenesis. Its site of origin is identified as the *foramen caecum* lying between the tuberculum impar and the copula of the tongue.

More than any other endocrine gland, the thyroid has great capacity for storage of its secretions. This is reflected in its histo-

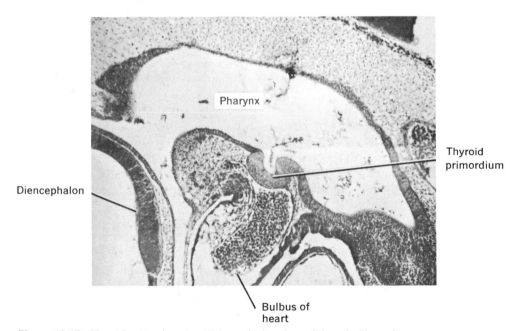

Figure 13-17 Thyroid primordium in a 72-hour chick embryo. (Photo by Torrey.)

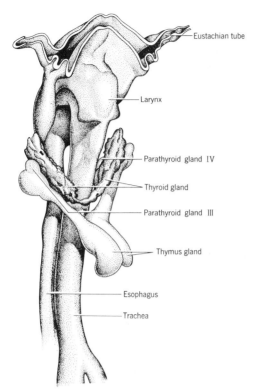

Figure 13-18 Pharyngeal region of a 23-mm human embryo. (After Weller.)

logical anatomy. The structural and functional unit of the thyroid in all vertebrates is the *follicle,* present in large numbers within the gland (Figure 13-19). Follicles are cystlike structures featuring a *secretory epithelium* consisting of a single layer of cells.

As a consequence of the secretory activity of the cells of the follicular epithelium, a sizable volume of *colloid* is deposited within the follicular lumen. The principal component of the colloid is a glycoprotein known as *thyroglobulin.* This is believed to be the storage form of the thyroid hormone whose tyrosine side chain is then iodinated to produce *triiodotyrosine* and *tetraiodotyrosine* (*thyroxine*), which pass into and circulate within the bloodstream. The bulk of the circulating hormone is actually thyroxine, so for all practical purposes thyroid hormone and thyroxine may be equated.

In warm-blooded vertebrates, thyroxine promotes greater oxygen consumption with attendant conversion of energy and production of heat. It also plays an important role in regulating early development and growth. These growth-regulating functions prevail in cold-blooded vertebrates as well (for example, the metamorphosis

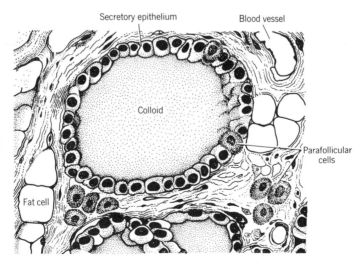

Figure 13-19 Microscopic anatomy of the thyroid gland.

of amphibians), but the general metabolic effects are not self-evident. The production of thyroxine is regulated by thyroid-stimulating hormone (TSH), secreted by the anterior lobe of the pituitary gland. The relationship is through a feedback system whereby an increase in TSH incites an increased thyroid output and increased thyroid hormone depresses the output of TSH.

There is recent evidence that the thyroid also produces a hormone other than triiodotyrosine and thyroxine (tetraiodotyrosine). It has been named *thyrocalcitonin* because its action appears to be that of lowering the level of blood serum calcium. In this respect it works antagonistically to parathyroid hormone. The evidence is not yet clear, but the aforementioned parafollicular cells are a probable source of thyrocalcitonin.

Derivatives of the Pharyngeal Pouches.
Beginning with the Amphibia, the first pair of pharyngeal pouches become intimately related to the developmental anatomy of the middle ear. Some consideration has already been given to the middle ear in conjunction with the history of the visceral skeleton, and we shall ultimately (Chapter 17) tie together all the threads of the story of the morphogenesis of the ear. Suffice it to say for the present that the distal end of each pouch expands to form the *tympanic cavity*, whose epithelial wall is wrapped about the ear ossicles. The proximal portion of each pouch retains its original connection with the pharynx and serves as a passageway known as the *eustachian tube* between the pharynx and middle ear (Figure 13-18).

The second pair of pharyngeal pouches tend to regress in air-breathing forms. It has been suggested that in mammals they contribute to one set of the several varieties of tonsils that arise in the pharyngeal region.

There are two types of glandular derivatives from the pharyngeal pouches found in varied form in all the vertebrates. The first is the *thymus gland;* the second is the category described by the term *epithelial bodies,* whose most important members are the *parathyroid glands* (Figure 13-18).

The thymus gland is found in all vertebrates except the Cyclostomata. It serves as a maturation center for a type of lymphocyte known as the T cell. The T-cell precursors originate, like other lymphocyte precursor cells, in the bone marrow. While in the thymus these cells mature and gain the ability to distinguish between "self" and "foreign" substances. The T cells are active in destroying certain types of tumors, and activating macrophages and antibody-producing (B-cell) lymphocytes. People born without a thymus gland, or people whose T cells are rendered nonfunctional (as by the virus causing the acquired immunodeficiency syndrome, AIDS) cannot get rid of infections and may die from common microorganisms. The thymus increases in bulk until the animal reaches sexual maturity and then steadily regresses, so that in old age virtually nothing but fat and connective tissue remain.

Several types of cells characterize the thymus (Figure 13-20), but most notable are lymphocytes (a feature of the lymphatic system), reticular cells, and macrophages. An object of speculation for many years, the mammalian thymus has recently been shown to play a major role in setting up the body's defenses against infection and the invasion of foreign tissues. It accomplishes this by serving as the original source of the lymphocytes that move out to colonize the spleen and lymph nodes. The function of the lymphocytes in these peripheral sites is to provide immunological barriers through the manufacture of antibodies. The thymus itself does not make antibodies; it is only the "seed bed" for the cells that perform this task and, once the spleen and lymph nodes have thus been seeded, the thymus regresses. There is also experimental evidence suggesting that the thymus produces a hormone serving to stimulate the lymph

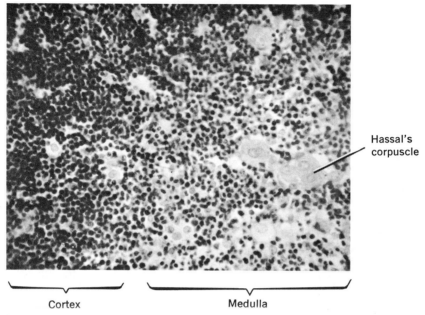

Figure 13-20 Microscopic anatomy of the thymus gland. (Photo by Torrey.)

glands to produce lymphocytes in their own right.

Like the thymus gland, the structures referred to collectively as epithelial bodies also involve various pouches. Most consistent in appearance are the parathyroid glands whose origin, form, and function are best known in mammals. So named because of their ultimate position alongside and more or less embedded in the thyroid gland, these are glands of internal secretion whose active principle, *parathormone,* regulates blood serum levels of calcium and phosphate ions by stimulating calcium resorption from bone and promoting phosphate excretion by kidney tubules. In the human embryo and other mammals, the parathyroid glands arise from the anterodorsal faces of the third and fourth pouches (Figure 13-18). Ultimobranchial bodies arise from the fifth pouches in most vertebrates, but no parathyroid-like function has ever been ascribed to them. It has been said that they may contribute tissue to the thyroid gland, within which they commonly become en-

gulfed, but this view is not universally accepted.

THE DIGESTIVE TUBE PROPER

From the rear of the pharynx back to its posterior exit at the anus, the alimentary canal is dedicated primarily to the processing of food. This, the digestive tube proper, has had a relatively conservative history and we shall do no more than point to its highlights and indicate the more notable functional and structural adaptations presented by various vertebrates.

There is a remarkable similarity among the digestive tracts of all vertebrates with respect to functional operations. The gut tube serves, first, for *transportation.* Food that is gathered by the mouth and pharynx is passed along the various sectors of the tube and the waste products are disposed of posteriorly. But the tract is more than a conveyor of food; it is also instrumental in the transfer of food resources to the tissues and cells of the entire body. To this end, it provides the necessary devices for physically breaking down raw food into particles of small size and chemically breaking it into absorbable molecular units. The *physical treatment* inaugurated by the jaws, teeth, and tongue (exclusive of filter feeders) is supplemented by contractions of the musculature of the tract, particularly the stomach, and by such special organs as the gizzard of birds. Fluid is also added to bring the food to a pasty condition. *Chemical treatment* is accomplished by enzymes produced by gland cells and gland masses. These enzymes hydrolyze carbohydrates to monosaccharides, proteins to amino acids, and fats to glycerol and fatty acids. Finally, there is *absorption* of the products of chemical breakdown for distribution by the circulatory system to all parts of the body.

These functional attributes are reflected in a basic histological architecture. Although highly variable from region to region, the digestive tube ordinarily consists of four layers, or tunics. These are depicted graphically in an illustration that includes a collation of some of the modifications that may be found along the length of the tube (Figure 13-21).

1. The innermost layer is the *mucosa,* and it features three regions. Lining the tube and therefore bounding its lumen is an *epithelial membrane.* It is this membrane that provides a variety of internal glands and is also the embryonic source of the liver and pancreas, which, although they lie outside the digestive tract, are connected to it by ducts representing the original outgrowths from the embryonic gut. It is the mucosal epithelium, too, that is thrown into a variety of folds that make for increased surface facilitating chemical digestion and absorption. Recall that, of all the components of the gut, only the mucosal epithelium is endodermal in origin; all the others have a mesodermal source. A *lamina propria* lies immediately be-

neath the epithelial membrane; it provides a variety of internal glands and is also the embryonic source of the liver and pancreas. Nodes of lymphatic tissue are also commonly found in the lamina, especially in the large intestine. A thin sheet of smooth *mucosal muscle* constitutes the outermost region of the mucosa and provides for independent, active movement of the entire mucosa.

2. The second tunic is the *submucosa.* It consists of a well-vascularized meshwork of connective tissue within which a plexus of nerve fibers and ganglion cells also lie. The nervous elements are components of the autonomic system.

3. The third layer is the *muscular tunic.* This is smooth muscle that is ordinarily disposed in two strata: an internal stratum with the muscle fibers oriented circularly and an external stratum with the fibers oriented longitudinally. The former serves to constrict the gut, the latter to shorten its length. These circular and longitudinal muscles are responsible for the peristaltic contractions that move food and waste along the tract and for the mixing movements of the stomach. Additional components of the autonomic nervous system lie between the two muscle strata.

4. The *serosa* (*visceral peritoneum*), consisting of coelomic epithelium and associated connective tissue, sheathes the digestive tract. (*See* Chapter 14.)

To speak of all the structural and functional permutations of the several sectors of the digestive tube of vertebrates would create an undesirable amount of detail. It is appropriate, however, to look to some of the major variations in gross anatomy (Figure 13-22) and also to translate the foregoing generalizations regarding functional operations and microscopic architecture into specific illustrations. To these ends, it is convenient to consider each of the subdivisions of the tract in order.

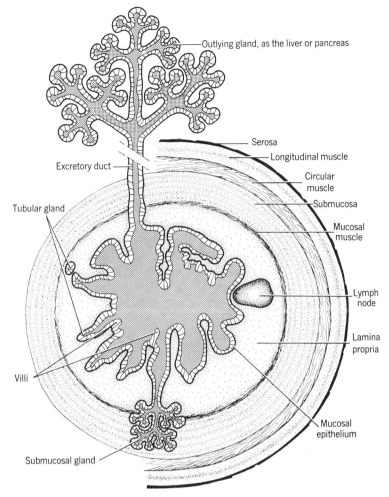

Figure 13-21   Basic four-layer plan of the digestive tract, including modifications and derivatives along its length. (Based on Fig. 7.1 [p. 148] in *Comparative Vertebrate Histology* by Donald I. Patt & Gail R. Patt. Harper & Row, 1969, by permission of the publisher.)

Esophagus. The esophagus may be defined as that segment of the gut leading from the pharynx to the sectors behind. Its limits are obscure in fishes, but in tetrapods it becomes circumscribed within the neck region and its length reflects the length of this region. Transportation of food is its essential function. Abundant mucus-secreting cells in the mucosal epithelium serve to lubricate the food which is moved along through the peristaltic contractions of the muscular tunic. Although the muscle is of the smooth variety in the majority of vertebrates, in many fishes and in mammals there is a notable amount of skeletal muscle. An interesting special adaptation of the esophagus is the saccular *crop* in birds, which serves as a food storage organ.

Stomach. Cyclostomes, chimaeras, lungfishes, and some teleosts lack a stomach, but all the other vertebrates have one. Despite many variations in form, the stomach consistently serves for storage of food, for the physical treatment of food, and for prelim-

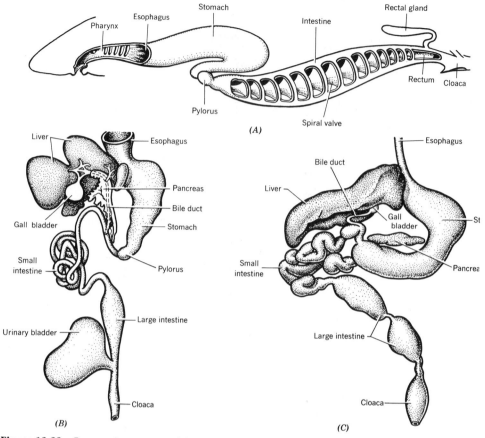

Figure 13-22 Comparative anatomy of the digestive tract. (*A*) Shark. (After Dean.) (*B*) Anuran amphibian. (*C*) Reptile. (*D*) Bird. (*E*) Mammal. (*A*, after Romer and Parsons, *The Vertebrate Body*, 4th ed., courtesy W.B. Saunders Co., Philadelphia. *B*, modified from Gaupp; *C*, from Potter; *D*, from Schimkewitsch.)

inary chemical treatment of food. The stomach may be essentially a straight, distensible sac or it may be bent and subdivided, with each especially named subdivision exhibiting distinctive functional attributes. Localized contractions of the stout muscular walls of the stomach churn and soften food for subsequent chemical treatment. The muscular tunic is usually fairly uniform throughout the stomach, but in crocodilians and birds the muscle is concentrated in a rear compartment, the *gizzard,* and especially designed grinding organ often containing sand or pebbles.

Extensive internal surface is provided by the elevation of mucosa and submucosa in longitudinal folds known as *rugae.* These are transitory, however, in the sense that they disappear when the stomach is distended with food. More significantly, the mucosal epithelium features numerous unicellular and compound tubular glands. Some of these secrete mucus; others produce a family of enzymes, collectively termed *pepsin,* that promotes a preliminary breakdown of proteins; still others secrete hydrochloric acid, which provides the acid environment favorable for the action of pepsin. A balanced production of hydrochloric acid is regulated by a hormone, *gastrin,* secreted by the mucosa according to the following scheme: the presence of food

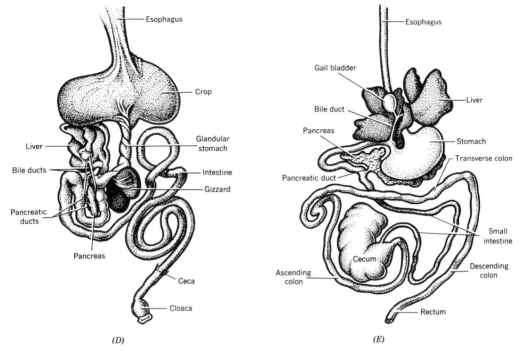

(D)   *(E)*

Figure 13-22 (*Continued*)

initiates local nervous stimulation of gastrin production, which, in turn, incites acid secretion; hydrochloric acid then serves as a feedback to inhibit gastrin secretion.

Intestine. It is in the intestine that the final events of digestion, completion of chemical treatment and absorption of food, take place. Once these events have been consummated, it is the intestine that resorbs excess water and passes waste material to the exterior. The anterior part of the intestine is largely dedicated to chemical processing and absorption, the posterior part to water recovery and waste disposal. Accordingly, the intestine is commonly delineated into *small intestine* (anterior) and *large intestine* (posterior). The demarcation, however, is very arbitrary; it is difficult to equate the sectors of the intestine of fishes with those of tetrapods and the final processing of food may be carried out in the large intestine. Moreover, the subdivi-

sions (and their names) of the mammalian intestine are based on human anatomy and cannot realistically be applied to other vertebrates.

The chemical treatment of food in the intestine involves a wide array of enzymes that fall into three groups: *carbohydrases* acting on carbohydrates, *esterases* and *lipases* acting on fatty materials, and *proteinases* acting on proteins. Many of these enzymes are products of the mucosal epithelium, but a sizable number derive from the pancreas which, as we shall see, is a special glandular outgrowth from the endodermal epithelium. These enzymes operate in the proximal part of the intestine (small intestine). But some may be carried rearward to complete their work in the large intestine. This sector is also populated with bacteria that break down food residues and, in the case of plant eaters, digest cellulose.

A primary requirement for efficient absorption is an extensive area of intestinal epithelium. This is provided in a number of ways and on three levels of magnitude. (1) In all vertebrates the mucosa is thrown into microscopically sized ridges and fingerlike projections (villi) (Figure 13-21). (2) Both the mucosa and submucosa may be elevated in folds of macroscopic size, as in the mammalian intestine. (3) Gross structural adaptations are utilized (Figure 13-22). Some vertebrates, notably teleostean fishes, feature numerous supplementary sacculations known as *caeca.* Sharks and a few other fishes possess a so-called *spiral valve,* a mucosal sheet traversing the intestine much as a spiral staircase descends the stairwell of a house. In vertebrates generally, however, extensive mucosal surface is provided by elongation and coiling of the intestine. Plant eaters tend to have longer intestines than meat eaters.

A number of intestinal hormones are involved in the regulation of the functional operations of the stomach, the gall bladder, the pancreas, and possibly the intestinal mucosa itself. At least they have been so demonstrated in mammals; little is known of their presence in other classes of vertebrates. One of these is *enterogastrone,* which inhibits the motility of the stomach and the secretion of hydrochloric acid. Another is *cholecystokinin* whose release is stimulated by the presence of fat in the intestine. This hormone elicits contraction of the gall bladder (when one exists) and the deposition of bile which is required for the emulsification of fat. A third is *secretin,* which stimulates the exocrine pancreas to release digestive enzymes. (There is evidence suggesting that the pancreatic juice elicited by secretin is low in enzyme content and that its enrichment depends upon a second hormone, *pancreozymin.*) It has also been claimed that other hormones are produced by the mucosa in the anterior intestine that activate enzyme production at lower levels, but the evidence is not complete.

THE LIVER

The liver is one of the two major organs associated developmentally and anatomically with the digestive tract. Its physiological operations, however, transcend digestion per se. It produces bile, which is both a secretion and an excretion. The secretory component consists of bile salts and other agents serving for predigestive emulsification of fats. The excretory components are primarily bile pigments derived from the hemoglobin of worn out red blood cells and surplus cholesterol and lecithin. It synthesizes glycogen from simple sugars and stores it in quantity. It converts other raw materials into fats and lipids. It is also the site of synthesis of plasma proteins, notably serum albumin, fibrinogen, and prothrombin, and undertakes the degradation of various toxins originating in the intestine. Therefore the liver is the most functionally versatile organ of the vertebrate body. This versatility is reflected in both its anatomy and its embryogeny.

Unlike other organs, the liver of any vertebrate has no characteristic shape. It tends to accommodate to whatever space is available in the peritoneal cavity; in slender animals it is slender and in plump animals it is plump. Yet, two developmental circumstances impose some regularity of form upon it. First, it originates embryonically at that level of the intestine where the ventral mesentery, largely destined to be eliminated elsewhere, is retained. Moreover, it grows into the ventral mesentery (Figure 13-23) and each plays a role in shaping the other (Chapter 14). Second, its site of origin is also that where the embryonic vitelline veins converge upon the rear of the developing heart and again there is a reciprocal relationship. The lobes and lobules of the liver are organized around the major blood vessels and the architecture of the hepatic portal system (Chapter 16) is conditioned by liver tissue.

It is upon its microscopic anatomy that

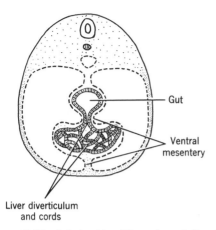

Figure 13-23 Relationship of the embryonic liver to the ventral mesentery.

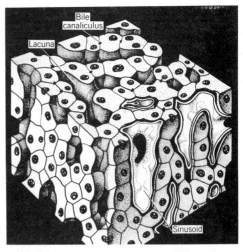

Figure 13-24 Stereogram of microscopic anatomy of human liver tissue. (After Elias and Sherrick, *Morphology of the Liver*, by permission of Academic Press.)

the great functional versatility of the liver rests. Although there are differences in detail in the several classes of vertebrates, the livers of all possess certain features in common. It is convenient to illustrate these essential features with the human liver, probably the most thoroughly studied of all, and then point to variants in other representatives. Human liver tissue is organized as a threedimensional, tunneled continuum of interconnected plates or walls of cells described by the term *muralium* (Figure 13-24). The individual plates are only one cell thick, but the geometric shapes (and volumes) of the individual cells vary with their location in the muralium. Although all of the cells are polyhedral in form, they may be five-, eight-, ten-, or twelve-sided. The spaces or tunnels between the plates are termed *lacunae* within which lie the blood *sinusoids* with which the liver is so generously endowed. The arrangement, then, is one wherein the faces of the polyhedral cells abut a maze of vascular channels, facilitating an efficient exchange of chemical products. The cells themselves are microscopic factories engaged in the fabrication of all the aforementioned products of the liver, including bile. The bile is ultimately deposited in *bile canaliculi,* which are merely tiny tubular spaces *between* the op-

posed walls of adjacent polyhedral cells. They all gradually converge toward coarser *hepatic ducts,* possessed of definitive walls of their own, which then lead to the *common bile duct* connected with the intestine. In most mammals, including humans, bile may be backed up and stored in a reservoir, the *gall bladder,* before exiting by the bile duct, but some species lack a gall bladder. The duct system itself is then used for the storage of bile.

THE PANCREAS

The pancreas is an organ common to all vertebrates. In the anamniotes it tends to be diffuse in distribution; in amniotes it is more compact. The gland ordinarily traces its embryonic origin to dorsal and ventral diverticula at the level of the developing liver. The former, the *dorsal pancreas,* arises from the dorsal side of the intestine; the latter, the *ventral pancreas,* arises from the base of the liver diverticulum, that is, from the primordium of the common bile duct. The dorsal and ventral pancreatic ru-

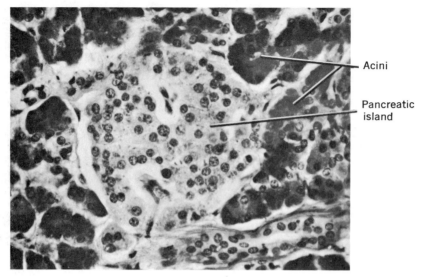

Acini

Pancreatic
island

Figure 13-25   Microscopic anatomy of the human pancreas. (Photo by Torrey.)

diments may remain completely independent, as in fishes and amphibians, but in amniotes the two units tend to fuse completely, with the dorsal pancreas ordinarily contributing the bulk of the adult gland. In humans, with the fusion of the two rudiments, the ventral pancreatic duct taps the dorsal one, which in turn largely atrophies, and the ventral duct, opening into the common bile duct, becomes the definitive adult *pancreatic duct.* In other mammals (for example, the pig and cow), the dorsal duct is retained as the definitive duct, and the ventral one regresses; in still others (for example, the horse and dog), both ducts are retained.

The pancreas plays a double-functional role that is reflected in its histogenesis. Its primordial cells rapidly form a branchwork of tubules at whose tips appear glandular *acini.* Meanwhile, clusters of cells bud from the tubules and proliferate as complex *pancreatic islands* interspersed throughout the duct system (Figure 13-25). The acini and the ducts into which they drain constitute the *exocrine* portion of the gland. Pancreatic fluid containing several digestive enzymes is produced in the acini and flows to the intestine through the duct system. The islands constitute the *endocrine* portion of the gland. They feature two main types of cells: (1) B (beta) cells, which secrete *insulin,* functioning to lower the level of blood sugar by increasing cellular uptake of glucose; and (2) A (alpha) cells, which secrete *glucagon,* functioning to raise the level of blood sugar. A cells are not very prominent in the mammalian pancreas, but are the predominant cell type in the pancreatic islands of birds and reptiles.

Chapter Fourteen

The Coelom and Mesenteries

The essential nature and manner of formation of the coelom and mesenteries was laid out in our earliest analysis of vertebrate embryogeny (Chapter 3 and Figures 3-1 to 3-4), and additional reference to it has been a necessary component of many discussions along the way.

A *coelom* is, by definition, a cavity bounded by an epithelium of mesodermal origin. This means that any space or cleft appearing within any division of the mesoderm whatsoever falls under this definition. Accordingly, the evanescent clefts that arise within the somites (Figure 12-7) and the lumina within the nephrotomes (Figure 15-1) are truly units of the coelom, and even the cavities of the blood vessels technically fall in this category. But *the* coelom, or body cavity, of a vertebrate is the space appearing within the hypomere (lateral plate), and it should be understood that this is the cavity under consideration in all the discussion that follows.

The coelom arises by the splitting of the lateral mesoderm (hypomere) on each side of the embryonic body and is, therefore, basically paired. If the embryo is a product of a mesolecithal egg, as an amphibian is, then it has a cylindrical form from the start and the two coelomic compartments originate in flanking positions. If, however, the early embryo is relatively flattened and is coextensive with extraembryonic areas, as in reptiles, birds, and mammals, and coelomic compartments are brought into flanking positions later as the embryo becomes constricted from its extraembryonic periphery and is rolled into a cylinder. In either case, the primitive coelom comes to consist of a pair of lateral compartments

311

extending from the level of the pharynx back to the level of the anal aperture. Although the coelom within the somites (myocoel) and nephrotomes (nephrocoel) is segmentally constituted, which may or may not suggest that the total coelom was originally metameric, the hypomeric compartments are never segmented in any vertebrate.

Externally, each coelomic compartment is bounded by the somatic layer of the hypomere which, abutting the developing ventrolateral body wall, provides the *parietal peritoneum* (Figures 3-4 and 14-1). Internally, the splanchnic layer of the hypomere on each side is applied to the wall of the gut and provides the investing *visceral peritoneum,* or *serosa* (Figures 3-4 and 14-1). Above and below the gut, the walls of the hypomere come together to form a double-layered sheet termed the *primary mesentery.* That part of the primary mesentery between the gut and the dorsal side of the coelom is designated the *dorsal mesentery;* that part between the gut and the ventral side is the *ventral mesentery* (Figure 14-1). If and when laterally developing organs (e.g., gonads and sexual ducts)

project into the coelom, they push the peritoneum ahead of them, and the peritoneum in turn converges behind them to form *lateral mesenteries.* In some instances, notably in the case of the kidneys, no mesenteries are established; the organs simply lie beneath the parietal peritoneum. They therefore lie outside the coelom and are termed *retroperitoneal organs.* Although we shall in due course have reason to consider certain lateral membranes as they participate in the subdivisions of the coelom, our immediate concern is with the general history of the dorsal and ventral mesenteries.

A moment's reflection on the preceding should bring the realization that actually none of the visceral organs lies within the coelom. Organs are either retroperitoneal, that is, they are situated external to the coelomic lining, or lie between the two layers of a mesentery. The mesenteries and peritoneum thus not only serve to support and/or cover organs (and in so doing provide access routes for nerves and blood vessels), but ordinarily wall off organs from the body cavity itself. Two notable exceptions to this isolation are the anterior funnels of the female sexual ducts (oviducts) which open

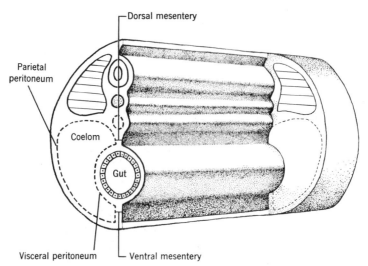

Figure 14-1 Stereogram of the trunk of a vertebrate embryo showing the basic architecture of the coelom and mesenteries.

to the coelom and the tapping of the coelom by the anterior kidney tubules of some vertebrates (p. 328 and Figure 15-5).

THE DORSAL AND VENTRAL MESENTERIES

We have seen that the coelom initially consists of a pair of lateral compartments separated by a midline partition, the primary mesentery. The gut lies between the two layers of the partition. Thus the mesentery is divided into dorsal and ventral components, and so long as the partition remains intact, the bilateral coelomic compartments are segregated. But in vertebrates generally the mesenteric partition is destined to be perforated to a greater or lesser extent in early embryonic life, in consequence of which the coelomic compartments become confluent. Generally speaking, it is the ventral mesentery that tends to be eliminated; with some exceptions, the dorsal mesentery remains largely intact.

With respect to the dorsal mesentery, it is descriptively convenient to speak of it in terms of subdivisions that are assigned with reference to the regions of the gut with which they are associated. That part of the dorsal mesentery supporting the stomach is identified as the *mesogaster;* that associated with the duodenum, the *mesoduodenum;* that associated with the colon, the *mesocolon;* and so on. Actually, the history of the various regions of the dorsal mesentery is quite conservative and largely reflects the maneuvers of the parts of the digestive tube to which it is attached. So it is that the subdivision and bending of the stomach, the lengthening and coiling of the intestine, and the outgrowth of various glandular accessories produce secondary foldings, fusions, and extensions of the dorsal mesentery with attendant anatomical complexity. The unraveling of these intricacies of the dorsal mesentery can be a challenging laboratory enterprise for students of developmental

anatomy, but it hardly seems necessary to undertake such an analysis here.

On the ventral side, as noted, the mesentery tends to become perforated and reduced in considerable degree. The developing heart, for example, lies originally in a segment of the ventral mesentery (mesocardium) whose ventral component (with reference to the heart) disappears almost as rapidly as it is formed and whose dorsal component disappears soon thereafter, so that the heart is left attached only by its vascular connections at each end. There are some exceptional instances of the ventral mesentery at other levels, but ordinarily the remainder of the mesentery becomes reduced to remnants. These are remnants, however, that are consistently retained in vertebrates and always involve the liver. One is the *falciform ligament,* which connects the anterior end of the liver with the ventral body wall. The other is the *gastroduodenohepatic ligament,* which runs between the liver and the stomach and duodenum. A third remnant appears in mammals as the *median ligament* between the urinary bladder and the ventral body wall. We shall have no occasion to deal further with the median ligament, but the hepatic portion of the ventral mesentery is inextricably linked to the history of the subdivision of the coelom to which our attention now turns.

THE PARTITIONING OF THE COELOM

To repeat, the coelom at first consists of a pair of continuous cavities extending the entire length of the trunk. With the extensive disappearance of the ventral mesentery, the two halves of the coelom become confluent beneath the gut. Although in theory the first stage in the evolution of the coelom would thus be exemplified by a vertebrate possessing a single, undivided body cavity, actually in the adults of all known vertebrates the coelom is again divided into at least two compartments. But the subdivision is now transverse rather than lengthwise, so as to provide a small anterior *pericardial cavity* containing the heart and a large posterior *pleuroperitoneal cavity* containing all the other viscera. The partition between these two coelomic cavities is termed the *septum transversum,* a structure whose embryogeny is linked to the development of the heart and liver.

The heart and that region of the coelom in which it lies have a bilateral origin. In essence, the heart originates in a pair of tubes, each lying in a separate coelomic chamber, which are extensions of vitelline veins from the yolk sac or its counterpart. The concomitant blending of these bilateral components creates the primordial heart tube and the pericardial cavity. Because the heart is connected from the beginning with the vitelline veins, there is no problem of venous drainage from this source. But the rise of the somatic veins calls for the creation of a route for them into the rear of the heart. This is provided by the extension of bridges of parietal peritoneum, *lateral mesocardia,* from the lateral body walls across the coelom to the serosa of the heart wall (Figure 14-2A). It is via the lateral mesocardia that the common cardinal veins receiving tributaries from the body reach the heart.

Simultaneously with the rise of the lateral mesocardia, the liver begins to take form. The liver is a diverticulum from the floor of the gut, and it so happens that it originates at the level of convergence of the vitelline veins on the heart, that is, immediately behind the lateral mesocardia. Growing downward from the gut floor, the liver lies within a persisting portion of the ventral mesentery. As the liver tissue increases in bulk, the ventral mesentery expands laterally to blend with the lateral body walls and also to fuse with the previously established lateral mesocardia (Figure 14-2B). The resulting rather thick membrane, a composite of the expanded ventral mesentery and lateral mesocardia, is the septum transversum.

The continued growth of the liver causes it to bulge posteriorly into the pleuroperitoneal cavity. In so doing, it carries the posterior wall of the septum with it, which thus becomes the serosa of the liver. By the same token, the ventral part of the septum, that is, the ventral mesentery, becomes longitudinally stretched and is identified as the falciform ligament. The septum transversum, therefore, not only provides the posterior wall of the pericardial cavity and the anterior wall of the pleuroperitoneal cavity, but also furnishes the falciform ligament and serosa of the liver (Figure 14-3). It should be added that whereas in amniote

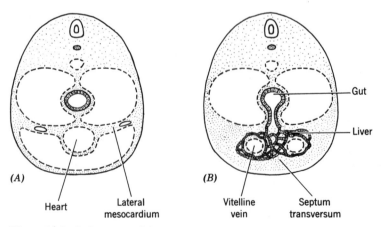

Figure 14-2 Embryogeny of the septum transversum.

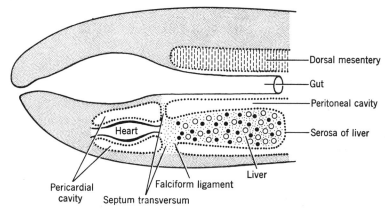

Figure 14-3 Coelom and mesenteries of a fish.

embryos (Figure 14-2*B*) and even in adults of elasmobranch and chondrostean fishes there are communications between the pericardial and pleuroperitoneal cavities over the top of the septum transversum, in fishes generally and in amphibians the dorsal passages are obliterated by the approximation and fusion of the gut and the dorsolateral body wall.

The conditions just described prevail not only in the Actinopterygii, but with slight modification also in choanichthyan fishes and amphibians, which exemplify a step toward the complications accompanying the elaboration of lungs. That is to say, in lung-fishes and Amphibia an anterior pericardial cavity is again partitioned off from a posterior pleuroperitoneal cavity by a septum transversum. The one difference is the projection of tubular or saccular lungs from the gut back into the pleuroperitoneal cavity (Figure 14-4). Although the lungs reside in no special compartments of their own—that is, they share the pleuroperitoneal cavity with other organs—they do, as they push out from the gut, carry a portion of the visceral peritoneum with them and thus, like every other organ, are encased in a serosa. In fact the serosa covering the lungs of urodeles is commonly extended to

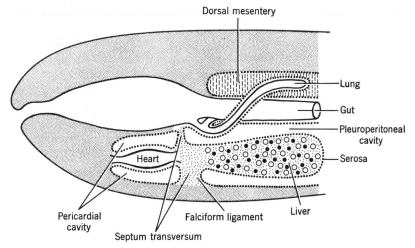

Figure 14-4 Coelom and mesenteries of a urodele.

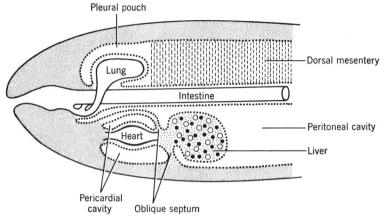

Figure 14-5   Coelom and mesenteries of a reptile.

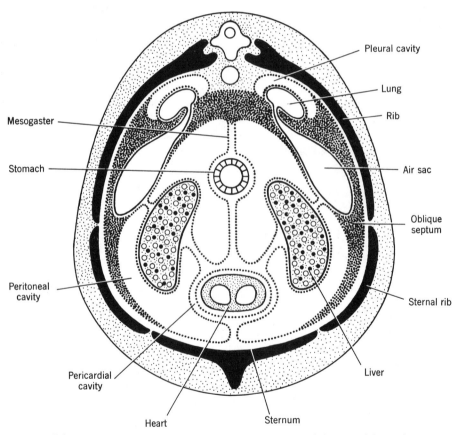

Figure 14-6   Transverse section of a bird showing mesenteries and division of the coelom.

form secondary mesenteries that tie the lungs to the lateral body walls, to the dorsal mesentery of the gut, and even to the liver.

In reptiles as a group, the coelom is also divided by a septum transversum into pericardial and pleuroperitoneal cavities. But in these forms, as an accompaniment of the reduction of the bulky pharyngeal apparatus, the pericardial cavity is no longer situated anterior to the pleuroperitoneal cavity. Rather, it has descended posteriorly

so as to lie ventral to the anterior part of the pleuroperitoneal cavity which, at the same time, pushes forward. In consequence of these relative movements, the ventral portion of the septum transversum is carried backward for a somewhat greater distance than is the dorsal part, and the septum assumes an oblique rather than vertical position (Figure 14-5). The lining of the pericardial cavity (parietal pericardium) is also split away from the body wall in whole or

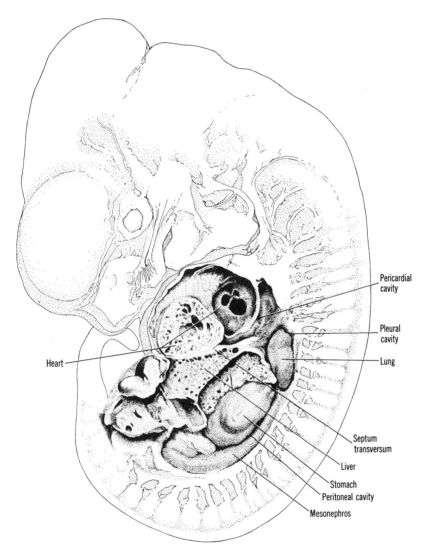

Figure 14-7 Lateral dissection of a $13\frac{1}{2}$-day-old rat embryo showing early continuity of the pericardial, pleural, and peritoneal cavities over the top of the septum transversum.

in part, so that the heart comes to lie in a separate pericardial sac (Figure 14-5). Most importantly, the anterior extension of the pleuroperitoneal cavity now becomes the site occupied by the lungs. In other words, this restricted portion of the coelom above the pericardial cavity will provide the individual *pleural cavities* housing the lungs. In some lizards a new incomplete septum partially separates the pleural cavities from the peritoneal cavity; in turtles and crocodiles the lungs are fastened close to the ribs

in a high dorsal position and are almost totally cut off from the peritoneal cavity.

The pleural cavities and lungs of birds (Figure 14-6) are walled off in essentially the same fashion as those of the Crocodilia, but conditions in birds are further complicated by the air sacs. Originating as outgrowths from the bronchi, the air sacs grow into the mesoderm constituting the oblique septum and from there push into the floors of the pleural cavities and also into the abdomen.

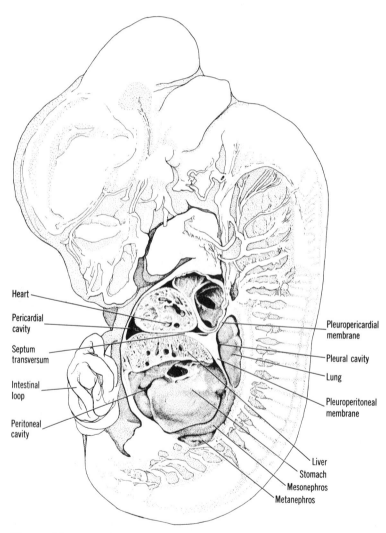

Figure 14-8   Lateral dissection of a 15-day-old rat embryo showing the partitioning of the coelomic cavities.

The first part of the story of the partitioning of the mammalian coelom involves events already described. As we have seen, the pericardial cavity is initially set off by the septum transversum, a composite membrane established by the fusion of the lateral bridges provided for the common cardinal veins entering the rear of the heart with the liver-expanded ventral mesentery. Because, of course, the rear of the septum is carried backward with the growing liver, it is the front wall alone of the original septum that serves as the preliminary partition between the pericardial cavity and the peritoneal cavity. But this is only a half-partition, for the coelom extends over it on each side of the gut and dorsal mesentery. That is, prospective pleural cavities lie dorsal to the pericardial cavity. The pericardial cavity initially communicates with the pleural cavities and these, in turn, with the peritoneal cavity behind (Figure 14-7). The situation might be likened to two rooms separated only by the lower half of a Dutch door. One room is the peritoneal cavity, the other is the pericardial cavity, and the door is the

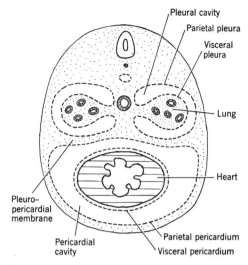

Figure 14-9 Cross section of a mammalian embryo with the lungs and pleural cavities dorsal to the heart and pericardial cavity.

septum. Thus the rooms are in communication over the top of the door, and because the lungs develop in the doorway itself, the area within which they lie is confluent with both rooms.

The next step is the closing off of the

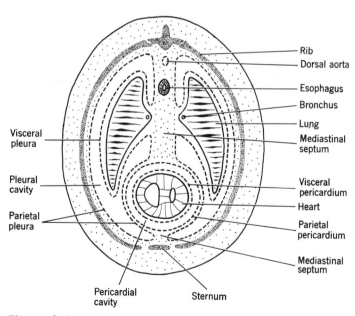

Figure 14-10 Cross section through the thorax of an adult mammal.

pleural cavities from the pericardial and peritoneal cavities. This is accomplished by two pairs of folds of parietal peritoneum that grow diagonally across the coelom from the lateral body walls to join the dorsal margin of the septum transversum. The more anterior pair of folds constitute the *pleuropericardial membranes,* which, as the name implies, cut off the pleural cavities from the pericardial cavity; the more posterior folds are the *pleuroperitoneal membranes,* walling off the pleural cavities from the peritoneal cavity (Figure 14-8).

As already emphasized, the pleural cavities and developing lungs are initially dorsal to the pericardial cavity and heart (Figure 14-9). But as the lungs increase in bulk, they grow ventrolaterally and accommodating space is provided by the concomitant expansion of the pleural cavities. As a consequence, the pleuropericardial membranes push ventrally, and the parietal pericardium is steadily split from the lateral body wall so that the heart, in its pericardial sac, becomes flanked by the pleural cavities (Figure 14-10). The onetime dorsal mesentery, which separates the pleural cavities in the midline, and the ventrally converged pleural peritoneum now constitute the *mediastinal septum* within which lie the gut, major blood vessels, and the heart in its pericardial sac (Figure 14-10).

With the incorporation of the pleuropericardial membranes within the mediastinal septum, the pleuroperitoneal membranes and septum transversum remain as the principal components of the partition that separates the peritoneal cavity from the thoracic (pleural and pericardial) cavity. This partition is the *diaphragm.* Unlike the membranous partitions in all the other vertebrates (except birds), the mammalian diaphragm becomes equipped with skeletal muscle derived from the lateral body walls. In terms of origins, then, the diaphragm consists of the following components: (1) anterior wall of the septum transversum, (2) dorsal mesentery, (3) pleuroperitoneal membranes, and (4) body wall musculature. The role played by the diaphragm and the rib basket in "breathing" has already been discussed (Chapter 13).

Chapter Fifteen
The Urogenital System

The urinary and reproductive systems are considered together because of their close developmental and anatomical association. They, in turn, assume an important relationship with the rear of the digestive tract, sharing with it a common outlet, the cloaca. For purposes of discussion, however, it is convenient to break the association into three parts—urinary, reproductive, and cloacal—whose interrelationships will then be considered.

THE URINARY SYSTEM

Physiological processes in cells and tissues result in certain products whose elimination (excretion) is necessary. Notable among these are various salts, the nitrogenous products of protein breakdown, and carbon dioxide and water resulting from carbohydrate metabolism. Many excretory devices are employed by vertebrates to dispose of these products. Carbon dioxide and some excess water are discharged via the respiratory organs; the sweat glands of mammals get rid of salts, some nitrogenous wastes, and water; and many marine fishes eliminate excess salt through their gills. But foremost as organs of excretion are the *kidneys,* which not only dispose of water and nitrogenous wastes, but operate to maintain a fine balance of water, salts, and other materials in the bloodstream. Our attention, therefore, centers on the kidneys and their associated ducts: the *urinary system.*

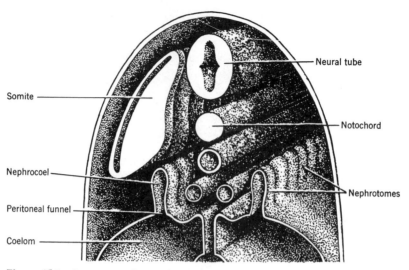

Figure 15-1   Stereogram of a vertebrate embryo showing nephrotomes.

THE STRUCTURAL AND FUNCTIONAL UNIT OF THE KIDNEY

The kidney, or *nephros,* of vertebrates is made up of many individual structural and functional units known as *nephrons.* The history of the kidney is thus the history of the developmental anatomy and physiology of this basic unit.

The development of the kidney begins with that portion of the embryonic mesoderm that has been designated as the meso-mere or intermediate mesoderm. Ideally this material becomes segmented, with each such segment being termed a *nephrotome* (Figure 15-1). Any given nephrotome contains a coelomic chamber, the *nephro-coel,* which opens to the adjacent general coelom by way of a so-called *peritoneal funnel.* It is the nephrotome that is the forerunner of a nephron.

The conversion of nephrotome to nephron involves the following events: from the dorsolateral wall of the nephrotome there arises a tubular outgrowth, *principal tubule,* which communicates with the nephrocoel via a *nephrostome;* the medial wall of the nephrotome comes to invest a tuft of arterial capillaries constituting the *glomerulus,* and the wall itself is then identified as a *renal (Bowman's) capsule.* Although it has been traditionally held that Bowman's capsule is established by a pushing of the glomerular tuft into the wall of the nephrotome, recent electron microscopy studies indicate that invagination does not occur. Rather, a cleft arises within the compact wall of the nephrotome, so as to produce an outer layer that becomes the capsular wall and an inner layer reflected over the surface of the glomerulus. The basic design of a nephron, as it results from these events, is shown in Figure 15-2.

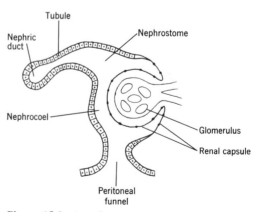

Figure 15-2   A nephron.

This excretory unit appears abruptly and full-blown on the evolutionary scene, having been identified in fossil imprint even in the oldest vertebrates known, the ostracoderms; however, little is known of its invertebrate origins. Whatever its evolutionary beginnings, it is this basic pattern of the excretory unit that, with variations, characterizes the kidneys of all vertebrates from cyclostome to mammal. Thus, by way of illustration, a single nephron in the human kidney (Figure 15-3*A*) consists of a greatly elongated principal tubule, differentiated into regions showing distinctive structural (and functional) details. At one end there is Bowman's capsule whose inner sievelike wall is intimately and intricately associated with the glomerular capillaries. At the other end the tubule, along with others, empties into a collecting system, which ultimately conducts the liquid waste (urine) through a drainage canal (ureter) into a storage reservoir (bladder). It should also be noted that the capillaries of the glomerulus communicate with a second network of vessels interlaced with the tubule proper. It is in conjunction with its elaborate and intimate blood supply that a nephron functions. Essentially three operations are in effect here, one having to do with the glomerulus and Bowman's capsule and the other two with the tubule proper and its capillary network.

The first operation involves a direct *filtration* process whereby noncolloidal solutions leave the glomerulus as a consequence of arterial pressure and pass into the tubule via the inner wall of the capsule. This wall consists of an epithelium of specialized cells called *podocytes,* because their cytoplasm is extended into long, footlike processes that subdivide into large numbers of smaller extensions which interdigitate with those of neighboring podocytes (Figure 15-3*B*). These cells are closely apposed to the basement membrane of the glomerular capillaries, which themselves are distended into very thin sheets containing many small perforations. The glomerular endothelium, basement membrane, and podocyte processes together constitute the filtration membrane through which all glomerular filtrate passes into the lumen of Bowman's capsule. Only the formed elements (for example, corpuscles) of the blood and large protein molecules are retained in the blood. All else, including nitrogenous wastes, salts, and food materials such as glucose, dissolved in a relatively large quantity of water, is passed into the nephron. It has been estimated, for instance, that the capsules of the roughly two million nephrons in the two kidneys of a human filter approximately 45 gallons of water solution per day. The obvious fact that nothing approaching this amount of final urine is produced daily, plus the additional fact that such unselective filtration alone would result in the loss of many valuable materials and do little to control the composition of the blood, directs attention to the other two operations. As the plasma filtrate passes down the tubule, the cells constituting the walls of the tubule go into action. Just what they do and how they do it varies from one vertebrate type to another. Moreover, there is still much that is not understood about the events that occur. In general, however, a two-way traffic between tubule cells and surrounding capillaries prevails. On the one hand, there is *reabsorption* of the bulk of the water, which is put back in the bloodstream, as are necessary food materials and salts; on the other hand, there is *secretion* whereby additional wastes are removed from the blood and added to the now much concentrated urine. In its regulation of this two-way traffic, the nephron exhibits truly remarkable capacities, creating what amounts to excretion thresholds for all the filterable elements of the blood. Substances in less than normal concentration in the blood are freed from the tubule and returned to the blood; those in greater than normal concentration in the bloodstream are removed in appropriate amounts. As one writer has summarized the work of a nephron, "the crude work is ... done by filtration from the glomerulus; the tubule

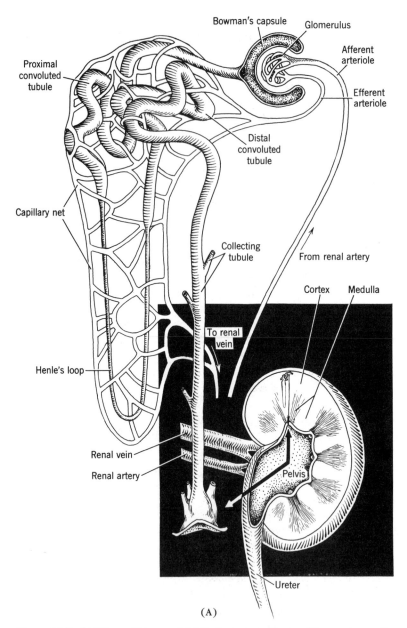

(A)

Figure 15-3 (*A*) Human kidney and detail of a single nephron and its vascular supply. (*B*) Detail of inner layer of capsule in contact with glomerular endothelium. This part of the capsule consists of *podocyte* cells (Po) whose cytoplasmic processes (↑) interdigitate within those of other podocytes to form a close cover over the glomerular surfaces.

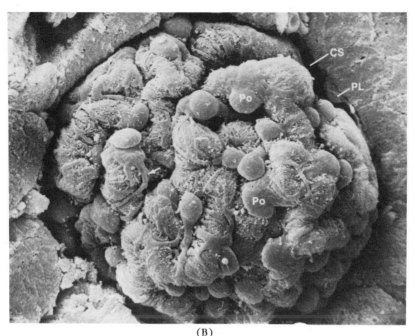

(B)
Figure 15-3 (*Continued*)

proper adds the necessary refinement to the process." The kidney tubules also play an additional role through the secretion of hormones in the regulation of blood pressure and sodium metabolism.

We have said that metabolic processes in cells and tissues result in a variety of byproducts, some of which must be disposed of, others conserved. A single substance may, in fact, at one time have to be voided and at another time conserved for metabolic purposes. Water is a case in point.

Although one of the refinements carried out by the tubule proper of the mammalian nephron is the restoration of all but surplus water to the bloodstream, the regulation of the water content of the body is not the same problem for all vertebrates.

We have come to think of many marine invertebrates as being in osmotic equilibrium with the sea, and thus facing no problem of water regulation. So it was presumably with the chordate ancestor of the vertebrates: its excretory system must have been little concerned, if at all, with water control. But the consensus is that the original vertebrates were freshwater forms, and this immediately posed the problem of excessive hydration. In keeping with the principles of osmosis, there would be a net flow of water into the body. Either this must be prevented or, once water has entered, the excess must be disposed of. One worker has made the interesting suggestion that the armor encasing the early fishes evolved as a protective waterproofing. But this alone was probably not enough, for the digestive and respiratory systems still afforded routes for the absorption of excessive water. Hence it has been presumed that the glomerulus and capsule originated as a device by which surplus water could be pumped out. Concomitantly, of course, the nephrons became capable of reabsorbing glucose, salts, and other materials, which otherwise would have been lost through the filtration system.

The return of some of the fishes to the sea and the evolution of land forms reversed the situation: water conservation rather than disposal became the problem, and the glomerulus became a liability rather than an asset. This difficulty appears to have been solved in a number of ways. Nearly all modern marine teleosts show more or less reduction in number and size of glomeruli, culminating in some forms whose nephrons are completely devoid of glomeruli. They also eliminate large amounts of salt through the gills, thus retaining fresh water. The elasmobranchs, on the contrary, employ an entirely different scheme. They have evolved a specialized segment of the nephron that reabsorbs urea (a nitrogenous waste) and returns it to the blood where it reaches concentrations of as much as 2.0% to 2.5%. This raises the osmotic pressure of the blood to a level slightly higher than that of sea water and brings about an inflow of water. Retaining their large glomeruli, marine elasmobranchs are in effect similar to freshwater teleosts in that they take in water and excrete copious amounts of dilute urine. Coelacanths (Crossopterygii) employ a similar device. Synthesized in the liver, high concentrations of urea are reached in the blood to maintain approximately the same degree of serum osmolality as that of sea water.

Terrestrial vertebrates (with the exception of Amphibia, which occupy a semiaquatic environment) also have a water conservation problem. The reptilian–avian line patterns itself after the marine teleosts by reducing the glomerular system, although no completely aglomerular kidney occurs. In the further interest of conservation, water is also reabsorbed through the wall of the cloaca and, in birds, by utilization of a segment of the nephron for water recovery. But more remarkably, it has been demonstrated that marine turtles, the marine iguana of the Galapagos Islands, and marine birds such as gulls drink sea water. Desalting of the water, and thus retention of fresh water, is accomplished in these forms by special glands in the head. These glands consist of thousands of well-vascularized branching tubules, which extract salt from the bloodstream and pour a concentrated salt solution into drainage ducts that either empty into the nasal cavities or pass directly to the exterior. It is interesting that the salt-secreting glands of the marine reptiles and birds have different embryonic origins. Those of birds are modified nasal glands, whereas those of reptiles are modified lacrimal (tear) glands.

The metabolism of protein calls for the elimination of nitrogen, which with relatively few exceptions is disposed of in the form of ammonia, urea, or uric acid. Free ammonia is extremely toxic and thus is excreted as such only when there is an abundance of water for its rapid removal. So it is that only those vertebrates living in fresh water and confronted with the problem of pumping out excess water that flush out nitrogen as ammonia. Animals that dispose of nitrogenous waste as ammonia are said to be *ammoniotelic*. But where water is at a premium, as it is for marine and terrestrial vertebrates, ammonia is converted to less toxic forms such as urea and uric acid. Urea-excreting animals are described as being *ureotelic;* uric acid-excreting animals are *uricotelic.*

It is generally agreed that ammoniotelism is the most primitive. But it also appears there has been an evolutionarily acquired adaptability of the cycle for protein degradation that can bring a shift to ureotelism. For instance, active lungfishes excrete ammonia but, when estivating, become uretelic; the aquatic larvae of the toad *Bufo* are 80% ammoniotelic, but are almost totally ureotelic as adults; and the African toad *Xenopus* is ammoniotelic while living in water, but becomes ureotelic when confined away from water. Urea not only offers the advantage of being relatively innocuous in moderate concentrations, but may even serve some useful physiological functions. It has already been noted that elasmobranch fishes retain urea for purposes of osmotic regulation, and in herbivorous mammals (cattle, for example) urea is secreted in the saliva and passes to the stomach where it serves as a source of nitrogen for the microflora in that organ.

Uric acid has the advantage of being highly insoluble. It can therefore be disposed of in solid form without the loss of water. So it is that uricotelism was one of the accompaniments of the emergence of the cleidoic egg (p. 146) and has always been successfully exploited by saurian reptiles and birds.

We have seen that the vertebrate nephron basically begins with a pressure filter (the glomerulus and renal capsule) and extends as a tube of variable length and structure, which carries the filtrate to the outside while reabsorbing useful substances from it and secreting additional substances into it. The real work of the nephron is therefore done in the principal tubule, and for this work a rich peritubular capillary network is essential. In the lower vertebrates the blood supply is venous and derives from the renal portal system (p. 394). In birds and mammals the renal portal disappears, and the peritubular network stems from the efferent glomerular arteriole (Figure 15-3A). It follows, then, that glomerular filtration can be eliminated in the interest of conserving water only in forms such as marine fishes whose peritubular blood supply is venous. When the capillary network is provided from the glomerular artery, the nephron is committed to glomerular activity, and other mechanisms for water recovery must be provided, notably through the tubule itself.

But water conservation is only one of the manifold operations of the tubule. It reabsorbs physiologically important solutes such as glucose, low-molecular-weight proteins, and chloride and secretes nitrogenous wastes, amino acids, and assorted salts. Many detailed studies involving analyses of the content of nephrons and the products ultimately discharged, tracer techniques, the clearance of various dyes and other indicators, and various chemical inhibitors have revealed much of the mechanism of nephron operation. Much remains to be learned, but the following generalizations appear to be valid: (1) all morphological variants of nephrons show a striking similarity in their physiological performances; (2) reabsorption and secretion depend on energy transfer requiring adequate oxygen supplies and the intervention of numerous cellular enzymes.

With this background of nephron structure and function behind us, we may now turn to the broader matter of the assembly of these basic units to form a total urinary system.

The original vertebrate nephros presumably consisted of nephrons that were alike in kind throughout the length of the organ, each opening independently to the exterior. But this arrangement is only hypothetical, for in all known vertebrates the nephrons empty into a common drainage duct (*nephric duct*) that passes back alongside the nephros and empties into the cloaca. This situation is further complicated in present-day vertebrates because of four variables. First, in the embryos of amniote vertebrates typical hollow nephrotomes are seldom formed; instead, nephrons differentiate without segmental arrangement within a continuous cord (*nephrogenic cord*) of intermediate mesoderm. Second, as the nephros develops embryonically, its entire length does not appear at one time; rather, the nephrons appear in sequence from front to rear, and the first-formed anterior tubules tend to disappear before the posterior ones arise. Third, the nephrons become progressively more complex from anterior to posterior. Fourth, the manner of establishing the nephric duct is not consistent in all vertebrates. Let us assess the results of these four variables.

THE NEPHROS OF FISHES AND AMPHIBIA

Embryonic development of the nephros is inaugurated within the most anterior mesomere. This intermediate mesoderm becomes segmented into nephrotomes, from each of which a nephron forms in the

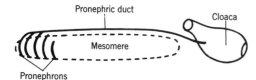

Figure 15-4 Developmental anatomy of the pronephros.

general fashion already described. The number of these first-formed nephrons varies with the species, but is always relatively small, usually in the order of three to five. These nephrons, because of their anterior position and by reason of being the first to appear, constitute the "head kidney" or *pronephros*. Accordingly, the nephrons themselves are termed *pronephric tubules* (*pronephrons*). As the first and most anterior pronephric tubule arises, it first extends dorsolaterally and then turns backward to join the one forming immediately behind. This one in turn joins the one behind it and so on, thus producing a common drainage duct designated the *pronephric duct*. The

pronephric duct, once initiated, extends itself backward along the still undifferentiated mesomere until it joints the cloaca (Figure 15-4).

With the exception of the hagfishes and some teleosts in which it persists throughout life, the pronephros has only a temporary existence. In the Chondrichthyes, for instance, it is present only during early embryonic stages and has little or no functional significance. But in those forms having an active free-living larval stage, such as amphibians and lampreys, it persists in the larva as a functional organ. The tubules not only show the general relationships to the circulatory system already described, but possess ciliated peritoneal funnels wide open so the adjacent coelom (Figure 15-5). This provides for the uptake of certain materials directly from the coelom as well as for traffic via the bloodstream. Direct demonstrations of the functional capacities of pronephric tubules have come from a variety of tests. For example, the pronephrons

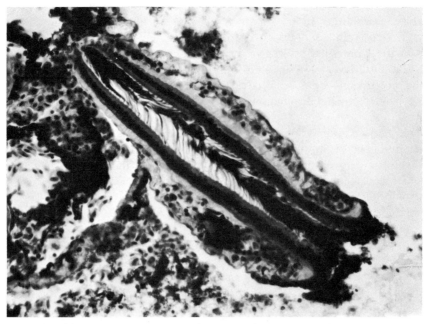

Figure 15-5 Photomicrograph of a portion of the pronephros of a larval lamprey showing uptake of coelomically injected carbon by a ciliated peritoneal funnel. ×175. (From Torrey, *J. Morphol.*, Vol. 63.)

of the larval lamprey will take up quantities of colloidal carbon, which may be injected within the coelom; through their ability to accumulate the dye phenol red, the pronephrons of frog larvae and certain teleosts have also revealed similar functional capabilities.

Whatever the length of its existence, the pronephros is supplemented and/or succeeded by a second generation of nephrons derived from the remainder of the mesomere. Although arising in basically the same manner and exhibiting the same fundamental structure, these later-forming nephrons tend to be longer and otherwise more complex in their makeup and they ordinarily lack peritoneal funnels. Unlike their pronephric forerunners, also, these nephrons fail to establish their own drainage duct. Rather, they join the already existing pronephric duct. Eventually, as noted, the earlier formed pronephros usually disappears, leaving this later generation of nephrons to constitute the final, definitive kidney. This organ, distinctive to fishes and amphibians, is known as the *opisthonephros,* or "back kidney," and the onetime pronephric duct that its nephrons have taken over is termed the *opisthonephric duct* (Figure 15-6).

THE NEPHROS OF REPTILES, BIRDS, AND MAMMALS

The embryonic initiation of the nephros of amniotes is customarily described as involving the establishment of a pronephros and a pronephric duct as in fishes and amphibians. Although this may be true for reptiles and birds (and even here there are reasons for questioning it), in mammals pronephric tubules rarely if ever appear. In humans, for example, that level of mesomere equivalent to the pronephros never gets beyond the point of conversion to a few rudimentary nephrotomes. Because definitive pronephric tubules are never provided, the nephric duct obviously must arise in some

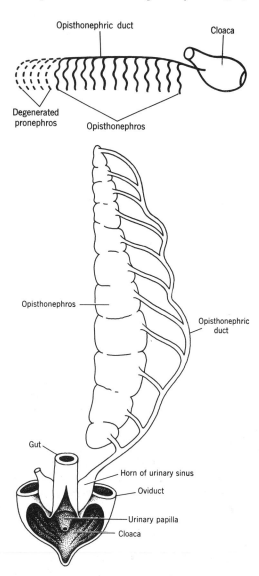

Figure 15-6 (*A*) Structure of the opisthonephros. (*B*) Ventral view of the excretory system of a female shark.

fashion other than by junction of the ends of pronephrons. Instead, we find that the original nephric duct is initiated as a solid rod, which splits off the dorsolateral side of the nephrogenic cord (Figure 15-7). For the human embryo this involves a distance represented by about four somites. Once established in this fashion, the solid "duct" frees itself from the parent mesomere and as a tapering rod extends itself backward

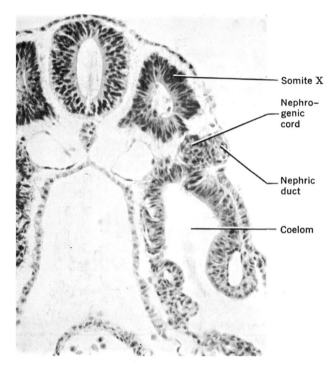

Figure 15-7 Origin of the nephric duct in a human embryo of 14 somites. × 150. (From Torrey, *Contributions to Embryology,* Vol. 35, Carnegie Institution of Washington.)

by independent terminal growth, ultimately contacting and fusing with the wall of the cloaca. In the meantime, the solid rod gradually hollows out and, within a few days of its original establishment, becomes tubular throughout.

This nephric duct is homologous with the pronephric or opisthonephric duct. In amniotes it is designated the *wolffian duct.* As it is being formed, the first of two generations of nephrons appears on the developmental scene. The first generation consists of a series of nephrons derived from the middle level of the mesomere. The details of their manner of development and final form vary from one category of animals to another, but in general they conform to the pattern described above. We may again use the human embryo as a specific illustration. Briefly, the nephrogenic cord paralleling the growing wolffian duct provides

an increasing number of serially arranged spherical bodies known as *nephric vesicles* (Figure 15-8). At first solid, these hollow out and send a *principal tubule* dorsolaterally to join the wolffian duct. The vesicle proper will provide the capsule surrounding the later-arriving glomerulus. Elongation and twisting of the principal tubule and the acquisition of a capillary network complete the nephron (Figure 15-9). These nephrons are known as *mesonephric tubules (mesonephrons),* and collectively they constitute the *mesonephros* or *wolffian body.* This kidney is well developed in reptiles and birds, but in mammals exhibits considerable variation. In the embryo rat (Figure 15-10), for example, it is quite rudimentary, and only a dozen or so abortive nephrons arise. At the other extreme, we may cite the pig embryo whose mesonephros is large and bulky and involves several

Nephric
duct

Principal
tubule

Somite XI

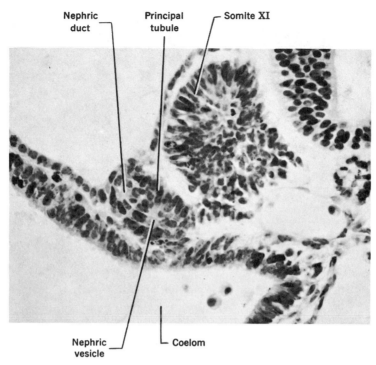

Nephric
vesicle

Coelom

Figure 15-8   Early differentiation of a mesonephron in a human embryo of 19 somites. ×400 (From Torrey, *Contributions to Embryology*, Vol. 35, Carnegie Institution of Washington.)

hundred long and convoluted nephrons (Figure 15-11). The human mesonephros lies between these two extremes.

The variable status of the mesonephros, especially in mammals, raises the question of the functional role it plays in the economy of the embryo. It is reasonable, of course, to infer function from exhibited structure. More convincing, however, are the direct experimental demonstrations of functional capacities of the mesonephrons.

The chick embryo has been a common subject for such tests. Solutions of indigo red and trypan blue injected into the vascular system ultimately turn up in the mesonephrons. Another approach has been that of cultivating fragments of mesonephros in a suitable culture medium to which an indicator such as phenol red has been added. The proximal portions of the tubules pick up the indicator and transport it to

their lumina; the distal portions resorb water. Still another method has been the direct identification of nitrogenous wastes deposited in the embryonic bladder, the allantois.

Similar experiments on mammals are

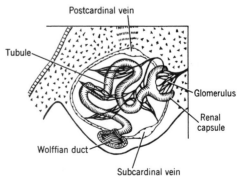

Figure 15-9   Fully differentiated human mesonephron.

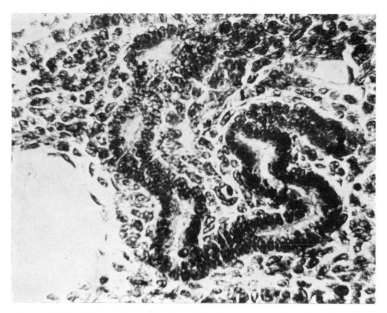

Figure 15-10   Photomicrograph of a rudimentary mesonephron in a rat embryo. ×400. (From Torrey, *Amer. J. Anat.*, Vol. 72.)

complicated by the intrauterine location of the embryo and its association with a placenta. Nevertheless, it has been possible to inject suitable indicators, either directly into the embryonic body or secondarily by transmission via the placenta from the maternal bloodstream. Such tests on embryos of the rabbit, cat, pig, and pouch-young opossum have provided positive demonstrations of the functioning of both the glomerular filter and the tubule proper. It is important to note, however, that the placenta itself also plays an important role in embryonic excretion, a role that appears to vary with the extent of development of the mesonephros. Or, conversely, the functional anatomy of the mesonephros appears to be correlated with the type of placenta at hand. The placenta of the pig is of the epitheliochorial variety (see Chapter 9), presumably the least efficient type and one of low excretory capacity; it has a relatively bulky mesonephros with full excretory powers. At the opposite extreme is the embryo rat with a rudimentary and, as far as has been determined, nonfunctioning

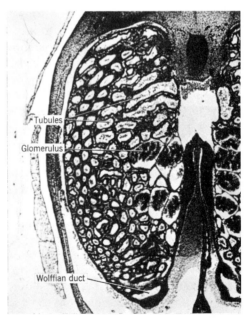

Figure 15-11   Photomicrograph of the mesonephros of a 10-mm pig embryo. Cross section. ×93 (Photo by Torrey.)

mesonephros; its maximally efficient hemochorial placenta must assume initially the entire burden of excretion. The human and all other mammalian embryos lie somewhere between these two extremes. It would be unwise to say which is cause and which is effect, but certainly there appears to be a correlation between placental type and degree of elaboration of the mesonephros.

Even as the mesonephros is attaining its maximum development, a second generation of nephrons is inaugurated. The history of this group is considerably more complicated than that of the former. Close to its entrance to the cloaca, a tubular outgrowth, the *ureteric diverticulum*, arises from the wolffian duct and pushes itself anteriorly into the still undifferentiated mesomere behind the mesonephros. The distal end of this diverticulum enlarges, and the mesomere coincidentally begins to condense around the enlargement. The proximal segment of the original diverticulum is the *ureter* or *metanephric duct*; the expanded distal end of the diverticulum is the *primitive renal pelvis*; the condensed mesomere around the pelvis is the *metanephric blastema* (Figure 15-12).

Subsequent events pertain primarily to the pelvis and blastema. The former exhibits a progressive series of subdiverticula, around each of which the blastema continues to condense (Figure 15-13). Within these blastemal condensations arise long, convoluted nephrons that, as they appear, open into the branches of the pelvis. These nephrons, then, are known as *metanephric (uriniferous) tubules* or *metanephrons*. Concomitantly, the pelvic diverticula elongate to form *collecting tubules*, all of which converge on a common chamber, the renal pelvis. The uriniferous tubules and collecting tubules together constitute the definitive kidney, or *metanephros*, of the late embryo and adult (Figure 15-3). The drainage duct is, as already noted, the ureter.

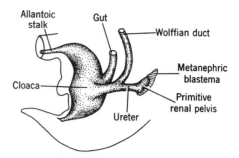

Figure 15-12   Origin of the metanephros in a 4-mm human embryo.

During the later part of embryonic life the mesonephros and metanephros function simultaneously. Gradually, however, the mesonephros regresses, and the metanephros assumes full responsibility for excretion. But this is not the end of the history of the mesonephros. It is destined to play a role in the development of the reproductive system, still to be considered.

CONCEPT OF THE HOLONEPHROS

Interpretations of vertebrate kidneys have for a long time been influenced profoundly by the view cultivated by the older German morphologists in the framework of the recapitulation principle (see Chapter 1). It has been held that the excretory system of amniotes (reptiles, birds, and mammals) consists of three sets of organs—the pronephros, mesonephros, and metanephros—which during ontogeny succeed each other in time and space, only the last being retained as the definitive adult kidney. Moreover, this ontogenetic succession has been presumed to parallel the phylogenetic order of events in the vertebrates as a group. Accordingly, the first kidney was a pronephros, which was succeeded by a mesonephros and then a metanephros. It is important to note, however, that for an equally long time there has existed an opposing view that looks upon the kidney as a single

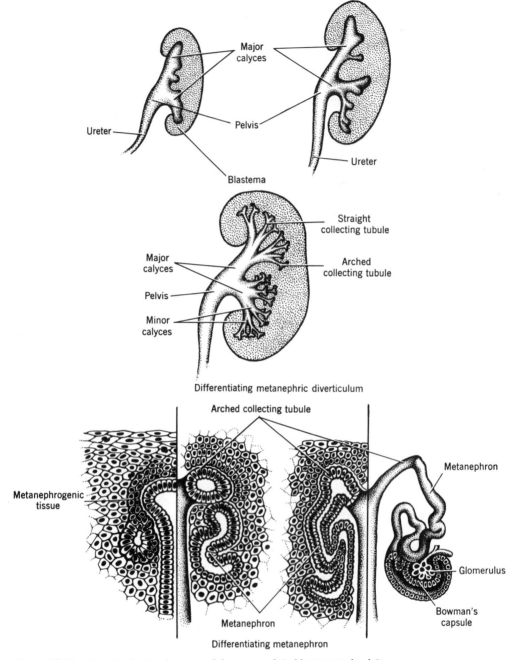

Figure 15-13   Stages in the development of the metanephric blastema and pelvis.

entity, that is, holonephros, parts of which may develop more or less separately, both temporally and spatially. By some curious circumstance, possibly because of its descriptive and pedagogical convenience or mere habit, the "tripartite" view of the German morphologists has tenaciously maintained its place in nearly all textbooks, even though the greater weight of anatomical and experimental evidence appears to favor the opposing view.

The tripartite concept requires that the three postulated organs be distinguishable and be set off one from another by differentiating criteria of one kind or another; the holonephric view denies the reality of such criteria, emphasizing instead the community of form at all levels, or at least the insensible gradation of one level of organization into the next. Facts such as the following support the holonephric assertions.

Although the original vertebrate nephros presumably consisted of nephrons alike in form throughout the length of the organ, this ideal is not realized in any known vertebrate. It is approached, however, in the embryo of the myxinoid agnath *Bdellostoma,* in which tubules arise from the nephrotomes over the distance of the 11th to 82nd somites, the distal end of each tubule uniting with the one behind to form the drainage duct. At the opposite ends of this organ the nephrons are distinctly different. Anteriorly, glomeruli are lacking, and the nephrocoels are wide open to the general coelom; posteriorly, the nephrocoels lack communication with the coelom, and well-developed glomeruli and capsules occur. But this distinction is apparent only at the extremes, for study of the whole reveals that one form grades into the other in an intermediate zone, within which assignment of nephrons to either category becomes impossible. It is only as *Bdellostoma* enters adulthood that the anterior aglomerular tubules, joined by two or three of the posterior glomerular ones, become a compact isolated pronephros.

A roughly comparable situation is found in *Hypogeophis,* one of the legless Amphibia. Nephrotomes of like structure are found throughout the trunk region of the embryo. Although this likeness is suggestive of linear uniformity, regional differences are destined to arise. Twelve nephrons, of which only eight or nine will develop fully, appear over the 4th to 15th somites. The first three of these units unite at their outer ends to form the nephric duct, which then grows back independently to join the cloaca. The remaining anterior tubules join the duct secondarily. There then occurs a gap, roughly from the 13th to 29th somites, where the nephrons are rudimentary, and fail to connect with the duct. This gap is followed caudally by a sequence of later-developing, duct-joining tubules.

These two cases have been presented in some detail so as to establish the important basic concept of progressing structural complexity along a continuum. The idea is that nephrons form ontogenetically in order from front to back, and originally all are alike. There follows a gradual increase in complexity, the first-formed anterior nephrons ultimately being set off as a discrete pronephros, but only for so long, apparently, as larval development and its accompanying physiological demands characterize the ontogenetic period. For, with a few exceptions, only the anamniotes with larval stages show the retention and elaboration of the anterior nephrons and their constitution as a pronephros. In the anamniotes without larvae, for example, the elasmobranchs, and in all the amniotes, the cranial end of the mesomere shows only the most rudimentary development; that is, a pronephros in the earlier sense no longer exists. At the same time the lower levels of the nephros show a steeper and steeper structural gradient. It has long been known, for example, that the posterior end of the nephros of adult sharks is considerably more complex than the anterior, and may in fact have a separate excretory duct. Similarly, in teleosts, there exists a whole series of types ranging from those with a nephros of uniform structure throughout to those in which a more complex rear portion, complete with separate blood supply and drainage duct, is isolated from the remainder. So it is also in amniotes, the chick embryo, for example, in which the last mesonephric tubules grade imperceptibly into the prospective metanephros behind. Although, therefore, the now well-established terms pronephros, mesonephros, and metanephros have considerable descriptive usefulness, the entities they describe may have no real significance other than as intergrading regions of a holonephros. The concept of the pronephros means little except in larval anamniotes; the metanephros has reality only as the end product of nephric sequence in amniotes. And the situation, as in fishes and Amphibia, of the simultaneous occurrence and/or structural continuity of the regions equivalent to mesonephros and metanephros, is better described by the term *opisthonephros*.

The essential wholeness of the nephros is reflected not alone by its morphology. Be reminded of the striking uniformity in physiological performance of all varieties of nephrons. To this may be added the revelations, provided by certain experimental analyses, of the interdependencies of the several parts of the developing nephric system.

As we have already seen, the original nephric duct grows back from the level of its origin to join the cloaca. In the normal course of events it is joined along the way by mesonephrons and thus becomes the wolffian duct. Suppose, now, that the backward extension of the duct alongside the prospective mesonephros-forming area is prevented. This can be accomplished readily in the chick embryo, for instance, by producing a minute wound or inserting a "roadblock" in the pathway of the duct. As a consequence, little or no development of the mesonephros will occur. This is interpreted to mean that the duct serves as an inductor of the mesonephric tubules; that is, differentiation of the mesonephrons depends on some kind of stimulus provided by the nephric duct. However, it is premature to accept the generalization that differentiation of vertebrate mesonephrons is *solely* dependent on induction by the primary nephric duct. Some inconsistencies in the results of the previously mentioned type of experiment have been reported. These may be a consequence of experimental error or it is possible that inductive influences emanate from such sources as the somites and the neural tube. Moreover, the presumed role played by the nephric duct in the induction of mesonephrons has been documented primarily by experiments on chick and amphibian embryos.

With respect to the metanephros, the facts of experimentation and observation are consistent and clear. If the elongating primary nephric duct in the chick embryo is blocked so that its posterior portion is lacking and as a consequence the ureteric diverticulum is also absent, even though a blastema appears, metanephrons fail to appear within it. This result conforms with those natural experiments leading to agenesis of a kidney in the human embryo. Congenital absence of a kidney is always accompanied by absence of the ureter on the corresponding side. The obvious deduction is that through failure of the primary duct to form properly, the ureteric diverticulum is lacking and no metanephrons are induced. Conversely, double kidneys are always accompanied by double ureters, suggesting that in the event of anomalous double ureteric diverticula duplicated inductions occur. The conclusion seems inescapable, then, that the ureter is essential to metanephric differentiation, a conclusion supported by experiments on the metanephros of the embryonic mouse.

If an intact metanephric primordium is cultivated in tissue culture, it undergoes characteristic morphogenesis. That is, the epithelial (pelvic) component forms a system of collecting ducts and the mesenchymal (blastemal) component forms coiled uriniferous tubules. But when the components are separated by the digestive enzyme trypsin and cultivated individually, neither shows ability to carry through morphogenesis. A reciprocal inductive interaction apparently normally operates. On the one hand, the branching of the primitive renal pelvis is dependent on some property of the nephrogenic blastema; on the other hand, tubule differentiation within the blastema is dependent on an inductive stimulus from the ureteric bud. In the first instance, the inductive capacity of the nephrogenic mesenchyme appears to be absolutely specific, for foreign mesenchyme is incapable of eliciting morphogenesis of the epithelial component. By contrast, the ureteric diverticulum shares its tubule-inducing capacity with some other tissues. Submandibular epithelium can induce kidney tubules, as can embryonic spinal cord.

Additional evidence for the "oneness" of the nephros is provided by another type of experiment performed on amphibian and chick embryos. If prospective mesonephros-forming mesoderm is transplanted to and replaces that area of the embryo that ordinarily would form a pronephros, it differentiates into nephrons of the pronephric variety. Similarly, if by suitable operative techniques prospective metanephric mesoderm (blastema) is brought into association with anterior wolffian duct rather than the ureter, it produces mesonephrons rather than metanephrons. The implication is that the mesomere has the general capacity to produce nephrons of a nonspecific type at all levels; that is, it is a holonephros. Its production of specific types at particular levels is in response to specific inductive stimuli at those levels: ureter for metanephrons, wolffian duct for mesonephrons. As for the pronephrons of fishes and Amphibia, the mechanisms directing determination and differentiation remain unknown.

THE REPRODUCTIVE SYSTEM

There is no chapter in animal history more varied than the evolution of the biological, psychological, and social characters of the individuals in sexually reproducing species that bear ova (females) and spermatozoa (males). On the biological level, organs are evolved for the reception and storage of the gametes, for their transfer from male to female, for protection of the development of the fertilized ovum without or within the body, and for the care of helpless young. These are endlessly varied in the different animal groups. As an accompaniment, there is the evolution of appropriate forms of behavior of the sexes with reference to such things as mating, nesting, and the care of young; in addition, there is the elaboration of social relationships within animal groups.

A treatment of all these matters is beyond our scope. Our attention is restricted to the history of the reproductive system proper. The system includes the *gonads* (testes in the male, ovaries in the female); the *sexual ducts* serving to transport the gametes and, in the case of the females of certain groups, housing and nurturing the embryo; and the organs of copulation, *external genitalia*. The history of these parts will in turn be seen to involve the excretory system and the cloaca.

We should remember that at the time of union of the gametes in the act of fertilization, the hereditary constitution of the new individual is established or, as we say, determined.

The mechanism for sex determination ordinarily operates through a pair of so-called *sex chromosomes,* present in all cells of the adult. In males, the pair of sex chromosomes consists of unlike mates, X and Y; in females, there is a pair of like X chromosomes.[1] Consequently, in gametogenesis

[1]Some vertebrates, notably the birds and certain fishes and urodele Amphibia, present the reverse situation: males are XX and females XY.

(see Chapter 4) half of the spermatozoa produced by the male will contain an X chromosome and half a Y chromosome, whereas the female will produce ova of one kind only, all carrying the X chromosome. An egg that is fertilized by an X-bearing sperm cell will thus be genetically determined for femaleness; and egg that is fertilized by a Y-bearing spermatozoon will be genetically determined for maleness.

Despite the fact that in terms of genetic makeup the newly inaugurated embryo is male or female from the moment of fertilization, development proceeds initially as though the embryo, to speak figuratively, were unaware of this. It prepares itself for both possibilities. Thus, in the early gonad, two discrete primordia appear. Whether a testis or ovary ultimately arises depends on which of the two primordia realizes its full development. So it is also with the sex ducts. Two sets are present, and a choice between them is made. As for the cloaca and associated genitalia, conditions are initially neutral, allowing for development in either direction.

Such a setup of discrete and neutral primordia obviously offers great opportunity for developmental variation in which those primordial destined to be discarded, according to the prospective sex, may be retained to a greater or lesser extent. Here lies the embryonic basis for the persistence of rudimentary structures belonging to the opposite sex, a persistence that may be of such an order as to produce those degrees of bisexuality that are termed *hermaphroditism*.

THE GONADS

The gonads arise from dorsal areas of coelomic epithelium between the hypomere and the mesomere, just medial to the embryonic kidneys. As a first step, the epithelium thickens and rapidly proliferates cells from its inner surface. These cells in turn multiply, and as a result there is built up a pair of *genital ridges* that project into the coelom (Figure 15-14). A genital ridge, on closer inspection, is seen to be a fairly compact mass of cells, described as a blastema, marked off rather sharply on the nephric

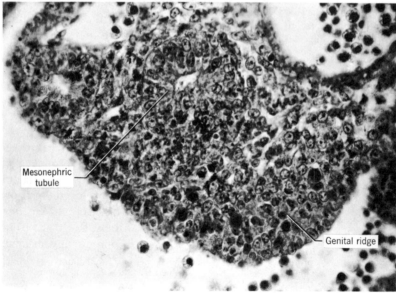

Mesonephric tubule

Genital ridge

Figure 15-14   Genital ridge of a 12-day-old rat embryo. ×250. (From Torrey, *Amer. J. Anat.,* Vol. 76.)

side, but blending imperceptibly with the epithelium on the coelomic side. Although the details differ in various vertebrate categories, the blastemal mass of the ridge becomes subdivided into strands and cords of cells. Coincidently, the coelomic epithelium is set off once more, becoming separated from the deeper cords by a layer of loosely assembled cells. The deeper-lying cords are termed *primary sex cords,* the coelomic epithelium is called the *germinal epithelium,* and the material lying between the sex cords and the epithelium is the fore-

runner of the *tunica albuginea* (Figures 15-15 *A, B*).

At this point the genital ridge is designated a *gonad.* For the present it is a bisexual gonad, which is to say that it presents two structural areas representing discrete primordia for both sexes. The usual practice is now to apply the term *cortex* in reference to the outer area and *medulla* to the inner area. In other words, cortex is equated with germinal epithelium and medulla with primary sex cords. The incipient tunica lies between cortex and medulla. Two addi-

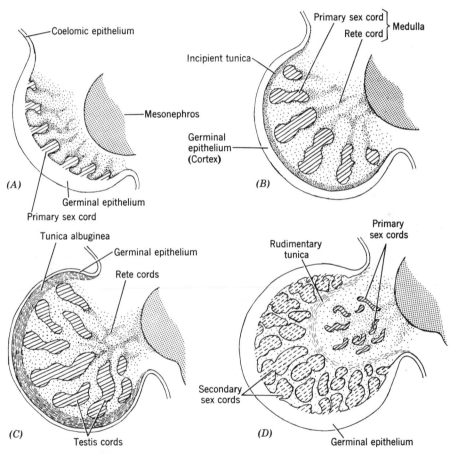

Figure 15-15 Differentiation of the medullary and cortical components of mammalian gonads. (*A*) Genital ridge with primary sex cords originating from germinal epithelium. (*B*) Sexually indifferent gonad with medullary component represented by primary sex cords and rete cords, and cortical component represented by germinal epithelium. (*C*) Embryonic testis. (*D*) Embryonic ovary. (From Burns, in *Analysis of Development,* Willier *et al.,* Eds., by permission of W.B. Saunders Company.)

tional features of the bisexual gonad also need noting. First, the inner ends of the primary cords are continuous with so-called *rete cords* which differentiate within the original blastema simultaneously with the sex cords themselves (Figure 15-15*B*). Second, both cortex and medulla contain *primordial germ cells,* which migrate into the genital ridge during the early stages of its formation (see Chapter 4).

If the gonad becomes a *testis* (Figures 15-15*C* and 15-16), the primary sex cords and the rete cords hollow out and become *seminiferous tubules* and *rete tubules,* respectively. The seminiferous tubules are lined by an epithelium within which the primordial germ cells lie. As these germ cells complete their maturation (Chapter 4), they are released, as spermatozoa, to the interior of the tubules (Figure 4-10). The rete tubules, joined with the seminiferous tubules from the very beginning, unite with nearby nephric tubules (see p. 328) to provide a channel to the exterior for the maturing sperm cells. In the meantime, the cortex (germinal epithelium) regresses to the status of a conventional coelomic epithelium, separated from the seminiferous tubules by a tunica albuginea, now a vascular, connective tissue sheath. Certain cells lying between the original sex cords give rise to special *interstitial tissues,* which become the source of male sex hormones.

If the gonad becomes an *ovary* (Figures 15-15*D* and 15-17), the germinal epithelium (cortex) continues to thicken and elaborates *secondary sex cords.* The earlier-formed primary and rete cords simultaneously degenerate, and the tunica albuginea is reduced to a rudimentary layer between the degenerate medullary area and the dominant cortex. The primordial egg cells lie within the secondary cords, and each becomes surrounded by other cells provided by the cords. This cellular investment of an ovum constitutes a *follicle,* whose developmental history within the ovary has already been reviewed in conjunction with the estrus cycle (Chapter 9).

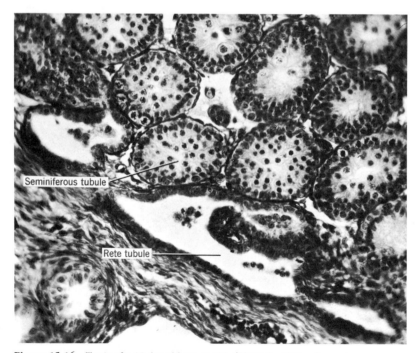

Figure 15-16 Testis of a 19-day-old rat. × 100. (Photo by Torrey.)

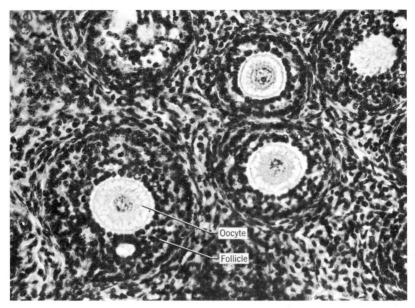

Figure 15-17   Ovary of a 19-day-old rat. × 100. (Photo by Torrey.)

It should be clear that the original gonad is a composite structure possessing male and female potentialities represented by medulla and cortex. Normal differentiation, in keeping with genetic sex determination, involves the gradual predominance of one component over the other. Now we must ask: What is responsible for directing this selective differentiation?

In Chapter 4 it was pointed out that the primordial germ cells make their initial appearance external to the site of the future gonads and migrate into or are transported to the genital ridge. Therefore, there exists the possibility that the germ cells themselves may in some way control gonad differentiation. But this turns out not to be the case, for experiments on amphibians and chick embryos, involving the elimination of the germ cells before they reach the gonad region, reveal that the gonads are not thereby prevented from forming. Moreover, it has been demonstrated that the sex type that materializes in a gonad bears no necessary relationship to the genetic constitution of the germ cells; in fact, the converse relationship obtains. That is, it has been possible to combine germ cells of one sex with

gonads of the other sex with the result that not only does gonad differentiation follow its genetically determined line, but the resulting gametes are sex-reversed.

With the germ cells ruled out, we must turn to other possibilities. Keep in mind that a gonad is initially bisexual and that differentiation is essentially a kind of competitive process by which one component ordinarily becomes predominant over the other. But it is a competition that can be swayed by a variety of unrelated circumstances. In fishes and amphibians, for instance, the two components are so well developed and persist for so long that the sex of the gonad may shift spontaneously with age. It is also known that temperature deviations may affect the competitive status of the two components. Thus, in amphibian embryos, high temperatures favor the medulla and lead to the transformation of genetic ovaries into testes, whereas low temperatures produce the converse result. Curiously enough, the gonads of mammalian embryos respond in just the opposite way: low temperature favors the medulla. The two components will also react to hormonal treatment, either directly through the injection of the chosen hormone into the embryo or indirectly through various types of grafts providing an association between heterologous gonads. Generally speaking, gonad differentiation can in this way be directed contrariwise to its genetic determination. Paradoxically, however, the reverse sometimes happens; for example, female hormone may completely masculinize female gonads. Finally, a literal realignment of the competitive status of cortex and medulla may be accomplished by the direct elimination of one component. Two classic experiments of this kind are based on the circumstance of the survival of the opposite component long after the dominant component has fully differentiated. The first pertains to adult male toads, which retain portions of the embryonic cortex in association with the testes. If the testes are removed, this cortical material, released as it were, proceeds to form ovaries capable of producing fertile eggs. The second provides the opposite result. In the female chicken only the left gonad becomes a functional ovary; the right gonad remains rudimentary and is composed chiefly of medullary tissue. But if the ovary is removed, the rudimentary right gonad will develop into a testis that may produce functional sperm.

From what has been said it should be clear that the primordial germ cells, notwithstanding their genetic makeup, are sexually indifferent or bipotential. Whether they become eggs or sperm depends on the gonadal environment within which they find themselves. The evidence for this derives from the just cited experiments on toads and birds and the earlier-described results of combinations of germ cells of one sex with gonads of another. It is not nearly so clear, however, how the competition between the components of the bisexual gonad is resolved. The most that can be said for the present is that the presence of these two structural components provides an uneasy balance that can be tipped by a variety of experimental agents and genetic factors.

THE SEXUAL DUCTS

The Generalized Sexual Ducts. Not only is the developing vertebrate embryo initially sexually indifferent with regard to the gonads, but in the beginning it also possesses both male and female sexual ducts (Figure 15-18). The potential male duct system is actually a part of the embryonic urinary system, namely, the original nephric ducts and certain of the nephrons. The female ducts are the *oviducts,* each of which opens anteriorly to the coelom by an *ostium* and courses backward to terminate in the cloaca. In elasmobranchs and urodeles the oviducts appear to arise by a lengthwise splitting of the nephric ducts. But in the majority of vertebrates they are formed by infoldings of the coelomic epithelium alongside the nephric ducts (Figure 15-19). Even in this circumstance the association with the nephric duct is close: both may be invested by a common sheath, and there is some indication that the nephric duct induces the formation of the oviduct. These, then, are the discrete primordia for the two sexual duct systems: nephric ducts and some of the nephrons for the male; oviducts

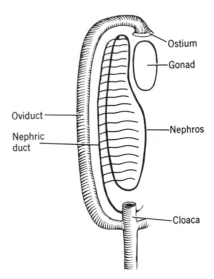

Figure 15-18 Generalized sexual ducts.

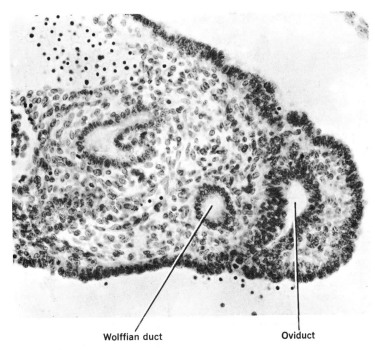

Wolffian duct   Oviduct

Figure 15-19   Origin of an oviduct by invagination of coelomic epithelium in a 14-mm human embryo. × 300. (From Faulconer, *Contributions to Embryology,* Vol. 34, Carnegie Institution of Washington.)

for the female. Sexual differentiation involves the retention and elaboration of one or the other and the degeneration in whole or in part of its opposite gender.

There are two major exceptions to these generalizations: (1) The Cyclostomata lack a sexual duct system; instead, the gametes are shed directly into the body cavity and from there make their way to the exterior via pores that pierce the body wall at the rear of the coelom. (2) The Teleostei exhibit a specialized system of ducts peculiar to themselves, which is considered later.

The Male Duct System. The history of the male sexual duct system may be considered to begin with the cyclostomes, even though, as we have just noted, they employ no ducts for the transport of sperm cells. They do, however, have nephric ducts draining the kidneys, and it is these ducts that have been taken over to a greater or lesser extent by all the other vertebrates for reproductive purposes. All that is required to add reproductive functions on the urinary system is the bridging of the gap between the kidneys and the nearby testes. Although there are variations in detail, this is accomplished essentially as follows. Recall that the seminiferous tubules of the testis are continuous inward, that is, toward the kidneys, with a network of canals known as rete tubules (rete testis) (Figure 15-16). These rete tubules tap certain of the nephrons of the adjacent kidney. There is thereby established a closed course for sperm cells to travel from seminiferous tubules to nephric duct and from there to the exterior. If no other adjustment is made, in these circumstances the nephric duct plays the dual role of transporter of sperm and urine. Presumably, this was the very situation in the first fishes following the cyclostomes, and so it is still in a scattering of modern vertebrates, notably sturgeons and gars, the Australian lungfish, and a few urodeles, for example, *Necturus* (Figure 15-20). But the total evolutionary history of urinary and male sexual ducts in the vertebrates has been one of the nephric duct playing one role and then another.

In the teleosts the nephric duct returns to its original status as the conveyor of excretory waste alone. By the same token, a completely new and separate sperm tube is provided for transportation out of the testis (Figure 15-21). In all the other vertebrates,

however, the nephric duct continues to be utilized in some degree for conduction of sperm. It will be recalled, for example, that the sharks and Amphibia have that type of kidney designated as an opisthonephros. Actually, only the posterior portion of this organ (the caudal opisthonephros) is excretory. The tubules of the anterior portion (cranial opisthonephros) function as genital ducts transporting sperm from the testis to the nephric (opisthonephric) duct. In all sharks and many Amphibia the nephric duct itself functions exclusively as a genital

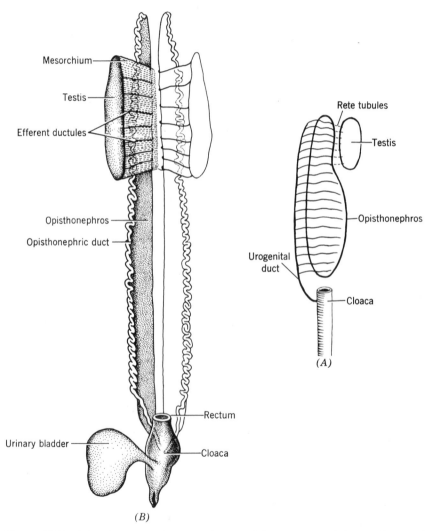

Figure 15-20 (*A*) Male urogenital system of the amphibian, *Necturus*. (*B*) Ventral view of the urogenital system of a male *Necturus*.

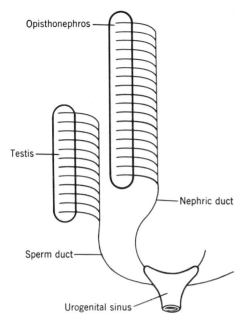

Figure 15-21 Excretory and genital ducts of a male teleost.

duct, the excretions from the caudal opisthonephros being carried away by one or more accessory ducts (Figure 15-22).

This is the stage setting for all the amniotes (reptiles, birds, and mammals). Here, as we have seen, the kidney is a metanephros drained by the ureter. During embryonic life the anterior end of the mesonephros (essentially the equivalent of the cranial opisthonephros of sharks and Amphibia) assumes the relationship to the testis described earlier, which is to say that certain mesonephrons are linked to the seminiferous tubules of the testis by rete canals. In this way the wolffian duct is taken over and retained as the sperm-conducting duct and henceforth is known as the *vas deferens* (Figures 15-23 and 15-37). The mesonephrons employed in the linkage with the testis are termed *vasa efferentia.* (It should be added that the remaining mesonephrons ordinarily degenerate.) The anterior end of the vas deferens and its associated vasa efferentia are commonly

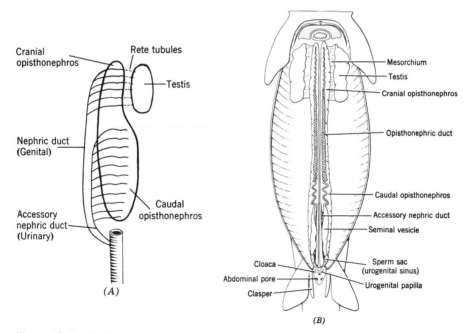

(A)

(B)

Figure 15-22 (*A*) Excretory and genital ducts of a male shark. (*B*) Ventral view of the urogenital system of a male shark, *Squalus.* (From Eddy, Oliver, and Turner, *Atlas of Outline Drawings for Vertebrate Anatomy,* John Wiley and Sons.)

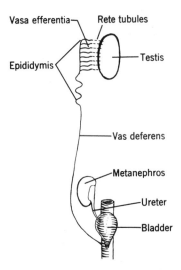

Figure 15-23 Urogenital system of a male amniote.

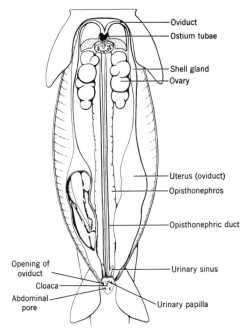

Figure 15-24 Ventral view of the urogenital system of a pregnant shark, *Squalus.* (From Eddy, Oliver, and Turner, *Atlas of Outline Drawings for Vertebrate Anatomy,* John Wiley and Sons.)

coiled and closely applied to the surface of the testis, providing a compact body termed the *epididymis* (Figure 15-37).

The Female Duct System. We have just seen the extent to which the nephros and nephric duct are taken over for reproductive purposes in males. In females, however, the nephric system either is retained in its original excretory capacity, as in elasmobranchs and amphibians (Figure 15-24), or with the advent of the metanephros is largely discarded (Figure 15-25). The point is that in contrast to the male, no part of the nephric system is salvaged and incorporated into the female duct system. The female sexual ducts are the oviducts that, as already noted, occur as a pair of tubes paralleling the nephric ducts. Each opens posteriorly to the cloaca and anteriorly to the coelom via an ostium. Also in contrast to the male, in which there is structural continuity between the testes and their ducts, in the female there is no such connection. Ova break through the walls of the ovaries, pass into the coelom, and then enter the oviducts through their ostia. The notable exception to this is again found in the Teleostei (Figure 15-29). Here the ovary is folded in such a fashion as to en-

close a special pocket of the coelom, the wall of which is extended as a funnellike tube to the exterior. Obviously, this tube is not the homologue of the conventional oviduct. Moreover, it is an arrangement that provides for the complete enclosure of the

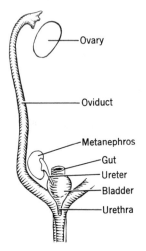

Figure 15-25 Female urogenital system of an amniote.

multitudinous eggs ordinarily produced by teleosts, rather than for their discharge into the coelom proper and from there into the oviducts as in vertebrates generally.

The oviducts serve as more than transportation tubes for the eggs. Specialized regions may perform as glands, providing albumen and/or shell investing the eggs. So it is in all the elasmobranchs where a *shell gland* (Figure 15-24) furnishes both albumen and a horny egg case. The eggs of Amphibia are invested by jelly as they traverse the oviducts. And in reptiles and birds one portion of the oviduct lays down albumen about the egg and another portion provides the shell. The oviducts, too, may serve as chambers within which the embryos are sheltered and nourished before they are born. Many elasmobranchs, for instance, possess an arrangement in which there is a close association between the lining of the oviduct and the embryo's yolk sac, a condition not unlike that of the mammalian placenta (Chapter 9). Many reptiles present a similar situation, and, of course, in eutherian mammals the oviduct is subservient to the maintenance of the embryo.

In most vertebrates the oviducts are present in the basic paired condition, each opening separately both anteriorly and posteriorly. Special exceptions are some sharks in which the two ostia are fused into one and birds in which the right oviduct, like the right ovary, is degenerate. Eutherian mammals represent a general exception. There is some measure of fusion of the two oviducts and, as an accompaniment of this, a regional differentiation of each oviduct. The upper (anterior) part of the oviduct occurs as a narrow, slender *fallopian tube.* It leads to a broader, muscular *uterus,* which in turn leads to a terminal vaginal portion. The vaginal regions are invariably fused in placental mammals to form a *vagina* proper (Figure 15-26). The fallopian tubes always remain separate, but the uteri show a variable history. In some they remain separate, providing what is known as a *duplex uterus* (Figure 15-26A). Or they

may be fused in part to a variable degree to form uteri of the *bipartite* (Figure 15-26B) or *bicornuate* (Figure 15-26C) types. And in primates there is a total fusion to form a *simplex uterus* (Figure 15-26D). When only partial fusion occurs, the single portion is usually termed the *body* of the uterus and the separate portions the *horns* (Figure 15-26C).

Differentiation of the Sexual Ducts. Remember that the vertebrate embryo starts off with a double set of discrete primordia, a nephric system for the prospective male and oviducts for the prospective female, one of which for each individual is destined for elaboration and the other for degeneration. We may very well ask again, as we did in conjunction with the gonads, what is known of the agency responsible for the choice of the one over the other?

Much experimentation has been directed to this question. Unfortunately, such generalizations as have come out of all this work are still hedged with qualifications and exceptions. However, we may with some security point to a few leading ideas—ideas, it should be added, that apply as much to the differentiation of the cloaca and external genitalia, still to be considered, as they do to the sexual duct system.

The guiding hypothesis behind much of the investigation that has been carried out is that once differentiation of the gonads has been consummated, however that may be accomplished, they in turn produce specific male and female hormones, which then stimulate and/or suppress the development of the appropriate duct system. In actual experimental practice, sex hormones can be added to embryos by injection of gonad grafting and subtracted by gonad excision. The general result is that male hormones accelerate differentiation of male structures and inhibit female structures in embryos of both sexes. Conversely, female hormones stimulate the development of female structures and inhibit male structures. So far so good. But as we noted before with the go-

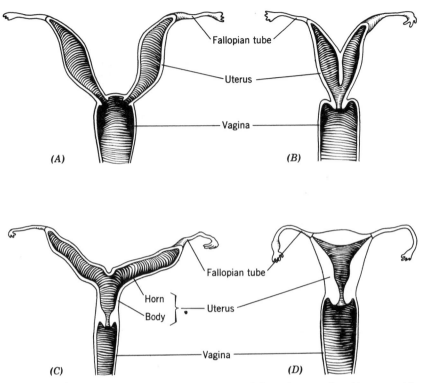

Figure 15-26 Types of uteri occurring in mammals. (*A*) Duplex type found in marsupials. (*B*) Bipartite type of some rodents. (*C*) Bicornuate type of carnivores. (*D*) Simplex type of primates. (After Wiedersheim.)

nads, sometimes both types of hormones paradoxically stimulate structures of the opposite sex, results that for the present confuse the issue. Some additional light, however, is provided by experiments involving the excision of the gonads from embryos, notably in mammals and birds. In mammals, the removal of the testes from a male results in the male ducts (and other parts) remaining rudimentary and a "release" of female parts that continue to develop. The Sertoli cells of mammalian testes have been shown to produce a hormone that specifically represses development of the mullerian ducts, the *anti-mullerian duct factor*. Removal of ovaries from a female does not lead to the converse effect, as one might expect, but again development is toward femaleness. This is interpreted to mean that male development depends on

the testes which, if present, also suppress female development; but the female pattern is essentially independent, developing without hormonal conditioning. In bird embryos the situation is reversed; the male system is the independent one.

Recently the active genetic site on the Y chromosome in human males has been isolated and its product characterized. XY males whose Y chromosome is deficient for this region do not develop testes or male secondary sex characteristics. In addition, phenotypically female XY humans are known in which a genetic defect prevents their cells from responding to testosterone; such individuals have rudimentary internal testes and no mullerian ducts. Estrogen from their adrenal glands apparently is sufficient to induce development of female secondary sexual characteristics.

THE CLOACA AND ASSOCIATED PARTS

In embryonic stages of all vertebrates, and in the adults of many, the urogenital ducts and intestine empty into a common chamber, the *cloaca,* which in turn opens to the exterior. Embryologically, the cloaca has a double origin (Figure 15-27). The bulk of it derives from the hindmost part of the intestine itself, which early shows a pronounced expansion. When the cloacal dilation first appears, the intestine ends blindly, its posteroventral endodermal wall lying in contact with the body wall ectoderm that pockets inward under the root of the tail. The ectodermally lined depression is the *proctodeum,* and the double-layered plate of tissue separating the endodermal intestine from the ectodermal proctodeum is the *cloacal membrane.* Eventually, the cloacal membrane ruptures, establishing a caudal outlet for the intestine and at the same time providing for incorporation of the proctodeum into the cloaca. Thus the lining of the definitive cloaca derives from both ectodermal and endodermal sources, although the latter predominates. The cloaca exhibits a variety of modifications among the vertebrates, and inseparable from its history are the histories of the urinary bladder and external genitalia.

THE URINARY BLADDER

At or near the rear ends of the nephric ducts there frequently is a reservoir for urine. This is the *urinary bladder.* Actually there are two basic varieties of bladders in vertebrates. One is found in fishes, in which the reservoir is nothing more than an enlargement of the posterior end of each urinary duct. Frequently the urinary ducts are conjoined, and a small bladder is formed by expansion of the common duct. The far more common type of bladder is that ex-

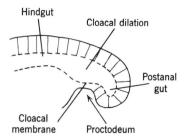

Figure 15-27   Origin and composition of the cloaca.

hibited by tetrapods. This is a sac that originates embryonically as an outgrowth from the ventral side of the cloaca. Present in all during embryonic life, it is exhibited differentially in adults. All Amphibia retain the bladder, but it is discarded by snakes, crocodilians, and a few lizards. All birds, with one or two exceptions, lack a bladder. It is, however, present in all mammals. Because much of the developmental history of the tetrapod bladder is linked to the history of the cloaca, it is considered further later.

THE CLOACA AND EXTERNAL GENITALIA

Although the embryos of all vertebrates feature a cloaca, as we have already indicated it is subject to modification or elimination in the adults of most groups, and with this goes a varied disposition of the parts associated with it. Thus the most convenient approach is to consider the cloaca and its accessories in each of the major vertebrate categories separately.

Fishes. Among fishes, a cloaca is retained in the adults of only the Elasmobranchii and Dipnoi (Figure 15-6*B* and 15-28). The cloaca is eliminated entirely in the Holocephali, the anal, urinary, and genital openings all being separate. So it is also with the Actinopterygii (Figure 15-29), where again the anal, genital, and urinary systems tend to open separately to the exterior, although in some teleosts the urinary and reproductive ducts may join in a common sinus (rep-

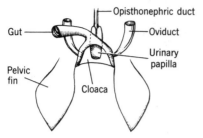

Figure 15-28   Cloaca and urogenital openings of a female shark.

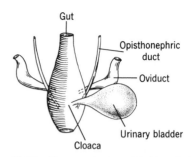

Figure 15-30   Cloaca and urinary bladder of a female amphibian.

resenting part of the embryonic cloaca) before making their exit (Figure 15-21).

Amphibia. All Amphibia have a cloaca (Figure 15-20B and 15-30). It receives not only the digestive tract, but the nephric ducts and, in females, the oviducts as well. And from its ventral side there emerges a urinary bladder. The nephric ducts do not empty directly into the urinary bladder.)

Reptiles and Birds. In reptiles and birds the cloaca is partly subdivided in such fashion that the intestine and urogenital ducts open into separate compartments, which then join in a common outlet (Figure 15-31). A prominent embryonic structure is the *allantois,* which sacculates from the floor of the cloaca. The history of the allantois has already been considered in conjunction with the extraembryonic membranes and should be familiar. In the present connection it need only be emphasized that the allantois serves as a urinary bladder for the embryo. In birds the entire

allantois is discarded at hatching, and ordinarily no adult urinary bladder is retained. This is also true for most reptiles, although some, notably the Chelonia and some lizards, retain the base of the allantois as an adult bladder and discard the remainder.

It is in the reptiles that one sees the evolutionary beginnings of certain cloacal accessories designed to facilitate the transfer of spermatozoa from the male to the female reproductive tract. By way of background, it may be recalled that in most water-dwelling vertebrates, eggs and sperm are shed directly into the water, and fertilization is thus external. But if, as in the Chondrichthyes, the egg is provided with a shell and thus must be fertilized before the shell is laid down, or if development of the embryo takes place within the uterus, then fertilization is necessarily internal. This requires a mechanism for transfer of sperm cells from male into female. So it is that

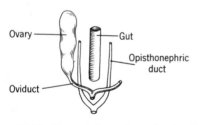

Figure 15-29   Urogenital openings of a female teleost.

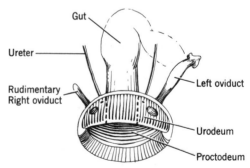

Figure 15-31   Cloaca of a female bird.

extensions (claspers) of the pelvic fins serve as intromittent organs in the sharks and rays, and similar structures involving the anal fin are found in some teleosts. These accessories, however, are not homologous to the cloacal devices found in amniotes. As for the birds, the proctodeal portions of the cloacas of male and female are simply everted and pressed together during copulation, so that spermatozoa are ejaculated directly into the urogenital portion of the female cloaca. The reptiles go further and come up with definitive accessory structures.

In male snakes and lizards a pair of tubular folds occurs in the wall of the proctodeum. These constitute what are termed *hemipenes* which, at the time of copulation, are turned inside out, extruded, and inserted into the female cloaca. Each hemipenis is grooved in such a way as to provide a channel for the spermatozoa. In turtles and crocodilians a somewhat different device occurs (Figure 15-32). Lying in the floor of the male cloaca is a pair of longitudinal ridges, *corpora cavernosa,* flanking a groove between them. Each corpus is composed of a vascular, spongy tissue, and the two corpora converge at the distal end of the groove in a spongy knob known as the *glans penis.* During copulation the corpora and glans become distended with blood, apparently as a consequence of a

stimulus provided by the secretion of adrenaline by the adrenal glands. This results in a protrusion of the glans and an expansion of the corpora, so that they meet each other and convert the groove between them into a tube that serves as a channel for sperm. This arrangement is not unlike that found in mammals.

Mammals. Among mammals, only the Prototheria retain a cloaca as adults. In all the others the cloaca is worked over in such a way as to be eliminated as such. The openings of the urogenital ducts are at the same time shifted, and the urinary bladder and external genitalia are established. Let us follow these events in some detail, with the human as the illustrative subject.

The original cloaca (Figure 15-33A) in the human embryo receives the hindgut dorsally, the rudimentary allantois ventrally, and the wolffian ducts, from which the ureters bud, on each side. Ventroposteriorly, cloacal endoderm meets proctodeal ectoderm in a cloacal membrane. A subdivision of the cloaca is inaugurated by a *urorectal fold,* which arises in the angle between the allantois and gut (Figure 15-33B). This fold works its way caudad until it contacts the cloacal membrane and, in so doing, divides the cloaca into a dorsal *rectum* and ventral *urogenital sinus* (Figure 15-33 B, C). The cloacal membrane is likewise divided into an *anal membrane* and a *urethral membrane* (Figure 15-33C). Both these membranes eventually rupture.

After its complete separation from the rectum, the urogenital sinus may be subdivided for descriptive purposes into a part anterior to and including the openings of the wolffian ducts, called the *vesicourethral division,* and a posterior *definitive urogenital sinus* (Figure 15-33C). There ensues a complicated and incompletely understood shift by which the wolffian ducts and developing ureters come to open separately into the vesicourethral division of the sinus. It appears to accompany an expansion of

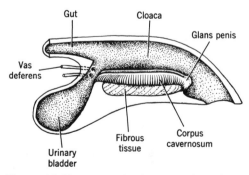

Figure 15-32 Longitudinal section through the cloaca and intromittent organ of a male turtle.

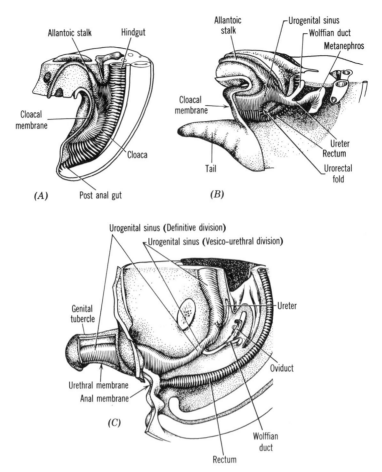

Figure 15-33 Subdivision and early differentiation of the human cloaca. (Drawn from Ziegler models.)

the anterior portion of the vesicourethral division, an expansion that, when completed, provides the *bladder*. As a result, the ureters come to open into the enlarging bladder while the wolffian ducts open some distance to the rear in that portion of the vesicourethral division destined to be the *primitive urethra*. Development so far being sexually indifferent (Figure 15-34), a female sexual duct system is also present and it opens into the definitive urogenital sinus. From this point on development takes divergent courses in the male and female except that, of course, the bladder and ureters opening into it are the same for both.

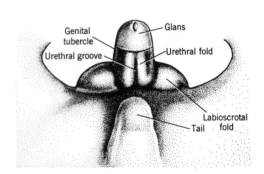

Figure 15-34 Sexually indifferent external genitalia on an 8-week-old human embryo.

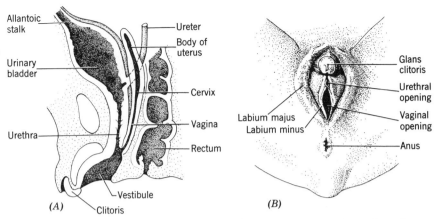

Figure 15-35 Differentiation of the female urogenital passages and external genitalia in a human fetus of about 12 weeks. (*A*) Longitudinal section. (*B*) Caudal view of genitalia.

If the individual becomes a female (Figure 15-35*A*), the definitive urogenital sinus shortens and expands to form the shallow *vestibule* into which the excretory and reproductive channels open separately. Specifically this refers to the opening of the *vagina,* the lower end of the sexual duct system, and the *urethra* running from the bladder and representing that part of the original vesicourethral division of the sinus previously designated the primitive urethra. The wolffian ducts, as prospective male ducts, degenerate in large part.

If the individual becomes a male (Figure 15-36*A*), the definitive urogenital sinus extends into the penis (whose origin is described later) as a continuation of the urethra running from the bladder. In other words, whereas the female urethra represents only the short segment of the vesicourethral division of the sinus extending from the bladder, in the male the urethra represents this segment plus the definitive sinus as well. Female and male urethras are therefore only partly comparable; that of the male is homologous to the female ur-

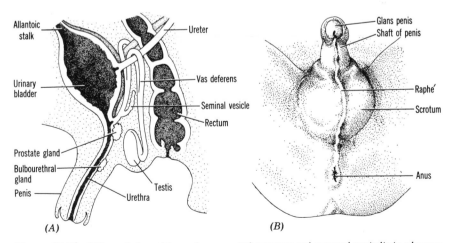

Figure 15-36 Differentiation of the male urogenital passages and external genitalia in a human fetus of about 12 weeks. (*A*) Longitudinal section. (*B*) Caudal view of genitalia.

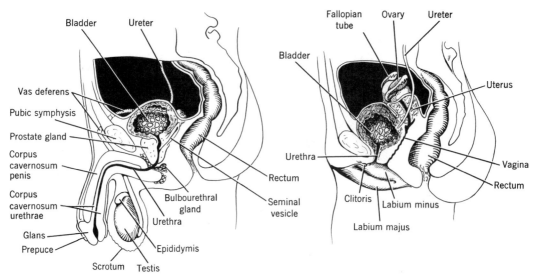

Figure 15-37 Male and female urogenital systems of adult humans.

ethra and vestibule combined. The base of the urethra of course receives the vasa deferentia (the onetime wolffian ducts), and the female duct system disappears. Special glands servicing the male duct system, that is, the *prostate* and *bulbourethral glands,* arise from the sides of the urethra.

The external genitalia have a common origin in the two sexes and all the parts in one sex have their exact homologues in the other (Figure 15-34). As the vesicourethral division of the urogenital sinus differentiates in the manner previously described, a conical elevation, the *genital tubercle,* appears at the site of the urethral membrane. On the caudal surface of this tubercle there

TABLE 15-1
UROGENITAL HOMOLOGIES

Male	Sexually Indifferent	Female
Testis	Gonad	Ovary
1. Degenerates	1. Cortex (Secondary cords)	1. Ovarian follicles
2. Seminiferous tubules and rete tubules	2. Medulla (Primary and rete cords)	2. Degenerates
3. Spermatozoa	3. Primordial germ cells	3. Ova
	Mesonephric tubules	
1. Vasa efferentia	1. Anterior group	1. Degenerate except for rudiments
2. Degenerate except for rudiments	2. Posterior group	2. Degenerate except for rudiments
Vas deferens	Wolffian duct	Degenerate
Degenerate	Oviducts	Fallopian tubes, uterus, and vagina
	Urogenital sinus	
1. Bladder and upper urethra	1. Vesicourethral division	1. Bladder and urethra
2. Lower urethra	2. Definitive division	2. Vestibule
1. Penis	1. Genital tubercle	1. Clitoris
2. Penis	2. Urethral folds	2. Labia minora
3. Scrotum	3. Labioscrotal folds	3. Labia majora

is a shallow median furrow, the *urethral groove,* whose margins are elevated as *urethral folds.* Essentially, the urethral groove corresponds to a portion of that ectodermally lined compartment we identified earlier as the proctodeum. The urethral membrane lies at the bottom of the groove and thus simultaneously provides a roof for the proctodeum and a floor for the definitive urogenital sinus which extends into the tubercle. The genital tubercle gradually elongates to form the cylindrical *phallus* at the end of which is a bulblike enlargement, the *glans.* At the base of the phallus, on either side, there develop the *labioscrotal folds.* The urethral groove runs along the caudal side of the phallus, extending in the potential male to the glans; in the female not quite to it. The definitive urogenital sinus keeps pace with this elongation of the tubercle and extends the length of the phallus. In the meantime, the urethral membrane (and anal membrane) ruptures, and the urogenital sinus comes to open to the exterior. Up to this point conditions are essentially duplicated in prospective males and females. From here on they show divergencies.

In the female (Figure 15-35*B*) the labioscrotal folds give rise to the *labia majora.* The phallus develops into the *clitoris* with a glans. The definitive urogenital sinus and the urethral groove remain open, as the vestibule into which the vagina and urethra open. The urethral folds become the *labia minora* flanking the vestibule.

In the male (Figure 15-36*B*) the phallus continues to elongate to become the shaft of the *penis* with a glans at the end. Progressive fusion of the urethral folds, except at the extreme terminal end, once more closes the urogenital sinus so as to create a tubular urethra within the penis. Coincidentally, paired columns of vascular connective tissue (corpora cavernosa penis) arise within the matrix of the penis. Their appearance plus the coming together of the urethral folds are certainly reminiscent of penile structure in reptiles. The principal difference is the addition of a third midline column of erectile tissue (corpus cavernosum urethrae) surrounding the urethra. The *scrotum,* into which the testes descend from the body cavity, results from a midline union of the labioscrotal folds.

Without our having specifically said so, it should be apparent that the story of development of the cloaca and its accessories is one of the differential development in the two sexes of what we earlier spoke of as neutral primordia. Here it is not a matter of competition, so to speak, between two distinct sets of primordia, but one of sexual divergence from a common starting point. The impetus for development in one direction or the other clearly comes from the gonadal hormones. Accordingly, in vertebrates generally, male hormones produce a male type of differentiation of the cloaca and external genitalia in embryos of both genetic sexes, and female hormones produce the converse effect.

To conclude on a summarizing note, a survey of the principal urogenital homologies with special reference to humans is provided by Figure 15-37 and Table 15-1.

Chapter Sixteen

The Circulatory System

The circulatory system of a vertebrate is composed of two parts: (1) the *blood vascular system,* consisting of vessels and chambers that transport blood, and (2) the *lymphatic system,* composed of delicate tubes and sinuses by way of which tissue fluids are returned to the blood vessels.

THE BLOOD VASCULAR SYSTEM

DEFINITIONS AND GROUND PLANS

In effect, the blood vascular system comprises transmission lines through which a liquid medium is pumped. The circulating medium is the *blood;* the pumping station is the *heart;* the transmission lines are the *blood vessels.* Those vessels that carry blood away from the heart are *arteries,* those in which blood flows toward the heart are *veins,* and the *capillaries* are minute vessels between the arteries and veins. Reduced to its physiological essentials, this is the system by which there are transported to the tissues and cells of the body those materials required for their maintenance and activity and by which metabolic by-products and wastes are removed. In addition, the circulating blood serves as a medium of transport for the secretions of the endocrine glands, and it also provides for uniformity of composition of interstitial fluids. To put it somewhat differently, the vascular channels are so laid out that the circulating blood picks up food and oxygen from the organs concerned with their absorption and transports these materials throughout the body; it gathers endocrine secretions at their sources and conveys them to their appropriate destinations; it carries waste materials to their elimination sites.

These fundamental physiological operations find expression in an anatomical plan, which, although it varies greatly in detail from one animal group to another, exhibits a measure of sameness in all the vertebrates. As a kind of preliminary preparation for the analyses of the subdivisions of the circulatory system to be dealt with separately later, let us first attempt to gain a general picture of the developmental anatomy of the total system. In so doing, it is well to keep two developmental principles in mind. First, the main vascular channels, that is, the blood vessels, will appear in conjunction with major centers of metabolic activity. The initial vascularization of these centers commonly takes the form of a diffuse network of channels out of which a main channel, the definitive blood vessel, is excavated. Here lies the source of the many interconnections and individual variations in the vascular systems of all vertebrates. Second, the final adult patterns of vessels are anticipated by transitory embryonic layouts, which, on the one hand, and manifestations of inherited embryogeny and, on the other, are accommodations to the functional requirements of embryos and larvae.

The primordial circulatory cycle in all vertebrates (Figure 16-1A) is one in which blood is pumped forward by the heart, via a *ventral aorta* lying beneath the gut, and passes through paired *aortic arches* running dorsally through the pharyngeal arches. The aortic arches unite dorsal to the pharynx to form a pair of *dorsal aortas* through which blood flows into a capillary network on the yolk sac. The vessels of this net then drain into channels leading to the rear of the heart, thus completing the circuit.

As development proceeds, *systemic arteries* branch from the dorsal aortas to supply the forming organs of the general body

and gut. Each branch breaks into capillaries in conjunction with the organ involved; capillaries then converge in *systemic veins* leading back to the heart. But at this point a marked dichotomy in the cycle appears between the Osteichthyes and Amphibia (Figure 16-1B) and the Chondrichthyes and all the amniotes (Figure 16-1C). In the former, blood to the yolk sac first passes through the capillary beds in the body and gut. That is, the yolk sac is supplied and drained by the systemic veins. In the latter, the yolk sac is supplied directly by arterial branches (*vitelline arteries*) from the aortas and is drained by separate *vitelline veins* passing to the heart. The systemic arteries and veins, in other words, bypass the yolk sac completely.

A little layer in development a new complication is introduced. This derives from the appearance of the liver. In all the vertebrates, the proliferating liver cords consistently invest the proximal portions of the veins draining the yolk sac: vitelline veins in the Chondrichthyes and amniotes, and analogous systemic veins in the Osteichthyes and Amphibia. In either case, the intervention of the liver serves to break up the venous channels, so that all blood from the gut and yolk sac passes through hepatic sinuses before reaching the heart. Such an arrangement in which veins are interrupted on their way to the heart constitutes a portal system, and this, the *hepatic portal system,* is of universal occurrence among the vertebrates (Figure 16-1 D, E). In all, venous blood from the yolk sac and digestive tract drains through liver sinuses before reaching the heart.

Still another complication appears in amniotes as an accompaniment of the development of the allantois. The allantois is supplied by direct branches of the aortas, the *allantoic (umbilical) arteries.* The *allantoic (umbilical) veins* that drain it initially pass directly to the rear of the heart; but shortly this circulation is also shunted to the liver whose sinuses it utilizes in vari-

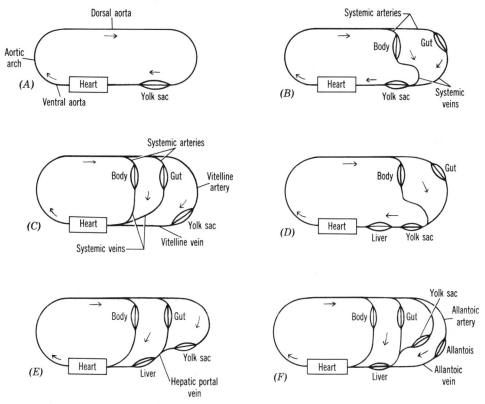

Figure 16-1 Basic circulatory arcs. (*A*) Primordial circuit in all vertebrates. (*B*) Early embryonic circuit in Osteichthyes and Amphibia. (*C*) Early embryonic circuit in Chondrichthyes and amniotes. (*D*) Intermediate embryonic circuit in Osteichthyes and Amphibia. (*E*) Intermediate embryonic circuit in amniotes. (*F*) Late embryonic circuit in amniotes. (Based on figures by Mossman.)

able degree in passing to the heart (Figure 16-1*F*). In reptiles and birds some anatomical bypassing of the liver sinuses is still effected, but in mammals there is extensive drainage of allantoic blood via the liver sinuses.

Two other circuits are also incorporated in vertebrate circulatory systems. One of these involves the kidneys. In embryonic stages of all vertebrates and in all adults except mammals and birds, more or less of the venous return from the rear of the body is routed through a capillary and/or sinusoidal net within the kidneys. This constitutes the so-called *renal portal system*. The other involves the aortic arches within the pharyngeal region. The key issue here, one to which we shall devote considerable at-

tention in due course, is the shift from a direct interposition of a gill circulation on the main line from the heart to the body in gill-breathing vertebrates to a side circuit in lung breathers.

The ground plan may be summarized as follows:

1. The primordial vascular circuit features vessels associated with the yolk sac. In the Osteichthyes and Amphibia the yolk sac is supplied by blood from the systemic veins; in the Chondrichthyes and all the amniotes the yolk sac is supplied directly by branches from the dorsal aortas. In either event, the yolk sac not only represents the site of formation of the first blood vessels and blood cells of the

embryo, but in the majority of verte-brates plays a major role in embryonic nutrition (see Chapter 9).

2. A hepatic portal system, that is, the interposition of the liver in the route of venous return from the digestive tract to the heart, is common to all vertebrates. This arrangement is linked to the multiple role played by the liver as an organ for storage of sugar, bile salt production, and urea synthesis, to name some of the more notable of its operations.

3. The allantois, a feature of the amniotes, is consistently supplied by arteries branching directly from the aortas. Its venous drainage initially passes directly to the heart, but eventually is shunted in some measure through the liver. In reptiles and birds, where the function of the allantois is primarily respiratory, a considerable bypassing of the liver sinuses prevails. The extensive utilization of the hepatic sinuses by the allantoic drainage of mammals appears to be correlated with the nutritive role played by the allantois in the placenta (see Chapter 9), thus, in effect, placing the liver in the same position of functional responsibility that it holds in conjunction with the hepatic portal system.

4. A renal portal system occurs in the embryos of all vertebrates and in all adults except birds and mammals. Thus, this circuit contributes to the demands of excretory requirements.

5. The aortic arches are either utilized within a gill respiratory system inserted directly between the heart and the body circulation, or modified in accompaniment with the development of a side circuit to the lungs.

THE BLOOD

The circulatory medium, blood, is a form of connective tissue consisting of a liquid matrix, *plasma,* within which formed elements, *blood cells,* are suspended.

The plasma is essentially water containing a variety of substances either in solution or in suspension. Some of its components are normally stable in amount. These are inorganic salts and an assortment of proteins, notably serum albumin, globulins, and fibrinogen. Serum albumin raises the osmotic pressure of the blood above that of tissue fluids, thus expediting transfer of materials through capillary walls; the globulins function primarily as antibodies in the defense against disease and infection; fibrinogen polymerizes under the influence of certain enzymes to form fibrin in blood clotting. Other substances are present in widely fluctuating amounts, depending on physiological circumstances. These include glucose, fats, amnio acids, and hormones plus metabolic wastes such as carbon dioxide and nitrogenous products.

With some rare exceptions noted below, the blood of all vertebrates contains two general kinds of cells: *erythrocytes,* or *red blood cells,* and *leukocytes,* or *white blood cells* (Figure 16-2). In mammals, still a third category, *blood platelets,* is found.

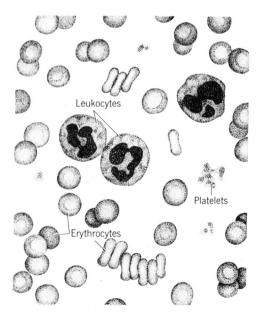

Figure 16-2 Types of human blood cells.

Attention has already been directed to the role played by hemoglobin in the transport of oxygen. In invertebrates, hemoglobin and other respiratory pigments are dissolved or suspended directly in the blood plasma, but in vertebrates hemoglobin is carried within the red blood cells. The only exceptions are certain Antarctic fishes, some larval eels, and some deep-sea fishes whose blood lacks hemoglobin. The presumed advantage to the intracellular transport of hemoglobin is that of preventing its loss by diffusion through capillary walls. Erythrocytes vary in size among the classes of vertebrates, those in lower classes tending to be larger than those in the higher classes. This is of little significance, however, for the smaller the cells, the greater the number. More importantly, the red blood cells of mammals, unlike those of all other vertebrates, are nonnucleated. (The nuclei are extruded early in the development of the cells.) The enucleate condition provides for maximum hemoglobin content per unit cell volume; however mammals pay a price for their red cells being unable to proliferate: constant replenishment is demanded.

Leukocytes are much fewer in number than erythrocytes but, at the same time, they present a great variety of form and function. Not only are there several types in the blood and extravascular tissues, but they vary considerably from one class of vertebrates to another. Their classification is a rather arbitrary one based on their reactions to standard cytological dyes. It is likely, too, that some of the types that have been described are only variants of a single category, for there is evidence that one type of cell may transform to another. However, they all appear to fall into two major groups: *lymphoid leukocytes* and *granulocytes.*

Lymphoid leukocytes are distinguished by having a simple, single nucleus and a nongranular cytoplasm. They are believed to encompass two categories of cells, *lymphocytes* and *monocytes.* Lymphocytes are found in all classes of vertebrates and are

especially numerous in the lower groups. They come in a variety of sizes and, in addition to circulating in the bloodstream, they have a ubiquitous distribution in the lymphatic system (p. 403) and throughout the connective tissues of the body. Their implication in the immunological defenses of animals now seems well established (p. 302), and it appears they may also give rise to other cell types, including erythrocytes. Monocytes are motile cells that more in extravascular areas and aggresively engulf cell and tissue debris.

Granulocytes, sometimes termed *polymorphonuclear leukocytes,* are large cells featuring abundant, highly granular cytoplasm and irregular, often lobed or multiple nuclei. They are commonly distinguished and classified on the basis of their reaction to cytological dyes. The cytoplasmic granules of some stain with acid dyes, some with basic dyes, and some with both. So they are termed, respectively, eosinophils, basophils, and neutrophils.

Except in reptiles, where they are relatively rare, neutrophils are the most abundant type of granulocyte in all the classes of vertebrates. The neutrophils are the major scavengers of the animal body. They are attracted to areas of infection both within and external to the blood vessels where they engulf invading microorganisms. Engorged and dying neutrophils at sites of heavy infection are the main ingredient of "pus." Basophils are the least numerous of the granulocytes. They are seldom found in most bony fishes and, with some exceptions, are rare in amphibians; even in the amniotes their numbers are small. They are believed to play some role in maintaining the physical and functional integrity of the endothelial walls of the blood vessels by way of endocrinelike secretions. They may also be involved in the blood clotting mechanism. In contrast to basophils, eosinophils have been found in most of the species of vertebrates that have been examined. Their function is actually unknown, but because their numbers increase dramatically in human individuals showing allergic sensitivity to given products, it has been suggested that the eosinophils may provide an antihistaminic defense against such sensitivity.

Because it is functionally distinct from the other categories of leukocytes, there is still another cell type to be considered. This is the *thrombocyte,* believed by many to be a derivative form of lymphocyte. Thrombocytes are found abundantly in all vertebrates except mammals. Their name derives from the fact that when a blood vessel is severed, the thrombocytes disintegrate as blood spews out, releasing a substance that is involved in the complicated transformation of protein fibrinogen to fibrin whose masses constitute a blood clot. In mammals, the thrombocytes are replaced by *blood platelets* whose role in the clotting mechanism is essentially the same. Blood platelets are enucleated fragments of certain cells produced in the bone marrow (p. 204).

Differentiation of most tissues of the vertebrate body occurs during embryonic life, and during adulthood there is adjustment only for normal wear and tear. The life span of blood cells, however, is a brief one, measured in months or days or even hours, and constant replenishment is required. Consequently, the organism maintains certain *blood-forming (hemopoietic) tissues* in which new blood cells are constantly formed. The first blood cells of the amniote embryo are derived from the blood islands in the extraembryonic mesoderm of the yolk sac. (See p. 142 and Figure 16-5.) But, later in embryonic life, other tissues and organs take over the task of blood cell formation. The most notable of such sites are the liver, the submucosa of the intestine (Figure 16-3), the intertubular tissue of the kidney (Figure 16-4), and the spleen (p. 404). These sites continue as the principal sources of new blood cells throughout adult life in lower classes of vertebrates. In the adults of higher classes, however, blood cell formation is centered in the spleen and other lymphoid organs and tissues and in

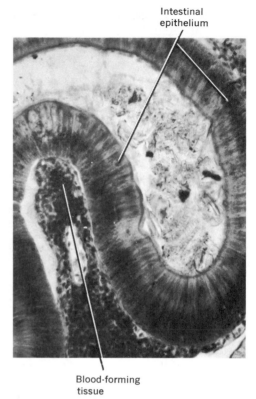

Intestinal epithelium

Blood-forming tissue

Figure 16-3 Blood-forming tissue in the intestine of *Ammocoetes.* (Photo by Torrey.)

the bone marrow. Reptiles exhibit a transitional condition wherein both major categories of blood cells are produced in varying proportions in the spleen and red bone marrow. In birds and mammals, then, there is a segregation of hemopoietic tissues: the red marrow becomes the center for erythrocyte and granulocyte production; the spleen and other lymphatic tissues provide lymphoid leukocytes.

THE HEART

The development of the heart is inaugurated concurrently with the origin of the primordial circulatory arc associated with the yolk sac. The starting point is the appearance of numerous clusters of mesodermal cells known as *blood islands* within

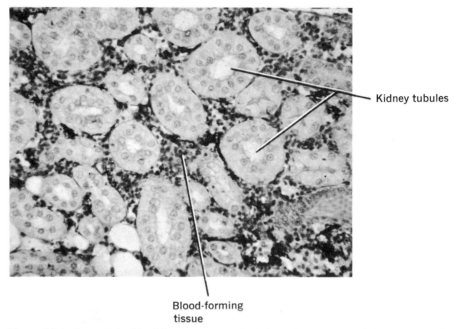

Kidney tubules

Blood-forming
tissue

Figure 16-4   Intertubular blood-forming tissue in the kidney of *Ammocoetes*. (Photo by Torrey.)

the splanchnic mesoderm of the yolk sac (Figure 16-5*A*). Fluid-filled spaces gradually appear within the blood islands, resulting in the separation of a peripheral layer of cells enclosing a central cluster (Figure 16-5*B*). Accumulation of the fluid and expansion of the spaces further define the peripheral layer, which may then be identified as the *endothelium* of a primordial blood vessel (Figure 16-5*C*). The enclosed cells represent the first blood cells, and the accumulated fluid is the plasma medium. The scattered, hollow vesicles formed in this fashion expand and coalesce so as to produce an irregular, anastomosing network of thin-walled vessels. The network is subsequently expanded by growth and by extension of previously established vessels. Meanwhile, the spreading yolk sac plexus makes contact with vessels arising within the body of the embryo. The usual pattern is that of the appearance of a pair of vessels ventral to the prospective pharyngeal region of the gut. Growing rearward and outward, these vessels communicate with the yolk sac plexus and constitute the *vitelline*

veins into which the plexus drains; growing forward and dorsad around each side of the pharynx as the first pair of *aortic arches,* they join newly formed vessels passing backward above the gut, the *dorsal aortas.* The latter then extend themselves as *vitelline arteries* to join the yolk sac plexus and thus complete the yolk sac arc (Figure 16-6).

Overlapping in development and supplementing this primordial vitelline circuit are the other embryonic and extraembryonic channels, with which we shall ultimately be concerned. For the present, however, let us center attention on the two intraembryonic vessels ventral to the pharynx. These are the forerunners of the heart.

ORIGIN OF THE HEART

Although the adult heart is an unpaired structure, it is in most vertebrates of bilateral origin. Exceptions to this generation are found in the lower fishes and in Amphibia in which the heart is from the very

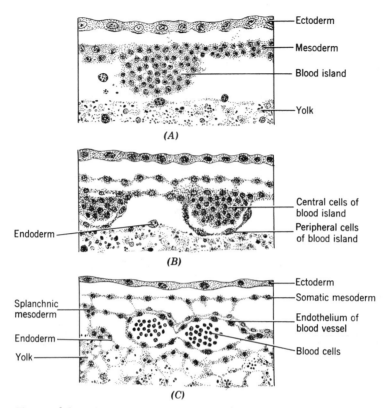

Figure 16-5   Differentiation of blood islands. Blood islands in extraembryonic blastoderm of (*A*) 8-hour, (*B*) 24-hour, and (*C*) 33-hour chick embryos. (From Patten and Carlson, *Foundations of Embryology,* courtesy McGraw-Hill Book Company.)

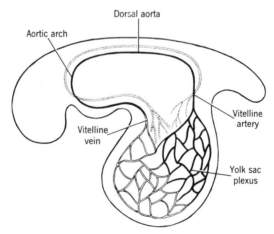

Figure 16-6   Primordial vitelline (yolk sac) arc.

start a single tube. But in all the other vertebrates, even allowing for some special circumstances of development in mammals, the heart derives from a pair of primordia. It is therefore appropriate to employ a single representative form, the bird embryo, to illustrate the fundamentals of the usual manner of origin of the heart.

It will be recalled (Chapter 8) that in the chick embryo of about 18 or 19 hours of incubation, the ectoderm immediately dorsal to the head process and prechordal mesoderm will have thickened to form a neural plate. Very shortly the plate becomes longitudinally depressed, and its thickened margins begin to elevate. The result is the conversion of the plate to a trough, the neural groove, flanked by neural folds. Meeting of the folds to produce the neural tube begins at about the level of the future midbrain and from there progresses forward and rearward. At about the same time that the neural groove and folds are being established, a crescentic depression appears in the ectoderm immediately in front of them. This depression is the *head fold,* which serves to delimit the anterior boundary of the embryonic head from the extraembryonic blastoderm beyond. Further downward and backward progression of the head fold, coupled with forward growth of the embryo itself, causes the head region to be completely undercut. The ectodermally lined space thus established between the head and the extraembryonic blastoderm is known as the *subcephalic pocket* (Figure 16-7).

Coincidentally with the inauguration of the head fold, the embryonic endoderm folds upon itself so as to produce an endodermal bay ventral to the neural tube. This endodermal pocket is the *foregut,* which opens posteriorly into the yolk-filled area under the blastoderm by way of the *anterior intestinal portal* (Figure 16-7). The foldings of ectoderm and endoderm correspond initially so closely that one gets the impression that both are parts of one

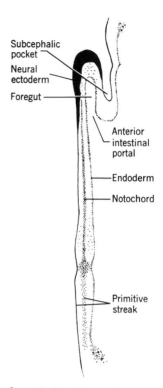

Figure 16-7 Chick embryo of approximately 24 hours incubation age in median sagittal section.

developmental process. It has been demonstrated experimentally, however, that they are actually independent events. This is revealed especially in later hours when the detachment of the foregut proceeds faster than that of the head. Consequently, the distance increases between the margins of reflection of the ectoderm and endoderm from the extraembryonic blastoderm onto the head and foregut, respectively, a circumstance closely linked with the initiation of the heart and the division of the coelom within which it will lie.

In a chick embryo with an incubation age of approximately 24 hours, the coelom on each side of the anterior intestinal portal shows a marked local enlargement. These dilated regions of the coelom, known as the *amniocardiac vesicles,* gradually push in from each side toward the midline, where they will eventually unite in a single cavity

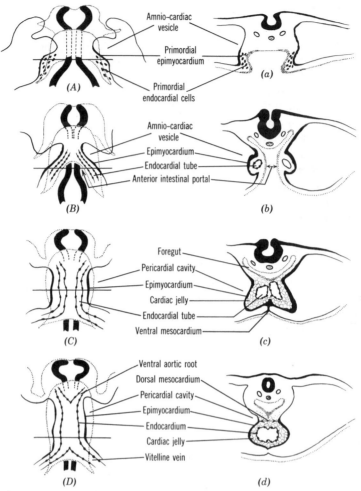

Amnio–cardiac vesicle
Primordial epimyocardium
Primordial endocardial cells
(A) (a)

Amnio–cardiac vesicle
Epimyocardium
Endocardial tube
Anterior intestinal portal
(B) (b)

Foregut
Pericardial cavity
Epimyocardium
Cardiac jelly
Endocardial tube
Ventral mesocardium
(C) (c)

Ventral aortic root
Dorsal mesocardium
Pericardial cavity
Epimyocardium
Endocardium
Cardiac jelly
Vitelline vein
(D) (d)

Figure 16-8   Diagrams of chick embryos showing the origin and subsequent fusion of the paired primordia of the heart, the establishment of the pericardial cavity, and the elongation of the foregut. *Left,* ventral views; *Right,* corresponding transverse sections at the levels indicated by heavy lines across ventral-view figures. Time interval: 25th to 30th hours of incubation.

(Figure 16-8). This serves to introduce a portion of the coelom into the progressively elongating space bounded in front by the ectodermal wall of the subcephalic pocket and behind by the endodermal wall of the intestinal portal. Accordingly, the ectodermal boundary is converted to a somatopleure and the endodermal boundary to a splanchnopleure. In the meantime, the lateral walls of the anterior intestinal portal rapidly converge, thus lengthening the foregut and moving the portal farther and farther caudad. Because the somatopleure remains practically stationary, the distance between the somatopleuric and splanchnopleuric boundaries is constantly increased, as is the extent of the intervening portion of the coelom—a portion that may now be identified as the *pericardial cavity.* Along with this process comes the inauguration of the heart.

The first visible indication of the embryonic heart is found in thickenings of the splanchnic mesoderm bounding the amniocardiac vesicles on each side of the anterior intestinal portal (Figure 16-8A, *a*). These thickenings are the primordia of the *epimyocardium,* which is destined to produce the external coat of the heart (epicardium or visceral peritoneum) and the muscular layer of the heart (myocardium). (Recent electron micrographic studies suggest that the epicardium and myocardium may be distinct from the very start.) Coincidentally, cells detach themselves from the epimyocardium and aggregate loosely between the splanchnic mesoderm and the adjacent endoderm of the intestinal portal. These cells constitute the primordium of the *endocardium,* the endothelial lining of the heart. The endocardial cells arrange themselves so as to produce a pair of thin-walled tubes (Figure 16-8B, *b*), each of which is continuous with the vitelline veins rearward and with the developing ventral aortas and first pair of aortic arches forward.

Each layer of epimyocardium next begins to bulge laterad into the coelom and each endocardial tube thus becomes enclosed in an epimyocardial semicircle. With the continued coming together of the walls of the intestinal portal to extend the length of the foregut, and the associated convergence of the amniocardiac vesicles, the endocardial tubes are progressively brought together immediately below the gut (Figure 16-8B–D, *b–d*). The epimyocardial halves likewise come together to form a single tube about the endocardium, although initially retaining continuity with the general splanchnic mesoderm above and below. Thus, the primitive heart, consisting essentially of a tube within a tube, remains suspended from the floor of the pharynx by a double mesodermal membrane, the *dorsal mesocardium,* and attached ventrally to the splanchnopleure by a *ventral mesocardium.* The ventral mesentery ruptures and disappears almost as rapidly as it is formed, bringing the two halves of the pericardial cavity into confluence. The dorsal mesentery is retained for a period, but it too ultimately disappears so that the heart tube

will be suspended hammocklike by its blood vessel connections at each end.

The origin of the heart from paired primordia is dramatically emphasized by experimental procedures that interfere with the normal midline fusion. It has been shown in chick embryos, for instance, that each of the two bilateral rudiments will produce a separate heart when an appropriately placed wedge prevents them from uniting. Double hearts also result when the paired primordia of the heart of the embryo rat are cultivated separately in tissue culture. And it has been demonstrated in Amphibia, where morphologically the heart appears to be single from the start, that a functioning heart will still develop after removal of a lateral half of the original primordium; or that two hearts will develop from one primordium split lengthwise.

When the heart tube is first established, it exhibits no notable subdivisions. It is necessary at this point, however, to indicate the divisions to be, for one important aspect of the fusion of the bilateral primordia, still to be considered, can be treated only in terms of the prospective divisions of the heart. Taking them in order from the intake end at the rear to the discharge end at the front, they are *sinus venosus, atrium, ventricle,* and *truncus arteriosus* (see Figure 16-13).

In the account of the fusion of the two primordia as given so far, the impression was probably left that union of the tubes occurred simultaneously along their lengths. Actually, the tubes unite sequentially from front to rear. The truncoventricular part is established first, then the atrium is added, and finally the sinus venous is added. So it is that in a chick embryo of 30 hours, fusion of the primordia is complete only in the truncoventricular region; the prospective atrium and sinus venosus are still double (Figure 16-9A). The definitive atrium is established at about the 38th hour (Figure 16-9B), and the sinus venosus some 10 hours later (Figure 16-9C, D). By 33 hours the ventricular portion of the heart is bent considerably to the right. This

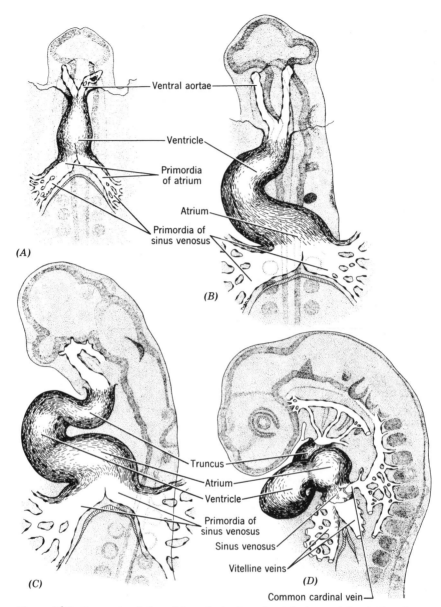

Figure 16-9 Progressive fusion of the paired primordia of the heart of the chick embryo. (*A*) At 9-somite stage (approximately 30 hours). Truncoventricular region established; primordia of atrium and sinus venosus still paired. (*B*) At 16-somite stage (approximately 40 hours). Atrium established. (*C*) At 19-somite stage (approximately 46 hours). Primordia of sinus venosus beginning to fuse. (*D*) At 26-somite stage (approximately 56 hours). Sinus venosus established. (From Patten and Carlson, *Foundations of embryology*, courtesy McGraw-Hill Book Company.)

bending, of which we shall have more to say later, is facilitated by the breakdown of the dorsal mesocardium at this level. For the present the atrial portion remains anchored, but with the later rupture of its mesocardium it too will be free to move forward somewhat, as the ventricle undergoes a spiral twist that carries it to the right and, topographically, somewhat caudad to the atrium.

To return to the main point of the moment, the heart tube forms sequentially from front to rear and as it forms, it begins to beat. The first contractions, therefore, are in the ventricular myocardium. The initial twitchings are seen at about 29 hours of incubation age and involve only localized patches on the right side alone. Gradually the areas of contraction spread until all the right side is involved, followed soon thereafter by a slow, rhythmical contraction of the entire ventricular myocardium. The few critical studies that have been made on the initiation of the first heartbeats in other vertebrates show, as might be expected, some deviations from the avian pattern. In the salamander, *Amblystoma,* for example, the first contractions appear in the middle of the prospective ventricle rather than on the right side as in the chick. Especially interesting is the situation in rat and rabbit embryos. Here the primordial heart tubes begin to pulsate prior to their fusion, with the left one showing activity some 2 hours in advance of the right. But these variations notwithstanding, the first contractions in all—bird, amphibian, and mammal—involve the ventricular myocardium.

With the later coming together of the bilateral primordia to form, in order, the atrium and sinus venosus, these divisions too begin to contract. As the atrium is established it not only starts to pulsate, but does so at a rate faster than that of the previously established ventricle. Still more remarkable is that the ventricle then increases its rate of contraction to keep pace. So it is again when the sinus venosus is established. It contracts at a still faster rate, and the

atrium and ventricle in turn step up their rates. What we are saying is that each new part of the heart that is added behind the first-formed truncoventricular region has a higher intrinsic contraction rate than the part immediately in front of it, and that the region with the highest rate, ultimately the sinus venosus, sets the pace for the entire heart. In effect, the pacemaker inaugurates contraction waves of a given frequency, which sweep through the tubular heart toward its anterior outlet.

That the region of the heart with the highest rate of contraction dominates the other regions has been demonstrated by a number of elegant experiments. If, for instance, the chick heart is transected into sinoatrial and truncoventricular halves, the sinoatrial division will continue to beat at essentially the rate of the intact heart, whereas the ventricle will revert to its original slower pace. A variation of this experiment consists of cultivating two such heart halves close together in a culture medium. In due time the parts are united by a bridge of myocardial tissue, and as soon as such material continuity is provided, the more slowly beating ventricular myocardium will once more step up to the rate of the sinoatrial region. Another demonstration of the role of the sinus venosus as a pacemaker is found in amphibians, where combinations of parts of the embryonic heart are assembled from two species with different heart rates. If a sinus venosus featuring one intrinsic rate is grafted to a ventricle with a different cardiac rate, the ventricle is driven at the rate of the donor pacemaker.

How each added heart region imposes its rate on the region in front and how the sinus venosus synchronizes the whole at this early time remain a mystery. Nerve fibers are yet to be supplied to the heart and the heart's own internal conducting system, which will ultimately govern the rhythmicity of its beat, is yet to be established. It has been suggested that progressive acceleration of heart rate may be correlated with increasing blood pressure. However, there are contradictory observations that suggest just the opposite: that heart rate is stepped up when blood pressure is reduced by experimental exsanguination.

Processes of Cardiac Morphogenesis. The events just described are the visible morphogenetic ones. But what do we know of the events and conditions that presage the primordial heart tubes? The answers to this question require a recollection of the early developmental history of the chick blastoderm (see Chapter 8).

It will be remembered that at the time the egg is "laid," that is, the zero hour of incubation, the blastoderm consists of two layers: an epiblast representing prospective ectoderm and mesoderm, and a lower endoderm. If a blastoderm in this state is cut into fragments and these fragments are cultivated in tissue culture, pulsating cardiac muscle will differentiate within those pieces taken from the edge of the blastoderm but not from those taken from the center. This result is interpreted to mean that at this early time cells capable of forming heart tissue are distributed around the periphery of the epiblast.

Next recall the pattern of migration of the prospective mesoderm. In regular order the areas destined to become lateral mesoderm, notochord, somites, and so forth, wheel toward the midline to converge upon and pass through the primitive streak on the way to their ultimate destinations. It has been demonstrated that the heart-forming cells are included in this mesodermal migration. Specifically, it has been shown that these cells first move from their initial position around the periphery of the epiblast toward the rear of the primitive streak, then pass through the streak, and finally assemble in two areas, one on each side of Hensen's node and the head process (Figure 16-10). Clearly, then, the bilaterality of the primordial heart is established very early. Recent studies have shown that the heart-forming areas are somewhat narrower than those pictured in Figure 16-

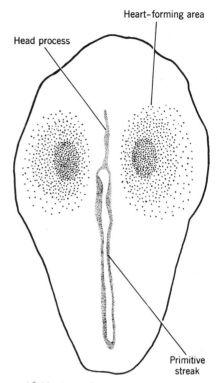

Figure 16-10 Heart-forming areas in a chick blastoderm of head process stage. Density of stippling indicates intensity of cardiogenic potency within each area. (After Rawles.)

10), but this refinement in no way alters the basic bilaterality of the primordia. Another refinement has been one of mapping the primordia. Each lateral heart-forming region consists of curved subdivisions of prescribed potentiality: an anteromedial band provides the truncus arteriosus, a midportion gives rise to the ventricle(s), and a caudolateral area furnishes the sinoatrial region (Figure 16-11).

Two subtle biochemical tools have provided a dramatic demonstration of the assumption by the heart cells of a specific chemical identity concomitantly with their migration and assembly in two population areas. One of these utilizes the principles of immunology. If an extract of heart muscle from an adult chicken is injected into a rabbit, antibodies will develop in the rabbit blood serum. By appropriate refinement

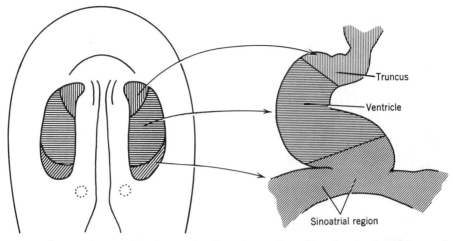

Figure 16-11 Embryonic fate of areas of cardiogenic mesoderm. (Based on data by Stalsberg and DeHaan, *Develop. Biol.,* Vol. 19.)

individual antisera specific to each of the principal proteins that characterize heart muscle can be prepared, and these will show clear-cut precipitin reactions when mixed with substances identical with or closely related to the proteins in the heart. These sera may then be mixed with extracts of whole blastoderms or selected pieces of blastoderm of varying ages, and through the resulting reactions the presence or absence of heart muscle proteins can be detected. Such tests reveal that the first heart muscle protein makes its appearance shortly after the inauguration of the movements of prospective mesoderm toward the primitive streak. With the completion of the movements through the streak, the protein is widely distributed, but soon thereafter becomes restricted to the heart-forming areas. It is not presently clear whether the aggregation of the cells with this distinctive biochemical property in two "organ-forming areas" is literally that, namely, the self-sorting of these cells from the general mesoderm, or the consequence of an elimination process whereby all the mesodermal cells except those in the heart-forming areas lose an original ability to synthesize cardiac muscle proteins. Be that as it may, through this delicate biochemical detection system it has been shown that the cells destined to

form the heart acquire their distinct chemical properties coincidentally with their migration to the general regions where the heart primordia are to appear.

Another approach has been through the utilization of various agents that have a selective effect on the chemical reactions that prevail during the biochemical differentiation of cells. It is known, for example, that the metabolic pathways followed in the differentiation of the heart and brain differ markedly and that this difference is manifested by the operation of different enzyme systems. A specific enzyme inhibitor may therefore be employed to pinpoint the time and place of a given metabolic pathway. One such inhibitor is sodium fluoride, which in low concentrations selectively disrupts the metabolism of the heart-forming cells and ultimately destroys them. Through its use the locations of the cells from the time of their diffuse distribution in early primitive streak stages to their ultimate assembly in the paired heart-forming areas can be established (Figure 16-12).

The differentiation and deployment of precardiac mesoderm are also known to be conditioned by the underlying endoderm. It has been demonstrated in amphibian embryos that if the endoderm is extirpated at the neurula stage of development, the heart

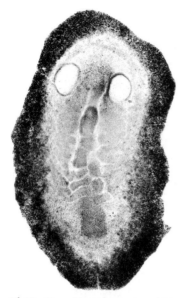

Figure 16-12 Heart-forming areas in a chick embryo appearing as vacuities following destruction of prospective heart cells with sodium fluoride. (Photo courtesy of L.M. Duffey.)

fails to form; or if fragments of precardiac mesoderm are cultivated in a culture medium, normal heart tissue does not result. Yet when a culture includes endoderm, vigorously beating, well-formed hearts develop. It is concluded, then, that the endoderm exercises an essential inductive influence on the precardiac mesoderm. Whether the same conclusion may be drawn for mammals and birds is not clear. There are no relevant experiments on mammalian embryos, and results have been conflicting when the epiblast of avian embryos has been cultivated in the absence of endoderm.

It is known in the case of the avian embryo, however, that the oriented movements of the precardiac mesodermal tissues from their original locations alongside the head process to their positions on each side of the anterior intestinal portal involve contact with the endodermal substratum. That these movements are correlated with and dependent on the folding of the endoderm to produce the foregut and the subsequent elongation of the foregut is indicated by two

types of experiments. (1) Any disruption of the pattern of folding of the endoderm produces a disruption in the migratory behavior of the precardiac tissue. (2) If the blastoderm is denuded of endoderm, the integrated sheets of precardiac cells fail to undergo migration. Instead, they organize into small cell clusters that differentiate into numerous pulsating vesicles of heart tissue.

EXTERNAL FORM OF THE HEART

The coming together of the paired primordia results fundamentally in the construction of a straight, cylindrical tube. But even as it is being formed, this cardiac tube becomes differentiated into four primary chambers—from rear to front, *sinus venosus, atrium, ventricle,* and *truncus arteriosus* (Figure 16-13). A confusion of terms is frequently encountered in reference to the anteromost division of the heart. Some writers refer to the truncus as the conus arteriosus. Actually, however, the conus is the narrowed portion of the ventricle leading to the truncus and properly should not be used synonymously with truncus. The term *bulbus cordis,* employed with reference to the mammalian heart, does correspond to the truncus. But the expression bulbus arteriosus refers to a frequently encountered enlargement of the ventral aorta and does not apply to the heart at all.

To avoid confusion, we shall restrict ourselves to truncus arteriosus and its mam-

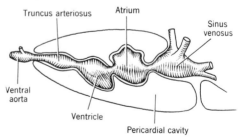

Figure 16-13 Hypothetical primitive arrangement of heart chambers. (After Goodrich.)

malian equivalent, bulbus cordis, in speaking of the most anterior division of the heart. The in-line arrangement pictured in Figure 16-13 is actually never seen in any adult vertebrate; in every instance the heart tube tends to form an S-shaped curve. It is commonly stated that this bending is an accommodation to the spatial limitations of the pericardial cavity. Indeed it does provide compactness with length, but the bending pattern is no mere consequence of mechanical forces deriving from a short coelomic compartment. No simple space restrictions would produce any regularity of bending pattern. This is something intrinsic to the cardiac tube itself, as revealed by tissue culture experiments in which an explanted embryonic heart tube, with plenty of room to extend itself lengthwise, still bends in characteristic fashion.

Recent studies involving cell counts in the embryonic chick heart suggest that this intrinsic behavior of the heart tube reflects an asymmetry of cell numbers on the two sides. A statistically greater number of cells are found on the right side at the anterior level of the S-shaped heart tube, whereas caudally the number is greater on the left side. That is, a higher cell count is correlated with expansion of the tube wall at the convexity of a bend, so the truncoventricular region swings to the right and rearward and the sinoatrial region swings to the left and forward.

The detailed twist pattern varies from one vertebrate to another. Thus in the

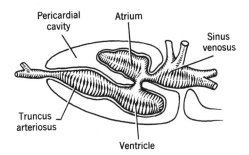

Figure 16-14 Heart chambers of an elasmobranch (shark). (After Goodrich.)

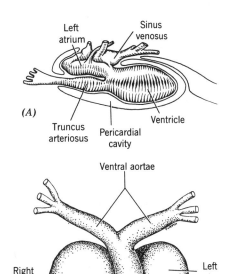

Figure 16-15 Amphibian heart. (*A*) View from left side. (After Goodrich.) (*B*) Ventral view. (Drawn from a Ziegler model.)

shark, the ventricle bends only moderately to the right. Most of its movement is ventrad and caudad, so that finally it lies immediately beneath the atrium (Figure 16-14). In Amphibia, the frog for example, not only does the ventricle move to the right and to the rear, but the atrium and sinus venosus move forward. The result is that the atrium comes to lie immediately above the truncus (Figure 16-15A). In the meantime the atrium will have dilated bilaterally, as an accompaniment of its internal subdivision (to be considered shortly), and the truncus thus lies in the shallow external groove separating the two halves of the atrium (Figure 16-15B). A similar pattern prevails in amniotes. As seen in the developing human heart (Figure 16-16), the ventricle first

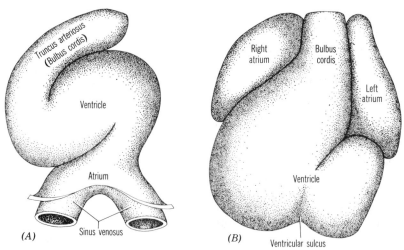

Figure 16-16   Ventral views of two stages in the development of the human heart. (Drawn from Ziegler models.)

swings to the right and then to a position behind the atrium; the truncus comes to lie between the lateral dilations of the atrium. Meanwhile, a distinct median furrow at the apex of the ventricle marks externally the subdivision of the ventricle into right and left chambers. It is to the internal partitioning of the heart that we now must turn.

EVOLUTION OF THE DOUBLE HEART CIRCUIT

A major circumstance in the morphogenesis of the heart is a redesigning of its internal architecture in company with the replacement of gill by lung respiration. Reduced to its essentials, the problem of developmental anatomy is as follows: In gill breathers, the gills lie in the main circulatory path between the heart and the various regions of the body (Figure 16-17); that is, the heart, gill, and systemic circulations are arranged in series. Now, a hydraulic engineer who wanted to design a lung breather would place the lungs in the same position in the circulatory arc as that occupied by gills and thereby achieve an equally simple morphological and functional layout. But it so happens that lungs (and those piscine air

bladders that serve as lungs) are set off the main line of the circulatory arc. Accordingly, some supplementing and complicating provisions are required. In front of the heart there must be a subdivision of the arterial channels, so that deoxygenated blood is routed to one destination, the lungs, and oxygenated blood to another, the body. There must also be separate pathways for the return to the heart of aerated blood from the lungs and deoxygenated blood from the body. Most importantly, there must be provision for keeping these two bloodstreams separated in the heart. In other words, the original "in-series" arrangement must be redesigned to provide two circuits, pulmonary and systemic, operating in parallel. The story of the establishment of the separate routes from and to

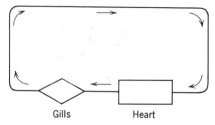

Figure 16-17   Schema of the circulatory arc of a gill breather.

the heart is one with which we shall deal in later considerations of the arteries and veins.

The developing mammalian heart exhibits the customary four chambers—sinus venosus, atrium, ventricle, and truncus arteriosus (bulbus cordis). First the atrium, and subsequently the ventricle as well, is divided into two compartments. The truncus is eliminated as such by being split into pulmonary and systemic channels; the two sides of the sinus are eliminated by reduction and absorption. These steps are presumed to be paralleled phylogenetically: the heart of a fish corresponds to the primitive four-chambers-in-series stage, that of an amphibian presents the divided atrium, and reptilian hearts show a progressively dividing ventricle. But a closer look at the facts indicates that this was not the phylogenetic order of events at all.

Fishes. The structure of the heart in the Chondrichthyes and actinopterygian Osteichthyes generally conforms to the basic pattern of four chambers in series. Such differences between them as do prevail pertain primarily to details of structure, number, and arrangement of the valves within and between the chambers, details that are beyond our concern. Unfortunately, the prime piece of evidence we need will likely never be available, namely, the status of the heart in the rhipidistian crossopterygians from which the original Amphibia, and thus tetrapods as a group, stemmed. They are known to us only as fossils. The other group of Crossopterygii, the coelacanths, whose internal anatomy is slowly being revealed as the number of specimens slowly increases, are apparently so specialized as to be little indicative of the basic conditions we should like to know. There are, however, important inferences to be drawn from the living relatives of the Crossopterygii, the Dipnoi, or lungfishes. The heart of the South American lungfish, *Lepidosiren,* is highly illustrative.

Starting embryonically as a conventional tube, the heart of *Lepidosiren* twists so as to bring the ventricle below and to the rear of the atrium and ventral to the sinus venosus (Figure 16-18). The atrium, ventricle, and truncus then become almost completely divided into right and left chambers. The embryological detail of this division is intricate, but its highlights can be delineated with profit. As for the atrium, a ridge

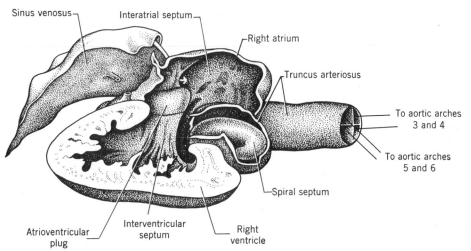

Figure 16-18 Heart of an adult lungfish, *Lepidosiren,* with wall removed to show right-hand chambers. Arrow indicates opening of sinus venosus iinto right atrium. Exits to aortic arches schematized. (After Robertson.)

derived from the endocardium of the rear wall of the chamber projects forward to join the so-called "atrioventricular plug" of which more will be said shortly. This serves to provide a partition that separates the original atrium into right and left chambers. Meanwhile, the ventricle also becomes subdivided, but its partition has a double origin. In part it derives from an endocardial ridge that projects forward from the rear wall toward the truncus. At almost the same time, however, numerous muscular buds, each covered with endocardium, arise from the lateral walls as well. These buds gradually coalesce to form a branching network (trabeculae) of muscles, which unite with the primary endocardial ridge. The significance of this double origin of the *interventricular septum* will become evident when we arrive at a consideration of the amniote heart.

To obtain an understanding of the atrioventricular plug, it is necessary to look ahead momentarily to a developmental feature of the heart of a higher vertebrate. In the mammalian heart, for example, the atrioventricular canal, the passageway between atrium and ventricle, becomes divided in two by the growth of dorsal and ventral endocardial cushions which fuse to form a partition (see Figure 16-21). The point is that in the lungfish only one of these cushions, the dorsal one, appears, and this becomes elaborated as a valvelike plug guarding the canal. In other words, the dipnoan atrioventricular plug is homologous only to the dorsal endocardial cushion of higher vertebrates.

The partitioning of the truncus is similar to that of the mammalian heart. As a consequence of the general torsion of the original cardiac tube, the truncus exhibits a marked spiral twist. Four longitudinal endocardial ridges—dorsal, ventral, right, and left—now appear. The dorsal and ventral ones are vestigial, but that on the right fuses with the one on the left to produce a septum dividing the truncus lengthwise. Because of the basic twist of the truncus, the septum also pursues a spiral course; hence

it is commonly referred to as the *spiral septum (valve)*.

In the adult dipnoan heart (Figure 16-18), the undivided sinus venosus receives the venous blood from the body and empties into the right atrium. The return stream from the bilobed air bladder (lungs) bypasses the sinus and empties aerated blood directly into the left atrium. Only because the atrioventricular plug is not fully efficient is there some mixture of these streams as they leave the atria, but thanks to the interatrial, interventricular, and spiral septa the two streams are still reasonably well segregated. Moreover, the outlets from the truncus are so connected to the aortic arches that those vessels (arches 3 and 4) leading to the head and general body circulation are supplied mainly with oxygenated blood and those (arches 5 and 6) leading to the lungs carry mainly deoxygenated blood.

Amphibia. If it may be granted that the Crossopterygii from which the Amphibia stemmed had a heart comparable to the one just described, then it is likely that the original Amphibia did also. Though no modern Amphibia have hearts like this, it must be remembered that present-day amphibians are highly specialized and compared with their forebears show evidence of considerable regressive evolution. The loss of dermal armor and the regression of bone with concomitant retention of cartilage are instances in point. Thus it may very well be that division of the chambers of the heart into right and left sides was accomplished with the Crossopterygii and has been a feature of tetrapods in general since their origin; modern Amphibia, with their incompletely divided hearts, may illustrate degeneracy rather than primitive intermediacy between fishes and amniotes.

The ventricle is undivided in all Amphibia. Atrium and truncus, however, show a variable status from one group to another. In the Anura the atrium is divided into two, the right one receiving via the sinus venosus the deoxygenated blood from the body and the left one receiving the pulmonary return. The truncus is likewise divided by a spiral septum, and the diverging streams of the truncus are in turn subdivided internally into three channels leading respectively to aortic arches 3, 4, and 6 on either side. Many physiological studies on the anuran heart have produced some conflicting conclusions. The consensus is, however, that a good degree of separation of the streams of aerated and nonaerated blood is achieved. It appears that the pulmonary and systemic streams reaching the ventricle from the left and right atria are kept reasonably well separated by ridges and pockets in the ventricular wall. The spiral septum in the truncus maintains this separation, so that oxygenated blood is sent to the head and body via aortic arches 3 and 4 and deoxygenated blood is sent to the lungs (and skin) via arch 6. Yet the separation is a somewhat inefficient one at best. The inference is that there has been a decline in physiological efficiency of a partitioning system that once included a ventricular septum.

A further decline, both anatomically and physiologically, is revealed by the Urodela and Apoda. In the terrestrial salamanders the truncus is much less well divided; in the aquatic urodeles and the Apoda the spiral septum is entirely absent. Moreover, in many urodeles the interatrial septum is perforated, and in the lungless salamanders the interatrial septum does not appear at all, representing a return to a sharklike or teleostlike condition.

To sum up, then, it is very possible that the essential division of the heart into right and left sides arose only once in vertebrate evolution and did so with the crossopterygian ancestors of the tetrapods, and the lungfish heart is reminiscent of this. Although the original amphibian heart was probably fully partitioned, modern Amphibia have not "held the line," so to speak. Their history, as with certain other structural features, has been one of decline of the cardiac septa. Only with the amniotes

has the partitioning of the heart been maintained and perfected; so it is to these vertebrates that we now turn.

Reptilia. The developmental anatomy of the reptilian heart involves many complex details, and its interpretation has given rise to considerable controversy. Accordingly, we shall confine ourselves to a few salient points that are significant for the understanding of conditions in birds and mammals.

The atrium is divided by an interatrial septum into right and left chambers, which receive the systemic and pulmonary returns. At the opposite end of the heart, not only is the truncus subdivided by a spiral septum into systemic and pulmonary channels, but this division involves a splitting of the entire truncus to produce separate vessels. Moreover, the systemic trunk becomes subdivided so that three channels, two systemic and one pulmonary, leave the ventricular region of the heart. Now comes the complex and still debated matter of the subdivision of the ventricle and the relationships of the atria and subdivisions of the truncus to it.

Three different situations occur in the three major groups of living reptiles—Squamata (snakes and lizards), Chelonia (turtles), and Crocodilia (Figure 16-19). The septum subdividing the ventricle in the Squamata and Chelonia is a horizontal one, which sets off a dorsal chamber from a ventral one (Figure 16-19A, B). These chambers, however, are almost surely the equivalent of left and right, having been brought into the dorsal–ventral relationship by the embryonic twisting of the cardiac tube; therefore, the septum is judged by some authorities to be the equivalent of the dipnoan primary septum. With respect to the outgoing channels, the pulmonary trunk leaves the right ventricle in both groups, but in the Squamata both systemic trunks run out of the left ventricle, whereas in the Chelonia one systemic trunk leaves the left ventricle and the other leaves the

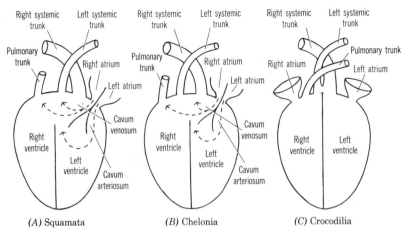

(A) Squamata   (B) Chelonia   (C) Crocodilia

Figure 16-19 Schemata of cardiac circulations in three categories of reptiles. (After Foxon.)

right (Figure 16-19A, B). As for the interventricular septum, it is not a complete one, permitting transfer of blood from one side to the other. This transfer is always from left ventricle to right ventricle, because in the hearts of Squamata and Chelonia both atria open into the left ventricle. However, because of the projection of the two flaps of the atrioventricular valve into the left ventricle, the left ventricle is subdivided imperfectly into two parts—a *cavum arteriosum* on the left and a *cavum venosum* on the right (Figure 16-19A, B). When the atria contract, the cavum venosum becomes filled with deoxygenated blood from the right atrium and the cavum arteriosum with oxygenated blood from the left atrium. Because the cavum venosum is nearer the incomplete interventricular septum, most of the deoxygenated blood in it flows past the edge of the septum into the right ventricle. Then when the ventricle contracts, its walls come in contact with the edge of the septum, thus temporarily segregating the two sides of the ventricle. The result is that the deoxygenated blood in the right ventricle is sent out the pulmonary trunk and, in the Chelonia, the right systemic trunk as well, whereas the oxygenated blood, mixed with some deoxygenated blood, in the left ventricle is sent through both systemic trunks

(Squamata) or the left systemic trunk (Chelonia).

The crocodilian heart presents an entirely different situation (Figure 16-19C). Here the interventricular septum is a complete one, providing for total separation of right and left ventricles. Still more significantly, this septum is a vertical rather than horizontal one and appears to be a new evolutionary development, which is not to be homologized with the interventricular septa of lizards, snakes, and turtles. It very likely corresponds to the dipnoan secondary septum. Furthermore, the right atrium now opens into the right ventricle and the left into the left ventricle. As in the Chelonia, the pulmonary trunk and one systemic trunk leave the right ventricle and the other systemic trunk leaves the left. (Where the right and left trunks cross each other and are in close contact they are in communication by a small aperture.) Although not in the direct line of ancestry of the birds and mammals, the crocodiles, as measured by the status of the heart, are closer to these warm-blooded amniotes than are either the Squamata or the Chelonia; that is to say, although not the direct progenitor, the crocodilian heart suggests conditions in birds and mammals, to which we may now turn.

Mammals and Birds. At first glance the fully formed, four-chambered hearts of mammals and birds appear to be the same, but important differences in the manner of partitioning of the primordial heart and the arrangement of channels at its discharging end in these two groups reveal that the similarity is more apparent than real. Let us first consider the mammalian heart and then point to the differences in the avian heart.

The human heart is employed for illustration. For descriptive purposes it is necessary to consider separately the several developmental events that take place.

As previously described (p. 374 and Figure 16-16), the ventricular portion of the original cardiac tube first swings to the right and then to a position behind the atrium. Subdivision of the atrium is foreshadowed by its lateral dilations between which the truncus (bulbus cordis) lies; a median furrow similarly marks the prospective subdivision of the ventricle. If, now, we were to prepare a frontal section at a level that would traverse the atrium and ventricle but miss the bulbus, a layout such as that pictured in Figure 16-20 would be seen. The

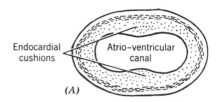
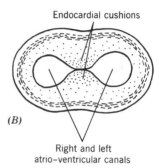

(A)

(B)

Right and left
atrio–ventricular canals

Figure 16-21 Partitioning of the atrioventricular canal by the meeting of endocardial cushions from above and below.

tured in Figure 16-20 would be seen. The prospective right and left atria, with the former receiving the sinus venosus, communicate with the ventricle via a constricted passageway termed the *atrioventricular canal.*

One of the earliest signs of subdivision is related to the atrioventricular canal. If the heart were sectioned transversely at this level, two ridges of loosely organized mesenchyme would be seen to develop in the walls of the canal. Each of these is known as an *endocardial cushion.* one projects downward into the canal from the dorsal wall and the other upward from the ventral wall (Figure 16-21A). These two ridges meet to form the definitive endocardial cushion that separates the canal into right and left channels (Figures 16-21B and 16-22).

Meanwhile, some of the interlacing muscular bands (trabeculae) at the apex of the ventricle become consolidated as a ridge that, capped with endocardial mesenchyme, steadily projects itself toward the endocardial cushion (Figures 16-22 and 16-23). This is the *interventricular septum.*

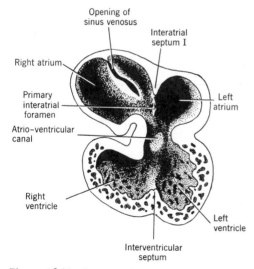

Figure 16-20 Diagram of a frontal section through the human heart to show the initiation of the partitioning of its chambers. (From Patten and Carlson, *Foundations of Embryology,* courtesy McGraw-Hill Book Company.)

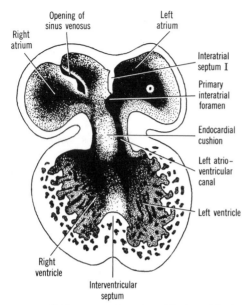

Figure 16-22 Progression of partitioning of the human heart. (This figure and Figures 16-23 through 16-25 from Patten and Carlson, *Foundations of Embryology,* courtesy McGraw-Hill Book Company.)

When it unites with the endocardial cushion, it partitions the ventricle into right and left sides (Figures 16-24 and 16-25).

While these events are taking place, the atrium is subdivided. The lateral dilations of the atrium initially communicate broadly through the *primary interatrial foramen* (Figure 16-20). A thin, crescent-shaped sheet of muscle, covered by endocardium, then begins to grow backward from the middle of the cephalic wall toward the atrioventricular canal. Because this is the first of two partitions destined to be formed, it is termed the *primary interactrial septum,* or simply *septum I* (Figure 16-22). As the septum expands, the primary interatrial foramen is steadily reduced and finally eliminated as the arms of the septum meet and fuse with the endocardial cushion in the atrioventricular canal (Figures 16-23 and 16-24). But just as it is about to fuse with the cushion, the more anterior part of the septum ruptures, thus reestablishing communication between right and left atria. Although in effect this represents no more

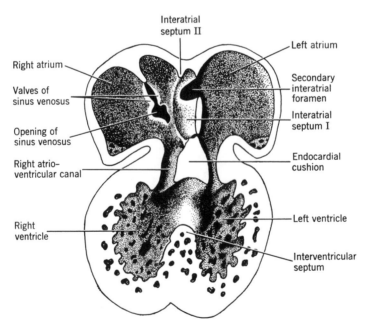

Figure 16-23 Beginning of the secondary interatrial septum and appearance of secondary interatrial foramen in the human heart. Embryonic age about 7 weeks.

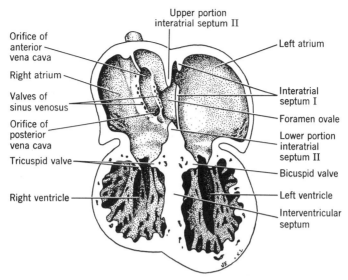

Figure 16-24 Internal configuration of the heart of a human embryo at 3 months.

than a partial restoration of the original communication between the atria, the new opening is called the *secondary interatrial foramen* (Figure 16-23).

It is at about the same time that a new septum, the *secondary interatrial septum,* or *septum II,* appears along the right side of septum I (Figures 16-23 and 16-24). Like septum I, septum II consists of a thin sheet of muscle covered by endothelium and is crescentic in shape. But unlike septum I, it never completes itself. As the limbs of the crescent join the endocardial cushion, an opening is retained, an opening so oriented that it lies alongside the entrance of the sinus venosus into the right atrium. This persistent passageway through septum II is called the *foramen ovale* (Figures 16-23 and 16-24).

This condition of two septa, each with an opening, is retained throughout fetal life and affords a means for passage of blood from the right atrium to the left (Figure 16-25). The fact that each septum overlaps the opening in the other guarantees that this will be a one-way passage. As already noted, the foramen ovale is adjacent to the opening of the sinus venosus (which later becomes the opening of the posterior vena cava upon the subsequent absorption of the sinus into the atrial wall). Thus, when the right atrium fills with blood, part of that blood may pass through the foramen ovale and, through its pressure, push aside the overlapping remnant of septum I and make its way by the secondary, interatrial foramen to the left atrium. Contraction of the left atrium and consequent buildup of pressure force the primary septum against the solidly anchored secondary septum, thus effectively closing the foramen ovale and preventing a return flow. The physiological significance of this arrangement and the establishment of the final interatrial septum are explored in a later consideration of the total fetal circulation and the changes that occur at birth.

Concomitantly with the partitioning of the atrium and ventricle, there comes a division of the truncus arteriosus into two separate channels. The process is inaugurated by the appearance of four longitudinal ridges of endocardial tissue, which pursue clockwise spiral courses along the length of the truncus (Figure 16-26). Two of these ridges are quite pronounced; the other two

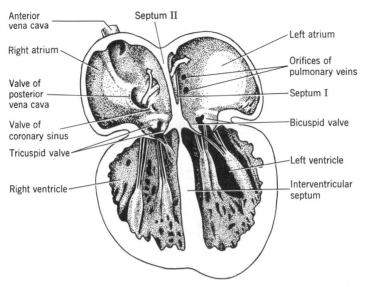

Figure 16-25 Internal configuration of the human heart in late fetal life. The divided arrow indicates the flow of a great part of the blood entering the right atrium from the posterior vena cava to the left atrium via the foramen ovale in the secondary septum and the secondary foramen in the primary septum; the remaining blood eddies back into the right atrium to mix with the anterior caval flow.

are of lesser thickness. For convenience they may be spoken of as major and minor ridges. In Figure 16-26 the major ridges are identified by the numbers 2 and 4, the minor ridges by 1 and 3. Division of the truncus is effected by the meeting of the major

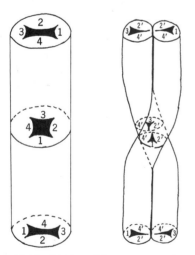

Figure 16-26 Division of the truncus arteriosus into pulmonary and systemic trunks. Numbers 1 and 3 indicate minor endocardial ridges; numbers 2 and 4, major endocardial ridges. Further explanation in text.

ridges and a coincidental constriction and subdivision of the total truncus. Each of the resulting two trunks will thus end up containing three ridges—two which are halves of the two major ridges and, the third, one of the minor ridges. By our numbering system, that is, both trunks contain halves of ridges 2 and 4, and one trunk contains ridge 1 and the other ridge 3. Because the original major ridges meet and become continuous with the interventricular septum, it follows that with the subdivision of the truncus one channel will run out of the right ventricle and the other out of the left ventricle. The channel out of the right ventricle constitutes the *pulmonary trunk,* which will lead cranially into the sixth pair of aortic arches from which the pulmonary arteries stem; the channel out of the left ventricle constitutes the *systemic trunk,* which will lead to the fourth pair of aortic arches. And because the original ridges pursued spiral courses, the pulmonary trunk coming out of the right ventricle will cross ventrad to the systemic trunk running out of the left ventricle (Figure 16-27). Finally, the three

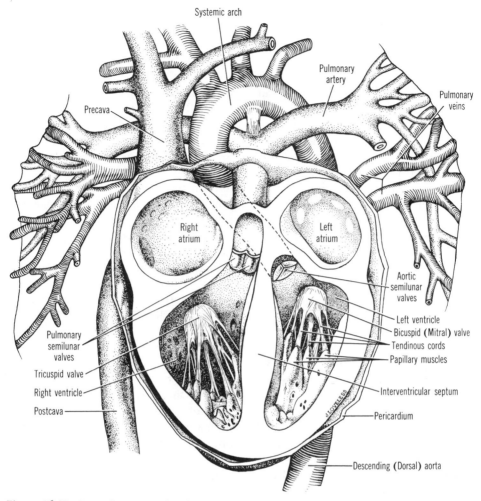

Figure 16-27 Internal anatomy of adult human heart.

ridges in each trunk will regress except near the exits of the trunks from the ventricles where they are not only retained, but enlarge and hollow out to form three thin-walled pockets, the *semilunar valves* (Figure 16-27).

The sinus venosus remains to be accounted for. It originally opens into the right atrium, and its inlet is guarded by a pair of flaplike valves (Figures 16-23 and 16-24). Gradually, however, the sinus is absorbed directly into the wall of the atrium and ceases to exist as a definitive entity. As a consequence, those vessels originally opening into the sinus (postcava, precava,

and coronary sinus) come to empty directly into the right atrium. During the process of absorption the left valve of the sinus, reduced in size, becomes fused with the secondary interatrial septum; parts of the right one are retained as valves guarding the orifices of the postcava and coronary sinus (Figure 16-25). Remember that the sinus venosus is the division of the heart that sets the pace, or rate, of rhythmic contractions of the entire heart. It continues to play this role, even though absorbed into the right atrium. The "pacemaker" is now identified with a cluster of specialized muscle, the so-called *sinoatrial node,* embedded in the muscular atrial wall from which emanate the specialized cardiac conduction fibers (p. 369).

The origin and nature of the semilunar valves guarding the exits of the pulmonary and systemic trunks have already been dealt with, as have the valves associated with the openings of the postcava and coronary sinus. Valves are also established in association with the right and left atrioventricular canals. Elevated folds of endocardium appear at the margins of these canals, folds that flatten and become anchored by fibrous chords to the ventricular muscles. Three such flaps appear on the right and constitute the *tricuspid valve* between the right chambers of the heart; two flaps make up the *bicuspid valve* between the left chambers.

The final form of the mammalian heart, as exemplified by that of humans, is pictured in Figure 16-27.

Rather than present a comparably detailed account of the morphogenesis of the heart of birds, only a few significant points of contrast are needed. These points are related to our introductory thesis: although the four-chambered hearts of birds and mammals appear to be very similar, there are certain differences in their morphogenetic histories that suggest that the similarity is more apparent than real.

The first issue has to do with the manner of origin of the interventricular septum. In mammals, as we have seen, the septum is inaugurated as a muscular elevation capped by a prominent ridge of endocardium, with the latter making the more significant contribution. In birds, however, the septum is a product of the coalescence of numerous muscular trabeculae, with the endocardium playing a minor role. These contrasting situations cause us to recall the double origin of the interventricular septum in the lung-fishes. Here, it will be remembered, the septum is a product of both an endocardial ridge and numerous muscle buds. The implication is that this double origin represents a "ground plan," which, carried into appropriate reptilian ancestors, was ex-

ploited differently by mammals and birds. The septum is derived mainly from the endocardial ridge in mammals and from muscle trabeculae in birds.

A second point of difference pertains to the manner of origin of the interatrial septum and the secondary openings through it. We have seen that the mammalian septum has a double origin, septa I and II, and that embryonic communication between the atria is effected by openings in each septum. In birds only one septum is formed, probably the equivalent of the mammalian primary septum. Several openings then develop in it, permitting a physiological exchange between right and left atria comparable to that in mammals. But here the similarity ends. There is no secondary septum operating with the primary one as a valvular mechanism, and at hatching the foramina are simply healed over.

A third difference relates to the subdivision of the embryonic truncus arteriosus. It should be recalled that in crocodiles there is first a subdivision of the truncus into pulmonary and systemic channels, and the latter in turn is split into two, the left one emerging from the right ventricle along with the pulmonary trunk and the right one emerging from the left ventricle. The avian truncus is likewise divided into pulmonary and systemic channels, and early in embryonic life there is also an incipient partitioning of the systemic trunk. But the left systemic channel disappears, leaving the pulmonary trunk alone to emerge from the right ventricle and the *right* systemic trunk to emerge from the left ventricle. In mammals, too, pulmonary and systemic trunks leave the right and left ventricles, respectively, but the systemic trunk is the counterpart of the *left* one. So although the avian arrangement could have been provided by the elimination of the left systemic trunk in a reptilian ancestor built to the crocodilian plan, the reciprocal elimination of the right trunk in a similar ancestor would not have created the mammalian arrangement. This would have resulted in both the pulmonary and systemic trunks leaving the right ventricle. The implication is that the discharging end of the mammalian heart, unlike that of birds, could not have been derived from a pattern like that of any extant reptile.

The fourth and final point is of least significance, but worth recording as a point of difference. Although the sinus venosus becomes considerably reduced, it is always retained in the adult bird as a discrete structure marked off externally from the right atrium by a sulcus and internally by the retention of its valves.

Summary. The vertebrate heart is basically a tube produced by the fusion of a bilateral pair of primordia. Through local dilations and constrictions the primitive cardiac tube becomes subdivided into four chambers: sinus venosus, atrium, ventricle, and truncus arteriosus, in that order from rear to front. Twisting of the tube brings a topographical reorientation of the chambers, so that the ventricle comes to lie to the rear of and beneath the atrium, but the basic serial arrangement is unaltered.

In gill breathers the heart remains in this form, receiving deoxygenated blood at the rear and pumping it out through the discharging end for aeration in the gills. But with the introduction of lungs, there is a separate return of oxygenated and deoxygenated blood to the heart and an anatomical separation (partial or complete) of the two streams of blood during their passage through the heart. Accordingly, some of the modern Dipnoi show a considerable measure of partitioning of the heart into right and left sides through the development of septa in the atrium, ventricle, and truncus.

Although the Dipnoi are not the immediate ancestors of the tetrapods, they are the nearest relatives of the Crossopterygii from which the original Amphibia stemmed, and thus the status of the dipnoan heart has great significance. The inference is that a subdivision of the heart into right and left sides was a feature of the first tetrapods and

has been retained in one form or another ever since. The lesser partitioning of the heart in modern Amphibia is indicative of specilization rather than evolutionary intermediacy between fishes and amniotes.

The hearts of modern reptiles present specializations that are difficult to interpret, particularly with respect to the interventricular septum. A current interpretation is one that takes as its starting point the double origin of the septum in the Dipnoi: a primary one consisting largely of an endocardial ridge and a secondary one derived by fusion of numerous muscle buds. Apparently, the interventricular septum in Squamata and Chelonia corresponds to the primary septum and that of Crocodilia to the secondary septum. In any event, the interventricular septa of birds and mammals appear not to be equivalent. That of birds is the secondary one; that of mammals, primary. Mammalian and avian hearts also show differences in the manner of development of the interatrial septum and the openings through it. In mammals two septa appear, each containing an opening; in birds there is only one with numerous perforations. The openings persist throughout embryonic life and provide a means of physiologically significant passage of blood from right to left atrium.

In looking for an ancestral pattern from which to derive the mammalian heart, particularly with respect to its discharge into the derivatives of the aortic arches, it seems necessary to turn to the dipnoan arrangement. Among modern reptiles, there are representatives of the diapsid line of evolution, a line that also produced the birds. The pattern of derivation of the heart of birds is therefore understandable. But the mammals came from the synapsid line of reptiles, of which there are no living descendants. For this reason and because of the specialization of modern amphibians, a great gap in the morphological history of the heart exists between crossopterygians and mammals.

THE ARTERIAL SYSTEM

A basic plan pervades the arterial system in all the vertebrates. Leaving the truncus of the heart, blood passes forward beneath the pharyngeal region of the gut via a *ventral aorta,* which just behind the level of origin of the thyroid gland forks into two. At intervals the ventral aorta(s) gives off paired *aortic arches,* which run dorsally through the pharyngeal arches and join on each side above the pharynx to form a pair of *dorsal aortas* (Figure 16-28). Anterior continuations of the ventral and dorsal aortas supply the head; behind the pharynx the two dorsal aortas unite to form a single *descending aorta,* which transports blood caudad, giving off branches en route to the various parts and organs of the body. Because the descending aorta and its branches (except for the vitelline arteries to the yolk sac and the allantoic arteries to the allantois and placenta to which special attention will be given) remain essentially standardized, the bulk of our concern is with the aortic arches.

THE AORTIC ARCHES

Although there may have been a greater number in ancestral vertebrates, with few exceptions the basic number of aortic arches in modern forms is six (Figure 16-

28). That is to say, six pairs of aortic arches appear in the embryo, although all six are not necessarily present at any one time and they are ultimately subject to greater or lesser modification, depending on the vertebrate. The basic scheme is that of a pair of aortic arches in front of each pair of pharyngeal pouches, thus corresponding to the pharyngeal arches. Accordingly, the first aortic arch is associated with the mandibular arch, the second with the hyoid arch, and so on.

In all the gill-bearing pharyngeal arches of fishes, the primary aortic arches become divided into two parts with an intervening capillary network. A given aortic arch thus consists of an *afferent branchial artery* bringing deoxygenated blood from the ventral aorta, a capillary system within the gill lamellae where aeration of the blood occurs, and an *efferent branchial artery* carrying oxygenated blood to the dorsal aorta (Figure 16-29). But many variations of this basic theme are present in the assorted categories of fishes. In the Chondrichthyes and Dipnoi, for instance, each efferent artery is subdivided so that each hemibranch is drained by a separate vessel. Moreover, in the sharks, there occurs a secondary shifting during ontogeny, whereby the efferent branchial arteries come to correspond to the gill chambers rather than to the gill arches (Figure 16-30). Another variation is found in the lungfish, *Protopterus* (Figure 16-31), in which the third and fourth arches

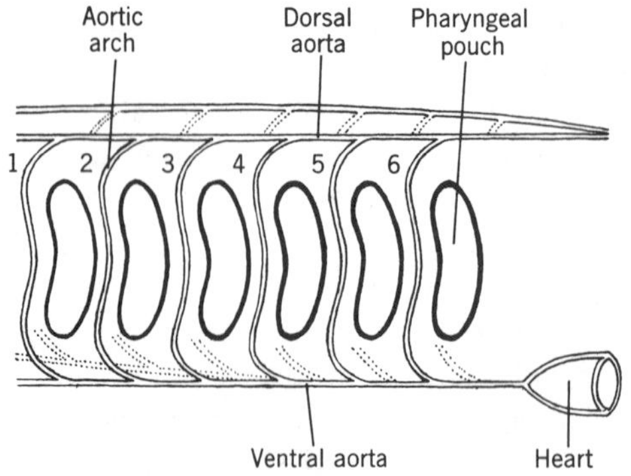

Figure 16-28 Ground plan of the aortic arches.

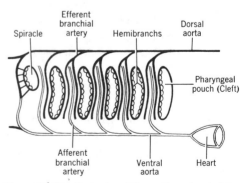

Figure 16-29 Basic plan of the aortic arches in fishes.

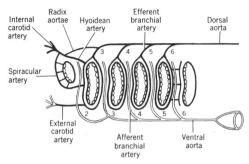

Figure 16-30   Aortic arches of elasmobranchs.

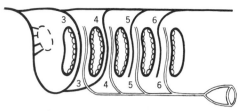

Figure 16-32   Aortic arches of teleosts.

run to the dorsal aorta without a break, because there are no gill filaments associated with the corresponding pharyngeal arches. As for the total number of aortic arches in fishes, the first one is consistently reduced and the second one variously modified. Thus in sharks (Figure 16-30) there are five afferent branchials representing the afferent portions of arches 2 to 6 and four efferent branchials representing the efferent portions of arches 3 to 6. The afferent half of arch 1 contributes to the external carotid; the efferent portions of arches 1 and 2 are incorporated into the vessels supplying the reduced gill in the first gill chamber (spiracle) and the hyoid arch. In teleosts and most other bony fishes only the last four definitive aortic arches are present, the first two having been reduced and incorporated into the cephalic blood supply (Figure 16-32).

The changeover from gill respiration to lung respiration in tetrapods is accompanied by two modifications: the aortic arches are no longer divided into afferent and efferent parts; there is a maturation of the pulmonary circuit. Taking the groups of tetrapods in order, let us first examine the Amphibia.

Only four of the basic six aortic arches show any developmental permanence in the Amphibia. Although the first two arches appear in early embryonic life, they shortly disappear and are then followed by arches 3 to 6 inclusive. However, the dorsal and ventral "roots" of the first two arches, that is, the dorsal and ventral aortas forward from the level of the third pair of aortic arches, are retained and extend into the head as the *internal* and *external carotids,* respectively (Figure 16-33). Branching off each sixth aortic arch is a *pulmonary artery* supplying a lung, hence the sixth aortic arches are often referred to as the pulmonary arches.

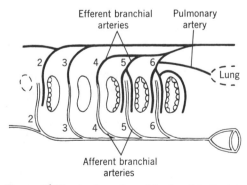

Figure 16-31   Aortic arches of the lungfish, *Protopterus.*

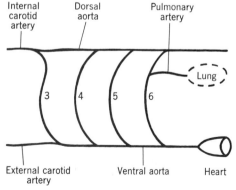

Figure 16-33   Basic plan of the aortic arches in Amphibia.

This pattern of four complete aortic arches actually prevails in many urodeles (Figure 16-33). But commonly in urodeles, and in the Apoda, the fifth arch is greatly reduced or even eliminated, leaving the aortic arch system to consist of arches 3, 4, and 6 (Figure 16-34). A third pattern is exhibited by adult Anura and results from the elimination of certain channels found in urodeles. One is the segment of the dorsal aorta on each side between the third and fourth arches, known as the "carotid duct"; the other is the so-called "arterial duct" (ductus arteriosus), which is the segment of the sixth arch between the level of exit of the pulmonary artery and the dorsal aorta (Figure 16-35). With the elimination of the carotid ducts, each third aortic arch becomes an integral part of the internal carotid, which stems together with the external carotid from that part of the ventral aorta running between arches 4 and 3 and is now designated the *common carotid artery.* The fourth arches remain as the *systemic arches,* each of which joins its fellow to form the dorsal aorta. The bases of the sixth arches become the *pulmonocutaneous arches,* which transport blood to the lungs and skin for aeration.

The following question of phylogeny may now be raised: Which of the aortic arch patterns seen in modern Amphibia is indicative of that in the first reptiles? The most likely answer is that it is one of those patterns in which the carotid and arterial ducts are still present, for although these vessels

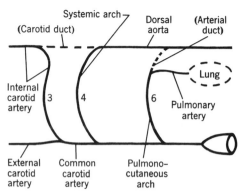

Figure 16-35 Aortic arches in the Anura.

are absent in most reptiles, they do remain in a few, leading to the view that their obliteration probably occurred independently in the Anura and amniotes in general. If it may be assumed that the original "stem reptiles" (Cotylosaurs) had an aortic arch pattern essentially like that of modern urodeles, then from this starting point the further history of the aortic arches can be traced. This history proceeded along two separate courses: one leading to modern reptiles and birds and the other to the mammals.

Although there are a few reptiles in which the arterial ducts are still present, these vessels are usually absent. That is, the pulmonary and systemic circuits are independent, as they are in the Anura. As we have already seen, this independence is furthered in all the amniotes by some measure of partitioning of the heart into right and left sides. This includes a splitting of the truncus along a lengthwise spiral course to produce a pulmonary trunk and a systemic trunk. A forward extension of the split into the ventral aorta results in the pulmonary trunk leading to the sixth aortic arches and the systemic trunk leading to the fourth aortic arches. We have observed, however, that in reptiles there is a subsplitting of the systemic trunk into two channels. One of these communicates with the right fourth aortic arch and the other with the left fourth aortic arch. The trunk leading to the right arch is the larger of the two and in all reptiles pro-

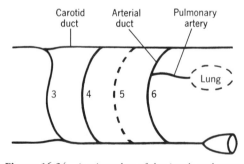

Figure 16-34 Aortic arches of the Apoda and many urodeles.

ceeds from the left ventricle; the smaller trunk leading to the left arch also leaves the left ventricle in the Squamata, but runs out of the right ventricle in crocodiles and turtles. In all reptiles the bilateral carotid system derives from a common stem, formed by the union of the bases of the third aortic arches, which branches from the right systemic arch. Figure 16-36 pictures a representative aortic arch pattern for reptiles.

Although in its early embryogeny the truncus of the avian heart gives evidence of triplication as in reptiles, only two channels are ultimately produced. One, the pulmonary trunk, runs as usual from the right ventricle and leads to the sixth aortic arches and lungs. The other, the systemic trunk, runs from the left ventricle to the dorsal aorta via the *right* fourth aortic arch (Figure 16-37). The left fourth aortic arch is eliminated. As in reptiles, the carotids branch from a common stem from the right systemic arch. This situation points to the contrast between the reptilian–avian line and the mammals. In reptiles and birds it is the right fourth aortic arch that either predominates (reptiles) or is employed exclusively (birds) in conjunction with the systemic circuit; in mammals, as we shall see shortly, it is the left.

Understanding of the aortic arches and their derivatives in mammals is furthered

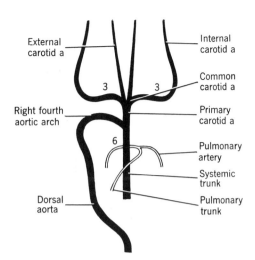

Figure 16-37 Ventral view of the aortic arches of birds.

by exploration of their embryogeny. Consistent with our past practice, the human is employed for illustration.

In the human embryo, as with mammals in general, the entire series of six pairs of aortic arches is not fully developed at any one time. The first pair, running through the substance of the mandibular arches, is followed shortly by the second pair associated with the hyoid arches (Figure 16-38). But as the third pair arrives on the developmental scene, the first arches begin to regress; similarly, the second pair declines as the fourth arches mature and increase in size. Inasmuch as the fifth arches never attain more than a transitory, rudimentary status, the sixth arches are the next and last to arise, and with their development the first and second will have disappeared entirely, except for remnants appropriated by the tissues of the mandibular and hyoid arches. With the first and second pairs of arches eliminated as definitive channels and the fifth never more than rudimentary, the third, fourth, and sixth arches are thus the only ones to play important roles in the formation of the adult channels (Figure 16-39). It should be noted, however, that the ventral aortic roots that originally served as feeders to the first two pairs of arches are retained as the external carotid arteries;

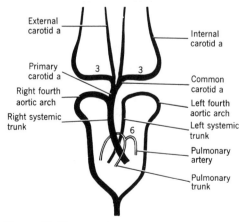

Figure 16-36 Ventral view of the aortic arches of reptiles.

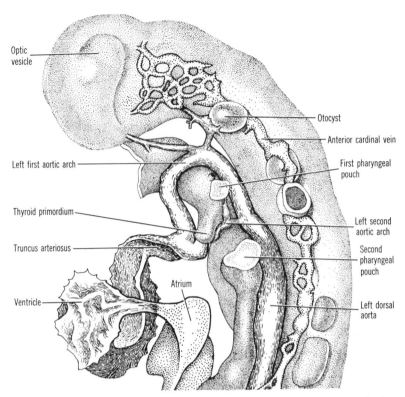

Optic
vesicle

Otocyst

Anterior cardinal vein

Left first aortic arch

First pharyngeal
pouch

Thyroid primordium

Left second
aortic arch

Truncus arteriosus

Second
pharyngeal
pouch

Atrium

Ventricle

Left dorsal
aorta

Figure 16-38   Left side of the forepart of a 22-somite human embryo showing the first two aortic arches. (After Congdon.)

likewise, the dorsal roots extend into the head as the internal carotid arteries.

Up to this point developmental events are the same on each side of the bilateral arch system; but now, in company with other modifications, the fourth and sixth arches show differential histories on opposite sides of the body. It is, therefore, necessary to turn to ventral-view figures, so that events on both sides may be viewed simultaneously (Figures 16-40 and 16-41). Figure 16-40 pictures diagrammatically the same layout as Figure 16-39. In addition to the vessels and relationships already spoken of, the pulmonary arteries extend from the sixth arches to the lungs and the truncus of the heart divides so that the pulmonary trunk leads to the sixth arches and the systemic to the fourth and third arches. Now come changes that alter the entire pattern.

First, on each side there is a disappear-

ance of the "carotid duct," that portion of the dorsal aorta between the junctions of the third and fourth arches. This results in each third arch coming to constitute the curved base of the internal carotid artery. The corresponding ventral aortic roots are retained, however, and become the common carotid arteries. At about the same time the right side of the remaining system becomes further altered. That segment of the right dorsal aorta between the right fourth arch and the descending aorta disappears, and, as a necessary accompaniment the right "arterial duct," the sixth arch beyond the origin of the pulmonary artery, also disappears. Blood from the heart can thus pass to the descending aorta only by way of the left fourth and sixth aortic arches. The left fourth arch consequently becomes greatly enlarged and will persist as the *arch of the systemic aorta*. Its re-

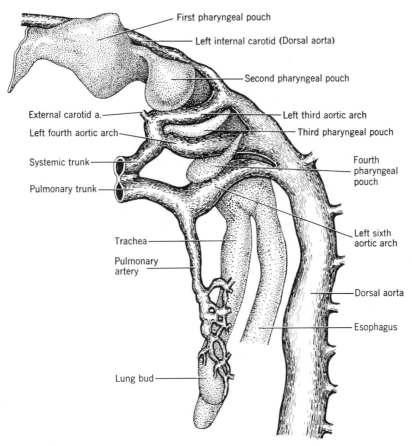

Figure 16-39 Left lateral aspect of pharynx and aortic arches of an 11-mm human embryo. (After Congdon.)

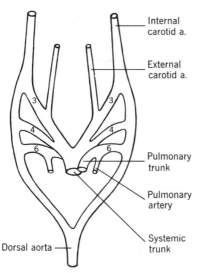

Figure 16-40 Human aortic arches, as seen in ventral view at approximately same state of development as that shown in Figure 16-39.

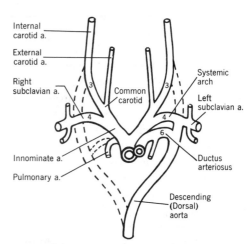

Figure 16-41 Derivation of the late fetal pattern of human aortic arches.

duced counterpart on the right becomes the base of the *right subclavian artery.* The right ventral aortic root proximal to the fourth arch, serving now as the common stem from which the right subclavian and right common carotid arise, is identified as the *brachiocephalic artery.* Its counterpart on the left simply becomes an integral part of the systemic arch from which the left common carotid and the independently formed *left subclavian* stem separately. (It should be remarked that there is great variation among mammals in the stemming of the carotids and subclavians from the systemic arch, so that the human pattern should not be interpreted as being indicative of the entire group.) A final important point relates to the arterial duct on the left. In contrast to that on the right, it is retained throughout embryonic and fetal life as the definitive *ductus arteriosus,* serving to shunt blood from the pulmonary circuit directly into the descending aorta. The significance of this physiological bypass and its final fate is dealt with in a later discussion of the fetal circulation and the changes occurring at birth.

BRANCHES FROM THE DESCENDING AORTA

The extensions of the dorsal aorta into the head as the internal carotid arteries have already been noted, as have the comparable extensions of the ventral aortas as the external carotids. Beyond our concern is a secondarily developed complex of vessels associated with the internal carotids and distributed to the neck region and brain. Caudally, the branches from the descending aorta fall into three general groups: (1) ventral, splanchnic arteries to the digestive tract and its derivatives; (2) lateral, splanchnic arteries to the derivatives of the mesomere and hypomere; and (3) dorsolateral, somatic intersegmental artries.

The primordial ventral, splanchnic branches are the embryonic *vitelline ar-*

teries supplying the yolk sac and, in amniotes, the *allantoic arteries* supplying the allantois. These are paired vessels and are more or less segmentally arranged. Through fusion of some stems and loss of the others, the vitelline vessels provide the adult with three midventral arterial trunks, the *coeliac, superior mesenteric,* and *inferior mesenteric,* distributed to the stomach and intestine. The paired allantoic arteries are present only during embryonic life, supplying the allantois and/or placenta.

The lateral, splanchnic branches provide the blood supply to the kidneys, suprarenal glands, and gonads.

The dorsolateral intersegmental branches supply the body wall musculature and the paired appendages.

THE VENOUS SYSTEM

The developmental anatomy of the veins— the vessels returning blood to the heart— is a complicated one, and the final pattern is highly variable from vertebrate to vertebrate. Yet within the welter of detail one can recognize four groups of veins that, despite their permutations and interconnections, maintain a measure of integrity from fish to mammal: (1) the *hepatic portal system,* consisting of veins draining the digestive system and running to the heart via the liver; (2) the *cardinals* (and their derivatives), draining the head, dorsal part of the body, and the appendages; (3) the *abdominals,* draining the ventrolateral parts of the body; and (4) in lung breathers, the *pulmonaries,* transporting blood from the lungs to the heart.

There are two possible ways to pursue the histories of these vessels. One is to trace each category separately in its changing form through the vertebrates; the other is to consider the four categories simultaneously in a succession of representatives of the major groups of vertebrates. Neither approach is fully effective, and therefore our approach is something of a compromise—

to pursue the developmental history of each set of vessels independently so far as necessary, but at the same time assemble them at appropriate intervals along the way into representative adult patterns.

EARLY ONTOGENY OF THE VEINS

The earliest of the veins to form in a vertebrate embryo are the *vitelline veins* (Figure 16-6). These vessels originate in the yolk sac, pass to the embryo via the mesentery supporting the gut, and enter the posterior end of the heart. Our concern is solely with the embryonic segments of

these vessels, for with the ultimate regression of the yolk sac, the extraembryonic segments will be eliminated. To get to the heart the vitelline veins must pierce the septum transversum (see Chapter 14) into which the liver is growing from the gut above. The liver thus comes to surround the proximal ends of the two vitelline veins (Figures 16-42A, B). In the meantime, the left vitelline vein acquires a tributary, the *subintestinal vein,* which originates in the tail as the *caudal vein,* loops around the anus, and runs forward on the ventral side of the gut (Figure 16-42B).

The vitelline veins first pass through the liver without interruption, but as the liver increases in bulk the veins are broken into

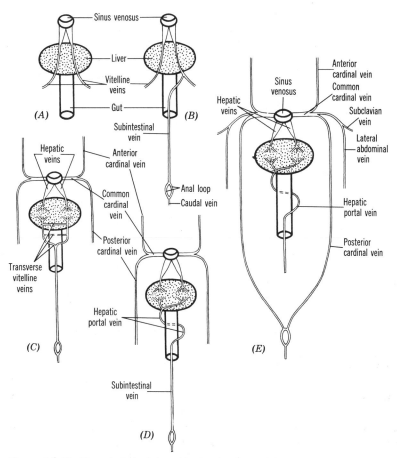

Figure 16-42 Ventral view of the early development of the veins of vertebrates as a group. Explanation in text.

a network of small vessels (Figures 16-42*C* and 16-47*B*). As a result, blood from the yolk sac and gut passes through the network in the liver and is then recollected in the proximal ends of the two vitellines that enter the heart. The latter vessels are now termed the *hepatic veins* (Figures 16-42*C* and 16-47*B*). At about the same time, the two vitelline veins become cross-joined above and below the gut (Figures 16-42*C* and 16-47*C*). With the subsequent elimination of sections of the original channels on each side, there is left a single, somewhat spiral vessel to empty into the liver. This is the *hepatic portal vein* (Figures 16-42*D* and 16-47*D*), which transports all blood from the digestive tract, and initially from the tail as well, to the liver. The hepatic veins complete the transport to the heart.

The cardinal system develops as follows. A pair of *anterior cardinals* (*precardinals*) arises on each side of the head and leads back toward the heart; a pair of *posterior cardinals* (*postcardinals*) runs forward from the rear (Figures 16-42*C, D*). Anterior and posterior cardinals on each side unite in a *common cardinal* (*duct of Cuvier*), which enters the sinus venosus of the heart. It should also be noted that, whereas the caudal vein originally drained into the subintestinal, the developing posterior cardinals tap the anal loop so that the caudal henceforth empties into the cardinals rather than into the hepatic portal system (Figure 16-42*E*). Lying as they do dorsolateral to the developing kidneys, the posterior cardinals initially receive numerous tributaries from the kidneys. But shortly a new set of cadinals develops ventromedial to the kidneys. These are the *subcardinal veins* (Figure 16-43*A*). The subcardinals and postcardinals connected not only at the front and rear of the kidneys, but through small vessels within the kidneys as well. Thus for a time blood in the caudal vein may either pass directly forward via the postcardinals or utilize the subcardinal route for a part of the way (Figure 16-43*A*).

There next occur an addition and a sub-traction. The addition consists of a pair of *lateral abdominal veins,* which course through the lateral body wall and join the common cardinals (Figure 16-43*B*). Each lateral abdominal receives an *iliac vein* draining the pelvic appendage, and a *subclavian vein,* draining the pectoral appendage. The subtraction involves those portions of the posterior cardinals between the anterior ends of the kidneys and the anterior junctions with the subcardinals. This means that blood from the tail can now travel by only one route: into the posterior portions of the postcardinals, now termed the *renal portal veins,* and from there through the many small vessels through the kidneys into the subcardinals, and on to the heart by the persisting proximal segments of the postcardinals and the ducts of Cuvier (Figure 16-43*B*).

Veins of Fishes. The pattern of veins whose progressive development we have just described, and whose final layout is pictured in Figure 16-43*B*, is representative of adult fishes in general and applies particularly to the elasmobranchs. There are, of course, many variations of this basic pattern from one fish to another. These variations are far too numerous to be reviewed in detail, but some of the more significant ones deserve to be noted.

In the elasmobranchs, for instance, the anterior and posterior cardinals feature numerous large sinuses; in fact, the cardinal system of these fishes is more in the nature of interconnected sinuses than of vessels of conventional small caliber. It will also have been appreciated that those vessels of the shark that are commonly designated in dissection manuals as "posterior cardinals" are actually composites of subcardinals and the proximal segments of the original postcardinals. The posterior segments of the postcardinals are identified with the renal portals. We see in many bony fishes a variety of connections between the renal portal and hepatic portal systems. The Teleostei also lack lateral abdominal veins, with the

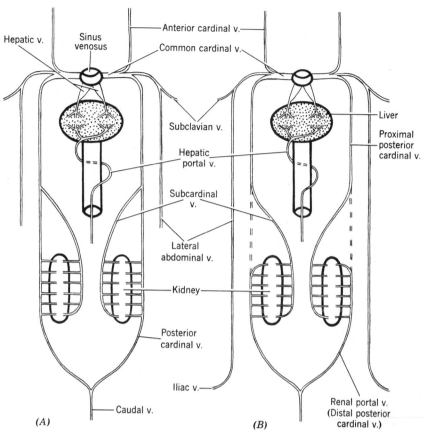

Figure 16-43   Ventral view of the veins of elasmobranch fishes. (*A*) Late embryo. (*B*) Adult. Explanation in text.

result that the subclavians empty into the common cardinals and the iliacs into the postcardinals (renal portals). Still another variation pertains to the drainage of the air bladder. In teleosts its veins usually join the hepatic portal vein, but with the utilization of air bladders as lungs by the Dipnoi, the veins enter the left atrium of the heart; in other words, they may appropriately be identified as *pulmonary veins.*

An additional special word needs to be said about the Dipnoi, for their venous pattern clearly foreshadows that of amphibians. The vessels of the Australian lungfish, *Neoceratodus,* may serve as an illustration (Figure 16-44). The more notable features are these: (1) The lateral abdominals are fused in a single, midline *ventral abdominal vein*

that enters the sinus venosus of the heart. Forward, the subclavians enter the common cardinals; rearward, the iliacs join the renal portals, and each one also communicates with the abdominal via a newly formed *pelvic vein.* (2) The caudal vein, instead of leading to the renal portals, drains into the subcardinal system. The originally paired subcardinals are fused into one for a part of their distance, but continue as a pair communicating with the posterior cardinals. However, the right sub-postcardinal channel is much larger than its counterpart on the left and to this major right-hand composite channel the term *postcava* is applied. It passes through the liver to empty directly into the sinus venosus, whereas the lesser sub-postcardinal channel on the left by-

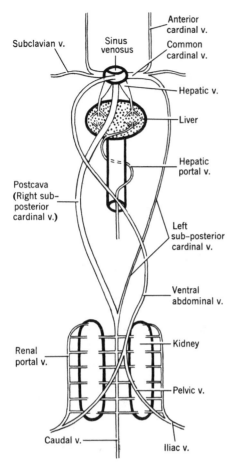

Figure 16-44 Venous system of the lungfish, *Neoceratodus*, ventral view. Explanation in text.

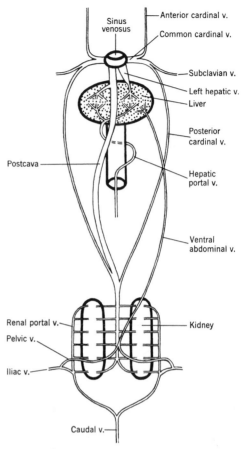

Figure 16-45 Venous system of a urodele.

passes the liver and enters the left common cardinal.

Veins of Amphibia and Reptiles. Only modest modifications in the dipnoan pattern just described are required to convert it to the amphibian type, especially that of urodeles (Figure 16-45). One involves the ventral abdominal vein. Whereas in *Neoceratodus* it leads to the sinus venosus, in amphibians it enters the liver and there joins the hepatic portal system. In effect, this serves to link the renal portal and hepatic portal systems together. Put in other terms, blood from the hind limbs passes to the heart either by the renal portal–postcaval route or by the pelvic–abdominal–hepatic portal route. Another modification pertains to the composition of the postcava. As in *Neoceratodus*, its rear portion is a product of the subcardinals, primarily the right one, but anteriorly it derives from a backward extension of the right hepatic vein (proximal portion of the original right vitelline vein) rather than through a utilization of the right postcardinal. This is a difference of considerable significance, for it presages the manner of development of the anterior segment of the postcava in all the other vertebrates. Moreover, it anticipates the arrangement in which all blood entering the liver via the portal vein leaves it by the postcava rather than passing to the heart independently. With respect to the posterior cardinals, they are retained in reduced form in urodeles, but usually are absent in adult anurans, so that the postcava is the sole channel from the kidneys to the heart. Cranial to the heart, the anterior and common cardinals with their tributaries remain basically unchanged, ultimately emptying to the sinus venosus, which enters the right atrium. Except in the lungless salamanders, the left atrium receives the pulmonary veins.

The venous system of reptiles shows little change from that of anuran amphibians. The anterior cardinals receive tributaries from the head and enter the sinus venosus; the postcava, again a composite of right subcardinal and right hepatic, courses forward from between the kidneys and passes through the liver on its way to the sinus venosus; the ventral abdominal vein remains as an important route from the hind limbs and tail, linking with the hepatic portal system. (During embryonic life the allantoic veins are also linked to the ventral abdominal. Full attention to this relationship is given in the analysis of mammalian veins.) The only significant change is a considerable reduction in the renal portal system. That is to say, the renal portal veins tend to communicate directly with the postcava by direct, major channels rather than through a maze of small vessels.

Veins of Birds. The venous system of a bird is essentially a replica of the reptilian pattern. The principal changes relate to the renal portal system and ventral abdominal. With regard to the former, the communication between the renal portals and the postcava is now so direct that the term *portal*, implying an intervening capillary network, is hardly appropriate. As for the ventral abdominal, it is questionable whether it actually exists, although the so-called *caudal mesenteric vein,* which connects the caudal vein with the portal, is a suspected homologue of the abdominal.

Veins of Mammals. The human continues to be the illustrative subject. An under-standing of the venous system of the human adult comes in part from the just described patterns in other vertebrates. But it is also necessary to trace certain preliminary ontogenetic events. To this end, the primordial venous plan must be laid out (Figure 16-46A). This is a plan that features bilateral symmetry. Paired *anterior cardinals* from the head region and paired *posterior cardinals* from the rear unite in *common cardinals,* which enter the sinus venosus of the heart. *Vitelline veins* from the yolk sac pass to the sinus venosus via the liver. Coming from the placenta is a pair of *umbilical (allantoic) veins* that, entering the embryo at the belly stalk, course through the lateral body walls to reach the sinus venosus. (The position of these veins in the body wall suggests a homology with the lateral abdominals of elasmobranchs.) The development of the venous system is primarily a changeover from a bilateral system to a unilateral one, and this in large measure involves a shift to the right, that is, a utilization of the right side of the original bilateral system. A summary of the two major steps in this process in the human embryo is provided by Figure 16-46. The evolutionary steps leading to the pattern seen in mammalian development are made difficult to follow by the addition of temporary channels such as the subcardinal sinus and the supracardinals. Throughout the gestation period, the tendency for blood to employ the right side of the shifting complex of vessels becomes greater and greater, so that there finally evolves a single, continuous channel from rear to front. This is the channel identified as the *posterior vena cava (postcava)* (Figures 16-46E, F). Although in its final form the postcava appears anatomically simple, it has been assembled piecemeal in an extremely tortuous fashion. Reading from rear to front, the postcava consists of the following components: caudal portion of the right supracardinal, right subsupracardinal anastomosis, right side of the subcardinal sinus, right subcardinal, and right hepatic.

As for the remnant vessels, the anterior

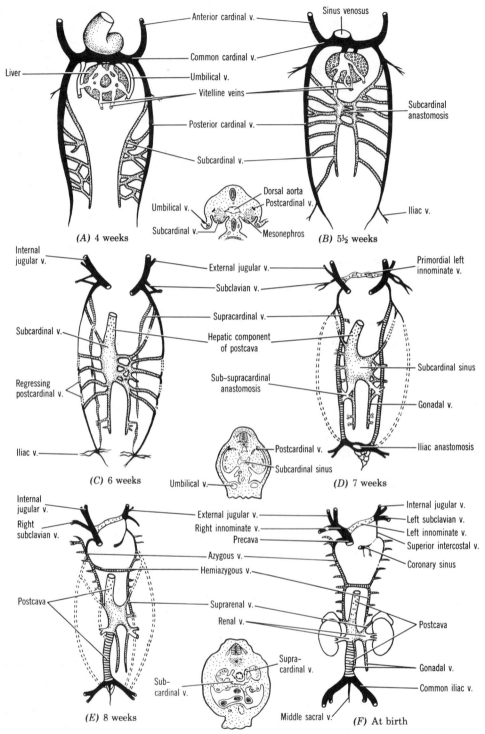

Figure 16-46   Ventral view of the development of the precaval and postcaval systems in the human embryo. The small cross-sectional diagrams indicate the relative dorsal–ventral positions of the principal vessels. (Slightly modified from Patten and Carlson, *Foundations of Embryology*, courtesy McGraw-Hill Book Company.)

portions of the supracardinals plus the anterior right postcardinal constitute the *azygous system* that empties into the precava, because the right common cardinal into which the postcardinal emptied is incorporated within the precava. The left side of the subcardinal sinus becomes incorporated within the precava. The left side of the subcardinal sinus becomes incorporated within the *left renal vein,* thus accounting for the connection of the *left suprarenal* and *left gonadal vein,* the one-time left subcardinals, with the left renal. By contrast, the *right gonadal vein* (rear of right subcardinal) and *right renal* and *suprarenal veins* (new vessels) empty directly into the postcava. As for the extreme caudal remnants of the postcardinals, they unite in a short common trunk, receiving the tributaries from the tail and pelvic appendages, and empty into the rear of the postcava, which is to say that there is no renal portal system.

Hepatic Portal System. The establishment of the hepatic portal system follows the pattern outlined earlier (p. 394) and requires no further elaboration. In fact, this generalization may be stated: in contrast to the maneuverings of the somatic veins, the hepatic portal exhibits in all the vertebrates consistency of development and final form. There are, however, some matters relating to its association with the umbilical veins that require special attention.

Umbilical Veins. As previously noted, when first established, the umbilical veins course through the lateral body walls and enter the sinus venosus (Figure 16-47A). But as the liver increases in bulk and the portal vein takes form, the liver contacts the lateral body walls. This provides for a diversion of the umbilical circulation into the liver (Figure 16-47B). At first a diffuse plexus, each channel into the liver becomes more and more direct, with the result that the links with the sinus venosus disappear (Figure 16-47C). Meanwhile, the right um-

bilical vein regresses, leaving the left as the sole route from the placenta; by way of it the entire venous return from the placenta is shunted into the liver. At first the flow of placental blood through the liver is through the maze of hepatic sinusoids, but gradually a principal channel is excavated. This is the *ductus venosus* (Figures 16-47C, D). As they leave the liver, the ductus venosus and hepatic veins become confluent. But, as we have seen, the right hepatic is one of the contributors to the posterior vena cava. The vena cava therefore becomes the common carrier into the heart for all blood originating in the systemic (rear), portal, and umbilical circuits. The ultimate fate of the umbilicals and ductus venosus are considered below.

CIRCULATION IN THE MAMMALIAN FETUS AND THE CHANGES OCCURRING AT BIRTH

Consideration of the circulation of the fetus provides an opportunity to review the more important developmental events that bring it into being and also to reestablish the homologies of the major vessels. Therefore, before turning to the very significant physiological aspects of the fetal circulatory plan, let us point to the highlights of its developmental anatomy (Figure 16-48).

The primordial heart tube, consisting of four chambers—sinus venosus, atrium, ventricle, and truncus arteriosus—arranged in series, is converted into two pairs of heart chambers arranged in parallel. These pairs are the right atrium and ventricle and the left atrium and ventricle. The truncus is eliminated as such, being split into the systemic trunk running from the left ventricle and the pulmonary trunk running from the right ventricle. The sinus venosus is also eliminated as a definitive chamber by absorption into the wall of the right atrium. All vessels originally terminating in the sinus venosus consequently come to empty

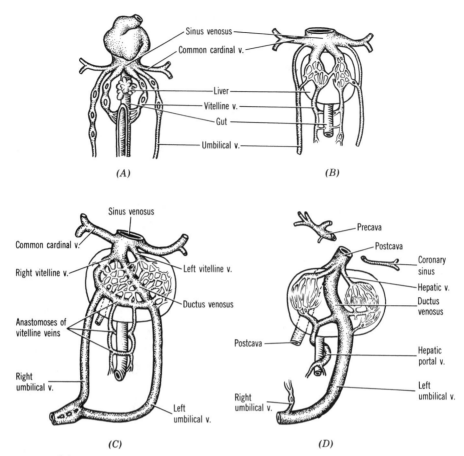

Figure 16-47 Ventral view of the development of the hepatic portal system and the relationship of the umbilical circulation to it. (From Patten and Carlson, *Foundations of Embryology,* courtesy McGraw-Hill Book Company.)

directly into the right atrium. The left atrium receives the pulmonary veins from the lungs. Most importantly, however, right and left atria communicate via the foramen ovale in the interatrial septum.

The systemic trunk leads to the systemic arch, which represents the left fourth aortic arch, and from there to the descending aorta. The right fourth aortic arch constitutes the base of the right subclavian which, with the right common carotid (ventral root of the third arch), stems from the brachiocephalic artery (ventral root of the right fourth arch). The common carotid provides the internal carotid (third aortic arch and extension of the dorsal aorta) and external carotid (extension of the ventral

aorta) supplying the head. On the left side the carotid system stems directly from the systemic arch, as does the independently formed subclavian.

The pulmonary trunk leads to the pulmonary arteries supplying the lungs. These arteries branch from the sixth pair of aortic arches, the left one of which is retained distally as the ductus arteriosus, which shunts blood from the pulmonary circuit to the descending aorta. After providing branches to the organs of the body, the descending aorta leads to a pair of umbilical arteries running to the placenta.

Blood returns from the head and trunk of the fetus by way of the precava and postcava, respectively, which empty into

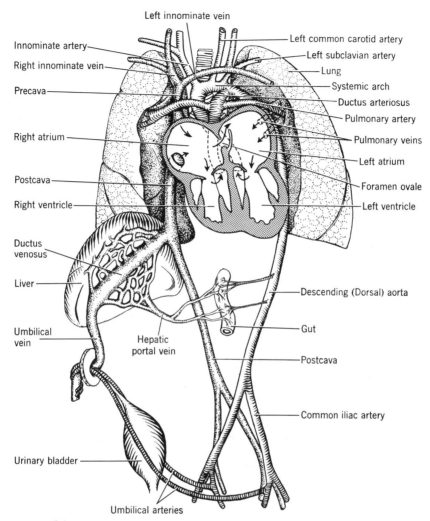

Left innominate vein

Innominate artery

Right innominate vein

Precava

Right atrium

Postcava

Right ventricle

Ductus venosus

Liver

Umbilical vein

Hepatic portal vein

Urinary bladder

Umbilical arteries

Left common carotid artery

Left subclavian artery

Lung

Systemic arch

Ductus arteriosus

Pulmonary artery

Pulmonary veins

Left atrium

Foramen ovale

Left ventricle

Descending (Dorsal) aorta

Gut

Postcava

Common iliac artery

Figure 16-48 Human fetal circulation immediately prior to birth. Arrows indicate routes into and out of the heart chambers, including passage through the foramen ovale.

the right atrium. The precava represents the onetime common cardinal and base of the anterior cardinal on the right side. The postcava is a composite vessel assembled from multiple sources: caudal portion of right supracardinal, right subsupracardinal anastomosis, right side of subcardinal sinus, right subcardinal, and right hepatic. Blood returns from the placenta by a single umbilical vein identified as the left one of an original pair of allantoic veins. Instead of going directly to the heart, the umbilical is diverted through the liver, where it is linked with the hepatic portal system (a product of the vitelline veins) by a major channel, the ductus venosus. The ductus venosus and associated hepatic vessels then join the postcava. Now we are prepared to explore the physiological implications and potentials of this anatomical layout.

For obvious reasons the human fetus is not very amenable to physiological experimentation; thus the interpretations that follow derive largely from studies on other mammals, notably the sheep. A highly rewarding technique with the fetal sheep has been that of injecting radiopaque substances into the bloodstream and then following their course through fluoroscopy or X-ray motion pictures. Recently the conclusions reached through these experiments have been brilliantly verified on the human fetus through the elegant technique of inserting a slender flexible tube through the blood vessels, directly into the heart. This permits the taking of blood samples from chosen locations for oxygen analysis and also the direct measurement of blood pressure.

In all these studies there has been a search for answers to the key questions created by these circumstances: the common receptacle, that is, the right atrium, for the precaval and postcaval streams; the communication between right and left atria via the foramen ovale; the linkage of the pulmonary and systemic circuits by the ductus arteriosus. As a consequence of these circumstances, to what degree does mixture of the caval streams take place in the right atrium? What is the nature and degree of flow through the foramen ovale? How effective a shunt is the ductus arteriosus? Is there any segregation of aerated and non-aerated blood so as to provide preferential treatment in regard to oxygen for one part of the fetal body over another? Here is the present status of the answers to these and related questions.

Aeration of the blood takes place, of course, in the placenta. The oxygenated blood from the placenta passes to the postcava by way of the ductus venosus. It is joined along the way by a small amount of deoxygenated blood arriving by the portal vein, and there is further "dilution" by postcaval blood coming up from the rear. But because of the relatively great volume of highly aerated blood in the umbilical vein, the stream finally collected and transported to the right atrium by the proximal part of the postcava is only slightly less oxygenated than that in the umbilical. By contrast, the stream brought in by the precava

is greatly deoxygenated. The question now is, Do these two streams, one poorly and the other well oxygenated, intermix or are they somehow kept separate? There is an anatomical answer to the query that is supported by X-ray motion picture studies. It will be recalled (p. 385) that with the absorption of the sinus venosus within the wall of the atrium, the right side of the original valve of the sinus is retained as a valve guarding the orifice of the postcava. This valve serves to direct the postcaval stream toward the foramen ovale and does it so effectively that the greater part of the postcaval flow passes through the foramen into the left atrium. There it is joined by deoxygenated blood brought in by the pulmonary veins from the still nonfunctioning lungs and then passes into the left ventricle. The remaining postcaval blood in the right atrium mixes with the precaval stream, none of which is diverted through the foramen ovale, and moves on to the right ventricle. Obviously, then, there is a considerable measure of segregation of the stream so that the blood in the left ventricle has a significantly higher oxygen content than that in the right.

The blood in the right ventricle passes out through the pulmonary trunk. A small part of it, increasing in amount during the latter part of fetal life, goes to the lungs, but most of it is diverted to the descending aorta by the ductus arteriosus. The blood in the left ventricle passes out through the systemic trunk and is distributed chiefly to the head and pectoral regions by the carotid and subclavian arteries. The remainder follows the descending aorta, picking up along the way the poorly aerated blood brought in by the ductus arteriosus. In other words, the cranial region of the body receives preferential treatment: its blood supply has a much higher oxygen content than that of the lower body. The less well aerated blood in the descending aorta, after some has been distributed to the viscera and lower limbs, moves out to the umbilical arteries and the placenta.

Two obvious changes occur in the blood vessels at the time of birth. The first is the abrupt contraction of the umbilical vessels, the arteries first and the vein shortly thereafter, thus cutting off placental circulation. The second is the going into action of the pulmonary circulation as the lungs take over. The caliber of the pulmonary arteries and veins steadily increases during the last weeks of fetal life, so that these vessels are in a sense ready and waiting to take over a greatly enlarged flow of blood.

With the decrease of blood volume in the right atrium, because of the cutoff of umbilical flow, and the increase of blood volume in the left atrium, because of increased pulmonary flow, there is a buildup of blood pressure in the left atrium relative to that in the right, which is sufficient to force septum I against septum II and effectively close the foramen ovale. The buildup in blood pressure in the left side of the heart also brings a brief reversal in flow of blood through the ductus arteriosus, after which the ductus constricts and the shunt between pulmonary and systemic circuits is closed.

Although physiological separation of the pulmonary and systemic circuits occurs rapidly at or shortly after birth, the anatomical changes that "fix" this separation require several weeks. In due time the ductus arteriosus and ductus venosus become reduced to ligaments, as do the intraabdominal remnants of the umbilical vessels, and a permanent atrial septum results from the fusion of its two fetal components. Any failure on the part of the ductus arteriosus or foramen ovale to be permanently sealed creates self-evident difficulties in the maintenance of physiologically necessary oxygen tensions.

THE LYMPHATIC SYSTEM

Because of the relatively high pressure in the closed blood vascular system, there is a slow, constant leakage of fluid and contained substances through the walls of the capillaries into the associated tissues. To prevent waterlogging of the tissues and to maintain the constancy of the cellular environment, it is essential that the interstitial fluids be returned to the bloodstream. The lymphatic system provides for this return. It also serves as the source of *lymphocytes*—about which we shall say more later—and as the route for the absorption of fats from the digestive tract.

The consensus is that cyclostomes and Chondrichthyes lack an independent system of lymphatic vessels. The functions of a lymphatic system in these vertebrates are presumably carried on by numerous thin-walled sinuses that drain into the veins. But bony fishes and all tetrapods are equipped with a distinct group of vessels that tend to run topographically parallel to the veins. They originate embryonically considerably later than the blood vessels, appearing first as fluid-filled spaces within the mesenchyme, spaces that subsequently coalesce and assemble themselves into an elaborate drainage system throughout the body.

Unlike the blood vascular system, the lymphatic vessels serve only to *return* fluid to the heart. There are no arteries in this system. The lymphatics arise as blindly ending capillaries into which fluid (*lymph*) diffuses; the capillaries unite to form larger vessels, which in turn combine to form vessels of still larger caliber; the largest vessels in most cases finally open into the great veins near the heart. In adult anurans the vessels are also expanded into large sinuses beneath the skin. Because the lymphatic system is not tied into the arteries and therefore does not in itself involve a complete circuit and have the benefit of arterial pressure behind it, the flow of lymphatic fluid is sluggish at best. To assist the flow, a few of the bony fishes, all of the amphibians and reptiles, and all bird embryos and some adults feature *lymph hearts.* These are local enlargements of the lymphatic vessels equipped with muscle, which rhythmically contract. They are also fitted with valves to prevent backflow as the hearts pulsate. Other controlling valves may be located elsewhere along the lymphatic channels. However, the embryonic lymph hearts of most birds become nonfunctional, and lymph hearts are never formed in mammals. The flow of lymph in these circumstances in dependent on the massaging action of the body musculature and the pressure of neighboring organs, with numerous valves serving to direct the flow.

Among other structures considered to belong to the lymphatic system are masses, even definitive organs, composed of *lymphoid tissue*. Three prominent instances are certain organs of pharyngeal origin: *tonsils, adenoids,* and the *thymus gland.* The role played by the thymus gland in the manufacture of lymphocytes and the production of antibodies for the defense of the body against infection has already been discussed (p. 302). It is possible that the tonsils and adenoids have a similar function. Mammals, but rarely other vertebrates, also feature large numbers of *lymph nodes* clustered at particular points along the lymphatic channels, notably in the groin and axilla, the neck, and the intestinal mesentry. They, too, manufacture lymphocytes and, as lymph filters sluggishly through them, they screen out bacteria and other materials. So it is that the presence of infectious organisms often causes the lymph nodes to become swollen and sore, and certain types of cancer cells that spread through the lymphatic vessels tend to aggregate in the nodes.

The largest lymphoid organ in the vertebrate body is the *spleen.* Except for the cyclostomes and Dipnoi, where it exists as a diffuse mass of tissue lying beneath the peritoneum of the gut, the spleen arises as a definitive organ from the mesenchyme of the mesentery supporting the stomach. But unlike the nodes and other lymphoid organs, the spleen is interposed in the blood vascular rather than lymphatic system. It and the central nervous system, in fact, are the only major organs not permeated by lymphatic vessels. Yet the spleen bears an important relationship to the lymphatic system, for after being colonized by lymphocytes provided by the thymus gland, it produces large numbers of lymphocytes in its own right and becomes one of the body's main lines of defense against infectious diseases. During embryonic life, the spleen is also a major source of erythrocytes, although in mammals it ultimately serves only as a storehouse for them. At the same time, worn out red blood cells are destroyed by the spleen, with one breakdown product of the hemoglobin of these cells, *bilirubin,* being transported to the liver and excreted in the bile. *Iron* is also recovered from the hemoglobin and returned to the general circulation, which carries it to the red marrow of the long bones for use in the synthesis of hemoglobin for new erythrocytes.

Chapter Seventeen

The Nervous System and Sense Organs

If one touches an amoeba with a finely pointed needle, it responds by putting out one or more extensions of its substance on the side opposite to the needle and moves away. A moment's reflection on this performance suggests that the needle prick provides a stimulus, which sets off some sort of signal that is transmitted across the body of the amoeba and, in turn, initiates a reaction. In other words, the single amoeboid cell receives the stimulus, transmits the resulting disturbance, and carries out the appropriate response.

The evolution of multicellular animals has been accompanied by a parceling out of these activities among special cells, tissues, and organs. It is no longer the organism as a whole that exhibits the three phases of response to a stimulus; there has been a division of labor, so to speak, whereby each phase is delegated to appropriately designed agencies. Accordingly, there are certain cells and accessory structures that specialize in receiving particular kinds of stimuli. These are known as *receptors* or *sense organs*. The excitations inspired within the sense organs are then transmitted by *conductors* to special organs of response. The latter, notably muscles and glands, are termed *effectors* because they carry out, or effect, a characteristic reply in some form of action. The conductors, the lines of communication between receptors and effectors, constitute the *nervous system,* whose complications provide for the coordination of actions featuring animal behavior and, in cooperation with the endocrine organs and cardiovascular system, ensure the stability of the body's internal environment.

405

THE NERVOUS SYSTEM

STRUCTURAL AND FUNCTIONAL UNITS

Segregation of nervous system functions and their allocation to specific cells and tissues are first demonstrated in the lowliest of multicellular animals, the sponges. Interestingly enough, however, it is an effector system that is first set off. Certain cells associated with the openings of the system of water-circulating channels respond directly to stimulation by contracting. These cells have been termed *independent effectors* and in most ways each performs like an amoeba. But there is one important addition. The reaction elicited by the stimulation of one such cell may be sluggishly transmitted to and across neighboring cells so as to evoke responses at a distance from the original stimulus. In this we see the beginning of a conducting system.

The divorce of reception and conduction from the effectors and their transfer to special cells first manifest in the coelenterates. The simplest and probably most primitive arrangement is found in the tentacles of sea anemones. Here we may identify numerous cells of a type featuring a cell body lying within (or immediately beneath) the covering epithelium and a fibrous extension attached to a muscle cell (Figure 17-1A). This type of cell, termed a *neurosensory cell,* is both a receptor and a conductor, because it plays the double role of receiving a stimulus and transmitting an impulse directly to the muscle cell. Much more commonly in coelenterates, however, an additional complication prevails: the segregation of reception and transmission. In essence, there exists a separate receptor or sensory cell, which upon stimulation inaugurates an impulse that is transmitted to the effector by a distinct conducting cell termed a *nerve cell* or *neuron* (Figure 17-1B). The ele-

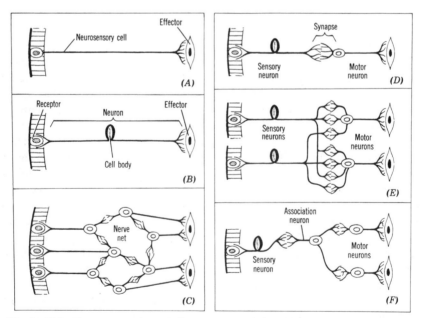

Figure 17-1 Receptor–conductor–effector systems. (*A*) Neurosensory cell. (*B*) Sensory cell and neuron. (*C*) Nerve net. (*D*) Elemental reflex arc. (*E*) Multiple reflex arc. (*F*) Sensory, motor, and association neurons. (After Simpson, Pittendrigh, and Tiffany, *Life,* by permission of Harcourt, Brace and Company.)

mental pattern pictured in Figure 17-1*B* is actually fictional for, instead of a single neuron between any given receptor and effector, there are always many of them, constituting a *nerve net* (Figure 17-1*C*) that is spread widely beneath the epidermis. Although the processes of the neurons making up the nerve net are not united, they do lie closely adjacent and impulses pass from one neuron to another. Furthermore, the impulses go in all directions equally well, so that excitations initiated by any receptor or group of receptors are propagated randomly through the nerve net and bring a diffuse response by the widespread muscle cells.

The nerve net type of nervous organization may occur locally in some organs in other animal phyla—it may even exist in the wall of the human intestine—but generally in higher animals a different structural and functional pattern prevails. Reduced to its essentials, it is a pattern involving a minimum of two neurons (Figure 17-1*D*): a *sensory neuron* carries impulses initiated by the receptor and passes them on to a *motor neuron,* which delivers the impulses to an effector. This simple chain from receptor to effector provides for direct and automatic response by the effector to a stimulus that impinges on the receptor. Such an automatic response is called a *reflex,* and the path followed by the nervous impulse that causes a reflex action is a *reflex arc.* It is unlikely that a reflex arc consisting of only two neurons actually exists. The usual arrangement is one in which a given sensory neuron can stimulate more than one motor neuron (Figure 17-1*E*). Because the response remains an automatic one, it is still a reflex, but now a single receptor can elicit a reaction from a whole set of effectors.

A further complication is found in the nervous system of the higher invertebrates and all the vertebrates. This consists of the introduction of *association neurons* between the sensory and motor neurons (Figure 17-1*F*). Through them an impulse may now be passed on to any number of different effectors, and it is by reason of the tremendous numbers and complexities of connections of the association neurons in vertebrates that great versatility of responses is provided. But before continuing with the analysis of the patterns of conduction pathways in vertebrates, it is essential that we examine neurons themselves a little more closely.

Neuron Structure. Despite their differences in size and shape, all neurons possess certain features in common. Any given neuron has a *cell body* containing a nucleus and a surrounding cytoplasm that is extended into a number of *processes* (Figure 17-2). The number and arrangement of the processes provide the basis for a useful classification of neurons. Of rare occurrence and confined almost exclusively to embryonic stages is the *unipolar neuron* having only a single process (Figure 17-2*A*). Somewhat more common is the *bipolar neuron,* in which one process projects from one side of the cell body and another one from the opposite side (Figure 17-2*B*). Bipolar neurons are found in adult vertebrates in the ganglia of the eighth cranial (auditory) nerve and, in modified form, in the retina of the eye (Figure 17-33). Still more common is the *pseudounipolar neuron* (Figure 17-2*C*), so designated because a single process arises from the cell body and then divides like the letter "T" to send a branch in each direction. It is believed that pseudounipolar neurons originate embryonically as bipolar neurons whose processes converge and fuse. Most of the nerve cells of the cerebrospinal ganglia (Figure 17-21) are of the pseudounipolar type. The great majority of neurons, however, are of the *multipolar* type, featuring several short, branching cytoplasmic extensions from the body and a single, often greatly elongated process (Figure 17-3).

It has been traditional to call all those processes that conduct nerve impulses *toward* the cell body *dendrites,* whereas the

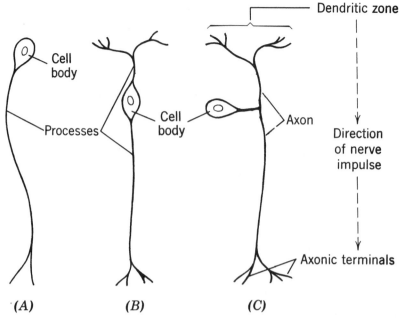

Figure 17-2   Three varieties of neurons. (*A*) Unipolar. (*B*) Bipolar. (*C*) Pseudounipolar.

fibers transmitting impulses *away* from the cell body are *axons*. But a more realistic interpretation is one that views the dendrites as constituting a zone of cytoplasmic extensions charged with the task of generating an electrochemical response to any outside stimulus, a response that as a nerve impulses is then generated at the base of an axon (axon hillock) and conducted away from the dendritic zone by the axon (Figures 17-2*C* and 17-3). In other words, the dendrites are only the stimulus-receiving component(s) of the neuron; the term axon applies to that part of the neuron that generates and propagates the nerve impulse, the reference point for the direction of propagation being the dendritic zone rather than the cell body.

The axon is uniquely designed for this function of transmission (Figure 17-3). It is characteristically uniform in caliber, and along its length it commonly, although not always, gives off collateral branches. The main stem, and to a lesser extent the collaterals, terminate in an *axon ending* con-

sisting of twigs of varying number and distribution. Except at the site of origin from the cell body and at the terminal arborization, the axon is encased by associated interstitial cells. The interstitial cells embracing axons of the cerebrospinal nerves are the so-called *Schwann cells*; those within the brain and spinal cord are the *glial cells* (p. 417). The common arrangement in peripheral nerves is that of the Schwann cells being arranged in a single layer constituting what has traditionally been termed the *neurilemma*, or *sheath of Schwann*; between the axons and neurilemma there lies a *myelin sheath* composed of alternate layers of fatlike material and protein. The myelin sheath is constricted and interrupted at regular intervals (*nodes of Ranvier*) and thus appears as a series of segments. The myelin sheath is therefore discontinuous at each node, but the neighboring Schwann cells so closely approximate each other that for all practical purposes the neurilemma is uninterrupted. Although the two sheaths are convention-

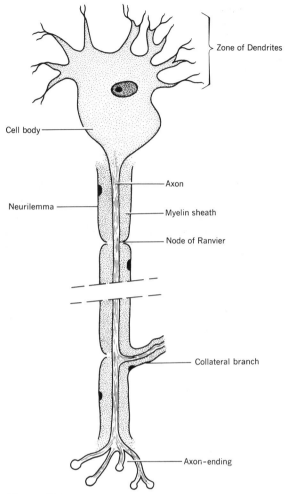

Figure 17-3   Structure of a multipolar neuron.

ally described as distinct entities, modern ultramicroscopic methods make it clear that myelin and neurilemma constitute an integral unit. This concept finds support in certain facts of embryogeny.

Neuron Development. Every neuron in the adult body traces its origin to a precursory cell in the embryo called a *neuroblast.* R. G. Harrison prepared cultures of neuroblasts and observed the formation of neuronic processes as direct protoplasmic extensions from the neuroblasts. This experiment, performed over 80 years ago, was a historically notable one, for not only did

it settle the issue of the manner of origin of neuron fibers, but it pioneered the now widely exploited technique of tissue culture.

Harrison's observations, which have been repeated and refined many times since and have even been recorded cinematographically, demonstrated that the outgrowth of a neuronic process has much in common with the extension of a pseudopodium by an amoeba. Accordingly, new axial cytoplasm is thought of as originating in the cell body from where it streams toward the tip of the steadily elongating process. But unlike an amoeba, which moves by the projection of pseudopodia, the nerve cell body remains anchored at whatever site in the embryo it occupies and the processes spin out peripherally. Sooner or later the free tips of the growing fibers attach themselves to appropriate receptors and effectors, and henceforth are passively towed along as these parts are displaced during growth. The growing tips, or growth cones, of axons form delicate microspikes that extend forward from the membrane. Treatment with the drug cytochalasin B causes withdrawal of the microspikes and cessation of growth. The extended axon maintains its new length by forming microtubules along its axis.

All newly sprouted axons are first devoid of sheathing. Prospective neurilemmal cells (or their equivalents) then migrate along the axons, multiply, and finally assemble themselves as a membrane. In so doing, each cell proceeds to wind itself as a double sheet around the axon and lay down a spiral of myelin substance. The nodes in the myelin represent the limits of deposition of a single sheath cell. That is, the length of a given segment of myelin is defined by the length of the wrapping Schwann cell, and the apposition of two consecutive Schwann cells marks the level of the node between two segments of myelin. The amount of myelin deposited varies greatly: some axons are provided with heavy coats, other with little or none. Except for the axon endings, however, there are probably no axons that are devoid of interstitial cell associations. In the case of the postganglionic fibers of the autonomic system (p. 444), for instance, no myelin is present, but groups of axons are embraced by a single series of Schwann cells. And the nonmyelinated axons within the central nervous system simply course through a matrix of glial cells rather than possessing definitive sheathing.

The integral nature of a neuron is also clearly demonstrated in the phenomena of degeneration and regeneration. If an axonic fiber is severed through accident or disease, or deliberately cut in an experiment, that portion of the axon so isolated from the cell body gradually degenerates. To be more specific, the severed axon always disintegrates and, if it is a myelinated axon, so does the myelin; but the Schwann cells survive. The breakdown of the axon is a manifestation of the fact that no part of a cell can survive when isolated from its nucleus; the breakdown of the myelin suggests that although it is a neurilemmal product, the axon has something to do with its maintenance. In due time, then, the central stump of the severed axon exhibits a growing sprout which follows the pathway marked by the persisting Schwann cells. The sprout eventually restores the axon to its normal length, and the axon may even reestablish functional connection with its appropriate end organ. The Schwann cells, in the meantime, replay their role of producing new myelin for the restored axon.

The Synapse. We have seen that the nervous system is composed of complex chains of neurons so arranged as to provide transmission of excitation from one neuron to another with resulting initiation of response by appropriate effectors. In the case of the neurons that constitute a nerve net, transmission is diffuse and largely unchanneled. In higher invertebrates and vertebrates, however, conduction is channeled in one direction only; that is, functional polarization prevails. Specifically, chains of neurons are always laid out in such a way that the branching axon ending of one neuron impinges upon, but does not fuse with, the dendritic zone of the succeeding neuron. Moreover, the nerve impulse passes only from the axon to the dendrites of the next neuron in line. However, this restriction in the direction of conduction is not dependent on any peculiar qualities of the axon itself, for it has been shown experimentally that axons can carry impulses in either direction. Rather, it is a consequence of the functional properties of the *synapse,* the site of association of axon endings with the dendritic zone (Figure 17-1*D*). The nervous excitation reaching the end of the axon elicits in the terminal arborization of the axon the formation of two chemical substances, in most cases. One of these is *noradrenalin* (norepinephrine) and the other is *acetylcholine.* (Both substances are not produced in any given synapse. Certain nervous circuits are said to be "adrenergic" because noradrenaline is the physiological agent; others, utilizing acetylcholine, are said to be "cholinergic." Still others secrete one of a variety of other compounds that act to inhibit the generation of a new impulse in the postsynaptic neuron, rather than stimulating one.) Thus, the neurons pass the "baton" from one to another. Because only the axon endings can secrete these chemical agents and only the dendrites are sensitive to those that are stimulatory, the conduction of impulses is necessarily unidirectional.

GENERAL ORGANIZATION OF THE NERVOUS SYSTEM

We have seen that the vertebrate nervous system is composed of large numbers of cellular units called neurons. Although of variable form, all ordinarily consist of a cell body, a conducting axon, and one or more stimulus-receiving dendrites; and they are polarized by virtue of the synapse. We have also seen that the neurons are not scattered at random, but are assembled in a morphological and functional pattern. The simplest possible functional combination of neurons is one in which a stimulus recorded by a receptor initiates an impulse that travels over a sensory neuron and from there to a motor neuron, which delivers the impulse to an effector (Figure 17-1*D*). But greater functional versatility is provided by the interposition of one or more association neurons between the sensory and motor links in the chain (Figure 17-1*F*). In this elementary scheme we have an introduction to the anatomical organization of the vertebrate nervous system.

The countless numbers of association neurons in the vertebrate body are assembled in a central switchboard, to use the analogy of a metropolitan telephone system, in and out of which run lines of communication with the parts of the body. The switchboard is the *central nervous system* and consists of the *brain* and *spinal cord.* The incoming and outgoing lines are composed of axons of sensory and motor neurons, which collectively make up the *peripheral nervous system.* These axons do not run singly, but are gathered together in cables known as *nerves.* That is, a nerve is a bundle of axons, each called a *nerve fiber,* invested by a connective tissue sheath, much as an electrical cable consisting of many independent wires is encased in a conduit. It is important to remember that a nerve fiber, no matter what its length, is part of a neuron and thus is connected to a cell body. In the case of motor nerves, the cell bodies of the neurons are located within the central nervous system (with an important exception to be noted later); in the case of sensory nerves, the cell bodies of the neurons are grouped in aggregations outside the brain and spinal cord, each such aggregation being termed a *ganglion.*

EARLY HISTOGENESIS OF NERVOUS TISSUES

The foregoing discussion, centered primarily on neurons, has taken the existence of embryonic nerve cells (neuroblasts) for granted. We must now turn back to the beginnings of embryonic development to establish the origin of these neuron precursors.

It will be recalled (Chapters 7 and 8) that one of the important by-products of the revelation of the morphogenetic movements involved in gastrulation has been the creation of maps of prospective organ regions (cf. Figures 7-1, 7-8, and 8-4). Accordingly, the positions of the materials destined to form the nervous system, notochord, somites, and so on have been laid out in the late blastula. But recall also that such a map is solely descriptive of topographic relationships. For instance, the designation of a given area as neural ectoderm is only a way of saying that this is the material destined to provide the nervous system; there is no implication of irrevocable determination for neural differentiation, as evidenced by the following experiment.

If small blocks of prospective neural ectoderm and epidermal ectoderm in an early amphibian gastrula are interchanged, it is found that each transplanted block differentiates in keeping with its new environment rather than in accordance with its normal prospects. That is, the bit of neural ectoderm acquires the qualities of epidermis and the bit of epidermal ectoderm differentiates as neural tissue. But if the experiment is repeated toward the end of gastrulation, an entirely different result is obtained. Epidermal ectoderm now pursues its normal course and appears as a patch of epidermis within the embryonic nervous system; likewise, neural ectoderm forms an island of neural tissue within the expanse of epidermis.

These experimental results clearly indicate that during gastrulation there is a restriction and channeling of developmental abilities. Whereas prior to gastrulation prospective neural and epidermal ectoderm are capable of forming tissues unlike those for which they are normally destined (can, in fact, even be made to transform into mesodermal tissues), at the end of gastrulation their abilities are restricted to their normal prospects. This is a way of saying that during gastrulation "determination" occurs. However, this is all that the experiments do tell us: that an original developmental versatility gives way to precise determination and

that this occurs during gastrulation. To learn of the agent responsible for bringing about determination, specifically determination of neural ectoderm, we must look to other aspects of gastrulation.

We have seen from our earlier considerations of gastrulation that the morphogenetic movements serve to bring the prospective endodermal and mesodermal parts into their normal positions in the interior of the embryo. The entire surface of the completed gastrula at the same time becomes covered by ectoderm, with the neural ectoderm in particular being disposed as a relatively broad band along the dorsal midline (Figures 7-8 and 8-4). A clue to the source of the influences bringing about determination in this neural area is found in the topography of the mesoderm. Like the neural ectoderm at the surface, that mesoderm destined to provide the notochord, somites, and prechordal plate also moves to a dorsal position and stretches out lengthwise. In other words, the neural ectoderm is accompanied on its undersurface by an axial strand of mesoderm. It would thus not be unreasonable to suspect the chordamesoderm of playing a role in neural determination. If this is true, the neural ectoderm should not acquire its qualities of determination if the chordamesoderm could somehow be prevented from assuming its normal position beneath the neural area.

A direct and relatively simple method of doing this very thing is possible with amphibian embryos. If a blastula is freed of its protective membranes and placed in a weak salt solution, the movements that normally carry the endoderm and mesoderm to the interior are reversed. The result is the production of a so-called "exogastrula," consisting of an endodermal–mesodermal mass connected by a narrow neck to an ectodermal mass (Figure 17-4). The former undergoes a remarkably normal histogenesis, except that the mesodermal tissues are enclosed by endodermal tissues. The ectodermal mass, however, proves to be in-

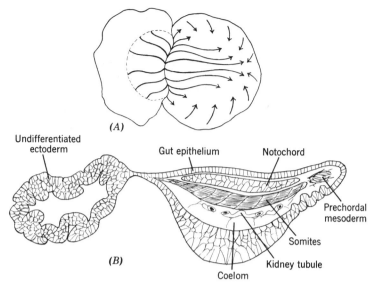

(A)

(B)

Undifferentiated ectoderm

Gut epithelium

Notochord

Prechordal mesoderm

Somites

Kidney tubule

Coelom

Figure 17-4 Exogastrulation in an amphibian. (*A*) Morphogenetic movements in exogastrulation. (*B*) Differentiation in an exogastrulated embryo. (After Holtfreter.)

capable of producing neural tissues; it forms only a mass of unspecialized epidermis. It therefore follows that neural differentiation is somehow dependent on an association with axial mesoderm. That is, the chordamesoderm exerts some kind of morphogenetic effect on the prospective neural ectoderm, an effect known as *induction.* This effect is then expressed in neural histogenesis.

Neural Induction. We have said that to speak of an area in the blastula as the one destined to produce neural parts is only to say that this is what the normal developmental fate of the area has been observed to be. But description is not explanation. We seek to know the causal interactions responsible for the results that we observe and describe.

The beginning of our understanding, incomplete as it remains to this very day, is found in an experiment performed over 50 years ago by the distinguished German embryologist Hans Spemann. Spemann found that if he constricted an amphibian blastula in such a way that the size of the prospec-

tive blastopore and the material immediately above it lay entirely in one half, this half-blastula would form a complete embryo while the other half remained inert. It was then demonstrated in a number of follow-up experiments that the crucial component was the material in the upper blastoporal lip, material that we now know to be prospective chordamesoderm and that traces back to the gray crescent of the undivided egg.

If a fragment of prospective chordamesoderm were to be grafted into the epidermal ectoderm or simply inserted within the blastocoel of an early gastrula, the graft itself would differentiate into characteristic mesodermal parts and induce the formation of a neural tube in the host epidermis (Figure 17-5). The final result, moreover, was more than just a collection of tissues; a fully organized secondary embryo was assembled in association with the primary host, with the original graft not only inducing a neural tube, but providinig a complete mesodermal axial system partly out of its own substance and partly by recruitment from the host. It was in deference to its remarkable

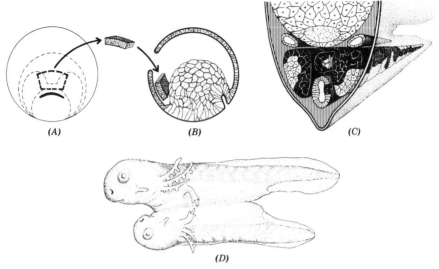

(A)   *(B)*   *(C)*

(D)

Figure 17-5   Induction of a secondary embryo by a chordamesodermal graft. A piece of dorsal blastoporal lip is taken from donor (*A*) and grafted into host gastrula (*B*). (*C*) and (*D*) show the induced secondary embryo attached to primary host. Tissues derived from graft shown in black; those provided by host shown in white. (From Holtfreter and Hamburger in Willier et al., *Analysis of Development,* courtesy W.B. Saunders Company.)

ability to direct the fabrication of a total embryo that Spemann coined the term *organizer* to describe the capacities of the chordamesoderm.

The investigations of many workers in the years immediately following Spemann's original experiments served to add many refinements. It was shown, for instance, that although the area that possessed organizing, or inducing, power corresponded to the prospective axial mesoderm, this power is greatest in the midline and gradually falls off on either side. Most importantly, too, it was demonstrated that the phenomenon is more than an experimental curiosity, such as the induction of a secondary embryo might suggest, but is a corporate feature of normal embryogeny; for as a consequence of any procedure, creation of an exogastrula or excision of prospective chordamesoderm, whereby axial mesoderm is prevented in whole or in part from associating itself with the neural ectoderm, there is a corresponding failure of neural differentiation. Significantly, also, neural induction has been shown to occur in vertebrates other

than amphibians and thus is probably a universal morphogenetic process.

In considering the manner of operation of the organizer, it is essential to realize that induction presents two aspects. The organizer not only elicits the differentiation of neural tissue, but somehow directs the creation of an organ with a definite architecture. That is, neural histogenesis is accompanied by organogenesis in which a neural tube featuring a brain and spinal cord of characteristic form comes into being. These two aspects of organizer action reflect different physiological mechanisms, for they can be separated from one another experimentally so that the former, elicitation of histogenesis, will occur either exclusively or with a minimum of organogenetic accompaniment. This is a way of saying that induction involves, first, a generalized kind of activation and, second, origination of architectural pattern.

The essential distinctness of the two aspects of induction is indicated by the results obtained when fragments of chordamesoderm are killed by heating or some other treatment and then are implanted beneath the prospective epidermis of a normal early gastrula. Neural tissue is still induced, but other than usually being rolled into a tube it exhibits little structural organization. Such architecture as may be manifested is almost surely an expression of the developmental powers inherent in the induced tissue itself rather than something imposed by the inductor. If this may not follow entirely from the experiment cited, there can be no other conclusion deriving from tests of unnatural inductors. It has been found, for example, that a great variety of embryonic and adult tissues, which in the living condition are ineffective as inductors, produce induction effects when killed. In fact, neural inductions may result from such an unlikely agent as methylene blue or a mechanical irritant such as silicious earth. One can therefore hardly avoid the conclusion that the first step in induction is the triggering of biochemical changes in an ectoderm that is already "cocked and ready to fire."

The nature of the inducing substance, although clearly soluble and diffusible, remains otherwise unknown. It is clear that distinct actions occur in different parts of the neural plate and that different areas of the chordamesoderm exert different influences.

At the conclusion of gastrulation the neural ectoderm is thin, single-layered, and undistinguishable from the adjacent epidermal areas. But it rapidly becomes thick and stratified so as appropriately to be termed the *neural plate* (Figure 17-6A). Beginning anteriorly and working steadily backward, the middle of the neural plate then buckles downward and its margins are concomitantly elevated, resulting in a *neural groove* bounded by *neural folds* (Figure 17-6B). The deepening of the neural groove and simultaneous elevation of the neural folds are caused by contractions of microfilaments in the apices of the neural plate cells, reducing the exposed surface area. A force is also generated that causes

the entire neural groove to elongate during neural fold closure. The groove continues to deepen, and the folds finally meet above it, thereby converting the original plate into a *neural tube* (Figures 17-6C, D). The meeting of the neural folds to create the tube is accompanied by a meeting of epidermal ectoderm, so that the tube is detached from the overlying epidermis.

With the inauguration of the neural groove and folds, a ridge of cells appears at the junction of each neural elevation and associated epidermal ectoderm. These cells, which probably migrate out of the margins of the neural ectoderm, constitute the *neural crest.* When the neural folds meet to create the neural tube, the neural crest is left as a wedge between the tube and the epidermis above (Figure 17-6D). The crest tissue subsequently separates into right and left liner halves, each of which moves out to occupy a position between the neural tube and the adjacent myotomes. Initially organized as continuous columns, the neural crests gradually break into segmental clusters corresponding to the myotomes. The myotomes clearly play a role in this segmentation, for any experimental disarrangement of the myotomes prior to crest segmentation results in a corresponding disarrangement of the neural crest. The removal of myotomes abolishes the segmentation of the neural crest; a grafting in of extra myotomes results in additional segments of neural crest.

With some exceptions to be noted elsewhere, the neural tube and paired segments of neural crest provide the material source of the central and peripheral nervous systems.

It follows, then, that whatever controlling agencies direct the formation of *segments* must exert a major influence on the formation of the nervous system. Recent work on the analysis of insect development has revealed a cascade sequence of gene activations and actions that progressively determine the axes of the embryo, the areas of segmentation, the orientation of the seg-

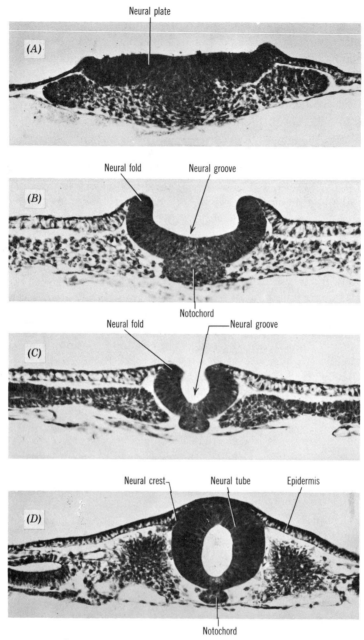

Figure 17-6 Photomicrographs of stages in the development of the neural tube of the chick. (*A*) Neural plate. (*B*) Early neural groove. (*C*) Late neural groove. (*D*) Neural tube and neural crest. All photos ×210. (Photos by Torrey.)

ments, and finally the individual character of each segment. Application of complementary DNA probe sequences from these insect genes to sections of mammalian embryos has shown the presence of at least some of this sort of gene activity in the establishment of segmentation, especially including the segmental portions of the hind brain, the *rhombomeres*. We may thus be close to both an understanding of the genetic control of segmental development in the vertebrate nervous system and a much better understanding of the phylogenetic relationship between the vertebrates and invertebrates.

At the beginning of its development, the neural tube is composed of cells that visibly, at least, appear to be of one kind. But, as they multiply, the cells enter two lines of specialization. One path leads to the production of *neuroglial cells* destined to produce a variety of nonnervous supportive and protective cells; the other leads to *neuroblasts* destined to become neurons. The term neuroglial cells is applied to the *ependyma* (p. 418) that lines the lumen of the tube and several types of so-called *glial cells* that not only will serve as a sheathing matrix for the processes of the nerve cells to come, but will be involved in the mediation of the normal metabolism of the neurons. It is to the neuroblasts within the neural tube that the motor neurons of the peripheral nerves and the countless association neurons constituting the "switchboard" of the nervous system trace their origin.

The segmentally arranged aggregations of neural crest outside the neural tube are termed *ganglia*. They, too, provide both nervous and nonnervous elements. The neuroblasts therein are the forerunners of the sensory neurons of the peripheral nerves (and some motor neurons as well; see p. 446). The nonnervous components of the ganglia, or their neural crest precursors, provide an even wider variety of materials than do their counterparts in the neural tube. In addition to supporting structures within the ganglia themselves, they are the major source of the Schwann cells investing both sensory and motor nerve fibers and, as noted elsewhere, furnish such seemingly unrelated materials as medullary tissue of the adrenal glands, pigment cells, and branchial skeleton.

Having laid the groundwork of the structural and functional units of the nervous system and their embryonic sources, we are now prepared to turn to the architectural history of its principal parts.

THE SPINAL CORD

As already noted, the original neural ectoderm is thin and composed of a single layer of cells, but as it is rolled into a tube its cells multiply rapidly and the wall of the definitive neural tube comes to consist of many cell layers. At one time it was believed that as the cells multiplied they lost their boundaries and merged into a syncytium, but closer study has revealed that, although they become delicate and inconspicuous, the cell membranes remain intact. When first established, that portion of the neural tube destined to become the spinal cord is considerably broader dorsoventrally than it is transversely, so that the cord is oval in outline and its lumen is slit-shaped (Figure 17-7*A*). The notochord seems to be involved somehow in the shaping of the early cord, for in the absence of the notochord the spinal cord and its lumen tend to become circular.

As an accompaniment of the oval configuration and slitlike lume, the cells of the embryonic cord are concentrated at the sides, leaving its roof and floor relatively thin (Figure 17-7*A*). Scrutiny of the cord at this time reveals that cell multiplication, as evidenced by the presence of mitotic figures, is going on only in the area adjacent to the lumen. Many of the new cells that are rapidly produced migrate peripherally and assemble themselves as a thickened stratum termed the *mantle layer* (Figure 17-7*B*). With the decline and ultimate termination of cell division, a residuum of cells bounding the central canal constitutes the *ependymal layer*. Processes from the ependymal and other nonnervous cells extend externally to provide an investing framework termed the *marginal layer*.

During the aforementioned decline of cell proliferation into the mantle layer, the walls of the ependymal layer gradually approximate and fuse dorsally. The central canal consequently becomes steadily reduced and assumes a roughly circular outline (Figures 17-7*C, D*). Meanwhile, the

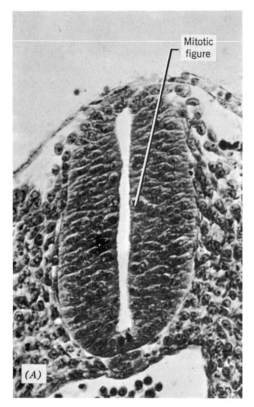

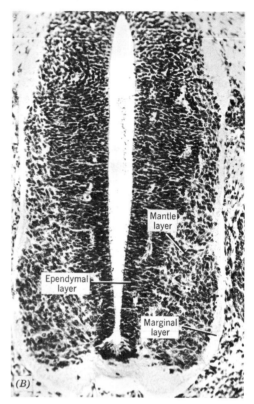

Figure 17-7 Development of the spinal cord. Photomicrographs of transverse sections through the cord of the embryonic rat at various ages: (*A*) 11 days, ×210; (*B*) 13 days, ×140; (*C*) 14 days, ×100; (*D*) 17 days, ×100. (*E*) Stereogram of the adult cord. (Photos by Torrey.)

cells of the mantle layer migrate and aggregate to create thickened masses dorsally and ventrally on each side. This gives to the mantle layer, as seen in cross section, a configuration roughly like the letter "H" (Figure 17-7*E*), whose thickened arms traverse the length of the cord. It is in the mantle layer that the neuroblasts become concentrated, since this layer represents the site of the cell bodies and dendrites of the neurons to come. Because these parts do not acquire myelin, the mantle layer assumes a characteristic gray color; hence it is termed the *gray matter.* By the same token, the previously mentioned longitudinal thickenings are referred to as *dorsal* and *ventral gray columns.* The two halves of the gray matter are connected by the *dorsal* and *ventral gray commissures* lying above and below the central canal.

With the growth of axons from the neuroblasts lying within the gray matter, the nonnervous meshwork of the early marginal layer is invaded by nerve fibers. Most of these fibers acquire myelin sheaths whose whitish color is responsible for the peripheral layer of the cord finally becoming called the *white matter.* The nerve fibers constituting the white matter are actually grouped in areas separated by the dorsal and ventral columns of gray matter. Thus, like the gray matter, the white matter of the cord is also arranged in longitudinal columns (Figure 17-7*E*). To avoid confusion in nomenclature, the white columns carry the name *funiculi* (the Latin equivalent of "cords") to distinguish them from the gray columns. A *dorsal funiculus* lies above and medial to the dorsal gray column on each side, a *ventral funiculus* lies below and me-

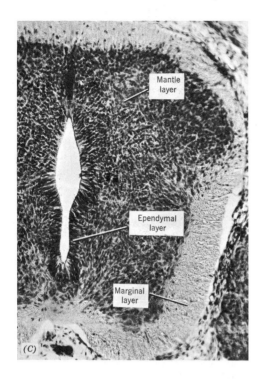

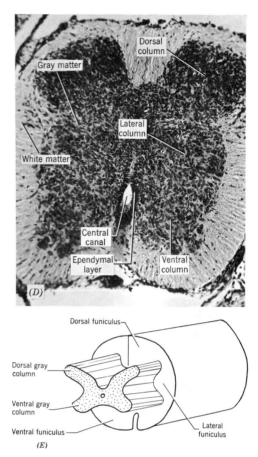

Figure 17-7 (*Continued*)

dial to each ventral gray column, and a *lateral funiculus* is present on each side between the dorsal and ventral columns. The white matter on the two sides of the cord is bridged by the *white commissure* lying immediately beneath the ventral gray commissure. It is by way of the nerve fibers in the funiculi that transmission up and down the cord and to and from the brain is expedited. Likewise, the white commissure provides communication between the two sides of the cord.

With the general architecture of the spinal cord before us, it is now possible to look more closely at its functional organization. First, however, it is necessary to speak briefly of the functional organization of the vertebrate body as a whole.

In a very broad sense the functions of the body may be resolved into two categories, *somatic* and *visceral*. The former relates to the operations carried on by the skin and its derivatives and the voluntary musculature; the latter relates to the operations of the other organ systems, that is, digestive, circulatory, and so on. Because the mediation of these operations requires an equipment of sensory and motor neurons, it follows that we have to deal with four functional types of nerve fibers: *somatic sensory, somatic motor, visceral sensory,* and *visceral motor*. The detailed arrangement of these fibers in the spinal nerves is examined later (p. 436). The somatic sensory fibers that carry impulses from somatic parts to the spinal cord connect synaptically with neurons lying in the upper portion of the dorsal gray columns. Visceral sensory fibers, transmitting from the visceral organs, connect with neurons in the lower portion of the dorsal columns. The cell bodies of the somatic and visceral motor neurons lie in the ventral columns, those of the somatic below and those of the visceral above. There are, therefore, four functional areas in the gray matter on each side of the spinal cord arranged in the following order from top to bottom: somatic sensory, visceral sensory, visceral motor, and somatic motor (Figure 17-8). In due time we shall see that the gray matter of the rear of the brain presents a similar pattern.

As just noted, the ventral gray columns consist largely of the cell bodies and dendrites of the motor neurons whose axons are distributed peripherally to the body. The cell bodies in the dorsal columns, however, belong for the most part to association neurons. The dendrites of these association neurons have synaptic association with the entering sensory fibers, but their axons follow a multiplicity of courses. Some may pass ventrally to form synapses directly with the dendrites of the motor neurons; others may connect with additional association neurons, which in turn communicate with the motor cells; and still others may run up and down the cord in the funiculi or cross to the opposite side in the white commissure, thus giving functional versatility and coordination to the total cord.

In cyclostomes and fishes the spinal cord tends to be fairly uniform in diameter along its length. The cord of tetrapods, however, features two local enlargements, one at the level where the nerves to the forelimbs take off, the other at the level of the hind limbs. These enlargements particularly reflect the ventral columns from which emanate the axons of the motor neurons supplying the limb musculature. We see in this an interesting correlation between the quantity of nervous supply and the peripheral "load" which that supply is obligated to bear. That this correlation is more than a coincidence is revealed by observations on "natural" and experimentally produced variations in the peripheral load. Congenital absence of a limb or its experimental removal from an embryo, for instance, is accompanied by an underdevelopment of the corresponding centers of the cord. Conversely, if an extra limb is grafted, say in a chick or amphibian embryo, an augmentation of the appropriate centers in the cord takes place.

THE BRAIN

A stripped down definition of the brain identifies it simply as the anterior part of the neural tube lying within the cranium. If that of *Amphioxus* is any criterion, the brain originally differed little architecturally from the spinal cord. But the morphogenetic history of the vertebrate brain is one of increasing complexity of gross anatomical form and internal organization.

From the very beginning of embryogeny, whether in amphibian or mammal, the prospective brain differs from the spinal cord behind. The forepart of the neural plate is considerably wider and thus, as the neural folds rise and meet, the diameter of the brain tube is far greater than that of the cord. Moreover, even before the neural

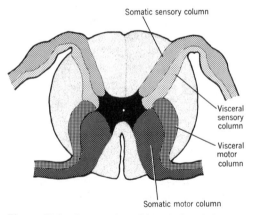

Figure 17-8 Cross section of the spinal cord showing the relative positions of the four functional areas of gray matter.

folds come together, it is possible to distinguish two subdivisions in the forming brain. These subdivisions manifest themselves with variable clarity from one vertebrate embryo to another, and over the years there have been differences of opinion as to the significance of the criteria that presumably distinguish them. There is fairly general agreement, however, that the primordial brain tube is initially divided into an anterior *prosencephalon* (*archencephalon*) and posterior *deutereoncephalon* (Figure 17-9). Although some morphological delineation between these two divisions is often provided by a fold in the floor of the forming brain, the principal distinction relates to the chordamesodermal induction system. As noted earlier (p. 413), the neural tube comes into being via the influence of regionally specific induction mechanisms. Specifically, that which is identified as the prosencephalon is the region of the brain induced by prechordal mesoderm, whereas the deuteroencephalon is induced by the anterior portion of the notochordal mesoderm (Figure 17-9).

The conventional story of the embryogeny of the vertebrate brain now tells of a subdivision of the deuteroencephalon by a constriction termed the *isthmus*. The brain is thus said to pass into a three-part stage consisting of the original prosencephalon (forebrain) in front, a *mesencephalon* (midbrain) in the middle, and a *rhombencephalon* (hind brain) at the rear (Fig-

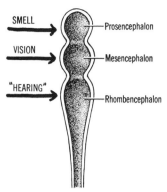

Figure 17-10 Three-part division of the brain in relation to smell, vision, and "hearing."

ure 17-10). Morphologically, however, the distinction between these three divisions is often so vague as to lead some workers to feel that this three-part organization has little or no reality. Such distinction of parts as may exist actually rests more on functional than anatomical grounds. This functional delineation has its basis in the site of termination of the sensory transmission lines coming in from the three major sense organs—nose, eye, and ear. Accordingly, the prosencephalon is associated with the sense of smell, the mesencephalon with vision, and the rhombencephalon with "hearing" (used in the broad sense to include both sound perception and equilibration).

The further embryonic history of the brain, for vertebrates generally, is one of the coming into play of a host of shaping processes. One of the earliest modifications is a subdivision of two of the three primary brain regions. The mesencephalon remains intact, but the prosencephalon becomes marked off into an anterior *telencephalon* and posterior *diencephalon,* and the rhombencephalon into a *metencephalon* and *myelencephalon.* As with the earlier three-part stage, these regions are often better defined functionally than anatomically; hence some experts are inclined to "play down" their reality. Yet they retain great descriptive usefulness, so we shall adhere to the convention of considering the basic

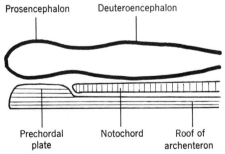

Figure 17-9 Subdivision of the brain into archencephalic and deuteroencephalic parts and the relationship of these to the inductor system.

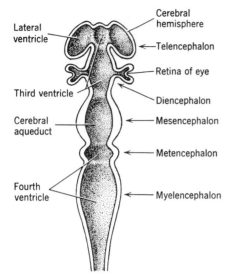

Lateral ventricle

Cerebral hemisphere

Telencephalon

Retina of eye

Third ventricle

Diencephalon

Cerebral aqueduct

Mesencephalon

Metencephalon

Fourth ventricle

Myelencephalon

Figure 17-11   Basic five-part anatomy of the vertebrate brain.

organization of the vertebrate brain, from fish to human, as one consisting of five parts: telencephalon, diencephalon, mesencephalon, metencephalon, and myelencephalon, in that order from front to rear (Figure 17-11).

In the lower vertebrates the five divisions of the brain are arranged essentially in a straight line. The brain tube in amniotes, however, tends to bend upon itself at an early period of its embryogeny. This is especially noticeable in birds and mammals, where *flexures* at three levels create sharp angles in the long axis of the embryonic brain. These flexures accompany a more general flexing of the entire head; but they probably also reflect morphogenetic propensities inherent in the brain itself. To illustrate, the human brain tube first exhibits a *cephalic flexure* whereby the forebrain is brought sharply ventrad from a fulcrum in the midbrain (Figures 17-12*A, B*). Soon thereafter, a *cervical flexure,* involving a bending of the entire head at its junction with the neck, brings the entire brain ventrad (Figures 17-12*A, B*). A later-appearing *pontine flexure* at the level of the metencephalon counteracts the effect of the other two flexures by bending the brain in the opposite direction (Figure 17-12*C*). The

cervical and pontine flexures eventually disappear, and the rear of the brain becomes straight once more, but the cephalic flexure is retained, so that the telencephalon and diencephalon remain permanently set at an angle with the regions behind.

Even as its five primary subdivisions are established, the brain wall exhibits a number of local outpocketings. Two of these, destined to produce the *cerebral hemispheres,* are identified with the telencephalon. Another pair will provide the retinas of the eyes (see p. 465) and are associated with the diencephalon (Figures 17-11 and 17-12). Such outpocketings are accompanied by continued cell multiplication, which also expresses itself in a number of other ways. Localized aggregations of differentiating neuroblasts bring thickenings of gray and white matter and a variety of folds, fissures, and invaginations. Conversely, the roof of the diencephalon and myelencephalon becomes thin, devoid of nervous tissue, and highly vascular.

The lumen of the original embryonic neural tube persists in the adult brain in the form of fluid-filled cavities adapted to the regional specilization of the brain walls (Figure 17-11). The telencephalic cavities that extend into the cerebral hemispheres are designated the *lateral ventricles.* These communicate with the lumen of the middle of the telencephalon and that of the diencephalon, which together constitute the *third ventricle.* The cavity of the mesencephalon, considerably reduced in amniotes, is termed the *cerebral aqueduct* and serves to connect the third ventricle in front with the *fourth ventricle* of the metencephalon and myelencephalon behind.

In the discussion that follows, we shall necessarily give considerable attention to the comparative morphology of the parts of the brain in the several classes of vertebrates. The series of illustrations constituting Figure 17-13 is dedicated to this end and will be referred to frequently as the need arises to point to external features and gross structures. It cannot be overempha-

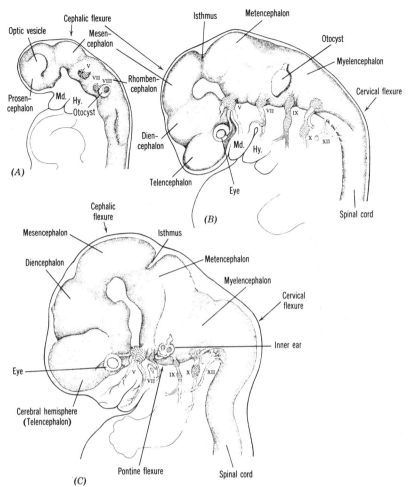

Figure 17-12   Three stages in the early development of the human brain: (*A*) 3½ weeks, (*B*) 5 weeks, (*C*) 7 weeks. (From Patten and Carlson, *Foundations of Embryology*, 2nd ed., © 1964. Used with permission of McGraw-Hill Book Company.)

sized, however, that this superficial anatomy is far less significant than the functional operations incorporated therein. As Professor A. S. Romer has put it, "an adequate understanding of the working of the brain can no more be gained ... than a knowledge of a telephone system can be had from an acquaintance with the external appearance and room plan of the telephone exchange building. What is important in a telephone system is the wiring arrangements and switchboards; in a brain it is the centers in which various types of activities occur and the tracts of fibers connecting these centers."

It was pointed out in our preliminary

discussion of the elementary functional anatomy of the nervous system how the interpolation of association neurons in a reflex arc contributes to the versatility of sensitivity to incoming sensory impulses and the initiation of motor responses. The brain represents only an elaboration of this principle. The brain may be said to differ from the spinal cord only in the magnitude of numbers of association neurons and a much greater variety of functional categories. Any given functional group of association neurons constitutes what is termed a *center,* whose role is to receive incoming "messages," sort, correlate, and integrate these messages, and relay them to appropriate

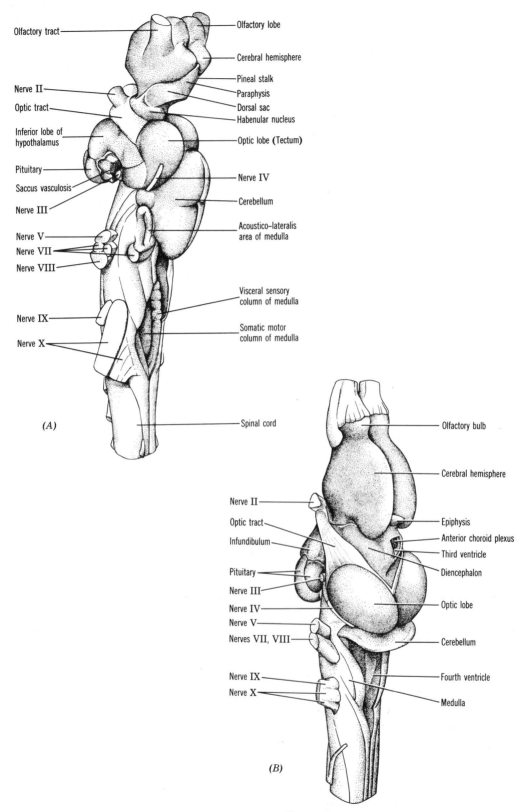

Figure 17-13 Comparative anatomy of the vertebrate brain. (A) Shark. (B) Frog. (C) Alligator. (D) Pigeon. (E) Rabbit. (Drawn from Mueller-Ward models.)

424

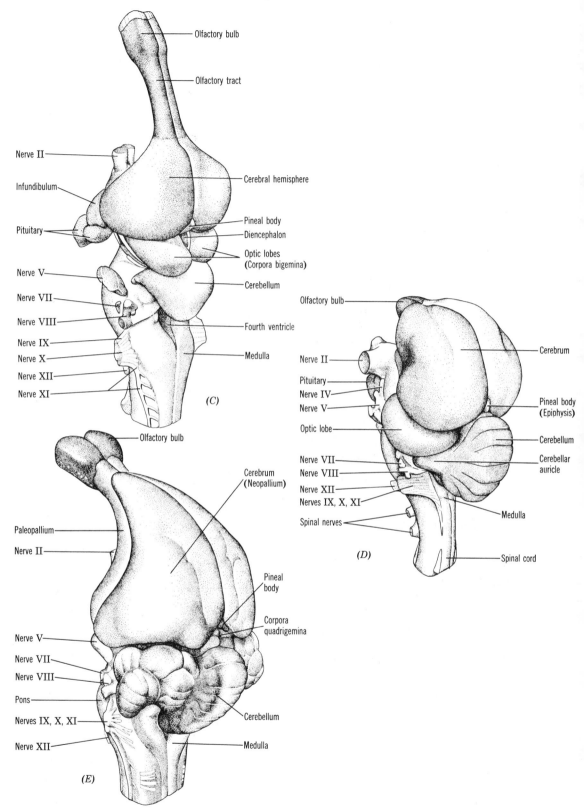

Figure 17-13 (Continued)

425

outgoing channels. This sorting and integration process, of course, requires a high degree of communication between the centers and/or the spinal cord. In some measure this communication is provided by randomly distributed fibers, but the more usual arrangement is for axonic fibers of like central connections to be assembled in bundles known as *tracts*, which run between the centers and the spinal cord. Although there are some fiber tracts in the brain running lengthwise for a distance and then crossing to the opposite side, bilateral integration is ordinarily provided by transverse fiber bundles termed *commissures*.

With these fundamentals of embryogeny and functional anatomy behind us, we are now prepared to consider separately the morphogenetic history of each of the five basic subdivisions of the brain.

The Myelencephalon. Because, in general, the rear of the brain is simpler in architecture than the parts in front, we shall begin our analysis with the last of the five divisions and work forward. The myelencephalon, or *medulla oblongata*, is, in fact, basically similar to the spinal cord with which it is continuous, except that its central canal (fourth ventricle) is greatly expanded and its roof consists of a thin, nonnervous epithelium equipped with an extensive blood supply (Figure 17-14). This vascular roof constitutes the *posterior choroid plexus*. It is in the thickened lateral and ventral walls of the medulla that we see the layout of nervous tissue already seen in the spinal cord; external white matter composed of myelinated fiber tracts and internal columns of gray matter. The latter are arranged like those in the spinal cord (reading from top to bottom): somatic sensory, visceral sensory, visceral motor, and somatic motor. In all embryonic vertebrates these columns are continuous fore and aft through the medulla, and so they largely remain in the adults of lower vertebrates. However, in higher vertebrates, notably mammals, the columns tend to be broken

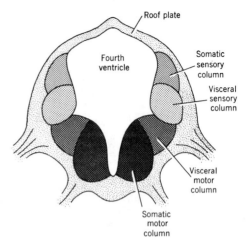

Figure 17-14   Cross section of the medulla, showing the distribution of sensory and motor columns. Compare with Figure 17-8.

into centers associated with specific functions. A consideration of the nature of these functions requires a preview of the cranial nerves.

The cranial nerves are those peripheral nerves that have connections with the brain. By conventional count there are 10 of them in anamniotes, plus 2 more for a total of 12 in amniotes. A detailed analysis of the fiber constitution of the cranial nerves will be presented shortly (p. 441). All the cranial nerves from number V rearward join with the medulla. Of these, numbers VI and XII contain somatic motor fibers only, and number VIII contains somatic sensory fibers only, whereas the others are a mixture of sensory and motor fibers. Speaking generally, now, the cells of origin of the purely motor fibers of nerves VI and XII lie in the somatic motor column of the medulla. The visceral motor fibers of nerves V, VII, IX, X and XI originate in the visceral motor column. The somatic sensory fibers of nerve VIII terminate along with the somatic sensory constituents of nerves V, VII, IX, and X in the somatic sensory column. The visceral sensory components of nerves VII, IX, and X terminate in the visceral sensory column.

To translate this anatomy into functional

operations, there must be some cognizance of the distribution of the nerves at issue. Nerve VI supplies one of the extrinsic muscles of the eye; nerve XII is associated with the tongue. Nerves V, VII, and IX are related to the head and visceral arches. Nerve X also supplies the visceral arches, but in addition relays visceral fibers to the heart, respiratory organs, and digestive tract and its associated organs, and somatic sensory fibers to the lateral line organs (p. 451) of fishes and amphibians. Nerve VIII is associated with the inner ear. Thus, generally speaking, the medulla provides the switchboard for the circuits between the receptors and effectors of the head, pharyngeal derivatives, and the principal visceral organs. When, as in mammals, the columns are broken into discrete centers, specific functions are found to be associated with specific centers. For example, the somatic motor fibers distributed to the tongue originate in a distinct center, the visceral motor fibers regulating respiratory rhythm originate in another, and the fibers to the salivary glands in still another; and on the sensory side, there is a special gustatory center for taste and an especially notable one related to hearing and equilibration. Through the fiber connections of nerve VIII with the inner ear and nerve X with the lateral line system, the medulla provides the center for the acousticolateralis system. Sensations from this system were primitively received in the somatic sensory column. However, the system is so well developed and special in nature in fishes that a conspicuous specific acousticolateralis area appears above the regular somatic sensory column. The lateralis system disappears in land vertebrates, but the acoustic center persists. And as we shall shortly see, this center is closely linked to the cerebellum (see later).

Some of the fiber tracts in the white matter of the medulla decussate to maintain integration of the two sides; others run posteriorly to the spinal cord and anteriorly to the forward divisions of the brain. Although the transmission lines between the medulla and other parts of the brain are extensive in vertebrates generally, similar communication between the spinal cord and the brain ranges from scanty in lower vertebrates to great in higher forms. Except for a limited number of sensory fibers that run from the cord through the medulla and as far forward as the midbrain, in fishes and amphibians the parts of the body under the jurisdiction of the spinal cord and its nerves tend to operate semiautonomously with little reference to the brain. It is only in amniotes, notably birds and mammals, that the trunk and tail come under the positive control of the higher brain centers, and thus the fiber tracts of the medulla become prominent.

The Metencephalon. In contrast to the myelencephalon whose roof remains thin and nonnervous and in conjunction with its blood supply serves as a site for the provision of cerebrospinal fluid, the dorsal side of the metencephalon becomes the elevated, prominent *cerebellum*. For vertebrates generally the cerebellum is concerned with the government of muscular coordination insofar as it relates to adjustments in body posture. Accordingly, there is seen a correlation between the extent of its size and the intricacy of bodily movements: in cyclostomes, amphibians, and the less active fishes and reptiles the cerebellum is of modest size, whereas in the more active fishes and in birds and mammals it becomes a conspicuous, intricately folded mass that overlaps the medulla behind and the mesencephalon in front. The cerebellum (like the cerebral hemispheres and the roof of the mesencephalon) also features a reversal of gray and white matter, so that the gray matter lies as a superficial cortex over the internal white matter.

The basic functional design of the cerebellum is that of a switching panel, or electronic computer, into which data are fed for sorting and correlation, and out of which appropriate commands are sent over the nervous circuitry. The cortical gray matter with its multitudinous cells is the "heart" of the panel, for it is here that all the data are ultimately received and all the "computations" are made. The white matter, then, provides the lines of communication in and out of the computer. It is in birds and mammals that this switching panel reaches its greatest capacity, for the cerebellar convolutions provide for a maximum of cortical substance and branching tracts of white matter in a minimum of cranial space (Figure 7-15).

The data fed into the cerebellum derive from two principal sources. One consists of the system of receptors associated with the skeletal muscles and tendons (p. 451), serving to record the state of muscle tension and the position of body parts. Sensory fibers from these receptors are directed into the cerebellum, via the spinal cord and medulla, where the information they bring in joins that from the other major source,

the acousticolateralis system (p. 451). Sensations relating to equilibrium and body movement are registered in the membranous labyrinth of the inner ear and, among fishes and aquatic amphibians, pressure and low-frequency vibrations of water are recorded by the lateral line organs. But proper muscular coordination does not come alone from the registering and integration of data on muscle tension and equilibrium. As we know from our own experience, body posture is also adjusted to what we touch, hear, and see. So the two primary sources of information are supplemented by signals from the receptors in the skin, from the eyes, and from the receptors for sound. Like those for equilibrium and position, the sensations of hearing are first centered in the medulla before being relayed to the cerebellum, and the circuits relating to touch basically follow a route through the spinal cord and medulla. Optic sensations, however, first

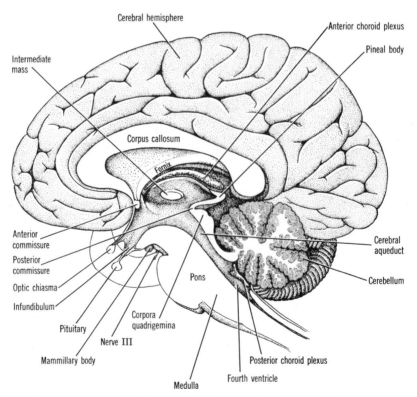

Figure 17-15 Sagittal section of the human brain (semidiagrammatic).

center in the mesencephalon (except in mammals) and are fed back to the cerebellum.

Although in vertebrates generally the cerebellum occupies a primary position as the control center for muscular coordination, special complications prevail in the mammalian brain. The great elaboration of the cerebral hemispheres in mammals has been accompanied by the rise within them of new centers of control. Now the sensations for touch, muscle tonus, hearing, and vision are relayed to the cerebral hemispheres (cerebrum) as well. With this has come an elaboration of prominent fiber tracts between the cerebellum and cerebrum, namely, the *pons* in the floor of the myelencephalon and metencephalon, the sidewalls of the mesencephalon, and the sidewalls of the diencephalon (Figure 7-15). The result is a complicated "feedback" system wherein sensations may first go to appropriate centers in the cerebrum and are then sent back to the cerebellum for final correlation and relay; or they may first go to the cerebellum, then to the cerebrum, and back once more to the cerebellum (Figure 17-16).

The Mesencephalon. It will be recalled that the brain is considered to have had a primary three-part organization based on functional operations. These functional criteria relate to the three principal senses, hearing, vision, and smell, whose nervous terminals are related to the rhombencephalon, mesencephalon, and prosencephalon, respectively. We have already observed how the sensory terminals of the acousticolateralis system are first assembled in the medulla and data are then transferred to the cerebellum where, after correlation with other data, coordinating motor responses emanate. And, as we shall see, the history of the telencephalon is linked to the olfactory sense. So it is that the mesencephalon is fundamentally linked to vision.

Although the retinas of the eyes are products of the diencephalon (see p. 464) and thus the fibers of the optic nerve might be

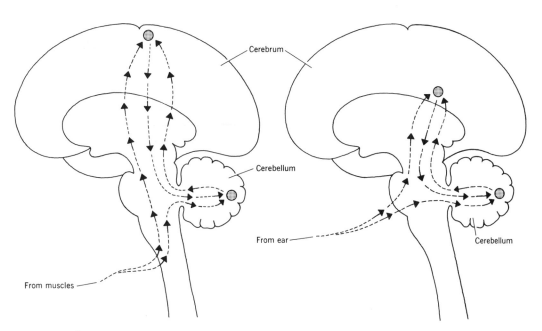

Figure 17-16 Wiring diagrams of two representative reverberation circuits to the cerebellum and cerebrum of a human. (From R.S. Snider, "The Cerebellum," *Scientific American* Vol. 199, No. 2. © August, 1958 by Scientific American Inc. All rights reserved. Courtesy W.H. Freeman and Co., San Francisco.)

expected to terminate in the diencephalon, in all vertebrates except mammals (and to a lesser extent in reptiles) the optic fibers pass without interruption upward and backward to the roof, *tectum*, of the mesencephalon (Figure 17-17*A*). If we ignore the mammals for the moment, the primary visual centers are located in the roof of the mesencephalon. The gross morphological expression of these centers is a pair of *optic lobes* (*corpora bigemina*), whose size variances reflect the sizes of the eyes and acuteness of vision (Figure 17-13). Thus in many teleosts and particularly in birds, the optic lobes are conspicuously large.

The history of the tectum of the mesencephalon, however, is not confined to vision alone. As we have already come to understand, coordinated behavior results

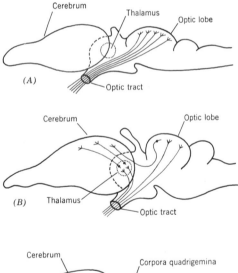

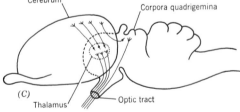

Figure 17-17 Wiring diagrams of the circuits to the optic centers in (*A*) amphibians, (*B*) reptiles, and (*C*) mammals. (After Portmann, *Vergleichende Morphologie der Wirbeltiere*, courtesy Benno Schwabe and Company, Basel.)

from a complete analysis and integration of all sensory data—touch, movement, hearing, smell, sight, and taste. Just as postural coordination results from a feeding of sensory data from all sources into the cerebellar computer, other body movements are governed by appropriate centers in the tectum of the mesencephalon, which also receives sensory data from many sources. So by way of fiber tracts in the sidewalls and floor, *tegmentum*, of the mesencephalon, the tectal region receives information from the olfactory organs via the diencephalon in front and from the acoustico-lateralis centers via the cerebellum behind, and sends out instructions to the appropriate motor columns. In fishes and amphibians, in fact, the tectum is the major site of control over body activity. It is also important in this respect in reptiles and birds, but in these forms there is a beginning of a shift in control to the cerebral hemispheres, a shift that reaches its maximum in mammals (Figures 17-17*B*, *C*).

Just as in the case of muscular coordination where, with the development of a complicated feedback system, the functions originally dominated by the cerebellum come more and more under the control of the cerebrum, the somatic sensations initially centering in the tectum of the mesencephalon are, in mammals, relayed to the cerebrum by way of the sidewalls of the diencephalon, and the final orders emanate from the cerebrum. This is especially true of the visual sense. Except for a few fibers that follow the original, primitive course to the mesencephalon, most of the visual sensations are shunted (via a relay in the diencephalon) into the cerebral hemispheres, and it is there that the principal visual centers of mammals are located (Figure 17-17*C*). The result is that morphologically the tectum of the mammalian mesencephalon is relatively reduced, taking the form of four small elevations, the *corpora quadrigemina* (Figure 17-13*E*). The mammalian tectum serves only in a limited way for visual and auditory reflexes, vision being centered in the anterior pair of quadrigeminal bodies, audition in the posterior pair. By contrast, the tegmentum becomes very prominent in mammals, for it is the site of the major motor pathways and relay stations between the medulla and cerebellum at the rear and the cerebral hemispheres in front.

The Diencephalon. Although relatively modest in size, the diencephalon is notable both for certain functions associated with it and for an assortment of accessories deriving as outgrowths from its walls. Of these accessories, the sensory portions of the eyes developing from the sidewalls of the embryonic diencephalon are especially noteworthy. However, the embryogeny of the eyes is treated separately (p. 464), so that the present analysis pertains to the functional differentiation of the diencephalon per se. In making this analysis it is convenient to consider the diencephalon in three parts: the roof or *epithalamus;* the sidewalls or *thalamus* proper; and the floor or *hypothalamus.*

Epithalamus. The third ventricle, it will be remembered, comprises the lumen of the diencephalon plus that of the midregion of the telencephalon. So although the bulk of the third ventricle lies within the diencephalon, its roof also includes a telencephalic component. The greater part of the roof of the third ventricle retains its original thin, epithelial character and, like the roof of the myelencephalon, acquires a vascular supply. This vascular membrane, often elaborately folded and invaginated into the third and lateral ventricles, constitutes the *anterior choroid plexus* (Figure 17-18). Important though this is in the physiological maintenance of the cerebrospinal fluid, the more interesting part of the history of the roof of the third ventricle pertains to a series of median evaginations from it. The most anterior of these is the *paraphysis,* a thin-walled sac extending dorsally from the telencephalic roof (Figure 17-18). Of unknown function and evolutionary significance, the paraphysis appears in the embryos of most vertebrates, but usually disappears in adult life.

Two additional median evaginations from the roof of the diencephalon appear immediately behind the choroid plexus (Figure 17-18). The anterior one is called

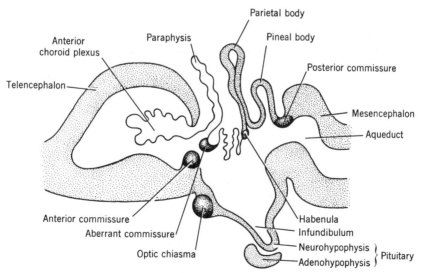

Figure 17-18 Sagittal section of a generalized vertebrate diencephalon and adjacent structures.

the *parietal body;* the posterior one the *pineal body* or *epiphysis.* These structures have had a variable and unique history in vertebrates, a history linked to the fact that in ancestral vertebrates the conventional bilateral eyes were supplemented by one or two median dorsal eyes. The fossil record, for instance, clearly reveals that the ostracoderms and most placoderms had a socket on the top of the head that undoubtedly bore an eye. It was present, too, in all the major groups of fishes, notably the crossopterygians, and in most ancient amphibians and reptiles, including the mammallike reptiles. Among modern forms, however, a median dorsal eye, of a miniature kind, is found only in lampreys and a few reptiles. The rhynchocephalian *Sphenodon* features the greatest development of such an eye; some lizards have a more rudimentary organ.

That two outgrowths are formed and that in lampreys both produce eyelike structures suggest that ancestral vertebrates may have had two median eyes. Yet the fossil evidence usually indicates the presence of a single eye alone. It is likely, therefore, that one outgrowth was exploited by some vertebrates and the other outgrowth by other vertebrates. In other words, the single median eyes of vertebrates are not homologous structures; in some the eye corresponds to the parietal body, in others to the pineal body. Thus, of the two structures present in lampreys, the pineal eye is the truly functional one, whereas it is the parietal eye that is represented in *Sphenodon* and lizards.

But aside from these exceptional instances, the dorsal evaginations from the diencephalic roof never form eyes in modern vertebrates. In most vertebrates, in fact, the parietal body either does not arise at all or has only a transitory embryonic existence. The pineal body, by contrast, does appear in all vertebrates (Figure 17-13), but seems to be glandular in nature.

As the foregoing indicates, the epithalamus is dedicated largely to nonnervous parts and functions. There is, however, in all vertebrates a pair of *habenular bodies* lying immediately in front of the parietal organ and serving as relay centers for olfactory stimuli being routed rearward from the cerebrum; and a posterior commissure lies adjacent to the mesencephalon (Figure 17-15).

Thalamus. The thickened sidewalls of the diencephalon, the thalamus, contain important relay and integration centers for sensory and motor stimuli passing to and from the cerebral hemispheres. Functionally, it is divided roughly into dorsal and ventral regions, the dorsal centers being concerned primarily with sensory impulses, the ventral centers with motor impulses. The ventral motor area serves as a pathway for motor tracts between the cerebrum and the brain divisions and spinal cord to the rear, so that in vertebrates generally it is consistently well developed. The status of the dorsal sensory area, however, varies with the degree of control exercised by the cerebrum. In the lower vertebrates, where the principal integrations are conducted in the cerebellum and tectum of the mesencephalon, the dorsal thalamus plays a relatively minor role. But with the gradual shift of control to association centers in the cerebrum, the dorsal thalamus shows a corresponding development of relay centers. Thus in mammals, where all somatic sensations are assembled in the cerebrum, the thalamus provides major centers for the relay of optic and auditory data.

Hypothalamus. The functional properties of the hypothalamus are significant indeed and relate to three principal anatomical parts (Figures 17-13 and 17-15). The most anterior of these parts is the *optic chiasma,* the site where the two optic nerves cross as they pass from the eyes on their way to the optic centers in the mesencephalic tectum and/or cerebrum. Behind the chiasma the hypothalamic floor features important centers tied in with the autonomic nervous system (p. 444). Although many visceral functions are handled by reflexes in the medulla and spinal cord, in all vertebrates it is the hypothalamus that exercises control over such truly involuntary actions as temperature regulation, sexual reactions, breathing rate, emotional responses, and the rhythm of sleep. The third area is the *infundibulum,* an evagination from the hypothalamic floor.

Embryonically, the infundibulum projects downward to assume an association with a diverticulum pushing upward from the roof of the embryonic mouth or stomodeum (Figures 13-4 and 17-19). The stomodeal evagination drew our attention earlier (p. 281) at which time it was identified as *Rathke's pouch,* the forerunner of the *adenohypophysis* or *anterior lobe of the pituitary gland.* The physiological roles played by the adenohypophysis through its variety of secretions—for example, the hormones directing the activities of other endocrine glands such as the thyroid, adrenal, and corpus luteum of the ovary—has been dwelt upon at numerous points elsewhere and need not be repeated. At issue now is the *neurohypophysis,* or *posterior lobe of the pituitary gland,* a product of the distal

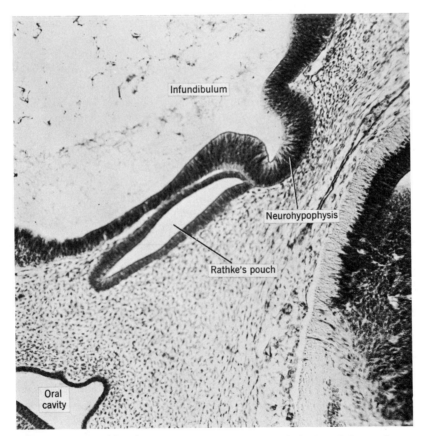

Figure 17-19 Photomicrograph of a sagittal section through the hypothalamus of a 10-mm pig embryo, showing the infundibulum and its relationship to Rathke's pouch. (Photo by Torrey.)

end of the infundibulum (Figures 17-18 and 17-19).

Two hormones, *vasopressin* (antidiuretic hormone) and *oxytocin,* have been identified with the neurohypophysis. It has been shown in representatives of all classes of vertebrates that vasopressin reduces the output of urine and so guards against excessive water loss. (That type of diabetes in humans, diabetes insipidus, characterized by excessive loss of water, is a consequence of a malfunctioning neurohypophysis.) Vasopressin also has the power to elevate blood pressure. Oxytocin produces contraction of the smooth muscle of the mammalian uterus and also of the contractile cells of the mammary glands to bring about "milk letdown." But unlike the adenohypophysis and other endocrine glands, the posterior lobe of the pituitary contains no glandular, secreting cells, and thus does not itself actually produce the substances attributed to it. Rather, its hormones are formed by clusters of neurosecretory cells located in the hypothalamus proper. The secretions then pass down the axons of nerve cells to the neurohypophysis. In other words, rather than being a secreting gland, the posterior pituitary is an organ in which products of a neurosecretory nature are stored and dispensed to the circulatory system.

The Telencephalon. The vertebrate telencephalon was originally only a center for olfactory sensations. Yet from this primitive beginning it pursued an evolutionary course marked by spectacular changes, culminating in the mechanism that provides for all human intellectual capacities. To trace this evolutionary history in detail is clearly impossible in a treatment of the scope to which we are limited, so we shall confine ourselves to major highlights. To this end, it is helpful to point first to some overall evolutionary trends and then, in this setting, to submit representative illustrations of changing form and function.

Beginning with a telencephalon con-

cerned solely with olfaction, we may note, first, a morphological subdivision of the telencephalon into bilateral halves, each of which, in turn, subdivides into a terminal *olfactory lobe* and a proximal, dorsolaterally expanded *cerebral hemisphere* (Figure 17-13). The olfactory lobes remain exclusively olfactory in function, whereas the hemispheres, although they retain olfactory functions as well, acquire other capacities. This, then, is the second general trend: a steady shift of dominance from association centers of other brain regions, notably the mesencephalon, into the cerebral hemispheres and the concomitant rise of new centers. A third trend is toward a shift of relative positions of gray and white matter, culminating in mammals with the gray matter largely occupying the surface of the hemispheres. Finally, as an accompaniment of the increase in number and importance of the association centers, the cerebral hemispheres increase steadily in bulk so as to tend to cover and envelop the brain region behind (Figures 17-13 and 17-15).

The cerebral hemispheres are primitively only areas into which olfactory sensations are assembled for ultimate routing to the hypothalamus and the hind brain for correlation with other sensory data. Such a primitive cerebral hemisphere, say in a cyclostome, presents a conventional external layer of white matter enclosing an internal layer of gray matter (Figure 17-20A). Looking forward to the time when the gray matter will assume a surface position, it may be termed a *pallium* (a Latin term meaning "cloak"), and because this is the pallium in its ancient, primitive form, it may be designated the *paleopallium.*

A more advanced stage is exemplified by the cerebral hemispheres of amphibians (Figure 17-20B). Now the gray matter is divided into three general regions. One area lies dorsolaterally and, because it remains largely olfactory in function, retains the designation paleopallium. A second region, termed the *corpus striatum,* lies ventrally and, in addition to receiving olfactory sen-

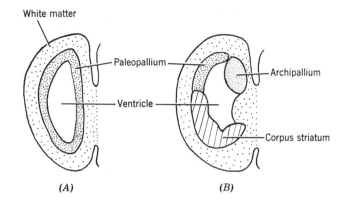

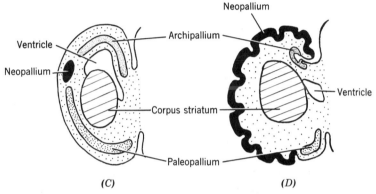

Figure 17-20   Cross sections through a cerebral hemisphere showing stages in the evolution of the cerebral cortex. (*A*) Primitive stage. (*B*) Amphibian. (*C*) Reptile. (*D*) Mammal. (Adapted from figures by Romer and Parsons, *The Vertebrate Body,* 5th ed., courtesy W.B. Saunders Company.)

sations, is linked to the thalamus and mesencephalon. Likewise linked to the diencephalon, as well as receiving olfactory sensations, is the dorsal (and median) *archipallium.*

Among reptiles the cerebral hemispheres not only exhibit an overall increase in size (reflecting an increase in internal complexity), but feature two important changes over amphibians. First, the corpus striatum moves inward to occupy a considerable area in the floor. (In birds the corpora striata enlarge greatly and, in fact, provide the bulk of the relatively large cerebral hemispheres; they become identified with the elaborate mechanical, instinctive behavior patterns that characterize these

vertebrates.) Second, the gray matter of the pallial areas tends to lie near the surface. In some reptiles, moreover, a small area of gray matter of a new type, the *neopallium,* makes its appearance (Figure 17-20*C*). It is in this reptilian neopallium that we see the first traces of the mammalian *cerebral cortex.*

The story of the mammalian cerebrum is essentially the story of the neopallium (Figure 17-20*D*). Now completely superficial, the neopallium becomes exceedingly bulky and often is extensively folded. Within its substance arise all the association centers, which not only attain dominance over the remainder of the brain but, in humans, are the source of all that we call intelligence or mental capacity. Conversely, the original archipallium, now called the *hippocampus,* and the paleopallium become relatively unimportant centers devoted exclusively to olfactory functions. Curiously enough, the corpus striatum, so important to the instinctive behavior of birds, remains a functional enigma in mammals. Finally, in eutherian mammals a massive new commissure, the *corpus callosum,* arises to connect the enlarged neopallia of the two sides (Figure 17-15). This broad cable (and lesser bridges) connects and coordinates the two halves of the brain.

SPINAL NERVES

As an introduction to the functional organization of the spinal nerves, it is desirable to review some points discussed earlier. Likening the vertebrate nervous system to a telephone system, we identify the central switchboard with the central nervous system (brain and spinal cord) and the incoming and outgoing lines with the peripheral nervous system. The peripheral nervous system consists of bundles of nerve fibers, that is, nerves, distributed to every region of the body. Those connected to the spinal cord constitute the spinal nerves, and those connected to the brain constitute the cranial nerves.

Now we have seen that the bodily functions with which the peripheral nerves are concerned fall into two general categories, somatic and visceral, and each category requires both sensory and motor equipment. Accordingly, we recognize four functional classes of nerve fibers: (1) *somatic sensory fibers,* carrying impulses from the skin and receptors in the skeletal muscles and tendons; (2) *visceral sensory fibers,* carrying sensations from the gut and other visceral organs; (3) *somatic motor fibers,* carrying impulses to the skeletal musculature; and (4) *visceral motor fibers,* carrying impulses to the visceral musculature and glands.

Embryonically, and thus in ultimate an-atomical terms, the fibers of the peripheral nerves are derived from two sources. Those of the motor neurons emerge from cell bodies lying within the brain and spinal cord; those of the sensory neurons ordinarily (there are exceptions) emerge from cell bodies clustered in ganglia lying alongside the brain and cord. With respect to the spinal cord and its nerves, the cell bodies of the somatic and visceral motor neurons lie in the ventral gray columns of the cord, whereas the proximal fibers of the somatic and visceral sensory neurons terminate in the dorsal gray columns of the cord. The cord, therefore, exhibits on each side four functional areas in the gray matter, corresponding to the four categories of nerve fibers (Figure 17-8). Regarding the manner in which the nerve fibers become appropriately distributed during embryogeny, the reader may be interested in the following brief discussion.

Orientation of Nerve Fibers. It has been pointed out elsewhere that nerve fibers originate as processes growing out from embryonic nerve cells (neuroblasts). Such outgrowths can be directly observed in plasma cultures of neuroblasts, under which conditions the developing fibers project in a completely random, chaotic fashion. Obviously, however, as fibers spin out from their neuroblastic sources in the central nervous system and the cranial and spinal ganglia of the normal embryo, their growth is not random but conforms to a pattern. There is thus posed the question of what directs the orientation of fibers so that they end up in association with their appropriate peripheral organs. To put it more simply, how do fibers get to where they are supposed to go?

At various times, three tentative answers to this question have been submitted. One of these postulates the existence of electrical fields in the periphery of the embryo, and the growing fibers are presumed to orient to the patterns present in these fields. Although there has been a limited experimental demonstration of the orientation of fibers in a culture medium through which an electric current is passed, there is no evidence for the existence of electrical forces of comparable density in the normal embryo. Moreover, the repeated observation that parallel fibers may grow in opposite directions militates against the likelihood that electric guidance is involved.

A second, and at first sight more likely, answer is that fibers are literally attracted to their destinations through the medium of chemical substances present in the parts to be innervated. Experimental results such as the following have been offered in support of this view. Specific outgrowth of sensory axons from the trigeminal ganglion to the nasal whisker pad epithelium in the mouse embryo has been demonstrated *in vitro.* If the forelimb rudiment of an amphibian embryo is removed prior to receiving its nerve supply and is grafted to an adjacent position on the side of the body, the outgrowing nerve fibers that normally would have supplied it continue to do so by deviating from their normal course. In the event that such a limb is grafted a greater distance away from its regular site, spinal nerve fibers ordinarily destined to go somewhere else detour and invade the limb. In fact, it has been demonstrated that almost any similarly grafted foreign object, eye vesicle, nasal sac, or even tumorous tissue will serve to "attract" nerve fibers. The behavior of regenerating nerve fibers in a severed adult nerve also suggests chemical attraction. If, for instance, a nerve is cut and its distal stump is deflected somewhat, the outgrowing fibers from the central stump alter their course and grow into the deflected stump. But, as we shall see shortly, that which apparently is chemical attraction turns out to be something else.

The third answer derives from the observation that nerve processes growing in a culture medium are unable to push through a pure liquid; they can proceed only along interfaces between liquids and

solids, two immiscible liquids, or liquid and gas. In other words, nerve fibers can and do grow only on or within some kind of structural framework. By cultivating growing neurons in colloidal media whose architecture can be altered at will, it has been shown that fibers assume patterns that conform to the submicroscopic architecture of the culture medium, especially responding to fibronectin and laminin proteins. It appears, therefore, that growing fibers are guided in their course by contact with surrounding structures. Translating this principle to the normal embryonic situation, it is the embryonic body that provides the architectural substratum to which nerve fibers orient. In the case of limbs particularly, they not only furnish within themselves a scaffolding for the guidance of fibers, but as areas of rapid growth they likely withdraw water from their basal surroundings, thus creating in the body proper stress patterns that converge on the limbs. Growing nerve fibers will thus be guided to the bases of the limbs by these stress patterns and, having entered the limbs, will orient to the limb patterns themselves. Fiber orientation, therefore, appears to be a matter of contact guidance rather than response to chemical or electrical factors. Similar direction has been shown for the growth of retinal axons into the optic tectum in *Xenopus.* Experiments on the regenerating optic nerve of goldfishes are difficult to evaluate except in terms of chemotropism. The optic nerve of the goldfish divides into two tracts, a lateral one made up of fibers from the dorsal half of the retina and a medial one made up of fibers from the ventral half of the retina. When these tracts are cut and crossed, the crossed fibers regenerate but recross against all mechanical biases and reenter their proper sites in the brain tectum.

When we consider the distance traversed and the devious routes followed by many nerves in reaching their end organs, we may momentarily wonder how contact guidance could operate effectively. But remember that when an organ first begins to acquire its nerve supply in the embryo, the distances are measured in millimeters, not centimeters or meters. The original pioneering fibers actually need to project only relatively minute distances before reaching and attaching to a peripheral receptor or effector. Henceforth, the fibers are passively towed along, as the parts they supply continue to grow and differentiate. Sometimes, as in the case of certain muscles, organs may become extensively displaced during development and the level of origin of the nerves they drag along with them provides a valuable clue as to their source.

The problem of guidance or orientation of nerve fibers actually concerns only the first pioneering fibers, for the later developing fibers simply follow the earlier ones. With the establishment, then, of the original delicate pattern of pioneering fibers, the buildup of the final innervation of an adult part proceeds by successive generations of fibers following the courses of these trailblazers, courses that in the meantime are shifted here and there as the organ in question attains its final form and position.

It was suggested earlier that under experimental conditions nerve fibers of a given kind can penetrate foreign parts with ease. But close histological scrutiny reveals that although nerves can be so deflected, they fail to establish proper functional connections, as they do when associated with the organs that they normally supply. We are therefore confronted with a new problem. Once the innervation of an organ has been expedited by the contact guidance of fibers into it, how does it happen that motor fibers connect with the correct effectors, and sensory fibers with the correct receptors? The answer to this question may yet depend on the large numbers of specific proteins found on growing axons and potential target organs.

Numerous pairs of spinal nerves are given off at regular intervals along the length of the spinal cord. In terms of gross anatomy, each spinal nerve originates in two roots, a *dorsal root* and a *ventral root* (Fig-

ure 17-21). Associated with the former is a *spinal ganglion.* The two roots then unite to form a common trunk. However, this main trunk promptly divides into a variety of branches or rami. A *dorsal ramus* is directed to the dorsolateral part of the body, a *ventral ramus* extends ventrolaterally, and one or two *communicating rami* run inward to the viscera. Each spinal nerve is typically a discrete structure, supplying a restricted area with a minimum of overlap with its neighbors. But in vertebrates with well-developed limbs, the branches of a number of successive nerves at the level of each limb are interwoven to form a *plexus,* so that a given limb region is supplied by fibers from several spinal nerves.

In terms of functional composition (Figure 17-21), the dorsal root embodies somatic and visceral sensory fibers exclusively; the ventral root embodies somatic and visceral motor fibers exclusively. But with the union of the roots, these fiber types come together, so that all four components are present in the main trunk of the nerve. Beyond this point, however, there is a new segregation of fibers in the rami. The dorsal and ventral rami carry mainly sensory and motor somatic fibers, whereas the visceral fibers are routed through the communicating rami. The functional distribution of the visceral fibers will be considered further in a subsequent discussion of the autonomic system (p. 444).

The foregoing description of a spinal nerve and its fiber components actually applies specifically to mammals and requires some qualification when we consider vertebrates as a whole. It is generally true that the dorsal root carries somatic and visceral sensory fibers, and the ventral roots carry somatic and visceral motor fibers. Yet in some reptiles and birds visceral motor fibers are carried by the dorsal root as well as the ventral, and in amphibians and fishes this is a common arrangement. The lampreys present still another situation, one judged to be primitive. In these forms the dorsal and ventral roots do not join, so that

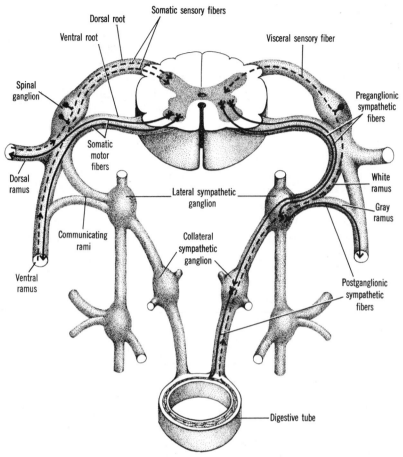

Figure 17-21 Anatomy and fiber composition of a spinal nerve. Somatic components pictured on left, visceral components on right.

separate dorsal and ventral nerves are maintained. Furthermore, the visceral motor fibers are found exclusively in the dorsal nerve: the dorsal nerve thus has three components, somatic sensory, visceral sensory, and visceral motor, whereas the ventral nerve has only one, somatic motor. We can only speculate as to how, starting with this primitive arrangement, the two roots come to unite and the visceral motor fibers tended to shift from the dorsal to the ventral root. It is interesting, however, that the arrangement of separate roots and the three-to-one parceling of the fiber components are reflected in the organization of the cranial nerves to which we now turn.

CRANIAL NERVES

According to firmly fixed convention, there are 12 cranial nerves, but this convention is based on humans and other mammals, and its application to the vertebrates as a group is inaccurate. In the first place, in anamniotes (fishes and amphibians) the 12th nerve is absent and the 11th is an integral part of the 10th. Furthermore, following the establishment of the nomenclature for the basic 12, a 13th anteriormost one was identified, and so as not to disrupt the time-honored numbering system, it was given the number 0. So anamniotes actually possess 11 cranial nerves, and amniotes have 13. Another difficulty stems from the fact that the second (optic) nerve is not a nerve at all, but a brain tract; it originates from cells within the retina of the eye and because the retina is a brain derivative (see p. 464), the optic "nerve" lies totally within the confines of the brain. Nevertheless, it retains its listing with the other true nerves. The following, then, is the list of cranial nerves in sequence from front to rear: (0) terminalis, (I) olfactory, (II) optic, (III) oculomotor, (IV) trochlear, (V) trigeminal, (VI) abducens, (VII) facial, (VIII) auditory, (IX) glossopharyngeal, (X) vagus, (XI) accessory, and (XII) hypoglossal.

There is reason to believe that primitively the peripheral nerves associated with the brain, like those associated with the spinal cord, consisted of separate dorsal and ventral units such as are presently exhibited in cyclostomes. Unlike the spinal nerves, however, the morphogenetic history of the cranial nerves did not involve a union of dorsal and ventral roots. Rather, there was a differential elimination of one root or the other, so that certain cranial nerves (for example, the oculomotor) came to represent onetime ventral roots alone, and others (for example, the trigeminal) represented dorsal roots alone. Recall, too, that primitively, as revealed by modern lampreys, the dorsal roots contained visceral motor fibers as well as somatic and visceral sensory fibers, whereas the motor roots were solely somatic motor in character. This parceling of fibers was retained in the evolving cranial nerves, so that those representing onetime ventral roots remain largely somatic motor in composition, whereas the dorsal root nerves are mixed.

Still another complication in the cranial nerve system, as compared to spinal nerves, is that in addition to the four regular varieties of components (somatic and visceral sensory and motor fibers), there occur fibers of restricted distribution and specialized function. One group of these transmits sensations from the three major sense organs of the body, the olfactory organ, eye, and acousticolateralis system, and so constitutes a category of *special somatic sensory fibers*. A second category of *special visceral sensory fibers* serves the organs of taste. Finally, the striated musculature of the pharyngeal arches is supplied by *special visceral motor fibers*. It follows, then, that seven distinct varieties of functional components are represented in the cranial nerves: (1) general somatic sensory, (2) general somatic motor, (3) general visceral sensory, (4) general visceral motor, (5) special somatic sensory, (6) special visceral sensory, and (7) special visceral motor. This is not to say that all seven components are present in all or even in any one of the cranial nerves. Rather, as Table 17-1 indicates, each nerve has its own characteristic composition. The table also reveals that the cranial nerves fall into three rather clearcut categories: (*a*) those identified with the three special types of sense organs; (*b*) those corresponding essentially to primitive motor roots; and (*c*) those corresponding to primitive dorsal roots and serving the pharyngeal region.

It has already been established as a basic principle that the cell bodies of motor neurons lie within the central nervous system, whereas those of sensory neurons are aggregated in ganglia external to the central nervous system. Obviously, then, those cranial nerves that are purely motor in composition have no ganglia; those that are either purely sensory or contain sensory components do have ganglia. However, special situations prevail in two of the latter. The cell bodies for the optic nerve lie within the retina, so that it is the retina that serves as a "ganglion" in this instance. Likewise, the fibers constituting the olfactory nerve originate in cells lying in the olfactory epithelium (p. 471), so that only in the loosest sense of the word can a ganglion be said to exist. The situation with respect to the olfactory nerve also points to another feature of the cranial ganglia. Although the gan-

TABLE 17-1
FIBER COMPONENTS OF THE CRANIAL NERVES

Nerve Type:	Special Sensory	Dorsal Root Nerves (Pharyngeal Nerves)					Ventral Root Nerves
Fiber Component:	Special Somatic Sensory	General Somatic Sensory	General Visceral Sensory	Special Visceral Sensory	Special Visceral Motor	General Visceral Motor	General Somatic Motor
O. Terminalis		F					
I. Olfactory	F						
II. Optic	F						
III. Oculomotor						(F)	F
IV. Trochlear							F
V. Trigeminal		F			F	(F)	
VI. Abducens							F
VII. Facial	F	(F)	F	F	F	F	
VIII. Auditory	F						
IX. Glossopharyngeal		(F)	F	F	F	F	
X. Vagus	F	F	F	F	F	F	
XI. Accessory						F	
XII. Hypoglossal							F

F = presence of fibers. Components shown in parentheses are of variable occurrence or negligible.

glia of the spinal nerves trace their origin to aggregations of neural crest cells, those of cranial nerves V, VII, VIII, IX, and X are developed from special thickenings of the embryonic epidermis termed *placodes.* As we shall see in due course (p. 452), the olfactory organ and the receptors of the acousticolateralis system have a similar origin.

The analysis of the cranial nerves may be concluded with a brief summary of the functional distribution of each. To this end, reference to Figures 17-22 and 17-23 is helpful. Whereas the terminal and olfactory nerves are attached superficially to the telencephalon, the optic nerve to the diencephalon, and the oculomotor and trochlear nerves to the mesencephalon, all the others are associated with the medulla.

Terminal Nerve (0). Originating from one or more ganglia, the fibers of this nerve come either from the *olfactory organ* or (in amphibians, reptiles, and mammals) from an accessory olfactory structure known as the *vomeronasal organ* and terminate in the telencephalon. Although associated with olfactory organs, its olfactory function has never been demonstrated. It may possibly represent a reduced and functionally obscure remnant of an anterior pharyngeal nerve that originally served the mouth area.

Olfactory Nerve (I). The fibers of the olfactory nerve originate in neurosensory cells that lie within the olfactory epithelium lining the nasal cavity, and pass inward to terminate in the olfactory lobe of the telencephalon. This nerve commonly consists of a dispersed array of short fibers rather than appearing as a definitive nerve trunk. The designation of the olfactory nerve as special somatic sensory is based on the embryonic origin of the olfactory organ from the somatic ectoderm of the head. But since the

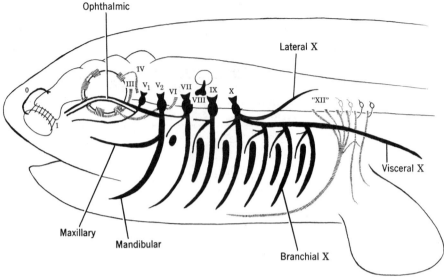

Figure 17-22   Cranial nerves of a fish.

sense of smell is associated with digestive functions, nerve I is sometimes classified as special visceral sensory rather than somatic.

Optic Nerve (II). As emphasized previously, the optic nerve is actually not a peripheral nerve at all, but a brain tract. The fibers originate from cell bodies in the retina of the eye (Figure 17-33), which embryonically is a brain derivative; hence its designation as a brain tract, referring to the fact that it lies within the confines of the brain rather than peripheral to it. But the brain, in the final analysis, is a derivative of the embryonic skin, so the "nerve" is arbitrarily classified as special somatic sensory.

Oculomotor Nerve (III), Trochlear Nerve (IV), Abducens Nerve (VI). These nerves supply the six extrinsic muscles of the eye. Because these muscles have the same source as the general axial musculature, their nerve supply is classified as general somatic motor. It will be recalled (p. 267) that the six eye muscles originate in three embryonic myotomes, each equipped with a nerve. One of these myotomes

provides four muscles (superior rectus, internal rectus, inferior rectus, and inferior oblique), so that all four muscles are serviced by one nerve, the oculomotor. The dorsal half of the second myotome provides the superior oblique muscle whose nerve is the trochlear. The posterior rectus muscle is a product of the ventral halves of the second and third myotomes, and its innervation, the abducens nerve, is that nerve originally serving the third myotome.

Trigeminal Nerve (V). This is the nerve that supplies the first visceral, or mandibular, arch and its associated parts. As its name signifies, it has three divisions: a *maxillary ramus* supplying the upper jaw, a *mandibular ramus* supplying the lower jaw, and an *ophthalmic ramus* passing through the orbit to the dorsal and lateral sides of the head. In fishes the ophthalmic ramus has superficial and deep divisions. The deep one originally was probably a separate and distinct nerve; but in higher vertebrates the deep ophthalmic branch either is missing or has been incorporated indistinguishably with the superficial division. The ophthalmic ramus is almost exclusively

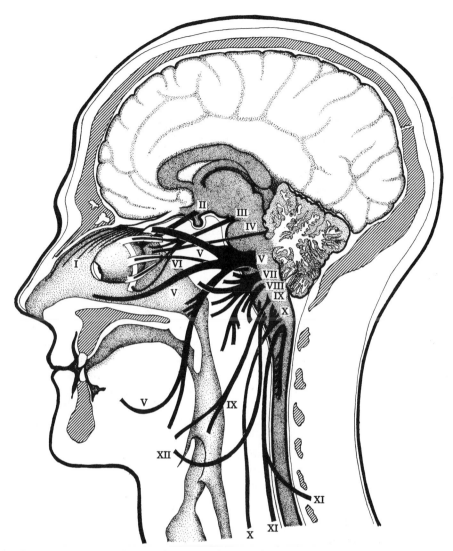

Figure 17-23 Cranial nerves of a human, shown as if projected upon a median section of the head. (From Neal and Rand, *Comparative Anatomy,* by permission of the Blakiston division of the McGraw-Hill Book Company.)

sensory and relays sensations from skin receptors in the head. Both the maxillary and mandibular rami carry somatic sensory elements distributed to the surface of the head, the jaws, and the mouth. Most of the motor fibers are in the mandibular ramus.

Facial Nerve (VII). The facial nerve is associated with the second visceral, or hyoid, arch and its associated parts and de-

rivatives. Its motor components supply the muscles of the hyoid arch, or its derivatives in higher vertebrates; its sensory components supply adjacent receptors, for example, lateral line and taste organs. With the loss of the lateral line system in terrestrial vertebrates, the somatic sensory components become relatively unimportant. Conversely, in mammals where we have noted (p. 276) the elaboration of the dorsal musculature of the hyoid arch provides the facial muscles, and the visceral motor fibers become widely distributed.

Auditory Nerve (VIII). The auditory nerve supplies the inner ear. In higher vertebrates the nerve has two main branches, each with a separate ganglion. A *vestibular nerve* serves the division of the inner ear concerned with equilibrium; a *cochlear nerve* serves the hearing organ. The classification of the auditory nerve as special somatic sensory relates to the fact that the inner ear is a product of the embryonic skin.

Glossopharyngeal Nerve (IX). This nerve supplies the third visceral (first branchial) arch and/or its derivatives and associated sense organs. Its motor fibers innervate some of the musculature of the throat and larynx (and part of the salivary glands, when present); its sensory fibers carry visceral sensations from the rear of the tongue and mouth, notably from some of the taste organs.

Vagus Nerve (X). As the nerve basically serving the last four branchial arches, the viscera, and lateral line system, the vagus nerve is one of the largest and most versatile of all the cranial nerves. Its *lateral division* disappears with the elimination of the lateral line organs in amniotes, and the *branchial division* becomes adapted to the working over undergone by the branchial arches. But in all vertebrates the *visceral division* serves as the main route of the fibers of the parasympathetic system (see Figure 17-24 and the discussion later).

Accessory Nerve (XI). Purely visceral motor in composition, the fibers of the accessory nerve in fishes and amphibians are incorporated into the vagus. Only in amniotes are these fibers set off as a distinct nerve; yet even in these forms the accessory continues to be distributed with the vagus.

Hypoglossal Nerve (XII). Also present only in amniotes, the hypoglossal is a purely somatic motor nerve that supplies the intrinsic muscles of the tongue.

THE AUTONOMIC NERVOUS SYSTEM

Because knowledge of the autonomic nervous system of vertebrates is by far the most complete for mammals, it is desirable to introduce the account with the mammalian situation. Some comparative observations on other vertebrates may then follow.

Many of the activities of the body are not under willful control, yet they are still subject to close regulation by the nervous system. It is the autonomic portion of the peripheral nervous system that is concerned with the government of these involuntary operations. Notable among the actions controlled by the autonomic system are rate and intensity of heartbeat, changes in size of blood vessels, movements of the digestive tract, secretions of various glands, expansion and contraction of the urinary bladder, and respiratory movements.

The autonomic system as a whole contains both sensory and motor neurons, but when we speak of the involuntary control of bodily activities, we have in mind the control exercised by the motor neurons. So for practical purposes the autonomic system is equated with the patterns and functions of its motor components.

These patterns and functions differ in a number of fundamental respects from those of conventional somatic motor neurons (and the visceral motor neurons serving the head and pharyngeal musculature). In the first place, unlike somatic motor impulses, which are carried without interruption by long axons running from the cord and brain directly to the appropriate effectors, the autonomic impulses are conducted by a two-neuron relay. The cell body of the first neuron in the set lies in the brain or cord; the cell body of the second neuron lies in

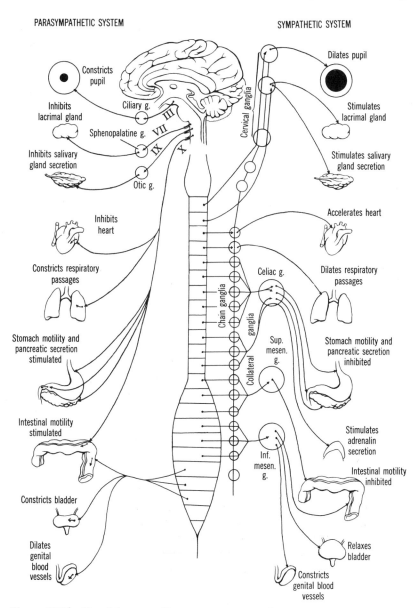

PARASYMPATHETIC SYSTEM

SYMPATHETIC SYSTEM

Constricts
pupil

Dilates pupil

Inhibits
lacrimal gland

Ciliary g.

Stimulates
lacrimal gland

Sphenopalatine g.

VII

III

Inhibits salivary
gland secretion

IX

X

Cervical ganglia

Stimulates salivary
gland secretion

Otic g.

Inhibits
heart

Accelerates heart

Constricts respiratory
passages

Celiac g.

Dilates respiratory
passages

Chain ganglia

Collateral ganglia

Stomach motility and
pancreatic secretion
stimulated

Sup.
mesen.
g.

Stomach motility and
pancreatic secretion
inhibited

Intestinal motility
stimulated

Stimulates
adrenalin
secretion

Inf.
mesen.
g.

Intestinal motility
inhibited

Constricts bladder

Dilates
genital
blood
vessels

Relaxes
bladder

Constricts
genital blood
vessels

Figure 17-24 Plan of the mammalian autonomic system. (Adapted from several sources, notably Johnson et al., *Principles of Zoology,* by permission of Holt, Rinehart and Winston.)

a ganglion somewhere outside the central nervous system (Figure 17-21). Accordingly, the axon of the first neuron, running from a cell body in the visceral motor column in the central nervous system to the outlying ganglion, is termed a *preganglionic fiber.* Its terminal fibrils synapse

with the dendrites of the second neuron in the ganglion, a neuron whose axon then leads, as a *postganglionic fiber,* to the effector. (Any given preganglionic fiber actually synapses with numerous postganglionic neurons. Thus, a single preganglionic fiber may elicit, via the multiple postganglionic relay, a diffuse response.) Preganglionic fibers are customarily myelinated; postganglionic fibers usually have little or no myelin. The outlying ganglia are, like spinal ganglia, derived from embryonic neural crest; but they are distinctive in being the *only motor ganglia* in the entire nervous system.

Another important characteristic of the autonomic system is that each internal organ under its jurisdiction receives a *double* set of fibers. One of these constitutes the *sympathetic system;* the other constitutes the *parasympathetic system.* Generally speaking, the two sets of nerves have antagonistic effects on the organ supplied. That is, impulses carried over one system cause a structure to react in one way, whereas the other system produces an opposite effect. The principal antagonistic effects of the two systems are noted in Figure 17-24. The final stimuli to a given organ are furnished, of course, by the postganglionic fibers in the two systems, and in both cases the effects are brought about by chemical substances (neurohumors) given off at the nerve terminals. But the differences in the effect produced by the two sets of fibers are expressions of differences in the chemical substances produced. Postganglionic sympathetic fibers, except those going to the sweat glands and uterus, produce substances similar to or identical with adrenalin. Hence they are said to be *adrenergic* fibers. Postganglionic parasympathetic fibers and the sympathetic fibers supplying the sweath glands and uterus liberate acetylcholine, and are thus said to be *cholinergic* fibers. It should be added that the preganglionic fibers in both systems liberate acetylcholine at their synapses with the dendrites of the postganglionic neurons.

The Sympathetic Nervous System. The preganglionic fibers of the sympathetic system originate in cell bodies lying in the visceral motor area of the spinal cord at the level of the thoracic and lumbar vertebrae.

(It is for this reason that the term *thoracolumbar outflow* is often applied to this portion of the autonomic system.) These fibers pass out through the ventral roots of the spinal nerves, in company with the somatic motor fibers, and then turn inward via the *white communicating rami* (so designated because the preganglionic fibers are myelinated) to enter adjacent ganglia (Figures 17-21 and 17-24). The adjacent sympathetic ganglia are differentiated into two groups: a series of fairly regularly spaced *lateral,* or *chain, ganglia* paralleling the cord, and a lesser number of irregularly spaced and deeper or more distal *collateral ganglia.* The preganglionic fibers first enter the lateral ganglia. Here they may synapse with the dendrites of postganglionic neurons, or they may pass *through* the lateral ganglia without synapse to link with postganglionic neurons originating in the collateral ganglia. In addition to the preganglionic fibers going to the lateral ganglia, there are fibers passing from one lateral ganglion to the next, so that the lateral ganglia are all chained together.

Some of the postganglionic fibers originating in the lateral ganglia turn back via the *gray communicating rami* (so designated because the postganglionic fibers are unmyelinated) and are distributed by the dorsal and ventral rami of the spinal nerves to the skin, where they innervate the smooth muscles of the blood vessels, the sweat glands, and the muscles of hairs (Figure 17-21). Other postganglionic fibers originating in the chain ganglia serve the iris of the eye, the lacrimal and salivary glands, and the mucous membranes of the mouth and nasal passages. Still others are distributed to the heart and respiratory organs. Those postganglionic fibers that originate in the collareral ganglia serve the digestive tube and its associated parts and the urogenital system (Figure 17-24).

It should be noted as an important exception that the medulla of the adrenal gland receives preganglionic fibers *directly,* without the intervention of a postganglionic link. The significance of this situation will be considered shortly.

The Parasympathetic Nervous System.
The preganglionic fibers of the parasympathetic system are associated with the third, seventh, ninth, and tenth cranial nerves, and certain spinal nerves at the level of the sacrum (Figure 17-24). (The term *craniosacral outflow* is often applied to this portion of the autonomic system.) They all terminate in ganglia lying near or within the organs innervated by the outgoing postganglionic fibers.

The preganglionic parasympathetic components of the oculomotor nerve (III) terminate in a minute *ciliary ganglion* situated in the rear of the orbit. Here they synapse with postganglionic neurons distributed to the sphincter muscle of the iris and ciliary muscles of the eye.

The preganglionic fibers in the facial nerve (VII) terminate in the *sphenopalatine ganglion* from which postganglionic fibers emerge to innerveat the lacrimal gland. (Other postganglionic fibers innervate certain salivary glands.)

The preganglionic fibers in the glossopharyngeal nerve (IX) terminate in the *otic ganglion* from which postganglionic fibers pass to the parotid gland and lining of the mouth.

Of the four cranial nerves involved in the parasympathetic system, the vagus (X) is by far the most important. Preganglionic fibers are distributed via its branches to all regions of the digestive tract (except the rear of the large intestine), to the heart, respiratory system, liver, gallbladder, and pancreas. The ganglia in which the preganglionic fibers terminate are all very minute and are located within the walls of the organs themselves; thus the postganglionic fibers are short. As for the sacral portion of the system, the preganglionic fibers course through the white rami of the third, fourth, and fifth sacral nerves which together form a com-

mon *pelvic nerve.* These fibers, like those of the vagus, pass directly to the organs they supply, namely, the lower part of the large intestine and the urogenital organs, there to synapse with short postganglionic neurons originating from ganglia within the organs themselves.

The Autonomic System and the Hypothalamus. Appropriate fiber tracts in the spinal cord and brain serve to link all the units of the autonomic system together and bring them under the control of major centers in the hypothalamus. We may thus reiterate the point made earlier (p. 433): the major function of the hypothalamus is to regulate the stability of the body's internal environment by exercising final control over all the activities with which the autonomic system is concerned.

The Adrenal Gland. The adrenal glands of mammals are compact organs capping the anterior surfaces of the kidneys. Each adrenal gland is composed of two distinct regions, an external *cortex* and an internal *medulla.* These two components have separate embryonic origins. The cortex is derived from the coelomic peritoneum (mesoderm) adjacent to the developing kidney. As the mesodermal cells multiply, aggregate, and differentiate, they are invaded by strands and cords of cells migrating out of the embryonic neural crest. It is this material from the neural crest that will constitute the medulla. Although the embryonic source of the two categories of tissue is the same in other vertebrates, it is only in mammals tht their delineation in an adrenal cortex and medulla prevails. In birds the two tissues are intermixed: in reptiles and amphibians they may be intermixed in discrete glands or appear as distinct and diffusely arranged clusters; and in fishes the two types of tissue tend to occur as separate clusters associated with the postcardinal veins and their tributaries.

The cortical and medullary components

of the adrenal gland are as unlike functionally as they are in material source. As many as 50 different substances have been identified in the cortex. Not all of these, however, are secretory products; most are intermediate compounds produced transitorily in the synthesis of *steroids*. The three major categories of steroid hormones produced by the mammalian adrenal cortex are (1) *mineralocorticoids,* which regulate sodium and potassium balance by promoting sodium resorption and potassium excretion by the kidneys; (2) *glucocorticoids,* which participate in the regulation of carbohydrate, protein, and fat metabolism; and (3) *corticosterones,* which complement or inhibit various gonadal hormones. The adrenal medulla, by contrast, produces only *adrenalin* and *noradrenalin.* These two medullary hormones are similar in some of their biologic actions, but they also show important differences. By way of illustration, adrenalin increases the blood flow through skeletal muscle, liver, and brain, whereas noradrenalin has no effect or decreases it. Adrenalin causes peripheral vasodilation; noradrenalin produces vasoconstriction. Both hormones increase heart rate, but adrenalin is the more potent; noradrenalin has been established as the principal neurohumor of adrenergic neurons. It is because of this last circumstance in particular that we have chosen to consider the adrenal gland in conjunction with the autonomic nervous system.

It was emphasized earlier (p. 446) that the adrenal medulla receives preganglionic fibers directly, without the intervention of a postganglionic link. Recall also that preganglionic fibers are always cholinergic, that is, they liberate acetylcholine at their terminals, whereas postganglionic sympathetic fibers (with some exceptions) are adrenergic, that is, they liberate noradrenalin. With respect to the adrenal medulla, therefore, there is a situation in which preganglionic fibers stimulate the production of adrenalin and noradrenalin by the medulla. In other words, the medullary cells are performing in the same way as do postganglionic neurons: they produce adrenalin. When we link to this circumstance the additional fact that the adrenal medulla and the autonomic ganglia whose cells provide the postganglionic fibers have a common source in embryonic neural crest, we reach the conclusion that the cells of the adrenal medulla and the postganglionic sympathetic neruons are homologous. They differ only in that the postganglionic fibers produce noradrenalin in tiny amounts affecting the structures immediately innervated by the fibers, whereas the medulla produces adrenalin and noradrenalin in volumes for wide distribution by the circulatory system.

Autonomic Nervous System of Lower Vertebrates. Observations are fragmentary, but there is evidence of the beginnings of an autonomic system in the cyclostomes. It will be recalled (p. 438) that in lampreys the dorsal and ventral roots of the spinal nerves do not unite. Sympathetic fibers emerge from the spinal cord and pass through both roots to scattered sympathetic ganglia. The ganglia are not connected with each other. In the hagfishes, in which the roots do unite, the sympathetic fibers apparently pass through the ventral roots only and are believed to exercise a vasomotor control. As for the parasympathetic system of cyclostomes, it is confined to fibers in the vagus nerve. In the hagfish, *Myxine,* these fibers regulate the gallbladder and intestine; in lampreys, they have a similar distribution with the addition of fibers to the heart.

The autonomic system of elasmobranch and teleost fishes is considerably more elaborate, although still falling short of the mammalian system. Its anatomical limitations are reflected in the fact that the two components of the system fail to exhibit the dual actions of excitation and inhibition with respect to many organs. None of the fishes has a sacral parasympathetic system. Nerves III, VII, IX, and X carry parasympathetic fibers in elasmobranchs as in mammals, but the teleosts lack the components in nerves VII and IX. All fishes have a sympathetic system, but it shows considerable variation. Elasmobranchs, for instance, have no supply to the head and skin—only to the abdominal viscera—and the chain ganglia are loosely connected. The teleosts, by comparison, have well-linked chain ganglia and a sympathetic supply to the head and skin as well as to the abdominal region. An important role played by the sympathetic fibers to the skin in many teleosts is the regulation of the chromatophores (p. 198).

Urodele amphibians have an autonomic system similar to that of teleosts. But anurans include the parasympathetic fibers in nerves VII and IX and also add a parasympathetic sacral flow. With that the basic pattern of the autonomic system of all the other tetrapods, including mammals, is established.

THE SENSE ORGANS

The nervous system, whose functional organization we have just been considering, is concerned essentially with sorting out and correlating the information that it receives and with relaying messages to their proper destinations so that the organism may carry out appropriate adjustments. The source of this information is the organism's environment, both its external surroundings and its internal milieu. The devices utilized to acquire this information are the *receptors* or *sense organs*.

We noted earlier (p. 406) that at the beginning of the evolution of the neurosensory system in multicellular organisms the functions of reception and transmission are combined in *neurosensory cells* (Figures 17-1A and 17-25A); that is, certain epithelial cells serve simultaneously to receive stimuli and to transmit the resulting disturbances to effectors. As a next step the bodies of the primitive neurosensory cells withdraw from the surface, but contact with the environment is still maintained by fibers extending into the epithelium or terminating just be-

neath it. Two additional categories of transmitting units, motor and association neurons, are then introduced, so that the original neurosensory cells become strictly sensory neurons whose naked peripheral terminals serve as *primary receptors* (Figure 17-25B). A further modification comes with the investment of the primary receptors by specialized supporting cells adapted to augment the stimuli to be received, thus creating *secondary receptors* (Figure 17-25C). Finally, there comes a specialization of many of the epithelial cells, so that they themselves become sensitive to stimuli. In these, the *tertiary receptors* (Figure 17-25D), the specialized cells, are the actual receptors, and the neurons associated with them serve only as conductors to the central nervous system.

Briefly, then, we may consider that there have been four steps in the evolutionary history of sense organs. Beginning with (1) primitive neurosensory cells whose cell bodies are the receptors, there have followed (2) primary receptors consisting of naked neuron terminals, (3) secondary receptors in which the fiber terminals are housed in specialized supportive cells, and (4) tertiary receptors featuring cells that are themselves sensitive to stimuli, thus relegating the sensory nerve fibers to a purely conducting function. The important common denominator in all these sense organs is the irritable cell. Whether it is an uncomplicated naked nerve fiber or a complex entity such as an ear or an eye with a host of accessories, in the final analysis the response to environmental change depends on receptor cells that are able to detect environmental energy.

Irritability—the capacity to respond to a stimulus—is, of course, one of the general properties of living stuff, and in some measure all the cells of the body are irritable. But the clear and intended implication in the foregoing is that in the division of labor among cells and tissues of the animal body the specific task of recording environmental changes and initiating the nervous impulses

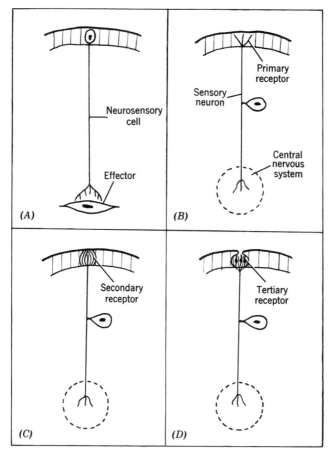

Figure 17-25 Stages in the evolution of receptors.

that, when relayed to the central nervous system, will lead to appropriate responses is allocated to the receptors. Accordingly, not only is sensitivity to environmental fluctuations identified with the receptors, but there is a segregation among receptors in terms of the particular kinds of stimuli to which they react. On this basis all the varieties of sense organs may be categorized under three main headings: *mechanoreceptors,* responsive to mechanical stimuli such as touch and pressure, motion, and the low-frequency vibrations of sound; *radioreceptors,* responsive to the high-frequency vibrations of heat and light; and *chemoreceptors,* responsive to chemical agents. So our review of the form and function of

the sense organs of vertebrates is organized under these three headings.

MECHANORECEPTORS

The varieties of receptors sensitive to mechanical stimulation range all the way from free nerve endings, primary receptors, through secondary receptors to specialized tertiary organs.

There may be noted, first, the widely distributed free nerve endings that penetrate the epithelial cells of the skin and are responsive to transient contact and/or intensive irritation. Fleeting contact may be recorded as the sense of *touch;* when touch is sustained, it becomes interpreted as *pressure;* extensive stimulation results in *pain.* It is not entirely clear whether any given primary receptor can report all these sensations, depending on the degree of irritation, or if there is functional segregation among such receptors. Probably touch and pressure are only degrees of sensation reported by the same receptors, whereas pain constitutes the sensation derived from the irritation of a completely separate array of free nerve endings with a much higher stimulus threshold. Although most such receptors are associated with the skin, there are internal ones located in tendons, joints, and skeletal muscle, serving to give information regarding muscle tension and the position and movements of various parts of the body. Other primary receptors are also found within the walls of the digestive tract and its derivatives. The peritoneal linings and mesenteries are likewise equipped, but primarily with pain receptors.

Several types of touch and pressure receptors of the secondary type are also present. Consisting of nerve endings encapsulated by cells of assorted form, they are located for the most part in the dermis of the skin, but they are also found in the tendons, periosteum, and mesenteries (Figure 17-26).

Of all the types of mechanoreceptors found in vertebrates, the most conspicuous are those tertiary organs constituting the *acousticolateralis system* whose components are the *lateral line organs* and the *ear.* The grouping of these sense organs in a single system is by no means capricious: they are both concerned with recording motion, changes in pressure, and low-frequency vibrations; their sensory cells share a common form and derive embryonically from intimately related sources. With respect to the latter, we have seen elsewhere (p. 438) that the ganglia of certain of the cranial nerves represent an important exception to the generalization that the central and peripheral nervous systems trace their origins to the neural tube and neural

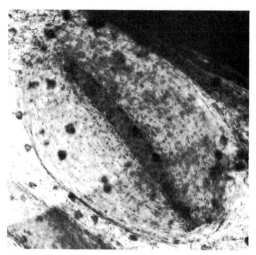

Figure 17-26 A pacinian corpuscle, a pressure receptor, in a mesentery. × 100. (Photo by Torrey.)

crest. The ganglia of nerves V, VII, VIII, IX, and X are products of local thickenings of the embryonic epidermis called placodes. In due course we shall see that the olfactory organs and the lenses of the eyes also come from epidermal placodes, as does the lateral line organs and sensory components of the ears. Each inner ear traces to a thickening of epidermis alongside the myelencephalon termed an *auditory placode.* Areas of the epidermis adjoining the auditory placode also become thickened and it is from these placodes that the lateral line organs arise. From this common starting point the two components of the acousticolateralis system then follow separate morphogenetic courses.

Lateral Line Organs. Through experimental techniques of vital staining, excision, and grafting applied to amphibian embryos, it has been demonstrated that cells move out of the primordial placodes and migrate along prescribed pathways over the head and body. Wherever lateral line organs form, clusters of these cells differentiate into specialized patches of epithelium termed *neuromasts,* consisting of sensory and supporting cells (Figure 17-27). Each sensory cell has a hairlike projection from

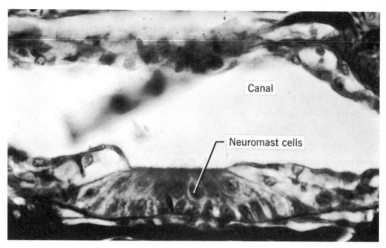

Figure 17-27 Photomicrograph of a lateral line organ of a catfish. ×480. (Photo by Torrey.)

its exposed end and is innervated by one of the terminal branches of a nerve fiber.

The lateral line system of sense organs is present in cyclostomes, bony and cartilaginous fishes, aquatic larval stages of all Amphibia, and in those adult amphibians that are aquatic. Except for some vestiges in adult turtles, no lateral line organs are present in reptiles, birds, and mammals. That it is a very antique system, however, is evidenced by the presence of grooves, with which lateral line organs were presumably associated, in the armor of ostracoderms and placoderms; such canal systems are also represented in fossil Chondrichthyes, Osteichthyes, and Amphibia.

In modern cyclostomes the system consists of linear rows of sense organs lying in individual shallow pits. This arrangement is probably the primitive one. Apparently the next evolutionary step was for the pits to blend into a continuous shallow groove; this groove, in turn, closed over to produce a canal opening at intervals to the surface by small pores. Modern holocephalans exhibit the groove arrangement, but the usual fish pattern is that of a main canal running along the side of the trunk and tail, plus a complex pattern of branching canals on the top and sides of the head. Ancient amphib-

ians appear also to have had such canals, but modern Amphibia retain, or have returned to, the discontinuous system of pit organs.

Whether discontinuous as in cyclostomes and amphibians, or spaced at intervals in the floor of a canal as in most fishes (Figure 17-27), the receptors along the sides of the body are innervated by the lateral branch of the vagus nerve and those on the head by branches of the facial and trigeminal nerves. Many suggestions have been made from time to time regarding the function of the lateral line system, but the best experimental evidence indicates that these receptors respond to water disturbances. These disturbances may be of many kinds, but the most effective stimulus is water flowing over the surface of the body. In fact, it has been shown that the neuromasts can discriminate between "head-to-tail" and "tail-to-head" water movement. The lateral line system, then, is one that responds to water movement. Although it is not sensitive to pressure waves and thus not to "sound" in the conventional sense, a sound source may produce local water movements to which lateral line organs are also responsive. So if such local disturbances are by definition considered to be "sound," then an animal can be said to "hear" with its lateral line system and thereby locate a source of disturbance at a distance.

The Ear. As it occurs in humans and other mammals, the ear consists of three portions: outer ear, middle ear, and inner ear (Figure 17-32). Of these three units, it is the inner ear alone in which the sensory receptors are found; the middle and outer ears are concerned only with receiving, amplifying, and transmitting sound waves. In other words, the definitive sense organ is the inner ear; the middle and outer ears are accessories. So it is that the inner ear is present in all vertebrates, whereas, beginning with the Amphibia, the accessories are added progressively.

The Inner Ear. We are prone to think of the inner ear as the organ of hearing, but it also serves to record direction of movement. In fact, its function as an organ of equilibrium and orientation is the basic one to which the detection and analysis of sound have been added. Accordingly, we find the function of equilibrium and the structures concerned therewith fundamentally unchanged from fish to human, whereas there has been a remarkable evolution of audition and the structures pertaining to it. The morphogenetic history of the inner ear, then, begins with the fishes

in which its form and function pertain primarily to equilibration. The evolutionary origin of the inner ear is entirely unknown. It has been suggested that an anterior portion of the lateral line system became embedded in the head and was modified to create the form and function to be described below. All the evidence—anatomical, embryological, and physiological—points to the intimate relationship between the inner ear and the lateral line system. But the transformation of lateral line to inner ear remains a great unfilled gap in the evolutionary history of the acousticolateralis system.

The inner ear of a representative fish (Figure 17-28) consists of a series of closed tubes and two sacs that constitute the so-called *membranous labyrinth.* The sacs are an upper *utriculus* (sometimes subdivided) communicating by one or more openings with a lower *sacculus.* A slender *endolymphatic duct,* usually terminating within the braincase in the an *endolymphatic sac,* runs dorsomedially from the sacculus. Three relatively long, narrow tubes, the *semicircular canals,* connect at both ends with the utriculus. These canals are set at right angles to each other, that is, they conform to the three planes of space; thus two are vertical and one is horizontal. The anterior canal is vertical and extends forward and outward; the posterior canal is also vertical and extends backward and outward; the horizontal

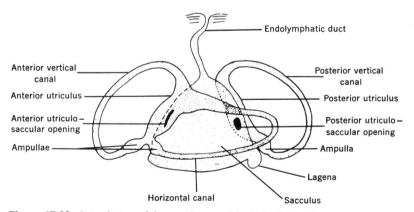

Figure 17-28 Lateral view of the membranous labyrinth of a shark.

canal extends laterally. Each canal exhibits an enlargement, an *ampulla,* at its lower end. Insignificant in fishes, but destined to play a major role in the evolution of hearing, is an outpocketing from the ventral floor of the sacculus known as the *lagena.*

The entire membranous labyrinth is embedded in the auditory region (capsule) of the skull in a set of channels that conform to the canals and sacs of the labyrinth. These channels constitute a *skeletal labyrinth,* which is lined with the membranous periosteum or perichondrium and contains a *perilymphatic fluid* serving as a cushion between the skeletal walls and the delicate membranous labyrinth within. The membranous labyrinth also contains a fluid, the *endolymph,* which plays an important part in the functioning of the ear.

The definitive receptors of the inner ear are identified with clusters of sensitive cells not unlike the neuromast cells of the lateral line organs. Three of these patches of sensory epithelium are associated with the utriculus and sacculus: the *utricular macula,* the *saccular macula,* and the *lagenar macula.* Each semicircular canal is likewise equipped with a sensory patch, called a *crista,* on the ampulla. Both maculae and cristae are supplied with fibers from branches of the auditory nerve (VIII).

In the case of the maculae the sensory cells bear hairlike processes, which project into an overlying mass of gelatinous material that contains crystals of calcium carbonate. These crystals may be dispersed, but commonly in bony fishes they are fused into one or more solid masses, often of considerable size, known as *otoliths.* Changes in the position of the head or changes in linear velocity during movement are believed to bring movements of the otoliths, thus stimulating the sensory hairs and providing information on the position of the body. Further information with respect to turning movements comes through the semicircular canals. Turning movements in the three planes of space cause displacement of the endolymph with resulting stimulation of the sensory cells of the cristae which, like those of the maculae, bear inwardly projecting sensitive hairs.

As emphasized earlier, the equilibrating apparatus remains essentially unchanged in all vertebrates. True enough, there are variations in architectural detail from one form to another, and among cyclostomes there are only one (hagfishes) or two (lampreys) semicircular canals (possibly a case of degeneracy rather than primitiveness). Also, otoliths are absent in most modern reptiles, and no birds and mammals possess them. But the history of auditory function is something else again. Among fishes, the capacity for hearing is at best of a very low order. The saccular macula may respond to water vibrations of low frequency, and several groups of fishes use the air bladder as a resonating chamber whose vibrations are carried forward to the ear by Weberian ossicles. The inner ear of amphibians is again essentially an organ of equilibrium, but with the reptiles elaboration and refinement of auditory capacities begin.

This elaboration is linked to the development of the *cochlea* whose origin involves three structures: the lagena, the perilymphatic space, and the so-called *basilar papilla.* We have previously identified the lagena as a small protuberance from the ventral floor of the sacculus, and the perilymphatic space as the fluid-filled skeletal chamber housing the membranous labyrinth. We also noted the macula associated with the lagena. The lagenar macula persists in all vertebrates except in metatherian and eutherian mammals, yet it is believed to be concerned only with the recording of directional movements. A second sensory area, the basilar macula, or papilla, appears in the lagena of tetrapods, and it is this structure that is concerned with hearing.

The development and final structure of the cochlea are complex indeed, but reduced to essentials they are as follows (Figures 17-30 to 17-32). The lagena elongates to form a narrow tube termed the *cochlear duct.* Because the lagena is a part of the membranous labyrinth, the cochlear duct is filled with endolymph. The basilar papilla also elongates and differentiates into an exceedingly intricate structure, the *organ of Corti,* running the length of the cochlear duct. The ultimate auditory receptors are hair cells in the organ of Corti. Keeping pace, the perilymphatic space investing the original lagena also elongates as a duct. Lying as it does within the perilymphatic duct, the cochlear duct in effect divides the perilymphatic duct into separate passages, one above and the other below the cochlear duct. The total cochlea thus consists of a triple-tube system composed of a central cochlear duct (*scala media*), a dorsal perilymphatic duct (*scala vestibuli*), and a ventral perilymphatic duct (*scala tympani*) (Figure 17-31). In reptiles and birds the cochlea is straight or moderately curved, but in mammals the greatly elongated cochlea is coiled in a tight spiral. A complete analysis of sound reception must wait upon the account of the form and function of the middle and outer ear to follow, but for now it may be said that sound vibrations brought in by the perilymphatic system set up pressure waves in the endolymph of the scala media that strike the organ of Corti and stimulate its hair cells. The hair cells then initiate impulses in the fibers of the cochlear branch of the auditory nerve, impulses that are transmitted to the auditory centers of the brain.

It was emphasized earlier (p. 451) that the ear and lateral line organs are interpreted as being components of a single system by reason of the basic similarity of their sensory cells and their common origin in epidermal placodes. The first embryonic evidence of the inner ear, then, is an epidermal thickening, the *auditory placode,* alongside the myelencephalon (Figure 17-29A). Very shortly the placode begins to invaginate to form a shallow *auditory pit,* opening broadly to the exterior (Figure 17-29B). The pit gradually deepens, and the

external opening narrows (Figure 17-29C). Finally, the opening closes, which is to say that a fully closed *auditory vesicle (otocyst)* is detached from the parent epidermis (Figure 17-29D).

The early embryogeny of the inner ear provides another illustration of embryonic induction. Suitable transplantation experiments reveal that the myelencephalon exercises some influence on the adjacent epidermis to bring the auditory placode into existence. However, the myelencephalon does not have full control over ear induction, for it has also been shown that the chordamesoderm, which itself induces the brain, plays an ancillary role in the stimulus of the ear vesicle as well. Apparently, then, induction of the inner ear proceeds in two steps: the initial stimulus comes from the mesodermal mantle and occurs near the end of gastrulation, and the induction is then reinforced by the myelencephalon.

It has also been shown that, once formed, the auditory vesicle in turn serves as an inductor of the cranial capsule that encloses it: if an ear vesicle is removed, the capsule does not form; if an ear vesicle is transplanted to a foreign site, local mesenchyme aggregates about it and produces a capsule. Here, then, is a case in which development of an important region of the primordial endocranium depends on an associated sense organ.

When first formed, the auditory vesicle is somewhat pear-shaped, with its pointed end directed upward (Figure 17-30). From the medial side of this upper pointed end an endolymphatic duct extends dorsad, eventually to terminate in an expanded endolymphatic sac. Meanwhile, the otocyst proper exhibits an unequal growth pattern consisting of three flattened folds and local constrictions and bulges (Figure 17-30), which ultimately will furnish the components of the membranous labyrinth. By thinning out and atrophy of their central areas, the folds are transformed into the three semicircular canals, each with an ampulla

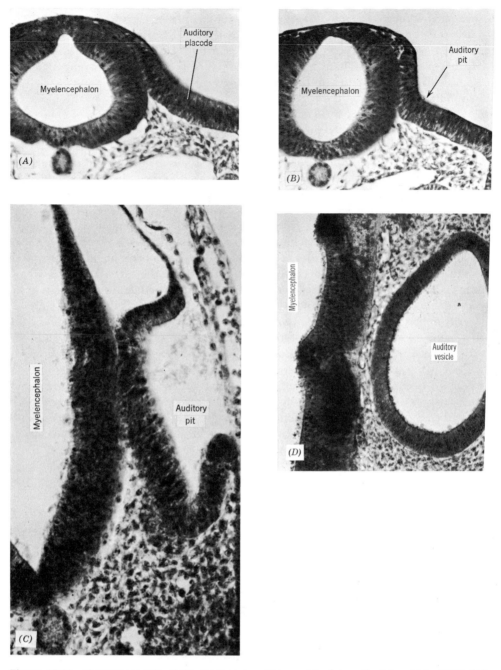

Figure 17-29 Photomicrographs of the developing auditory vesicle (otocyst) of the chick. ×230. (*A*) Auditory placode, 33 hours. (*B*) Shallow auditory pit, 38 hours. (*C*) Deep auditory pit, 48 hours. (*D*) Auditory vesicle, 72 hours. (Photos by Torrey.)

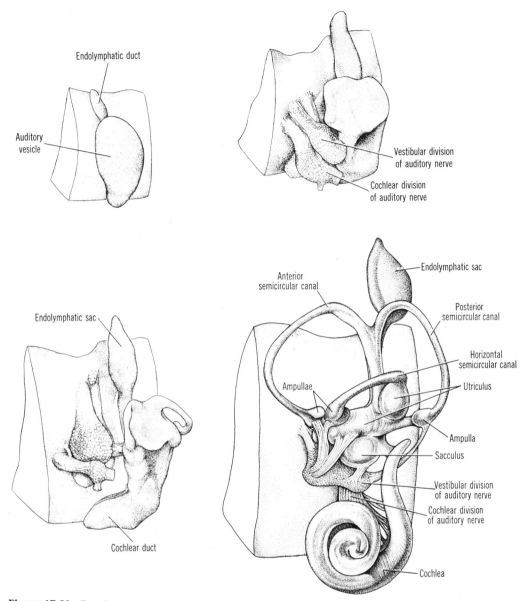

Figure 17-30   Development of the membranous labyrinth of a human. See text for explanation. (Drawn from Ziegler models.)

at one end. That portion of the auditory vesicle into which the semicircular canals open becomes set off as the utriculus, distinct from the sacculus below. An outgrowth from the floor of the sacculus provides the lagena, destined in amniotes to become greatly elongated and to provide

the cochlear duct within which the organ of Corti will form.

The differentiation of the sensory cells that constitute the maculae and cristae and of the ganglionic cells destined to provide the fibers of the auditory nerve goes hand in hand with the differentiation of the membranous labyrinth. In fact, these cells trace their origin to the wall of the auditory vesicle itself. The medioventral wall of the vesicle furnishes the cells that organize themselves as the cristae on the ampullae and the utricular and saccular maculae, one of the latter of which (the basilar papilla) will furnish the organ of Corti. Other cells from this same source will become the auditory ganglion; that is, they will provide the neurons of the auditory nerve.

The Middle Ear and Outer Ear. Recall that in many teleost fishes the swim bladder is connected to the inner ear by way of a tubular extension or a chain of bones. The swim bladder with its extension or bony connections may then quite properly be considered a kind of middle ear, for it provides a mechanism for the transfer of pressure waves to the inner ear. The inner ear *hears*. But the middle ear of tetrapods involves the first, or spiracular, pouch from the pharynx rather than an air bladder. Now there is evidence that some Devonian Rhipidistia (Crossopterygii) possessed one or more diverticula from the first pouch that, filled with air, might like a swim bladder also have served as a kind of middle ear. This is significant, for the first Amphibia came from rhipidistian stock and were thus equipped with the pharyngeal appurtenance necessary for a middle ear. The original Amphibia, however, were confronted with a new problem: emergence to dry land and the acoustic impedance of air. The middle ear is a pressure-sensitive device, and a sound source in air produces only a small fraction of the pressure developed in an underwater source. A new mechanism for transmission (and amplification) of sound pressure was therefore required. The basis for this was also provided by the Rhipidistia.

In the early consideration of the crossopterygian skull, the point was made that the basal hinge between the lower and upper jaws involves two replacement bones, articular and quadrate, with the former abutting upon the latter. The quadrate in turn, however, does not articulate directly with the neurocranium. Rather, there is interposed between it and the auditory region of the neurocranium the dorsalmost unit of the hyoid arch, the hyomandibula (Figure 11-14). But in the primitive Amphibia the quadrate bone becomes an integral part of the sidewall of the neurocranium, and because the articular bone abuts it, the jaws are attached directly to the neurocranium. This arrangement is accompanied by an alteration of the hyomandibula. In the first place, the first pharyngeal pouch, which in fishes typically opens to the exterior as the spiracle, remains closed. That is, the area of contact between the distal end of the pouch and the surface of the skin does not perforate but is retained as a *tympanic membrane* (eardrum). The hyomandibula then becomes modified in form and, as a *stapes* (*columella*), attaches externally to the inner surface of the tympanic membrane, crosses the pouch cavity, and terminates in an opening through the auditory capsule to the perilymphatic cavity of the inner ear (Figure 11-16). In this position it serves to transmit vibrations impinging on the tympanic membrane across to the inner ear. Associated as it is now with audition, the expanded terminus of the first pouch is designated the *middle ear,* or *tympanic, cavity,* and its narrow connection with the pharynx is termed the *eustachian tube.* (Not all modern Amphibia have retained this full complement of middle ear parts. They all have a stapes (columella), but the tympanic membrane and cavity and eustachian tube have been lost by present-day Apoda and urodeles and even some anurans.)

The arrangements just described for primitive amphibians are found essentially in reptiles and birds as well. But with the creation of a new suspension design for the jaws of mammals, involving an articulation between two dermal bones, the quadrate and articular bones are reduced and added to the sound-transmitting system. Because in amphibians and reptiles the articular—

quadrate joint lies near the tympanic membrane, the articular now makes contact with the membrane and is identified as the *malleus.* The quadrate, originally interposed between the articular and hyomandibula (stapes), becomes the *incus,* lying between the malleus and the stapes (Figures 17-31 and 17-32). Expressed somewhat differently, the reptilian stapes, which bridged the tympanic cavity and inserted on the tympanic membrane at one end and in an opening to the perilymphatic space at the other end, becomes considerably shortened in mammals (in fact, the mammalian stapes may correspond only to the proximal portion of the reptilian stapes), and the incus and malleus are inserted between its distal end and the tympanic membrane. Because, with the development of the organ of Corti, the original perilymphatic space has been divided into dorsal and ventral ducts, it also happens that the stapes inserts in the opening (oval window) to the dorsal duct alone.

When present in amphibians, the tympanic membrane lies flush with the body surface, and so it is in most reptiles (except in snakes in which the cavity, membrane, and eustachian tube have been altogether lost). In certain lizards and in crocodilians, however, the tympanic membrane lies at the base of a shallow canal opening to the exterior. This, the *external auditory mea-tus,* is in effect the deepened pharyngeal groove matching the first pouch. In mammals the meatus becomes deep and tubular. Moreover, it becomes bounded by a projecting *pinna* supported by elastic cartilage and when, as in many mammals, the pinna is provided with well-developed voluntary muscles, it may be moved so as to "collect" sound waves. It is to the pinna and external auditory meatus that the term *outer ear* applies.

The structure of the mammalian ear may be summarized in terms of the mechanism of sound reception (Figures 17-31 and 17-32). Pressure waves in the air pass inward via the external auditory meatus and cause the tympanic membrane to vibrate. This vibration is transmitted mechanically by the three middle ear bones (malleus, incus, and stapes, in that order) to the membrane that covers the opening (oval window) to the scala vestibuli. Vibration of the oval membrane sets up pressure waves in the fluid of the scala vestibuli, which are then conducted to the scala media. The vibrations in the fluid of the scala media strike the organ of Corti and stimulate its hair cells, causing them to initiate impulses in the fibers of the cochlear nerve. The pressure waves ultimately return to the tympanic cavity via the scala tympani and the membrane in the round window.

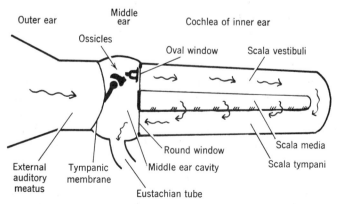

Figure 17-31 Functional anatomy of the mammalian ear. See text for explanation. Arrows indicate pathway of vibrations.

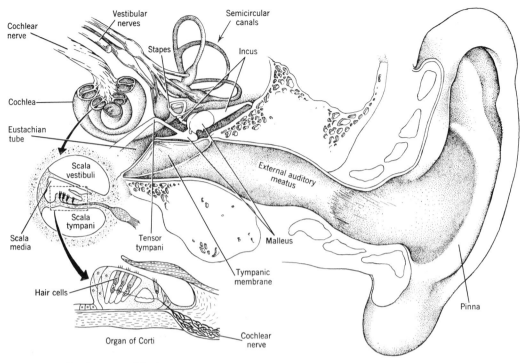

Figure 17-32 Anatomy of the human ear.

RADIORECEPTORS

Although vertebrates, as animals and plants generally, are constantly subjected to radiation of many kinds and degrees, so far as is known there are special receptors only for that relatively narrow band of radiation encompassing heat and light. And of these two varieties of radiant energy, only light is consistently perceived; the known instances of special receptors for heat are scanty indeed.

Among mammals, temperature receptors are widely distributed in the skin and mucous membranes, in the conjunctiva of the eye, and in the heart. These may be free nerve endings, that is, primary receptors, or encapsulated nerve endings, secondary receptors.

Among fishes, the so-called *pit organs* distributed over the head and trunk and the *ampullar organs* on the heads of elasmobranchs are said to be temperature recep-

tors. The most thoroughly documented cases of temperature receptors are the pit organs of the pit vipers, for example, the rattlesnake. The tissues within these pits are extensively innervated and are known to enable the snakes possessing them to detect warm objects, say mammalian prey, at a distance of several feet.

Specific sensitivity to light is widespread in the animal and plant worlds, sometimes through the medium of particular cells or cell clusters, sometimes through a highly organized mechanism complete with a light-concentrating lens and light-responsive epithelium. The light receptors, the *eyes,* of vertebrates, however, are constructed in a manner distinctive unto themselves and have no known homologue in any other animal phylum.

It will be recalled that in their evolutionary history the vertebrates have exhibited two general varieties of visual organs: unpaired median dorsal eyes and paired lat-

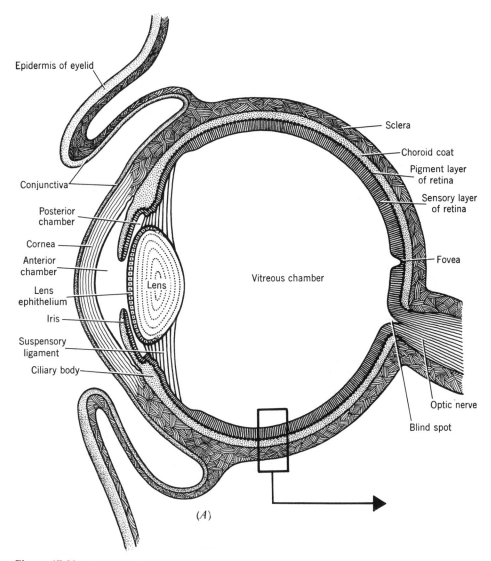

Epidermis of eyelid

Conjunctiva

Posterior chamber

Cornea

Anterior chamber

Lens ephithelium

Iris

Suspensory ligament

Ciliary body

Sclera

Choroid coat

Pigment layer of retina

Sensory layer of retina

Fovea

Vitreous chamber

Lens

Optic nerve

Blind spot

(A)

Figure 17-33   Anatomy of the human eye.

eral eyes. The former have already been dealt with briefly (p. 432), and because the predominant visual organs of vertebrates are the paired eyes, the account that follows is devoted exclusively to them.

The Structure of the Eye. The eyes of vertebrates exhibit many variations in structural and functional specifications, yet they are all built to a common plan. The human eye, about which more is known than that

of any other vertebrate, may therefore serve as an introductory example (Figure 17-33).

The human eye is a hollow sphere whose wall is composed of three coats, only the outer one of which is complete. It is the outer coat that provides the essential support of the eye and gives it shape. This coat, termed the *fibrous tunic,* is divided into two regions: the *cornea,* the transparent, exposed portion of the tunic; and the *sclera,* the remainder of the tunic to which the

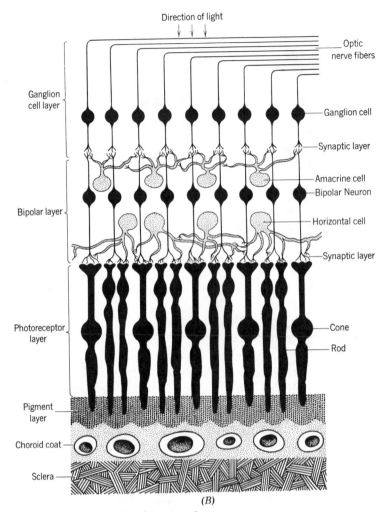

Figure 17-33 (*Continued*)

extrinsic muscles moving the eye are attached. The oneness and continuity of the sclera and cornea are, however, more apparent than real. The sclera is a product of a mesodermal envelope that encases the embryonic eye, whereas the cornea is essentially an area of the skin of the head and thus has both ectodermal and mesodermal components. In the adult this head skin is doubled back on itself to form the underlinings of the eyelids, and its epidermal component continues over the cornea as the *conjunctiva.*

The middle one of the three coats is the *uvea,* which consists of three regions. (1)

Closely adherent to the sclera is a vascular, pigmented *choroid layer.* (2) At the level of transition between the sclera and cornea, the uvea becomes a relatively thickened, muscular *ciliary body.* (3) From the ciliary body the uvea turns inward and away from the sclera and joins with the margin of the innermost coat (retina) to form the *iris,* which is also equipped with muscle. The opening in the center of the iris is the *pupil.*

The innermost coat is the *retina,* which abuts the entire extent of the uvea. That part of it in contact with the choroid layer is fairly thick and is the region sensitive to light. The remainder is thin and nonsensory and contributes to the ciliary body and iris. The structure of the light-sensitive portion of the retina is complex indeed and it is beyond our province to pursue its functional anatomy in detail: however, attention must be given to an irreducible minimum of facts about it.

The retina consists of two strata, an outer *pigmented layer* lying against the choroid coat and an inner *sensory (nervous) layer* containing the definitive light receptors and a succession of neurons in synaptic series (Figure 17-33). The cellular components of the nervous retina are, in turn, arrayed in three layers. Reading from the outside inward, these are a layer of *photoreceptors,* a layer of *bipolar neurons,* and a layer of *ganglion cells.* Between the three layers of cells are two *synaptic layers* in which the processes of the cells make close contact.

As we have often noted previously, sensitivity to stimuli is associated with especially designed cells. In this instance the sensory cells are responsive to light, hence their designation as photoreceptors, and are of two kinds: long and narrow *rods* and shorter, thicker *cones.* The free ends of the rods and cones are oriented toward the outer pigmented layer of the retina. Centrally, the rods and cones send processes into the outer synaptic layer where they form synapses with the bipolar neurons. At this level of the bipolar layer there also lie *horizontal cells* whose processes connect with the terminals of neighboring rods and cones, as well as with other horizontal cells, and thereby provide horizontal integration of the photoreceptors. In the inner synaptic layer, each bipolar neuron terminal forms synapses with one or more ganglion cells and additional horizontal integrative units known as *amacrine cells.* The axons of the ganglionic cells course over the inner surface of the retina, finally converging at one point to form the optic nerve. The nerve then pierces the choroid and sclera and makes it way to the brain. Where the optic fibers converge, neurosensory cells are necessarily lacking, so that this point represents a *blind spot* in the visual field. Conversely, there is an adjacent area where visual acuity is greater than in any other region of the retina. This is the *fovea centralis,* a highly specialized depression that contains many thousands of closely packed cones, but is devoid of rods. There are no blood vessels in this tiny area, and the ganglionic nerve fibers diverge so that sharply focused light falls directly on the cones.

The careful reader will have detected that the order of arrangement of the receptors and neurons of the retina is such that the rods and cones are furthest removed from the light entering the front of the eye. In other words, except for the fovea, light must pass through the layers of ganglionic and bipolar neurons to reach the rods and cones. By the same token, stimuli thus set up must pass back through these layers of cells to reach the optic nerve. Because of this circumstance the retinas of vertebrates are said to be "inverted," and one of the problems of the evolutionary origin of the eye is that of accounting for the origin of the inverted retina.

In the interior of the eye, just behind the iris, lies the *lens.* Clear and glassy in appearance and biconvex in form, it is held in place by a *suspensory ligament* composed of fibers radiating from the ciliary body. The large cavity behind the lens contains a transparent, semigelatinous substance, the *vitreous body* or *humor.* The space between the lens and the cornea is divided by the iris into *anterior* and *posterior chambers,* both filled by a clear watery fluid, the *aqueous humor.*

A useful practice in summarizing the form and function of the eye is to compare it with a camera. Light is excluded or permitted to enter by the eyelids, the equivalent of the camera shutter. Once admitted, the amount of light is regulated by a variable aperture: the diaphragm between the lens elements of the camera, the pupil of the eye, whose diameter is governed by expansion and contraction of dilator and constrictor muscles in the iris. Focusing of the light rays is accomplished by a lens system, which in the eye consists of the cornea and lens, and an inverted image is projected on a sensitive screen, photographic film or the retina. To prevent the blurring of images by internal reflection, the inner walls of the camera are painted black; the interior of the eye is darkened by the pigmented choroid layer. Finally, accommodations to varying distances of objects are made in the camera by substituting one lens for another, and by varying the distance between the lens system and the retina in the eye.

Everyone is surely aware of the fact that it is impossible to photograph a scene in color if the illumination is poor; the recording of color requires bright light. To extract the maximum information from a scene, a photographer uses sensitive black-and-white film for twilight and color film for full daylight or its artificial counterpart. But it is not necessary to change the "film" in the eye camera. The retina contains two "emulsions," one for color and insensitive to dim light, the other for twilight. These are the rods and cones. The rods are the "emulsion" for twilight vision, the cones for daylight vision.

The rods contain a pigment known as *rhodopsin,* or *visual purple,* and its presence is necessary for any vision in dim light. Conversely, its absence is a requisite for vision in bright light. Thus, if an abrupt transition from bright to dim light is made (say, in entering a motion picture theater during daylight hours), some time is required for the restoration of the pigment before good vision is possible. (Vitamin A is known to play an important role in the synthesis of rhodopsin.) The reverse, the destruction of rhodopsin, is required when going from dim to bright light.

Over a century ago it was demonstrated that all colors can be created by suitable mixtures of red, green, and blue primary colors. This led to the suggestion that the retina contains three different kinds of light-sensitive substances, and that the information provided by the excitation of each sub-

stance is transmitted to the brain and there combined to reproduce the colors of the outside world. The general validity of this concept now seems well established. Through refined methods of single-cell microspectrophotometry not only has it been demonstrated that the postulated substances are present in the cones, but also that it is indeed likely that there are three types of cone, each of which contains principally one of the three pigments. One of these pigments, *erythrolabe,* is sensitive to light in the red–orange range of the spectrum. The second is *chlorolabe,* sensitive to the yellow–green range. The third one is less well documented, but presumably responds to the blue range; it is called *cyanolabe.* An interesting spinoff of the studies of these retinal pigments pertains to individuals with red–green color blindness. Such individuals are actually of two kinds: red-blind and green-blind. It has been shown that individuals in the first category lack the red-sensitive pigment, whereas those in the second category lack the green-sensitive pigment.

With these basic facts of architecture and function behind us, we are now prepared to turn to some of the more conspicuous variations of form and function in the eyes of vertebrates generally. First, however, it is desirable to look into the embryogeny of the eye.

The Embryogeny of the Eye. The various parts of the eye originate from three embryonic sources: (1) the retina and optic nerve are products of the brain wall; (2) the lens and part of the cornea are provided by the epidermal ectoderm of the head; (3) all the other parts are furnished by neighboring mesenchyme.

Even before the prosencephalon becomes subdivided into telencephalon and diencephalon, a pair of saclike evaginations appear from its lateral walls. Each of these is an *optic vesicle* (Figure 17-34A). The vesicle gradually expands and, in so doing, displaces the intervening mesenchyme and

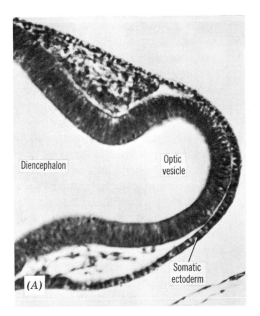

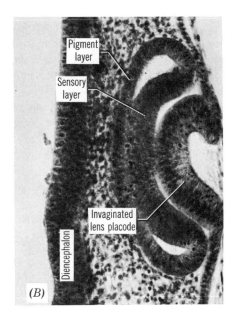

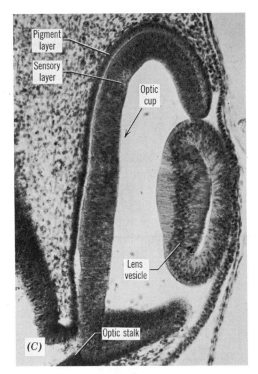

Figure 17-34 Photomicrographs of sections through the developing eye of the chick embryo. (*A*) Optic vesicle of 33-hour chick; × 190. (*B*) Optic cup and early lens of 48-hour chick; × 230. (*C*) Optic cup and lens of 72-hour chick; × 190. (Photos by Torrey.)

makes contact with the inner surface of the epidermis. At the same time, the proximal portion of the vesicle becomes relatively constricted so as to constitute an *optic stalk,* serving to connect the vesicle proper with the brain. As the vesicle enlarges, it also projects itself upward and backward, so that eventually the stalk runs out of its ventral side. Meanwhile, the subdivision of the prosencephalon is consummated and the stalk is found joining the diencephalon.

The external wall of the vesicle now flattens and pushes inward so as to convert the vesicle to a double-walled *optic cup* (Figures 17-34*B, C*). The invaginated layer, that is, the inner layer, of the optic cup is considerably thicker than the external wall. Pigment granules shortly appear in the external layer; hence it is designated the *pigment layer.* The thicker internal layer is designated the *sensory layer,* for it is here that the three strata of cells of the retina—rods and cones, bipolar neurons, and ganglionic neurons—will differentiate. The pigment and sensory layers eventually fuse. The opening of the cup to the exterior represents the future *pupil;* thus the rim of the cup, to which other materials will be added, is the prospective *iris.*

It is important to note that the invagination by which the optic vesicle is transformed into a cup involves not only the distal wall of the original vesicle but its ventrolateral wall as well. Because the proximal part of the original vesicle is represented in the optic stalk and the invagination of the vesicle is continued ventrally, there results a groove on the undersurface of the stalk leading into the cavity of the cup proper. This is the *choroid fissure* (Figure 17-35). Expressed somewhat differently, the ventral projection of the invagination of the vesicle creates a gap in the ventral edge of the iris, a gap continuous with a trough on the underside of the stalk. For a time the choroid fissure remains open and furnishes a route for blood vessels and migrating mesenchyme on their way to the interior of the cup. The fissure eventually closes, so that the blood vessels serving the interior of the eye lie within the optic stalk. Ultimately, too, the axons growing out of the ganglionic cells of the retina converge upon the optic stalk and, as the fibers constituting the optic nerve, course through the walls of the stalk to the brain. So it is that the blood vessels to and from the retina and other internal parts actually lie within the optic nerve.

Even as the optic vesicle is transforming into an optic cup, the somatic ectoderm immediately overlying the cup undergoes an associated change. This consists of a thickening to form a *lens placode.* In birds and mammals the placode then pockets inward (Figure 17-34*B*) and finally pinches away from the parent ectoderm to form a *lens vesicle* lying in the pupil (Figure 17-34*C*). In amphibians and bony fishes only

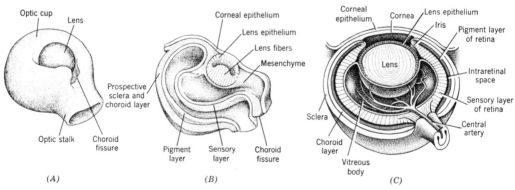

Figure 17-35 Stereograms of the developing eye. (Drawn from Ziegler models.)

the inner part of the epidermis is involved in lens formation, and in this instance the prospective lens is first nipped off as a solid bud which then becomes vesicular. In each case, the vesicle differentiates into a definitive *lens* in the following manner. The cells of the inner wall of the vesicle increase greatly in height, thus bringing an overall thickening of the inner layer and eventually eliminating the lumen of the vesicle (Figure 17-35). These cells transform into transparent *lens fibers.* The cells of the outer wall, on the contrary, remain a low columnar type and constitute a permanent *lens epithelium.* The interior of the fully formed lens remains wholly nonvascular, but blood vessels derived from the supply entering via the optic stalk do invest its surface. That the sensory layer of the retina plays a role in the histogenesis of the lens is attested to by the following experiment. If the lens of a 5-day-old chick embryo is surgically reversed so that its epithelium faces the neural retina, the elongation of those lens cells that have already differentiated is arrested and the epithelial cells differentiate into a new set of lens fibers. This internal reorganization results in a complete reversal of the polarity of the lens.

With the exception of the outer surface of the cornea, all the remaining tissues of the eye are derived from mesoderm. A double layer of mesenchyme invests the lens and optic cup. The outer layer will become the *sclera* and part of the *cornea;* the inner, the *choroid layer, ciliary body,* and part of the *iris.* The mesenchyme between the lens and the surface ectoderm, originally solid, hollows out to form the *anterior* and *posterior chambers.* The layer of mesenchyme left in front of the anterior chamber, and continuous with the sclera, combines with the overlying somatic ectoderm to form the *cornea.* Mesenchyme also invades the chamber behind the lens where, with supplements of ectodermal cells provided by both lens and retina, the *vitreous body* differentiates.

Optic Induction. Through the conventional methods of vital staining, notably on amphibian embryos, the materials destined to provide the optic cups have been shown to occupy a midline position at the front of the open neural plate. Moreover, it has been demonstrated that determination of this localized area for eye formation comes about under the influence of the subjacent prechordal mesoderm. In other words, induction of the eyes occurs hand in hand with that of the neural plate itself. The initial determination is of a very general sort only, for if the eye rudiment in the neural plate stage, or even at the later optic vesicle stage, is split, each part is capable of forming a complete eye. Only with advancing development is there an increasing restriction of the developmental potentialities of the parts of the optic cup. The bilateral deployment of the midline rudiments has also been shown to be under the influence of the prechordal mesoderm, because if for any reason the prechordal mesoderm fails to engage in its customary bilateral differentiation, a midline "cyclopean" eye results.

That the rise of the lens vesicle simultaneously with the differentiation of the optic cup is more than a coincidence has been revealed by experimental studies on a number of vertebrates. If the optic vesicle is removed before it reaches the epidermis, the lens ordinarily fails to develop. Or if the epidermis over the vesicle is removed and replaced by epidermis from some foreign source, say the flank of the body, this foreign epidermis forms a lens. It is concluded, therefore, that the optic cup furnishes an inductive stimulus causing epidermis to differentiate into a lens. The nature of the stimulus is unknown. This is suggested by the fact that if a cellophane membrane is inserted between the vesicle and epidermis, lens development is stopped, whereas a thin slice of agar or a millipore filter in the same position allows for the passage of the hypothetical chemical agent and lens development proceeds. An interesting situation paralleling that found in the induction of the neural tube (p. 412) has also been found with respect to the passage of cytoplasmic particles (ribosomes). At the time when the optic vesicle comes into contact with the prospective lens epidermis it contains large amounts of ribonucleic acid, whereas the epidermis contains none. Then as the lens vesicle develops, ribonucleic acid accumulates within it while that in the optic cup declines. One possible interpretation is that lens induction is a consequence of the passage of ribonucleic acid from vesicle and cup to the lens epidermis. Yet it is also possible these are only coincidental and not necessarily related events: the induction process may call for the inductor to use up its ribonucleic acid, whereas the rise of ribonucleic acid in the inductee may be only a manifestation of its biochemical response.

Experimental analyses of lens development in some species of amphibians reveal that more than the optic cup may be involved in induction. In these instances a lens may form even following a prior removal of the optic cup. Originally interpreted as an exception to inductive requirements, it presently appears that the mesoderm of the head replaces the optic cup in whole or in part as the inductor.

The creation of a transparent cornea out of the skin overlying the pupil also comes via inductive stimuli, the source of which is the remainder of the eye. No cornea forms if the eye is extirpated, but if the prospective cornea is replaced by foreign skin, the graft transforms into typical cornea. Both the lens and retina appear to work together, for each, acting alone, can elicit corneal differentiation when transplanted under foreign skin.

In summary, the development of the eye involves a chain of inductive events. (1) The prechordal mesoderm, a primary inductor, incites the differentiation of the forebrain and the optic area within it. (2) The optic cup (sometimes supplemented by head mesoderm) then serves as a secondary inductor, eliciting the formation of the lens. (3) The lens, acting together with the retina, serves as a tertiary inductor of the cornea.

Comparative Functional Anatomy of the Eye. The human eye served as the example on which our earlier description of the functional anatomy of the eye was based. Now let us consider some of the variations that attend the eyes of other vertebrates.

First, not all vertebrates, not even all mammals, share the human's capacity for perception of color. Any animal that does see color must have cones containing visual pigments. Lacking these requisites, some animals have no color vision at all; others, with minimal equipment, have no more than rudimentary color vision. An especially interesting arrangement is found in some reptiles and birds where a single photopigment is combined with oil droplets of various colors which serve as differential filters of the light impinging upon the pigment. Among mammals, only Old World primates have color perception approaching that of humans.

It should be appreciated, too, that the retina is more than just the biological equivalent of a photographic emulsion. By way of the nerve impulses generated by the biochemical transformation of the visual pigments under the impact of light energy, a kaleidoscopic pattern of sensations is sent to the brain. The mental image that we call "seeing" is the consequence of the processing and integration of these sensations and this takes place in some measure in the retina itself as well as the brain. The relative extent of processing in the retina and brain varies from species to species and, curiously enough, there is no correlation with taxonomic position. For instance, though one might expect just the opposite, the retina of the frog is considerably more complex than that of primates. The greater complexity lies in the synaptic layers where, in contrast to the human retina, the integrative connections of the horizontal and amacrine cells are far more elaborate. The implication is that in those species exhibiting synaptic complexity like that of the frog, the retina plays a greater role in image processing than it does when it is more simply organized. At the same time, there is a correlation between the degree of complexity of the retina and the nature of the visual centers of the brain. Recall (p. 430) that those centers are shifted from their original location in the tectum of the mesencephalon to the cerebral hemispheres. It appears that vertebrates have evolved two styles of visual systems. One involves an intensive processing of visual data in the retina, with the mesencephalon playing a lesser role; the other involves intensive processing in the cerebral cortex, with the retina playing a lesser role.

In order that the patterns projected on the retina may be recorded accurately, they must be received with the greatest possible sharpness and refinement. It is to this end that all the parts of the eye other than the retina are primarily dedicated. Especially important is the lens system which, when functioning properly, ensures that the picture of the object to be perceived falls upon the retina in perfect focus. This ability of the eye to bring objects at various distances into focus is termed *accommodation.*

The cornea and the lens are the two components of the lens system. Because its refractive index is virtually the same as that of water, the cornea plays no part in accommodation in aquatic vertebrates, so that the task is performed by the lens alone. Terrestrial vertebrates, however, utilize both cornea and lens, with the general trend being toward an increase in importance of the cornea. The cornea operates in accommodation through changes in the degree of its curvature: the greater the curvature, the shorter the distance of focus. Alterations in the curvature of the cornea are brought about by the contraction and relaxation of the muscles of the ciliary body. Contraction of these muscles exerts a force on the peripheral margin of the cornea, thus causing

it to bow outward. Both reptiles and mammals have the ability to adjust the cornea in this fashion, and birds do so to a remarkable degree.

Accommodation through adjustments in the lens may be accomplished in either of two ways: by changing its position forward and backward or by changing its shape. In all anamniotes the method employed is that of moving the lens. Lampreys and teleosts maintain the lens for the most part in a forward position for near vision and move it backward for distant vision. These movements are ordinarily mediated by special muscles within the eye. In all amniotes (except snakes) the shape of the lens is changed: it is flattened for focusing on distant objects and its curvature is increased for focusing on near objects. The muscles of the ciliary body are always concerned, but the general mechanism varies between mammals on the one hand and reptiles and birds on the other. In reptiles and birds a ring of processes extends from the ciliary body inward to contact the periphery of the lens and, as the ciliary muscles contract, these processes push against the lens and force it to bulge. When the ciliary body relaxes, the lens flattens. But in mammals the lens is suspended by the fibers of the suspensory ligament, which under strain hold the lens in a flattened condition. When the ciliary body contracts, it pulls the choroid coat forward, thus relieving the tension on the suspensory ligament and allowing the lens to relax and become more rounded.

The quality of the image that impinges on the retina also depends in measure on the amount of light permitted to enter the eye camera. The iris plays a major role here. Although in some fishes the iris remains in a fixed position, and thus the diameter of the pupil remains unchanged, in vertebrates generally the iris is equipped with muscle under the control of the autonomic system and by its expansion and contraction the size of the pupil may be regulated. The iris-diaphragm is closed down in bright light, opened in dim light. As an added protection against excessive bright light, the closed down pupils of nocturnal animals are often slitlike rather than circular. Conversely, to make most efficient use of minimum light at night, many nocturnal animals are equipped with a reflective surface, *tapetum,* in the choroid layer serving to reflect unused light back into the retina.

The point was made earlier that the fibers of the optic nerve, originating in the ganglionic layer of the retina, transfer sensations to the brain for the creation of the final mental image. In all vertebrates except mammals, all the fibers in each optic nerve cross to the opposite side, so that the object "seen" by the right eye is recorded in the left side of the brain, and vice versa. Because in the majority of vertebrates the eyes are directed laterally and the two fields of vision are largely different, the brain thus reports two entirely separate views. If, however, the eyes are directed forward, as they are in birds of prey and many mammals, the two fields of vision overlap in considerable measure and the mental pictures likewise overlap, with resulting *stereoscopic vision.* In mammals, moreover, a proportion of the fibers in each optic nerve fails to cross to the opposite side, and so duplicated pictures are created on each side of the brain. Perfection of this arrangement occurs in primates where, with the two fields of vision practically identical, stereoscopic vision is especially refined.

CHEMORECEPTORS

Throughout the history of the vertebrates the chemical senses, taste and smell, have been intimately associated, so much so that the physiological distinction between them is easily confused. Human experience clearly reveals the extent of this confusion, for that which we refer to as "flavor" in foods, for instance, is a combination of the sensations of taste and smell. However, taste and smell actually have only one thing in common: to be tastable or smellable, substances must be capable of going into solution. Smellable substances must be volatile at ordinary temperatures and must be soluble in fat solvents; substances are tastable only when they are dissolved in water. Otherwise, taste and smell differ in many ways.

In the first place, these sensations are associated with two very different kinds of receptors, *taste buds* and the *olfactory organs.* Second, the cranial nerves involved in taste are the facial, glossopharyngeal, and vagus, whereas only the olfactory nerve is concerned with smell. Third, the taste buds are capable of producing only four sensations, sourness, sweetness, bitterness, and saltiness, whereas a countless array of odors, may be detected through smell. Finally, the sensitivity of the olfactory receptors is infinitely greater than that of taste buds. To detect a substance by taste, it must be present in concentrations several hundred or thousand times greater than for smell.

Taste Buds. The taste buds of vertebrates are oval or barrel-shaped clusters of elongated cells sunk within some epithelial covering (Figure 17-36). The outermost cells of a taste bud are only supportive in function; the innermost cells are the true receptors. Each receptor cell ordinarily has a fine bristle projected toward the surface, and the terminals of the nerve fibers that serve a bud lie between the receptor cells. Taste buds are for the most part confined to the mouth, but among the lower vertebrates their distribution may be wide indeed. In fishes they are commonly scattered over the surface of the head, and even the trunk, as well as the mouth. The common catfish, for example, has an extensive spread of taste buds over the body and they are especially concentrated on the barbels, or "whiskers," featured by this fish. In amphibians the buds have a general distribution within the mouth and pharynx, and sometimes also on the head. Reptiles and birds

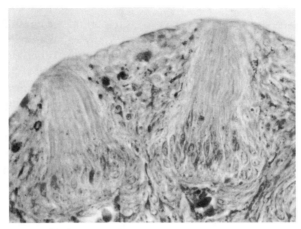

Figure 17-36   Taste buds of a catfish. × 180. (Photo by Torrey.)

have them largely in the pharynx, whereas in mammals they tend to be concentrated on the tongue, particularly in association with localized elevations known as papillae. Although all the taste buds in a given animal are structurally alike, curiously enough there is a physiological distinction among them: the separate sensations of sour, sweet, bitter, and salt are recorded by separate taste buds. Thus in humans, sourness is largely recorded by the taste buds on the lateral margins of the tongue, saltiness and sweetness by the buds at the tip of the tongue, and bitterness by the buds at the base of the tongue. Apparently, any given taste bud is physiologically specialized to respond only to the specific ions responsible for each of these sensations.

Olfactory Organs. If it is correct to interpret the evolutionary history of the telencephalon to the effect that this region of the brain was primitively devoted to olfaction, then the olfactory organs are ancient attributes of vertebrates. The earliest stage in their evolution is likely still exemplified in sharks and teleosts where the organs are paired, blind *nasal sacs* located well anteriorly on the head, a situation paralleled in early embryonic stages of amphibians and amniotes where the nasal organs are initially blind sacs created by the invagination of epidermal *olfactory placodes* (Figure 17-

37). Each sac is lined by a considerably folded epithelium, which contains *olfactory cells* interposed among nonsensory supportive cells (Figure 17-38). Additional evidence for the ancientness of the olfactory epithelium lies in the fact that the olfactory cells are of that primitive kind designated neurosensory cells. That is, these cells are both receptors and conduc-

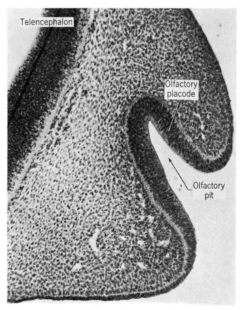

Figure 17-37   Olfactory pit (invaginated olfactory placode) of a 10-mm pig embryo. × 140. (Photo by Torrey.)

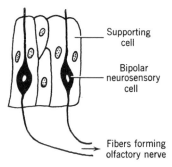

Supporting cell

Bipolar neurosensory cell

Fibers forming olfactory nerve

Figure 17-38 Cellular components of the olfactory epithelium of a mammal.

tors: the exposed ends of the olfactory cells are stimulated by the dissolved substances that make contact with them, and the sensations created are relayed directly to the brain by fibrous extensions of the cells, fibers that collectively constitute the olfactory nerve.

We have already commented on the later history of the nasal sacs (p. 282 and Figure 13-3). Originally possessed of anterior openings (external nares) only and subservient to olfaction, they acquire openings to the mouth (internal nares) and assume a major role in aerial respiration. Also, beginning with the reptiles and culminating in mammals, there is a separation of food and air passages and olfaction becomes ancillary to respiration. The olfactory epithelium in each nasal passage also becomes deployed in two distinct regions, one termed a *nasal organ* and the other a *vomeronasal organ* (*Jacobson's organ*).

The vomeronasal organ is seen in its simplest form in urodele amphibians, where it consists of a groove in the ventrolateral floor of the nasal sac opening into the roof of the mouth. In other amphibians the organ is almost completely separated from the nasal sac, and in lizards and snakes it is entirely so and thus appears as a blind sac connected to the mouth. It serves in these reptiles to record the odors of substances taken into the mouth and transferred to the organ by the tongue. By contrast, the vomeronasal organ is absent in turtles and crocodilians and is vestigial or lacking in birds as well. Among mammals the vomeronasal organ may be a distinct structure connected to the mouth by a duct piercing the palate, reduced to a small structure buried in the floor of the nasal cavity, or absent entirely (for example, humans).

The nasal organ is an area of olfactory epithelium that lies in the main air passage, or nasal cavity, primarily in its roof. It reaches its maximum development in mammals, where its area is greatly expanded through the elaboration of scrolls of bone (turbinals) from the lateral wall of the nasal chamber.

General References

Arey, L.B. 1974. *Developmental Anatomy,* 7th ed. Saunders, Philadelphia.

Balinsky, B.I. 1981. *An Introduction to Embryology,* 5th ed. Saunders, Philadelphia.

Bentley, P.J. 1982. *Comparative Vertebrate Endocrinology,* 2nd ed. Cambridge University Press, New York.

Carlson, B.M. 1988. *Patten's Foundations of Embryology,* 5th ed. McGraw-Hill, New York.

Gardner, E. 1975. *Fundamentals of Neurology,* 6th ed. Saunders, Philadelphia.

Goodrich, E.S. 1930. *Studies on the Structure and Development of Vertebrates.* MacMillan, London. (Republished by Dover, New York, 1958).

Gorbman, A., et al. 1983. *Comparative Endocrinology.* John Wiley & Sons, New York.

Ham, A.W., and D.H. Cormack. 1979. *Histology,* 8th ed. J.B. Lippincott, Philadelphia.

Hildebrand, M. 1988. *Analysis of Vertebrate Structure,* 3rd ed. John Wiley & Sons, New York.

Hildebrand, M., Bramble, D.M., Liem, K.F., and Wake, D.B., eds. 1985. *Functional Vertebrate Morphology.* Harvard University Press, Cambridge, MA.

Jarvik, E. 1980. *Basic Structure and Evolution of Vertbrates,* Vols. I and II. Academic Press, New York/London.

Kent, G.C. 1983. *Comparative Anatomy of the Vertebrates,* 5th ed. C.V. Mosby, St. Louis, MO.

Leake, L.D. 1975. *Comparative Histology: An Introduction to the Microscopic Structure of Animals.* Academic Press, New York.

Romer, A.S., and Parsons, T.S. 1986. *The Vertebrate Body,* 6th ed. Saunders, Philadelphia.

Saunders, J.W. 1982. *Developmental Biology: Patterns, Problems and Principles.* MacMillan, New York.

Trinkaus, J.P. 1984. *Cells into Organs: The Forces That Shape the Embryo,* 2nd ed. Prentice-Hall, Englewood Cliffs, NJ.

Turner, C.D., and Bagnara, J.T. 1976. *General Endocrinology,* 6th ed. Saunders, Philadelphia.

Villee, D.B. 1975. *Human Endocrinology: A Developmental Approach.* Saunders, Philadelphia.

Wessels, N.K. 1977. *Tissue Interactions in Development.* W.A. Benjamin, Menlo Park, CA.

Young, J.Z. 1981. *The Life of Vertebrates,* 3rd ed. Oxford University Press (Clarendon), London/New York.

Supplemental Reading

Chapter 1

Alberch, P. 1980. Ontogenesis and Morphological Diversification. *Am. Zool.* **20**:653–667.

Fink, W. 1982. The conceptual relationship between ontogeny and phylogeny. *Paleobiology* **8**:254–264.

Futuyma, D.J. 1986. *Evolutionary Biology,* 2nd ed. Sinauer, Sunderland, MA.

Gould, S.J. 1977. *Ontogeny and Phylogeny.* Harvard University Press, Cambridge, MA.

Hailman, J.P. 1976. Homology: Logic, information and efficiency. In R.B. Masterson, W. Hodos, and H. Jerison, eds., *Evolution, Brain and Behavior: Persistent Problems,* pp. 181–198. Erlbaum, Hillsdale, NJ.

Chapter 2

Bramble, D.M., and Wake, D.B. 1985. Feeding mechanisms of lower tetrapods. In M. Hilde-

grand et al., eds. *Functional Vertebrate Morphology,* pp. 230–261. Harvard University Press, Cambridge, MA.

Carroll, R.L. 1988. *Vertebrate Paleontology and Evolution.* W.H. Freeman, New York.

Colbert, E.H. 1980. *Evolution of the Vertebrates.* 3rd ed. John Wiley & Sons, New York.

Crompton, A.W., and Parker, P. 1978. Evolution of the mammalian masticatory apparatus. *Am. Scientist* **66**:192–201.

Edwards, J.L. 1977. The evolution of terrestrial locomotion. In M.K. Hecht, P.C. Goody, and B.M. Hecht, eds., *Major Patterns in Vertebrate Evolution,* pp. 553–557. Plenum Press, New York.

Feduccia, A. 1980. *The Age of Birds.* Harvard University Press, Cambridge, MA.

Frazzetta, T.H. 1968. Adaptive problems and possibilities in the temporal fenestration of tetrapod skulls. *J. Morphol.* **125**:145–158.

Gans, C., and Northcutt, R.G. 1983. Neural crest and the origin of vertebrates: A new head. *Science* **220**:268–274.

Gans, C., et al., eds. 1969. *Biology of the Reptilia.* Academic Press, New York.

Kemp, T.S. 1982. *Mammal-like Reptiles and the Origin of Mammals.* Academic Press, New York/London.

Northcutt, R.G., and Gans, C. 1983. The genesis of neural crest and epidermal placodes: A reinterpretation of vertebrate origins. *Q. Rev. Biol.* **54**:1–28.

Panchen, A.L., ed. 1981. *The Terrestrial Environment and the Origin of Land Vertebrates.* Academic Press, New York/London.

Romer, A.S. 1942. Cartilage: An Embryonic Adaptation. *Am. Nat.* **76**:394–404.

Schmidt-Nielsen, K. 1972. *How Animals Work.* Cambridge University Press, New York.

Thomas, R.D.K., and Olson, E.C. 1980. *A Cold Look at the Warm-Blooded Dinosaurs.* Westview Press, Boulder, CO.

Vaughn, T.A. 1986. *Mammalogy,* 3rd ed. W.B. Saunders, Philadelphia.

Wever, E.G. 1978. *The Reptile Ear. Its Structure and Function.* Princeton University Press, Princeton, NJ.

Chapter 3

Arey, L.B. 1974. *Developmental Anatomy,* 7th ed., Saunders, Philadelphia.

Saunders, J.W. 1982. *Developmental Biology: Patterns, Problems and Principles.* MacMillian, New York.

Trinkaus, J.P. 1984. *Cells into Organs: The Forces That Shape the Embryo,* 2nd ed. Prentice-Hall, Englewood Cliffs, NJ.

Wessles, N.K. 1977. *Tissue Interactions in Development.* W.A. Benjamin, Menlo Park, CA.

Chapter 4

Billett, F.S., and Adam, E. 1976. The structure of the mitochondrial cloud of *Xenopus laevis* oocytes. *J. Embryol. Exp. Morphol.* **33**:697–710.

Clermont, Y., and Leblond, C.P. 1955. Spermiogenesis of man, monkey and other animals as shown by the periodic-acid–Schiff technique. *Am. J. Anat.* **96**:229–253.

Coggins, L.W., and Gall, J.G. 1972. The timing of meiosis and DNA synthesis during early oogenesis in the toad, *Xenopus laevis. J. Cell Biol.* **52**:569–576.

Fawcett, D.W. 1970. A comparative view of sperm ultrastructure. *Biol. Reprod.* **2**(Suppl): 90–127.

Fawcett, D.W. 1975. The mammalian spermatozoon. *Dev. Biol.* **44**:394–436.

Gilbert, S.F. 1988. *Developmental Biology,* 2nd ed. Sinauer, Sunderland, MA.

Macgregor, H.C. 1980. Recent developments in the study of lampbrush chromosomes. *Heredity* **44**:3–35.

Packard, G.C., and Packard, M.J. 1980. Evolution of the cleidoic egg among reptilian antecedents of birds. *Am. Zool.* **20**:351–362.

Poccia, D. 1986. Remodeling of nucleoproteins during gametogenesis, fertilization and early development. *Int. Rev. Cytol.* **105**:569–576.

Wasserman, W.J., and Smith, L.D. 1981. Calmodulin triggers the resumption of meiosis in amphibian oocytes. *J. Cell Biol.* **56**:219–225.

Chapter 5

Bavister, B.D. 1980. Recent progress in the study of early events in mammalian fertilization. *Dev. Growth Differ.* **22**:385–402.

Bestor, T.M., and Schatten, G. 1981. Anti-tubulin immunofluorescence microscopy of microtubules present during the pronuclear movements of sea urchin fertilization. *Dev. Biol.* **88**:80–91.

Bleil, J.D., and Wassarman, P.M. 1986. Autoradiographic visualization of the mouse egg's sperm receptor bound to sperm. *J. Cell Biol.* **102**:1363–1371.

Cherr, G.N., *et al.* 1986. In vitro studies of the golden hampster sperm acrosome reaction: Completion on zona pellucida and induction by homologous zonae pellucidae. *Dev. Biol.* **114**:119–131.

Conklin, E.G. 1905. The orientation and cell-lineage of the ascidian egg. *J. Acad. Sci. Philadelphia* **13**:5–119.

Cross, N.L., and Elinson, R.P.C. 1980. A fast block to polyspermy in frogs mediated by changes in the membrane potential. *Dev. Biol.* **75**: 187–198.

Gerhart, J., *et al.* 1986. Amphibian early development. *BioScience* **36**:541–549.

McGrath, J., and Solter, D. 1984. Completion of mouse embryogenesis requires both the maternal and paternal genome. *Cell* **37**:179–183.

Sadleir, R.M.F.S. 1973. *The Reproduction of Vertebrates.* Academic Press, New York.

Surani, M.A.H., *et al.* 1986. Nuclear transplantation in the mouse: Hereditable differences between parental genomes after activation of the embryonic genome. *Cell* **45:**127–136.

Wassarman, P. 1987. The biology and chemistry of fertilization. *Science* **235:**553–560.

Wassarman, P. 1988. Fertilization in mammals. *Sci. Am.* **259:**78–84.

Chapter 6

Gilbert, S.F. 1988. *Developmental Biology.* Sinauer, Sunderland, MA.

Odell, G.M., *et al.* 1981. The mechanical basis of morphogenesis. I. Epithelial folding and invagination. *Dev. Biol.* **85:**446–462.

Trinkhaus, J.P. 1965. Mechanisms of morphogenetic movements. In R.L. DeHaan and H. Ursprung, eds., *Organogenesis.* Holt, Rinehart & Winston, New York.

Chapter 7

Black, S.D., and Gerhart, J. 1985. Experimental control of the site of embryonic axis formation in *Xenopus laevis* eggs centrifuged before first cleavage. *Dev. Biol.* **108:**310–324.

Black, S.D., and Gerhart, J. 1986. High frequency twinning of *Xenopus laevis* embryos from eggs centrifuged before first cleavage. *Dev. Biol.* **116:**228–240.

Boucaut, J.-C., *et al.* 1985. Evidence for the role of fibronectin in amphibian gastrulation. *J. Embryol. Exp. Morphol.* **89**(Suppl.):211–227.

Byers, T.J., and Armstrong, P.B. 1986. Membrane protein redistribution during *Xenopus* first cleavage. *J. Cell Biol.,* 2176–2184.

Conklin, E.G. 1931. The development of centrifuged eggs of ascidians. *J. Exp. Zool.* **60:**1–119.

Conklin, E.G. 1932. The embryology of *Amphioxus. J. Morphol.* **54:**69–151.

Gurdon, J.B. 1968. Changes in somatic nuclei inserted into growing and maturing amphibian oocytes. *J. Embryol. Exp. Morphol.* **20:**401–414.

Keller, R.E. 1986. The cellular basis of amphibian gastrulation. In L. Browder, ed., *Developmental Biology: A Comprehensive Synthesis,* Vol. 2, pp. 241–327. Plenum Press, New York.

Keynes, R.J., and Stern, C.D. 1988. Mechanisms of vertebrate segmentation. *Development* **103:**413–429.

Kimelman, D., *et al.* 1987. The events of the midblastula transition in *Xenopus* are regulated by changes in the cell cycle. *Cell* **48:**399–407.

Nakatsuji, N., *et al.* 1985. Fibronectin visualized by scanning electron microscope immunocytochemistry on the substratum for cell migration in *Xenopus laevis* gastrulae. *Dev. Biol.* **107:**264–268.

Newport, J.W., and Kirschner, M.W. 1984. Regulation of the cell cycle during *Xenopus laevis* development. *Cell* **37:**731–742.

Trinkaus, J.P. 1984. *Cells into Organs: The Forces That Shape the Embryo,* 2nd ed. Prentice-Hall, Englewood Cliffs, NJ.

Whittaker, J.R. 1987. Cell lineages and determinants of cell fate in development. *Am. Zool.* **27:**607–622.

Chapter 8

Azar, Y., and Eyal-Giladi, H. 1981. Interaction of epiblast and hypoblast in the formation of the primitive streak and the embryonic axis of the chick, as revealed by hypoblast rotation experiments. *J. Embryol. Exp. Morphol.* **61:**133–141.

Duband, J.L., and Thiery, J.P. 1982. Appearance and distribution of fibronectin during chick embryo gastrulation and neurulation. *Dev. Biol.* **94:**337–350.

Fisher, M., and Solursh, M. 1977. Gylcosaminoglycan localization and role in maintenance of tissue spaces in the early chick embryo. *J. Embryol. Exp. Morphol.* **42:**195–207.

New, D.A.T. 1959. Adhesive properties and expansion of the chick blastoderm. *J. Embryol. Exp. Morphol.* **7:**146–164.

Solursh, M., and Revel, J.P. 1978. A scanning electron microscope study of cell shape and cell appendages in the primitive streak of the rat and chick embryo. *Differentiation* **11:**185–190.

Spratt, N.T., Jr. 1963. Role of the substratum, supracellular continuity, and differential growth in morphogenetic cell movements. *Dev. Biol.* **7:**51–63.

VaKaet, L. 1984. The initiation of gastrular ingression in the chick blastoderm. *Am. Zool.* **24:**555–562.

Chapter 9

Beaconsfield, P., *et al.* 1980. The placenta. *Sci. Am.* **243**(2):94–102.

Borland, R.M. 1977. Transport processes in the

mammalian blastocyst. *Dev. Mammals* **1**:31–67.

Dyce, J., *et al.* 1987. Do trophectoderm and inner cell mass cells in the mouse blastocyst maintain discrete lineages? *Development* **100**:685–698.

Fleming, T.P. 1987. Quantitative analysis of cell allocation to trophectoderm and inner cell mass in the mouse embryo. *Dev. Biol.* **119**:520–531.

Gulyas, B.J. 1975. A reexamination of the cleavage patterns in eutherian mammalian eggs: Rotation of the blastomere pairs during second cleavage in the rabbit. *J. Exp. Zool.* **193**:235–248.

Levy, J.B., *et al.* 1986. The timing of compaction: Control of a major developmental transition in mouse early embryogenesis. *J. Embryol. Exp. Morphol.* **95**:213–237.

McIntire, J.A., and Faulk, W.P. 1979. Trophoblast modulation of maternal allogenic recognition. *Proc. Natl. Acad. Sci. USA* **76**:4029–4032.

Mintz, B. 1970. Clonal expression in allophenic mice. *Symp. Int. Soc. Cell Biol.* **9**:15.

Rodger, J.C., and Drake, B.L. 1987. The enigma of the fetal graft. *Am. Scientist* **75**:51–57.

Chapter 10

Bagnara, J.T., and Hadley, M.E. 1973. *Chromatophores and Color Change.* Prentice-Hall, Englewood Cliffs, NJ.

Feder, M.E., and Burggren, W.W. 1985. Skin breathing in vertebrates. *Sci. Am.* **253**:126–142.

Maderson, P.F.A., ed. 1972. The vertebrate integument: Symposium. *Am. Zool.* **12**:12–171.

Montagna, W., and Parakkal, P.F. 1974. *The Structure and Function of Skin,* 3rd ed., Academic Press, New York.

Osborn, J.W., and Crompton, A.W. 1973. The evolution of mammalian from reptilian dentitions. *Brevoria* **399**:1–18.

Sengel, P. 1976. *Morphogenesis of Skin.* Cambridge University Press, London.

Spearman, R.I.C., and Riley, P.A., eds. 1980. *The Skin of Vertebrates.* Academic Press, London/New York.

Chapter 11

Butler, P.M., and Joysey, K.A., eds. 1978. *Development, Function, and Evolution of Teeth.* Academic Press, New York/London.

Currey, J. 1984. *The Mechanical Adaptations of Bone.* Princeton University Press, Princeton, NJ.

Fleischer, G. 1978. Evolutionary principles of the mammalian middle ear. In *Advances in Anatomy, Embryology and Cell Biology,* Vol. 55, Pt. 5.

Herring, S.W. 1972. Sutures—A tool in functional cranial analysis. *Acta Anat.* **83**:222–247.

Jollie, M.T. 1977. Segmentation of the vertebrate head. *Am. Zool.* **17**:323–333.

Kummer, B. 1976. Biomechanics of the mammalian skeleton. Problems of static stress. *Fortschr. Zool.* **24**(3):57–73.

Laerm, J. 1982. The origin and homology of the neopterygian vertebral centrum. *J. Paleontol.* **56**:191–201.

Maderson, P.F.A. 1967. A comment on the evolutionary origin of vertebrate appendages. *Am. Nat.* **101**:71–78.

Maisey, J.G. 1980. An evolution of jaw suspension in sharks. *Am. Museum Novitates* **2706**:1–17.

Moore, W.J. 1981. *The Mammalian Skull.* Cambridge University Press, Cambridge.

Moss, M.L. 1968. The origin of vertebrate calcified tissues. In T. Orvig, ed., *Current Problems of Lower Vertebrate Phylogeny,* pp. 360–371. New York, Wiley–Interscience, New York.

Panchen, A.L. 1977. The origin and early evolution of tetrapod vertebrae. In *Problems in Vertebrate Evolution,* Linnean Soc. Symp. Ser. No. 4, pp. 289–318. Academic Press, London/New York.

Patterson, C. 1977. Cartilage bones, dermal bones, and membrane bones, or the exoskeleton versus the endoskeleton. In *Problems in Vertebrate Evolution,* Linn. Soc. Symp. Ser. No. 4. Academic Press, London/New York.

Williams, E.E. 1959. Gadow's arcualia and the development of tetrapod vertebrate. *Q. Rev. Biol.* **34**:1–32.

Chapter 12

Bourne, G.H., ed., 1972. *The Structure and Function of Muscle,* 4 vols. Academic Press, New York.

Doran, G.A. 1975. Review of the evolution and phylogeny of the mammalian tongue. *Acta Anat.* **91**:118–129.

English, A.W., and Wolf, S.L. 1982. The motor unit. *J. Am. Phys. Ther. Assoc.* **62**:1763–1772.

Grundfest, H. 1960. Electric fishes. *Sci. Am.* **203**:115–124.

Hiiemae, K.M., and Crompton, A.W. 1985. Mastication, food transport, and swallowing. In M. Hildebrand et al., eds., *Functional Vertebrate Morphology,* pp. 262–290. Harvard University Press, Cambridge, MA.

Lanyon, L.E., and Rubin, C.T. 1985. Functional adaptation in skeletal structures. In M. Hildebrand et al., eds., *Functional Vertebrate Morphology,* pp. 1–25. Harvard University Press, Cambridge, MA.

Lauder, G.V. 1980. On the relationship of the myotome to the axial skeleton in vertebrate evolution. *Paleobiology* **6**:51–56.

Murray, J.M., and Weber, A. 1974. The cooperative action of muscle proteins. *Sci. Am.* **230**:58–71.

Chapters 13 and 14

Duncker, H.R. 1978. General morphological principles of amniote lungs. In J. Piiper, ed., *Respiratory Function in Birds, Adult and Embryonic,* pp. 2–15. Springer-Verlag, New York.

Elias, H., and Sherrick, J.C. 1969. *Morphology of the Liver.* Academic Press, New York.

Gans, C. 1970. Respiration in early tetrapods— The frog as a red herring. *Evolution* **24**:723–734.

Hughes, G.M., ed. 1976. *Respiration of Amphibious Vertebrates.* Academic Press, New York/London.

Laurent, P. 1982. Structure of vertebrate gills. In D.F. Houlihan et al., eds., *Gills,* pp. 25–43. Cambridge University Press, New York.

Liem, K.F. 1985. Ventilation. In M. Hildebrand et al., eds., *Functional Vertebrate Morphology,* pp. 185–209. Harvard University Press, Cambridge, MA.

Moog, F. 1981. The lining of the small intestine. *Sci. Am.* **245**:154–176.

Pang, P.K.T., and Epple, A., eds. 1980. *Evolution of Vertebrate Endocrine Systems.* Texas Tech Press, Lubbock.

Perry, S.F. 1983. Reptilian lungs: Functional anatomy and evolution. *Adv. Anat. Embryol. Cell Biol.* **79**:1–81.

Randall, D.J., Burggren, W.W., Farrell, A.P., and Haswell, M.S. 1981. *The Evolution of Air Breathing in Vertebrates.* Cambridge University Press, London.

Schummer, A., Nickel, R., and Sack, W.O. 1979. *The Viscera of the Domestic Mammals,* 2nd ed. Springer-Verlag, New York.

Stevens, C.E. 1977. Comparative physiology of the digestive system. In M.J. Swensen, ed., *Dukes' Physiology of Domestic Animals,* 9th ed., pp. 216–232. Cornell University Press, Ithaca, NY.

Young, J.A., and van Lennep, E.W. 1978. *The Morphology of Salivary Glands.* Academic Press, New York.

Chapter 15

Donehoe, P.K., *et al.* 1984. Molecular dissection of mullerian duct regression. In R.L. Trelstad, ed., *The Role of Extracellular Matrix in Development.* Alan R. Liss, New York.

Ferguson, M.W.J., and Joanen, T. 1982. Temperature of egg incubation determines sex in *Alligator mississippiensis. Nature* **296**:850–853.

Josso, N., and Picard, J.-Y. 1986. Anti-mullerian hormone. *Physiol. Rev.* **66**:1038–1090.

Maynard-Smith, J. 1979. *The Evolution of Sex.* Cambridge University Press, London.

Page, D.C., *et al.* 1987. The sex-determining region of the human Y chromosome encodes a finger protein. *Cell* **51**:1091–1104.

Pang, P.K.T., Griffith, R.W., and Atz, J.W. 1977. Osmoregulation in elasmobranchs. *Am. Sci.* **17**:365–377.

Poole, T.J., and Steinberg, M.S. 1984. Different modes of pronephric duct origin among vertebrates. *Scanning Electron Microsc.* **1**:457–482.

Saxen, L., *et al.* 1986. Sequential cell and tissue interactions governing organogenesis of the kidney. *Anat. Embryol.* **175**:1–6.

Segal, S.J. 1985. Sexual differentiation in vertebrates. In Halvorson and Monroy, eds., *The Origin and Evolution of Sex.* pp. 263–270. Alan R. Liss, New York.

Zackson, S.L., and Steinberg, M.S. 1986. Cranial neural crest cells exhibit directed migration on the pronephric duct pathway. Further evidence for an *in vitro* adhesive gradiant. *Dev. Biol.* **117**:342–353.

Chapter 16

Burggren, W.W. 1987. Form and function in reptilian circulations. *Am. Zool.* **27**:5–19.

Golde, D.W., and Gasson, J.C. 1988. Hormones that stimulate the growth of blood cells. *Sci. Am.* July 1988:62–70.

Holmes, E.B. 1975. A reconsideration of the phylogeny of the tetrapod heart. *J. Morphol.* **147**:209–228.

Kirby, M.L. 1987. Cardiac morphogenesis: Recent research advances. *Pediat. Res.* **21**:219–224.

Linask, K.K., and Lash, J.W. 1986. Precardiac cell migration: Fibronectin localization at mesoderm–endoderm interface during directional movement. *Dev. Biol.* **114**:87–101.

Parsons, T.S. 1968. Functional morphology of the heart of vertebrates. *Am. Zool.* **8**:177–229.

Webb, G.J.W. 1979. Comparative cardiac anatomy of the reptilia. III. The heart of crocodilians and an hypothesis on the completion of the interventricular septum of crocodilians and birds. *J. Morphol.* **161**:221–240.

Chapter 17

Arnold, A.P. 1980. Sexual differences in the brain. *Am. Sci.* **68**:165–173.

Eakin, R.M. 1973. *The Third Eye.* University of California Press, Berkeley.

Glickstein, M. 1988. The Discovery of the Visual Cortex. *Sci. Am.* Sept. 1988:118–127.

Hudspeth, A.J. 1985. The cellular basis of hearing: The biophysics of hair cells. *Science* **230**:745–752.

Karfunkel, P. 1972. The activity of microtubules and microfilaments in neurulation in the chick. *J. Exp. Zool.* **181**:289–302.

Levine, J.S. 1985. The vertebrate eye. In M. Hildebrand, *et al.,* eds., *Functional Vertebrate Morphology,* pp. 317–337. Harvard University Press, Cambridge, MA.

Loring, J.F., and Erickson, C.A. 1987. Neural crest cell migration pathways in the trunk of the chick embryo. *Dev. Biol.* **121**:220–236.

Nilsson, S. 1983. *Autonomic Nerve Function in Vertebrates.* Springer-Verlag, New York.

Northcutt, R.G. 1984. Evolution of the vertebrate central nervous system: Patterns and processes. *Am. Zool.* **24**:701–716.

Popper, A.N., and Ray, R.R., eds. 1980. *Comparative Studies of Hearing in Vertebrates.* Springer-Verlag, New York.

Ralph, C.L., *et al.* 1979. The pineal complex and thermoregulation. *Biol. Rev.* **54**:41–72.

Weston, J.A. 1970. The migration and differentiation of neural crest cells. *Adv. Morphog.* **8**:41–114.

Weston, J.A., *et al.* 1984. The role of the extracellular matrix in neural crest development: A reevaluation. In R.L. Trelstad, ed., *The Role of the Extracellular Matrix in Development.* Alan R. Liss, New York.

Glossary

Most biological terms either are common Latin or Greek words or are compounded from Latin and Greek roots. The following list is intended not only to provide identification of the roof sources of the principal terms used in this book, but to indicate how truly descriptive every term is. Moreover, the alert student will shortly discover that many roots are employed again and again in varying combinations. So as the student's familiarity with these roots grows, new terms either take on instant meaning or become challenges for analysis rather than exercises in memory.

Although brief definitions or other identifications are provided herewith, each term is also defined or discussed in the body of the text. And this glossary should not preclude the use of a good medical dictionary or an unabridged Webster.

Abdo′men (L. *abdere*, to hide). That portion of the body cavity containing the greater part of the digestive system.

Abdu′cens (L. *ab*, away, + *ducere*, to lead). Ordinarily with reference to the sixth cranial nerve.

Abduc′tor (L. *abducere*, to draw away). Muscle serving to draw a part away from the axis of the body.

Acanthod′ii (G. *akantho*, spiny, + *eidos*, shape). Group of Paleozoic fishes featuring spiny fins.

Acetab′ulum (L. *acetabulum*, vinegar cup). Cup-shaped depression on the innominate bone, in which the head of the femur fits.

Ac′inus (L. *acinus*, grape). Referring to a minute saclike structure.

Acous′tic (G. *akoustikos*, pertaining to hearing). Relating to hearing or the perception of sound.

Ac′rosome (G. *akron*, tip, + *soma*, body). Anterior extremity of the head of a spermatozoon.

Ac′tinop′terygii (G. *aktis*, a ray, + *pteron*, feather, wing). Extensive group of fishes, having the projecting part of the paired fins supported only by dermal rays.

Adduc′tor (L. *adducere*, to bring toward). Muscle serving to draw a part toward the axis of the body.

Ad′enohypoph′ysis (G. *aden*, gland, + *hypophysis*, an undergrowth). The anterior lobe of the hypophysis (pituitary).

Adre′nal (L. *ad*, to, + *ren*, kidney). Endocrine gland near or upon the kidney.

Adrener′gic (*adrenalin* + G. *ergon*, work). Relating to nerve fibers that liberate adrenalin.

Aes′tivate (L. *aestivare*, to reside during the summer). To enter a state of dormancy because of external stress such as extreme periods of warm weather or lack of food. Certain lungfishes aestivate.

Affer′ent (L. *ad*, to + *ferre*, to bear). Bringing to or into.

Agna′tha (*a-* not, + G. *gnathos*, jaw). Division of craniate vertebrates, comprising those without jaws.

Al′bicans (L. *albicare*, to be white). White in color.

Albu′men (L. *albumen*, white of egg). White of an egg such as that of birds.

Alisphen′oid (L. *ala*, wing, + G. *sphen*, wedge, + *eidos*, form). One of the sphenoid bones; in the human skull, the greater wing of the composite sphenoid bone.

Allan′tois (G. *allas*, sausage, + *eidos*, appearance). Extraembryonic saclike extension of the hindgut of amniotes, serving excretion and respiration.

Allele (G. *allēlōn*, reciprocally). In Mendelian heredity, one of a pair of alternative genetic characters.

Al′losaur′us (G. *allos*, other, + *sauros*, lizard). Large, carnivorous theropod dinosaur of more than 30 feet, so named because it is distinguishable from any known dinosaur by its vertebrae.

Alve′olus (L. *alveolus*, a small chamber, or cavity.) Any small chamber, lobular or pitlike.

Am′ia (G. *amia*, a kind of tunny or mackerel-like fish). Genus of holostean fish containing the living bowfin.

Am′mocoe′tes (G. *ammos*, sand, + *koite*, bed). Larval form of the lamprey; occurs in streams with sandy substrate.

Am′nion (G. *amnion*, fetal membrane). Inner membrane surrounding the embryo.

Amniogen′ic (G. *amnion*, fetal membrane, + *genesis*, origin). Relating to the formation of the amnion.

Amphib′ia (G. *amphibios*, leading a double life). Class of vertebrates, intermediate in many characters between fishes and reptiles, which live part of the time in water and part on land.

Am′phiox′us (G. *amphi*, both, + *oxys*, sharp). Collective name for members of the Cephalochordata. These small, fishlike creatures are "sharp" at both ends, and are often called lancelets.

Ampul′la (L. *ampulla*, a flask). Saccular dilation of a canal.

Anal′ogy (G. *analogos*, conformable). Bearing reference to similarities in function of two organs or parts in different species of animals.

Anam′nio′ta (G. *an*, without, + *amnion*, fetal membrane). Vertebrates that lack an amnion during development.

Anap′sida (*an-* not, + G, *apsis, hapsis, -idos*, juncture, mesh, net). Subclass of reptiles of primitive structure in which the skull lacks temporal openings.

Anastomo′sis (G. *anastomōsis*, cause to communicate). Natural communication between two blood vessels or other parts.

Ang′ular (L. *angulus*, an angle or corner). Relating to a part occupying an angle or corner.

Anky′losaur′us (G. *ankylos*, curved, + *sauros*, lizard, named because of the strongly curved ribs). Comparatively moderately sized, armored, ornithischian dinosaur.

Antag′onist (G. *anti*, against, + *agonizomai*, I fight). That which opposed or resists the action of another.

Anthra′cosaur′ia (G. *anthrax*, coal, + *sauros*, lizard). Order of the Paleozoic labyrinthodont amphibians that contains the ancestors of the reptiles.

An′trum (G. *antron*, a cavity). Any closed or nearly closed cavity.

Anu′ra (*an-* not, + G. *oura*, tail). Amphibia without a tail.

A′nus (L. *anus*, fundament). Rear opening of the digestive tract.

Aor′ta (G. *aeirē*, I lift up). Main trunk of the arterial system.

Ap′atite (G. *apatē*, deceit). Name given to a group of minerals of the general formula $Ca_{10}F_2(PO_4)_6$.

Apo′da (G. *apodos*, footless). Any of several different groups of animals that lack limbs or feet. Refers especially to one group of Amphibia.

Appendic′ular (L. *appendere*, to hang upon). Relating to appendages or limbs.

Ap′sidospon′dyli (G. *hapsis*, union, + *spondylos*, vertebra). Amphibia with vertebrae assembled from two parts.

Aq′ueduct (L. *aquaeductus*, a conduit). Relating to a canal or passageway.

Ar′chaeop′teryx (G. *archaios*, ancient, + *pteryx*, wing). First known bird from the Jurassic Period.

Ar′chaeor′nithes (G. *archē*, beginning, + *ornithes*, birds). The original birds.

Archenceph′alon (G. *archē*, beginning, + *enkephalos*, brain). Anterior division of the primitive brain.

Archen′teron (G. *archē*, beginning, + *enteron*, intestine). Primitive, or embryonic, digestive tube.

Archipal′lium (G. *archi*, first, + *pallium*, cloak). Olfactory cortex.

Archosau′ria (G. *archi*, old, + *sauros*, lizard). Subclass of the reptiles that includes the ruling reptiles of the Mesozoic, including the dinosaurs and their allies.

Ar′cual (L. *arcua*, a bow). Relating to an arch.

Arcua′lia (L. *arcua*, a bow). Primordial elements from which a vertebra is formed.

Ar′tery (L. from G. *arteria*, an air conveyer; the arteries were once believed to be air tubes). Blood vessel that transports blood away from the heart.

Artic′ular (L. *articulaire*, to connect). Relating to a joint.

Ar′tiodac′tyla (G. *artios*, even in number, + *dakylos*, finger). Division of the ungulate or hoofed animals having toes even in number, two or four.

At′las (G. *Atlas*, mythological Titan who supported earth on his shoulders). First cervical vertebra.

A′trium (L. *atrium*, antechamber, entrance hall). Usually refers to heart chamber, but also applied to other organs.

Au′ditory (L. *audire*, to hear). Relating to the perception of sound.

Au′ricle (L. dim. of *auris*, ear). Relating to the external ear or a receiving chamber of the heart.

Autonom′ic (G. *autos*, self, + *nomos*, law). Self-controlling, independent of outside influences.

Autosty′lic (G. *autos*, self, + *stylos*, pillar). Type of jaw suspension in which the jaws articulate directly with the cranium.

A′ves (L. pl. of *avis*, a bird). Class of birds.

Ax′ial (L. *axis*, axle of a wheel). Relating to the central part of the body as distinguished from the appendages.

Ax′on (G. *axōn*, axis). Principal process of a nerve cell.

Az′ygous (G. *a*, not, + *zygos*, yoke). Unpaired anatomical structure.

Ba′leen (L. *balaena*, a whale). Hard keratinous material hanging in the mouth in toothless whales that serves to filter plankton from sea water.

Bap′tor′nis (G. *baptos*, dipped under water, + *ornis*, bird). Diving toothed bird of the Cretaceous Period.

Bar′bule (L. *barbula*, beard). Minute barb or fringe.

Ba′sal (L. *basis*, footing or base). Relating to a base.

Ba′si- (L. *basis*, base). Prefix pertaining to the base.

Basibran′chial (L. *basis*, base, + *branchia*, gill). Median-ventral component of a gill arch.

Basihy′oid (L. *basis*, base, + G. *hyoeides*, Y-shaped). Median-ventral component of the hyoid arch.

Basioccip′ital (L. *bases*, base, + *occipitis*, back of the head). Relating to the basal component of the occipital bone.

Basisphe′noid (L. *basis*, base, + G. *sphen*, + *eidos*, likeness). Basal member of the series of sphenoid bones.

Batoid′ea (G. *batis*, a skate or ray). Order of the class Chondrichthyes containing the skates and rays.

Bicus′pid (L. *bi*, two, + *cuspis*, point). Having two points or prongs.

Bigem′inal (L. *bi*, two, + *geminus*, twin). Referring to a doubling or twinning.

Bile (L. *bilis*, gall). Fluid secreted by the liver.

Blaste′ma (G. *blastema*, a sprout). Primitive cell aggregate from which an organ develops.

Blas′tocoel (G. *blastos*, germ, + *koilos*, hollow). Cavity in the blastula.

Blas′tocyst (G. *blastos*, germ, + *kystis*, bladder). Hollow sphere of early mammalian development.

Blas′toderm (G. *blastos*, germ, + *derma*, skin). Primitive cellular plate at the beginning of embryogeny.

Blas′todisc (G. *blastos*, germ, + L. *discus*, plate). Plate of cytoplasm at the animal pole of the ovum.

Blas′tomere (G. *blastos*, germ, + *meros*, part). One of the cells into which the egg divides during the cleavage phase of development.

Blas′topore (G. *blastos*, germ, + *poros*, opening). Opening into the gastrocoel.

Blas′tula (G. *blastos*, germ). Early embryo, the cells of which are commonly arranged in the form of a hollow sphere.

Bra′chial (I. *brachialis*, arm). Relating to the arm.

Bran′chial (G. *branchia*, gill). Relating to gills.

Branchiomer′ism (G. *branchia*, gill, + *meros*, part). Segmentation corresponding to the visceral arches.

Branchio′stoma (G. *branchia*, gill, + *stoma*, mouth). Genus of lancelet or amphioxus superseding the old generic name *Amphioxus*.

Bron′chus (G. *bronchos*, windpipe). One of the two branches given off the trachea and leading to a lung.

Bron′tosaur′us (G. *brontos*, thunder, + *sauros*, lizard). Tremendous sauropod dinosaur.

Buc′cal (L. *bucca*, cheek). Relating to the mouth; the surface toward a cheek.

Bul′bous (L. *bulbus*, a bulbous root). Relating to any structure of globular form.

Cae′cum (L. *caecus*, blind). Any structure ending in a cul-de-sac.

Calcito′nin (L. *calx*, lime, + G. *tonos*, tone). Referring to the level of blood plasma calcium.

Callo′sum (L. *callosus*, hard). Relating to something thick or hard.

Cam′brian (named for the noun referring to an inhabitant of Cambria; a Welshman). The first geologic period of the Paleozoic Era; the period in which the first vertebrates appeared.

Can′cellous (L. *cancellus*, grating or latticework). Relating to spongy bone.

Ca′nine (L. *caninus*, relating to a dog). Ordinarily refers to a tooth type, but there are other references as well.

Cap′illary (L. *capillaris*, relating to hair). Microscopic blood vessel intermediate to arteries and veins.

Capit′ulum (L. dim. of *caput*, head). Small head or rounded extremity of a bone.

Cap′sula (L. *capsa*, a chest). Capsule in any sense.

Car′bonif′erous (L. *carbon*, charcoal + *ferous*, bearing or yielding). Geologic period in the late Paleozoic characterized by the adaptive radiation of amphibians and large deposits of organic carbon (coal and oil) from plants.

Car′diac (G. *kardia*, heart). Relating to the heart.

Car′dinal (L. *cardinalis*, hinging, important). Of special importance.

Car′inate (L. *carina*, the keel of a ship). Term referring to all birds that possess a keeled sternum; all birds exclusive of the ratites.

Carniv′ora (L. *caro*, *carnis*, flesh, + *vorare*, to devour). Order of flesh-eating mammals that possess teeth and claws adapted for attacking and devouring prey.

Car′otenoid (L. *carota*, carrot, + G. *eidos*, like). Referring to a group of pigments having a yellow color.

Carot′id (G. *karoo*, cause to sleep profoundly). Relating to blood vessels supplying the head, compression of which produces unconscious state.

Car′pus (G. *karpos*, wrist). Collection of bones constituting the wrist.

Car′tilage (L. *cartilago*, gristle). Gristlike skeletal tissue.

Cau′dal (L. *cauda*, tail). Relating to the tail or rear.

Caverno′sus (L. *caverna*, cave). Cavern or cavity.

Ca′vum (L. *cavum*, hollow). Any hollow, hole, or cavity.

Cell (L. *cella*, a small chamber). Structural and functional unit of plant and animal organization.

Cen′ozo′ic (G. *kainos*, new or recent, + *zoe*, life). Current geologic era, following the Mesozoic, known as the age of mammals.

Cen′triole (G. *kentron*, a point). Central granule in a centrosome.

Cen′trum (L. *centrum*, center). Central structure of any kind.

Cephal'ic (G. *kephale*, head). Relating to the head.

Ceph'alochorda'ta (G. *kephale*, head, + *chorde*, string, cord). Subphylum of Chordata in which the notochord extends into the head.

Cer'ato- (G. *keras*, horn). Prefix denoting composition of horny substance or resembling a horn in shape.

Cer'atop'sia (G. *keras* or *keratos*, horn, + *ops*, face). Horned, herbivorous, ornithischian dinosaurs.

Cer'ebel'lum (L. dim. of *cerebrum*, brain). Brain mass deriving from the roof of the metencephalon.

Cere'brum (L. *cerebrum*, brain). Brain mass deriving from the roof of the telencephalon.

Cer'vical (L. *cervix*, neck). Relating to the neck.

Ceta'cea (L. *cetus*, whale). Order of completely aquatic, mostly marine, mammals of the subclass Eutheria, including the whales, dolphins, and porpoises.

Chala'za (G. *chalaza*, a sty). Suspensory ligament of the yolk in a bird's egg.

Chelo'nia (G. *chelone*, a tortoise). Order of reptiles comprising the tortoises of turtles.

Chev'ron (Fr. *chevron*, rafter). Shaped like a gable; V-shaped bones in the vertebral column.

Chias'ma (G. *chiasmaa*, figure of X). Describing the crossing of fibers.

Chirop'tera (G. *cheir*, hand, + *pteron*, wing). Order of placental mammals, characterized by a modification of the forelimbs that enables them to fly, for example, bats.

Cho'anich'thyes (G. *choane*, funnel, + *ichthyes*, fish). Group of fishes with nostrils opening into the mouth.

Cholecystoki'nin (G. *cholē*, bile, + *kystis*, bladder, + *kineo*, move). Hormone liberated by the intestinal mucosa causing contraction of the gallbladder.

Choliner'gic (G. *cholē*, bile, + *ergon*, work). Relating to nerve fibers that liberate acetylcholine.

Chon'drich'thyes (G. *chondros*, cartilage, + *ichthyes*, fish). Fishes with a cartilaginous skeleton, for example, sharks.

Chon'dro- (G. *chondros*, cartilage). Combining form meaning cartilaginous.

Chon'droblast (G. *chondros*, cartilage, + *blastos*, germ). Embryonic cartilage-forming cell.

Chon'droclast (G. *chondros*, cartilage, + *klastos*, broken in pieces). Cell concerned in the destruction of cartilage.

Chondrocra'nium (G. *chondros*, cartilage, + *kranion*, skull). Cartilaginous skull.

Chon'drocyte (G. *chondros*, cartilage, + *kytos*, cell). Cartilage cell.

Chondros'tei (G. *chondros*, cartilage, + *osteon*, bone). Order of fishes characterized by a largely cartilaginous skeleton.

Chorda'ta (L. *chorda* fr. G. *chorde*, a string). Phylum of animals with a notochord, transient or persistent, that is, in the adult or at a stage in their development.

Cho'rion (G. *chorion*, a skin). Outer membrane surrounding the embryo.

Cho'roid (G. *chorion*, skin, + *eidos*, like). Relating to a coat or membrane.

Chro'matophore (G. *chroma*, color, + *phorus*, bearer). Pigment-containing cell.

Chro'mosome (G. *chroma*, color, + *soma*, body). Deeply staining rod-shaped or threadlike body in the cell nucleus; bearer of genes.

Cil'iary (L. *ciliaris*, resembling an eyelash). Relating to cilia or hairlike processes.

Clav'icle (L. *clavicula*, a small key). Collarbone.

Cleav'age (Ger. *kleiben*, to cleave). Denoting division or splitting.

Cleido'ic (G. *kleidō*, lock up). Referring to eggs, such as those of birds, that are insulated from the environment by albumen, membranes, and shell.

Glei'thrum (G. *kleithron*, a closing bar). Bone external to and beside the clavicle in the pectoral girdle.

Cli'toris (G. *kleitoris*). Organ composed of erectile tissue; the homologue in the female of the penis.

Cloa'ca (L. *cloaca*, sewer). Combined urogenital and rectal receptacle.

Coch'lea (L. *cochlea*, snail shell). Spiral canal and its contents in the inner ear.

Coe'lacan'thini (G. *koilos*, hollow, + *akantha*, spine). Group of crossopterygian fishes.

Coe'liac (G. *koilia*, belly). Relating to the abdomen.

Coe'lom (G. *koiloma*, a hollow). Cavity bounded by mesodermal epithelium.

Coelu'rosaur (G. *koilos*, hollow, + *oura*, tail, + *sauros*, lizard). Small, carnivorous therapod dinosaurs thought by many to have given rise to birds.

Col'lagen (G. *kolla*, glue, + *gennao*, I produce). An albuminoid present in connective tissues.

Collat'eral (L. *con*, together, + *lateralis*, relating to the side). Accompanying or running by or from the side of.

Col'lum (L. *collum* [gen. *colli*], the neck). Relating to the neck or a collar.

Columel'la (L. dim. of *columna*, column). A small column. Usually refers to a bone in the middle ear.

Com'missure (L. *commissura*, connection). Passing from one side to the other.

Comp'sogna'thus (G. *kompsos*, elegant or ornate, + *gnathos*, jaw). Small coelurosaurian dinosaur possibly close to the ancestry of birds.

Conduc'tor (L. *conducere*, to lead). That which conveys or transmits.

Conjuncti'va (L. fem. *conjunctivus*, to connect). Mucous membrane covering the front surface of the eye and lining the eyelids.

Constric'tor (L. *constringere*, to draw together). Anything that binds or squeezes a part.

Cop'ula (L. *copula*, yoke, joining). Relating to a part connecting two structures.

Cor'acoid (G. *korax*, raven, + *eidos*, appearance). Shaped like a crow's beak.

Cor'dis (L. *chorda*, a string). Relating to any cordlike structure.

Cor'nea (L. *corneus*, horny). The transparent front portion of the eye.

Cor'neum (L. *corneus*, horny). Relating to a horny layer.

Cor'nu (L. *cornus*, pl. *cornua*, a horn). Horn-shaped anatomical structure.

Cor'onary (L. *coronarius*, a crown; encircling). Denoting various encircling anatomical parts.

Cor'onoid (G. *korone*, a crow, + *eidos*, resembling). A process shaped like a crow's beak.

Cor'pus (L. *corpus*, the body). Any body or mass.

Cor'tex (L. *cortex*, bark, rind). Outer portion of an organ or part.

Cos'moid (G. *kosmos*, orderly arrangement, + *eidos*, form). Relating to formal pattern.

Cos'tal (L. *costa*, rib). Relating to a rib.

Cotyle'donary (G. *kotyledon*, the hollow of a cup). Referring to aggregations of villi on the chorionic surface of a placenta.

Cot'ylosauria (G. *kotyle*, cup, + *sauria*, lizard). Order of late Paleozoic and early Triassic Anapsida comprising the most primitive reptiles having short legs and massive bodies.

Cra'nium (G. *kranion*, skull). Bones of the head collectively.

Creta'ceous (L. *creta*, chalk). Last geologic period of the Mesozoic Era.

Cris'ta (L. *crista*, a crest). Outwardly projecting fold or ridge of a membrane.

Crocodil'ia (L. *crocodilus*, crocodile). Order of reptiles including the crocodiles, gavials, alligators, and related extinct forms.

Crossop'terygii (G. *krossoi*, fringe, + *pterygion*, wing). Fishes ancestral to terrestrial vertebrates.

Cten'oid (G. *kteis*, a comb, + *eidos*, form). Type of scale featuring toothlike projections.

Cu'mulus (L. *cumulus*, a heap). Collection or heap of cells.

Cuta'neous (L. *cutaneous*, skin). Relating to the skin.

Cu'ticle (L. *cuticula*, dim. of *cutis*, skin). Outer horny layer of the skin.

Cy'cloid (G. *kyklos*, circle, + *eidos*, form). Type of scale featuring circular growth rings.

Cyclostoma'ta (G. *kyklos*, circle, + *stoma*, mouth). Fishlike forms, possessing a cartilaginous skeleton, a circular mouth without jaws, and dorsal and caudal fins but no paired fins.

Cy'nogna'thus (G. *kunos*, dog, + *gnathos*, jaw). Dog-sized mammallike reptile or therapsid.

Cys'tic (G. *kystis*, bladder or cyst). Relating to a bladder or cyst.

Cy'to- (G. *kytos*, cell). Combining form indicating cell.

Cytotro'pholast (G. *kytos*, cell + *trophē*, nourishment, + *blastos*, germ). Inner or cellular layer of the trophoblast.

Decid'ua (L. *deciduus*, falling off). Endometrium of the pregnant uterus.

Def'erens (L. *de*, away, + *ferens*, carrying). Applying to a duct carrying away.

Den'drite (G. *dendritēs*, relating to a tree). One of the receiving, branching processes of a nerve cell.

Den'tal (L. *dens*, a tooth). Relating to the teeth.

Den'tary (L. *dens*, a tooth). Tooth-bearing bone, specifically in the lower jaw.

Den'ticle (L. *dentriculus*, a small tooth). Projection from a hard surface.

Den'tine (L. *dens,* a tooth). Substance proper of a tooth.

Depres'sor (L. *depressus,* to press down). Muscle serving to lower or pull down a part.

Dermat'ocranium (G. *derma,* skin, + *kranion,* skull). Collectively the superficial bones of the skull without cartilaginous precursors.

Der'matome (G. *derma,* skin, + *tome,* cut, section). One of the embryonic skin segments.

Der'mis (G. *derma,* skin). Inner, or lower, layer of the skin.

Dermop'tera (G. *derma,* skin, + *pteron,* wing). Flying lemurs.

Deu'teroenceph'alon (G. *deuteros,* second, + *enkephalos,* brain). Posterior division of the primitive brain.

Devo'nian (named for Devonshire, England, where the strata were initially studied). Geologic period of the Paleozoic Era called the "Age of Fishes" during which most fish groups radiated and the first amphibians appeared.

Di'aphragm (G. *diaphragma,* a partition). Partition between the abdominal and thoracic cavities.

Diaph'ysis (G. *diaphysis,* a growing through). Shaft of a long bone.

Diapoph'ysis (G. *dia,* through, + *apophysis,* an offshoot). Upper articular surface of the transverse process of a vertebra.

Diap'sida (G. *di,* double, + *apsis,* arch). Reptiles with two temporal openings on each side.

Di'atry'ma (G. *dia,* "through," + *tryma,* a hole, referring to large foramina that penetrate the bones). Large flesh-eating Eocene bird commonly termed terror crane.

Di'enceph'alon (G. *dia,* through, + *enkephalos,* brain). Posterior of the two subdivisions of the primitive forebrain.

Dig'it (L. *digitus,* finger). Finger or toe.

Dila'tor (L. *dilatare,* to expand). Muscle whose function is to pull open an orifice.

Di'nosaur (G. *deinos,* terrible, + *sauros,* lizard). Dominant group of prehistoric reptiles.

Diplo'docus (G. *diploos,* double, + *dokos,* beam; named for the bifid spines on the vertebrae). One of the largest of the sauropod dinosaurs.

Dip'loid (G. *diploos,* double, + *eidos,* resemblance). Full number of chromosomes in the fertilized ovum and in all cells, except the mature germ cells.

Dip'noi (G. *dipnoos,* with two breathing apertures). A group of fishes that have, in addition to gills functioning in the usual manner, a lung or pair of lungs.

Discoid'al (G. *diskos,* disc + *eidos,* appearance). Resembling a disc or plate.

Dis'tal (L. *distalis,* distant). Farthest from the center or median line.

Dor'sal (L. *dorsalis,* back). Relating to the top or back.

Dor'sum (L. *dorsum* [gen. *dorsi*], back). Upper or back of any part.

Duc'tus (L. *ductus,* to lead). Tubular structure giving exit to or conducting any fluid.

Duode'num (L. *duodeni,* twelve). First division of the small intestine.

Ec'toderm (G. *ektos,* outside, + *derma,* skin). Outermost of the three primary germ layers.

Ec'totherm (G. *ectos,* outside, + *thermos,* heat). Cold-blooded vertebrate that derives its major heat from outside the body, from the ambient environment.

Edenta'ta (L. *endentalus,* toothless). Order of placental mammals without incisor teeth; the other teeth are poorly developed and devoid of enamel.

Effec'tor (L. *efficere,* to bring to pass). Organ that reacts by movement, secretion, or electrical discharge.

Ef'ferent (L. *ex,* out, + *ferre,* to bear). Conducting outward or centrifugally.

Elas'mobran'chii (G. *elasmos,* plate, + *branchia,* gills). Group of fishes comprising the modern sharks and the rays and their extinct allies.

Em'bolom'erous (G. *embolos,* wedge, + *meros,* part). Referring to vertebrae with two-part centra.

Embryo (G. *embryon,* embryo). Animal in an early stage of growth while in the egg or maternal body; denoted in humans from implantation to 8 weeks after conception.

Embryog'eny (G. *embryon,* embryo, + *gennao,* I produce). Origin and growth of the embryo.

Embryol'ogy (G. *embryon,* embryo, + *-logia,* discourse). Study of the origin and development of the embryo.

En'docar'dium (G. *endon,* within, + *kardia,* heart). Lining of the cavities of the heart.

Endochon'dral (G. *endon,* within, + *chondros,* cartilage). Within a cartilage or cartilaginous tissue.

En'docrine (G. *endon*, within, + *krinein*, separate). Denoting internal secretion.

En'doderm (G. *endon*, within, + *derma*, skin). Innermost of the three primary germ layers.

En'dolymph (G. *endom*, within, + L. *lympha*, water). Fluid contained within the membranous labyrinth of the inner ear.

Endome'trium (G. *endon*, within, + *metra*, uterus). Mucous membrane lining the uterus.

Endomys'ium (G. *endon*, within, + *mys*, muscle). Connective tissue sheathing of a muscle fiber.

Endoskel'eton (G. *endon*, within, + *skeletons*, dried up). Internal framework of the body.

En'dostyle (G. *endon*, within, + *stylos*, pilar). The subpharyngeal gland of the lower chordates that secretes mucus for food entrapment; homologous to the thyroid gland of vertebrates.

Endothe'liochor'ial (G. *endon*, within, + *thēlē*, nipple, + *chorion*, skin). Placenta type involving contact between chorionic epithelium and uterine blood vessels.

Endothe'lium (G *endon*, within, + *thēlē*, nipple). Layer of flat cells lining serous cavities and blood vessels.

En'dotherm (G. *endo*, within, + *thermos*, heat). Warm-blooded vertebrate that derives the majority of its body heat from internal thermogenesis.

En'terocoel (G. *enteron*, intestine + *koilos*, hollow). Coelom originally in communication with the lumen of the gut.

En'terogas'trone (G. *enteron*, intestine, + *gaster*, belly). Hormone from intestinal mucosa that inhibits gastric secretion and motility.

En'teropneus'ta (G. *enteron*, intestine, + *pneustos*, to breathe). Order of hemichordate "worms."

E'ocene (G. *eos*, dawn + *kainos*, new). Second epoch of the Tertiary Period of the Cenozoic Era.

Epen'dymal (G. *ependyma*, an outer garment). Referring to the layer of cells lining the central canal of the spinal cord.

Ep'i- (G. *epi*, upon). Prefix denoting above or on top.

Ep'iblast (G. *epi*, upon, + *blastos*, germ). Outer or upper layer of a double-layered embryo.

Epicar'dium (G. *epi*, upon, + *kardia*, heart). Visceral peritoneum enveloping the heart.

Epider'mis (G. *epi*, upon, + *derma*, skin). Outer epithelial portion of the skin.

Epidid'ymis (G. *epi*, upon, + *didymos*, twin). First, convoluted, portion of the excretory duct of the testis, passing from above downward along the posterior border of this gland.

Epiglot'tis (G. *epi*, upon, + *glotta*, tongue). Structure that folds back over the aperture (glottis) of the larynx.

Ep'imere (G. *epi*, upon, + *meris*, part). Dorsal region of the mesoderm on each side of the neural tube.

Epimys'ium (G. *epi*, upon, + *mys*, muscle). Fibrous envelope surrounding a muscle.

Epiph'ysis (G. *epi*, upon, + *physis*, growth). Separately developed terminal ossification of a long bone; the pineal body.

Epithal'amus (G. *epi*, upon, + *thalamos*, chamber). Dorsal portion of the thalamus of the brain.

Epithe'liochor'ial (G. *epi*, upon, + *thēlē*, nipple, + *chorion*, skin). Placenta type involving contact between chorionic epithelium and uterine epithelium.

Epithe'lium (G. *epi*, upon, + *thēlē*, nipple). Cellular layer covering a free surface.

Eryth'rocyte (G. *erythros*, red, + *kytos*, cell). Red blood corpuscle.

Er'ythrophore (G. *erythros*, red, + *phoros*, carrying). Cell containing red pigment.

Esoph'agus (G. *oisō*, I shall carry, + *phagēton*, food). Portion of the gut between the pharynx and the stomach.

Es'trogen (G. *oistros*, mad desire, + *gennao*, produce). Substance that produces estrus.

Es'trus (G. *oistros*, mad desire). Period of sexual excitement in the female.

Eth'moid (G. *ethmos*, sieve, + *eidos*, form). Resembling a sieve; relating to the ethmoid bone.

Euryap'sida (G *eurys*, broad, + *apsis*, arch). Reptiles with single temporal opening on each side, bounded below by postorbital and squamosal bones.

Euryp'terid (G. *eurys*, broad, + *pteron*, wing). Name given to the large predaceous, aquatic arthropods of the Paleozoic that were contemporaneous with the first fishes.

Eus'thenop'teron (G. *eusthenes*, stout or strong, + *pteron*, wing, in reference to the large, fleshy fin). Genus of rhipidistian crossopterygian fish close to the ancestry of the Amphibia.

Euther'ia (G. *eu*, well, + *therion*, a wild beast). Subclass of Mammalia embracing the placental mammals.

Ev'olu'tion (L. *evolutus*, *evolvere*, to roll out).

Doctrine maintaining that all forms of animal or plant life have been derived by gradual changes from simpler forms.

Ex′ocrine (G. *exō*, out, + *krinō*, I separate). Denoting external secretion of a gland.

Exten′sor (L. *extendere*, to stretch′out). Muscle serving to straighten a part.

Extrin′sic (L. *extrinsecus*, on the outside). Originating outside of the part on which it acts.

Fac′et (Fr. *facette*, face). Small smooth area on a bone or other firm structure.

Fa′cial (L. *facies*, face). Relating to the face.

Fal′ciform (L. *falx*, sickle, + form). Having a curved or sickle shape.

Fas′cia (L. *fascia*, a band or fillet). Fibrous tissue investing an organ or area.

Fe′mur. Thigh bone.

Fer′tilization (L. *fertilis*, to bear). Union of ovum and spermatozoon.

Fe′tus (L. *fetus*, fruitful). Unborn or unhatched vertebrate; in humans, from 2 months after conception to birth; before 8 weeks it is called an embryo.

Fib′ula (L. *fibula*, a brooch). External and smaller of the two bones of the hindleg.

Fis′sipe′dia (L. *fissus*, from *findere*, to cleave, + *pedis*, foot). Suborder of the Carnivora containing the terestrial carnivores.

Flex′or (L. *flectere*, to bend). Muscle serving to bend a joint.

Fol′licle (L. *folliculus*, a small bag). Vesicular body in the ovary containing the ovum; any crypt or circumscribed space.

Fora′men (L. *foramen*, an aperture). Aperture or perforation through a bone or membrane.

Fos′sa (L. *fossa*, trench or ditch). Depression below the level of the surface of a part.

Fo′vea (L. *fovea*, a pit). Any depression or pit.

Frondo′sum (L. *frons*, branch or leaf). Shaggy surface.

Fron′tal (L. *frons*, brow). Relating to the front or forehead.

Funic′ulus (L. dim. of *funis*, cord). Small, cord-like structure such as a bundle of nerve fibers.

Fur′cula (L. *furca*, forked). Fused clavicles that form the V-shaped "wishbone" in birds.

Gam′ete (G. *gametes*, husband; *gamete*, wife). Germ cell, ovum or spermatozoon.

Gam′etogen′esis (G. *gametes, gamete*, husband, wife, + *genesis*, origin). Production of gametes.

Gam′etogo′nium (G. *gametes, gamete*, husband, wife, + *gonus*, a begetting). Primordial reproductive cell that produces gametes.

Gang′lion (G. *ganglion*, swelling under the skin). Aggregation of nerve cells along the course of a nerve.

Gas′ter (G. *gaster*, belly). General reference to the stomach; the main body of a muscle.

Gastra′lia (G. *gaster*, belly). Special reference to skeletal parts supporting the abdomen.

Gas′trin (G. *gaster*, belly). Hormone that is produced in pyloric division of the stomach and excites secretion by glands of cardiac stomach.

Gas′trocoel (G. *gaster*, stomach, + *koilos*, hollow). Cavity within the gastrula.

Gas′trula (L. dim. of G. *gaster*, belly). Embryo in the stage of development following the blastula, consisting of a sac with a double wall.

Gene (L. *gennao*, I produce). Hereditary unit located in a chromosome.

Gen′ital (L. *genitalis*, pertaining to birth). Related to reproduction or generation.

Germinati′vum (L. *germen*, bud, germ). Relating to multiplication, development, or growth.

Glans (L. *glans*, acorn). Mass capping the body of a part.

Glen′oid (G. *glene*, a socket). Resembling a socket.

Gli′a (G. *glia*, glue). Supporting and binding substance.

Glob′ulin (L. *globulus*, globule). Simple protein present in blood, milk, and muscle.

Glomer′ulus (L. *glomus*, a skein). Tuft formed of capillary loops at the beginning of each uriniferous tubule in the kidney.

Gloss′al (G. *glossa*, tongue). Relating to the tongue.

Glos′sopharyn′geal (G. *glossa*, tongue, + *pharyngis*, throat). Relating to the tongue and pharynx, specifically to the ninth cranial nerve.

Glot′tis (G. *glottis*, aperture of the larynx). Opening to the larynx and trachea.

Glu′cagon (G. *glykys*, sweet). One of the pancreatic hormones concerned with regulating blood sugar.

Gna′thosto′mata (G. *gnathos*, jaw + *stoma*, mouth). Vertebrates possessing jaws.

Gon′ad (G. *gone*, seed). Reproductive gland, ovary or testis.

Gona′dotropic (G. *gone*, seed, + *trophē*, nourishment). Referring to a hormone that influences the gonads.

Granulo′sa (L. *granulosus*, full of grains). Layer or region having a granular appearance.

Gy′rus (G. *gyros*, circle). Rounded elevation on the surface of the brain.

Haben′ular (L. *habenula*, a strap). Relating particularly to the stalk of the pineal body.

Had′rosaurs (G. *hadros*, bulky, + *sauros*, lizard). Aquatic, duck-billed dinosaurs of the order Ornithischia.

Hap′loid (G. *haplous*, simple, + *eidos*, resemblance). Reduced number of chromosomes in the gamete.

Haver′sian (named for English anatomist, Havers). Relating to bone structure described by Havers.

He′mal (G. *haima*, blood). Relating to the blood or blood vessels.

Hem′i- (G. *hemi*, half). Prefix signifying one half.

Hem′ichorda′ta (G. *hemi*, half, + *chorde*, string). Phylum of animals closely related to echinoderms and chordates.

Hem′isphere (G. *hemi*, half, + *sphaira*, ball). Lateral half of the cerebrum or cerebellum.

Hem′ochor′ial (G. *haima*, blood, + *chorion*, skin). Placenta type involving contact between chorionic epithelium and maternal blood.

He′moglo′bin (G. *haima*, blood, + L. *globus*, globe). Respiratory pigment of the blood.

Hepat′ic (G. *hepar, kepatikos*, relating to the liver). Relating to the liver.

Hermaph′rodi′tism (G. *hermaphroditos*, son of Hermes/Mercury, + Aphrodite/Venus). Seeming occurrence of both male and female generative organs in the same individual.

Hes′peror′nis (G. *hesperos*, western, + *ornis*, bird, so named because of its discovery in western United States). Cretaceous toothed bird that superficially resembled a loon.

Het′erocer′cal (G. *heteros*, different, + *kerkos*, tail). Type of tail characteristic of sharks in which the upper lobe is larger and has caudal vertebrae extending into it.

Het′erodont (G. *heteros*, different, + *odont*, tooth). Having teeth of varying shape.

Hippocam′pus (G. *hippocampus*, sea horse). Elevation on the floor of the lateral ventricle of the brain.

Histogen′esis (G. *histos*, tissue, + *genesis*, origin). Origin of a tissue; formation and development of the tissues of the body.

Hol′o- (G. *holos*, entire). Prefix, signifying entire or total.

Hol′oblas′tic (G. *holos*, whole, + *blastos*, germ). Denoting the involvement of the entire egg in cleavage.

Hol′obranch (G. *holos*, complete, + *branchion*, gill). Complete gill with filaments on each side of the supporting arch.

Hol′oceph′ali (G. *holos*, entire, + *kephale*, head). Small group of fishes included in the Chondrichthyes.

Holoneph′ros (G. *holos*, entire, + *nephros*, kidney). Single rather than multiple kidneys.

Holos′tei (G. *holos*, entire, + *osteon*, bone). Group of fishes having a well-developed bony skeleton and approaching teleosts in structure.

Homeos′tasis (G. *homoios*, similar, + *stasis*, standing still). Condition of bodily equilibrium with regard to temperature, fluid content, and so on.

Ho′mocer′cal (G. *homos*, same, + *kerkos*, tail). Type of tail characteristic of most modern bony fishes in which each lobe is about the same size.

Ho′modont (G. *homos*, same, + *odont*, tooth). Having teeth all alike in form.

Homoi′otherm (G. *homos*, same, + *thermos*, heat). Organism that maintains a constant body temperature.

Homol′ogy (G. *homos*, same, + *logos*, ratio). Similarity in structure and origin of two organs or parts in different species of animals.

Ho′motypy (G. *homos*, same, + *typos*, type). Correspondence in structure between two parts or organs in one individual.

Hor′mone (G. *hormon*, I rouse). Chemical substance formed in one organ or part and carried in the blood to another organ or part which it stimulates to functional activity.

Hu′merus (L. *humerus*, shoulder). Bone of the upper arm.

Hu′mor (L. *humor*, fluid). Any fluid or semifluid substance.

Hy′alin (G. *hyalos*, glass). Of a glassy, translucent appearance.

Hy′oid (G. *hyoeides*, U-shaped). Second visceral arch; tongue bone.

Hy′omandib′ula (G. *hyoeides*, U-shaped, + L.

mandibular, jaw). Uppermost segment of the hyoid arch.

Hyosty′lic (G. *hyoeides,* U-shaped, + *stylos,* pillar). Type of jaw suspension wherein the hyomandibula is inserted between the jaws and cranium.

Hy′per- (G. *hyper,* above). Prefix denoting excessive or above normal.

Hy′po- (G. *hypo,* under). Prefix denoting beneath something else or a diminution or deficiency.

Hy′poblast (G. *hypo,* under, + *blastos,* germ). Inner or lower layer of a double-layered embryo.

Hypocen′trum (G. *hypo,* beneath, + L. *centrum,* center). Ventral vertebral centrum.

Hy′poglos′sal (G. *hypo,* beneath, + *glossa,* tongue). Beneath the tongue, usually denoting 12th cranial nerve.

Hy′pomere (G. *hypo,* under, + *meris,* part). Most ventral subdivision of mesoderm.

Hypoph′ysis (G. *hypophysis,* an undergrowth). Endocrine gland lying at the base of the brain.

Hypothal′amus (G. *hypo,* under, + *thalamos,* chamber). Ventral portion of the thalamus of the brain.

Hy′racoi′dea (G. *hyrax,* shrew mouse). Small order of African mammals (conies) thought to be closely related to the elephants.

Ich′thyoptery′gia (G. *ichthyos,* fish, + *pterygion,* fin). Group of extinct marine reptiles with fishlike bodies.

Ich′thyor′nis (G. *ichthys,* fish, + *ornithos,* bird). Cretaceous toothed bird, similar superficially to living gulls and terns.

Ich′thyosau′ria (G. *ichthys,* fish, + *sauria,* lizard). Group of Mesozoic marine fishlike reptiles.

Ich′thyoste′ga (G. *ichthys,* fish, + *stege,* roof). Genus of the first known amphibian from the Devonian Period.

Ich′thyostegal′ia (G. *ichthys,* fish, + *stege,* roof). Order containing the first known amphibians from the Devonian Period.

Il′ium (L. *ilium,* flank). Dorsalmost element of the pelvic girdle.

Im′par (L. *impar,* unpaired). Relating to a single or unpaired part.

Inci′sor (L. *incidere,* to cut into). One of the cutting teeth.

In′cus (L. *incus,* anvil). Middle one of the three middle ear bones.

In′fra- (L. *infra,* below). Prefix denoting a position below a designated part.

Infundib′ulum (L. *infundibulum,* funnel). Funnel-shaped structure or passage.

Innom′inate (L. *in-,* negative, + *nomen,* name). "Nameless" in reference to blood vessels and hip bone.

Insectiv′ora (L. *insectum,* insect, + *vorare,* to devour). Order of placental mammals whose chief food consists of insects; embraces the moles and shrews.

In′sulin (L. *insula,* island). One of the pancreatic hormones concerned with regulation of blood sugar.

Integ′ument (L. *integumentum,* covering). Membrane (skin) covering the body.

In′ter- (L. *inter,* between). Prefix meaning between or among.

Intercal′ary (L. *intercalare,* to insert). Occurring between two others.

Intersti′tial (L. *inter,* between, + *sistere,* to set). Relating to space or structures within parts.

Intes′tine (L. *intestinum*). Digestive tube passing from the stomach to the anus.

Intrin′sic (L. *intrinsecus,* on the inside). Belonging entirely to a part.

Invag′ination (L. *in,* in, + *vagina,* a sheath). Process in which one part is inserted into another.

Involu′tion (L. *involvere,* to roll up). Turning of a part around a margin or edge.

Irid′iophore (G. *iris,* rainbow, + *phoros,* carrying). Cell containing crystals that reflect and disperse light.

I′ris (G. *iris,* rainbow). Pigmented disklike diaphragm of the eye.

Is′chium (G. *ischion,* hip). Posteroventral bone in the pelvic girdle.

Iso- (G. *isos,* equal). Prefix signifying equal.

Isolec′ithal (G. *isos,* equal, + *lekithos,* yolk). Denoting a uniform distribution of yolk within an egg.

Is′thmus (G. *isthmus,* Isthmus). Constricted connection or passage between two larger parts of an organ or other anatomical parts.

Ju′gal (L. *jugum,* yoke). Connecting; relating to the malar bone.

Jug′ular (L. *jugulum,* throat). Relating to the throat or neck.

Juras′sic (named for the Jura Mountains of France). Middle period of the Mesozoic Era.

Ker′atin (G. *keras,* horn). Hard, relatively insoluble protein or albuminoid, present largely in cutaneous structures.

La′brium (L. *labium,* a lip). Lip-shaped structure.

Lab′yrin′thodon′tia (G. *labyrinthos,* labyrinth, + *odontos,* tooth). Division of extinct amphibians having an infolding of enamel and dentine in the teeth.

Lacertil′ia (L. *lacertus,* a lizard). Suborder of the Squamata to which the lizards belong.

Lac′rimal (L. *lacrima,* tear). Relating to the eye and tear ducts.

Lacu′na (L. *lacus,* a hollow). Small depression, gap, or space.

Lae′ve (L. *lavare,* to wash). In reference to a smooth surface.

Lage′na (L. *lagena,* a flask). Extension from the sacculus of the inner ear; extremity of the cochlear duct.

La′gomor′pha (G. *lagos,* hare, + *morphe,* form). Group of mammals including rabbits and hares.

Lamel′la (L. dim. of *lamina,* thin plate). Thin sheet, layer, or scale.

Lam′ina (L. *lamina,* thin plate). Thin plate or flat layer.

Lar′ynx (G. *larynx,* upper part of windpipe). Box composed of several cartilages at the upper end of the trachea.

Lat′eral (L. *latus,* side). One the side.

Lat′imer′ia (named for its discoverer, Marjorie Courtenay-Latimer). Genus containing the only living crossopterygian fish.

Lep′idosaur′ia (G. *lepis,* scale, + *sauros,* lizard). Subclass of diapsid reptiles containing some ancestral forms and the members of the Squamata and Rhynchocephalia.

Lepisos′teus (G. *lepis,* scale, + *osteon,* bone). Genus of living holostean fish containing the gars or gar pikes.

Lep′ospon′dyli (G. *lepos,* scale, + *spondylos,* vertebra). Amphibians with vertebrae consisting of hourglass bone cylinders enclosing the notochord.

Leu′kocyte (G. *leukos,* white, + *kytos,* cell). White blood corpuscle.

Leva′tor (L. *levator,* that which lifts). Muscle serving to raise a part.

Lig′ament (L. *ligamentum,* a band or bandage). Band or sheet of fibrous tissue connecting two or more skeletal parts.

Lin′eal (L. *linea,* a line). Referring to a line, strip, or streak.

Ling′ual (L. *lingua,* tongue). Relating to the tongue.

Lip′ochrome (G. *lipos,* fat, + *chroma,* color). Pigment in the orange–yellow range.

Lip′ophore (G. *lipos,* fat, + *phorus,* bearer). Cell containing lipochrome.

Liss′amphib′ia (G. *lissos,* smooth + *amphibia,* living a double life in water and land). Subclass of amphibians containing all the living forms.

Lo′phophore (G. *lophos,* ridge, + *phoros,* bearing). Tentacle-bearing arm of pterobranchs.

Lum′bar (L. *lumbare,* apron for loins). Relating to the loins, or the part of the back and sides between the ribs and pelvis.

Lu′teum (L. *luteus,* yellow). Yellow in color.

Ly′caenops (G. *lykaina,* shewolf, fem. of *lykos,* wolf, + *ops,* appearance). Genus of highly advanced therapsid reptiles, very mammallike in appearance.

Lymph (L. *lympha,* clear springwater). Fluid circulating in the lymphatic channels.

Lym′phocyte (L. *lympha,* + G. *kytos,* cell). White cell present in the lymph.

Lyt′ic (G. *lysis,* loosening). Relating to solution or destruction.

Mac′rolec′ithal (G. *makros,* large, + *lekythos,* yolk of egg). Referring to a large amount of yolk stored in the egg.

Mac′rophage (G. *makros,* large, + *phagō,* I eat). Large blood cell that engulfs and destroys other cells.

Mac′ula (L. *macula,* a spot). Colored spot or patch of distinctive tissue.

Ma′jus (L. comp. of *magnus,* great). Denoting large or major.

Mal′leus (L. *malleus,* hammer). Largest of the three middle ear bones.

Mamma′lia (L. *mamma,* breast). Highest class of living organisms; includes all the vertebrate animals that suckle their young.

Mam′mary (L. *mamma,* breast). Relating to the mammary glands.

Mandib′ular (L. *mandibula,* a jaw). Relating to the jaw; specifically, the lower jaw.

Marr′ow (Anglo-Saxon, *mearh*). Any soft or gelatinous material.

Marsu′pia′lia (G. *marsypos,* belly). Order of mammals featuring an abdominal pouch for nurture of young.

Masse′ter (G. *maseter,* masticator). One of the muscles for chewing.

Maxil′la (L. *maxilla,* jawbone). Upper jaw.

Mea′tus (L. *meatus,* passage). Canal or channel.

Mediasti′num (L. *mediastinus,* being in the middle). Septum between two parts of an organ or a cavity.

Medul′la (L. *medulla,* marrow, pith). Inner or central portion of an organ or part.

Meio′sis (G. *meiōsis,* lessening). Cell divisions of gametes resulting in halving of the number of chromosomes.

Mel′anin (G. *melas, melanos,* black). Dark pigment in the skin.

Mel′anophore (G. *melanos,* black, + *phorus,* bearer). Cell containing melanin.

Men′inx (pl. **menin′ges**) (G. *meninx,* membrane). Membranous envelope of the brain and spinal cord.

Menstrua′tion (L. *menstruare,* discharge). Periodic discharge from the uterus.

Meroblas′tic (G. *meros,* part, + *blastos,* germ). Denoting the involvement of only a restricted cytoplasmic area at the animal pole of the egg in cleavage.

Mes-, meso (G. *mesos,* middle). Prefix signifying middle.

Mesenceph′alon (G. *mesos,* middle, + *enkephalos,* brain). Midbrain; the second of the three primitive divisions of the brain.

Mes′enchyme (G. *mesos,* middle, + *enchyma,* infusion). Embryonic connective tissue.

Mes′entery (L. *mesenterium,* membranous support). Double layer of peritoneum enclosing an organ.

Mesocar′dium (G. *mesos,* middle, + *kardia,* heart). Mesentery supporting the heart.

Mes′oderm (G. *mesos,* middle, + *derma,* skin). Middle layer of the three primary germ layers.

Mes′olec′ithal (G. *mesos,* middle, + *lekthyos,* yolk of egg). Referring to a moderate amount of yolk stored in the egg.

Mes′omere (G. *mesos,* middle, + *meris,* segment). Middle region of the mesoderm.

Mes′oneph′ron (G. *mesos,* middle, + *nephros,* kidney). Tubular unit of a mesonephros.

Mes′oneph′ros (G. *mesos,* middle, + *nephros,* kidney). Second stage in the development of the amniote kidney.

Mes′ozo′ic (G. *mesos,* middle, + *zoe,* life). Geologic era after the Paleozoic and before the Cenozoic characterized by a great adaptive radiation of reptiles and the origin of both birds and mammals.

Meta- (G. *meta,* after). Prefix denoting behind or after something else in a series.

Metacar′pus (G. *meta,* after, + L. *carpus,* wrist). Part of the hand between the wrist and fingers.

Metam′erism (G. *meta,* after, + *meros,* part). Segmentation, resulting in a series of homologous parts.

Met′aneph′ron (G. *meta,* after, + *nephros,* kidney). Tubular unit of a metanephros.

Met′aneph′ros (G. *meta,* after, + *nephros,* kidney). Last stage in the development of the amniote kidney.

Met′apleu′ral (G. *meta,* after, + *pleura,* side). Finlike tissue extensions along the sides of *Amphioxus.*

Metatar′sus (G. *meta,* after, + L. *tarsus,* ankle). Part of the foot between the ankle and toes.

Met′ather′ia (G. *meta,* after, + *therion,* wild beast). Group of mammals coextensive with Marsupialia.

Met′enceph′alon (G. *meta,* after, + *enkephalos,* brain). Anterior of the two subdivisions of the primitive hind brain.

Mi′crolec′ithal (G. *mikros,* small, + *lekythos,* yolk of egg). Referring to a small amount of yolk stored in the egg.

Mi′nus (L. *minus,* less). Referring to absence or relative smallness.

Mi′ocene (G. *meion,* less, + *kainos,* new). Fourth epoch of the Tertiary Period of the Cenozoic Era.

Mitochon′drion (G. *mitos,* thread, + *chondros,* granule). Granule or filament in the cytoplasm of a cell containing enzymes governing metabolism.

Mo′lar (L. *molaris,* a mill). Grinding tooth.

Mon′otre′mata (G. *monos,* single, + *trema,* a hole). Egg-laying mammals.

Morphogen'esis (G. *morphe,* shape, + *genesis,* production). Origin and development of form.

Morphol'ogy (G. *morphe,* form, + *-logia,* discourse). Study of the form or structure of organisms.

Muco'sa (L. *mucosus,* mucus). Membrane containing or secreting mucus.

Mu'cus (L. *mucus,* juice). Clear, viscid secretion.

My'elenceph'alon (G. *myelos,* marrow, + *enkephalos,* brain). Posterior of the two subdivisions of the primitive hind brain.

My'elin (G. *myelos,* marrow). Fatty sheath of a nerve fiber.

My'ocar'dium (G. *myos,* muscle, + *kardia,* heart). Musculature of the heart.

Myocom'ma (G. *myos,* muscle, + *komma,* separation). Partition of connective tissue separating muscle segments.

Myom'erism (G. *myos,* muscle, + *meros,* part). Segmental or metameric arrangement of muscles.

Myosep'tum (G. *myos,* muscle, + L. *saeptum,* barrier). Connective tissue partition between two adjoining muscle segments.

My'otome (G. *myos,* muscle, + *tomos,* cutting). Prospective or actual muscle segment.

Myx'inoid'ea (G. *myxrinos,* slime fish). Typical hagfishes, distinguished by having on each side only a single external gill opening.

Na'ris (L. *naris,* nostril). Nostril.

Na'sal (L. *nasalis,* nose). Relating to the nose.

Neo'gnath'ae (G. *neso,* new, + *gnathos,* jaw). All birds belonging to the Neornithes exclusive of the Palaeognathae.

Neopal'lium (G. *neos,* new, + L. *pallium,* cloak). Pallium of the brain cortex of recent origin.

Neor'nithes (G. *neos,* new, + *ornithes,* birds). All recent and living birds.

Neot'eny (G. *neos,* recent, + *teinein,* to stretch). Reproductive capacity in a morphologically immature condition.

Neph'rocoel (G. *nephros,* kidney, + *koiloma,* a hollow). Cavity of a nephrotome.

Neph'ron (G. *nephros,* kidney). Tubular unit of the kidney.

Neph'rostome (G. *nephros,* kidney, + *stoma,* mouth). Funnel-shaped opening by which a kidney tubule communicates with the nephrocoel.

Neph'rotome (G. *nephros,* kidney, + *tomos,* slice). Segment of mesomere, forerunner of a kidney tubule.

Neu'ral (G. *neuron,* nerve). Relating to the nervous system.

Neurilem'ma (G. *neuron,* nerve, + *lemma,* husk). Delicate sheath surrounding the myelin substance of a nerve fiber or the axis of a nonmyelinated fiber.

Neu'roblast (G. *neuron,* nerve, + *blastos,* germ). Embryonic nerve cell.

Neurocra'nium (G. *neuron,* nerve, + *kranion,* skull). Part of the skull enclosing the brain.

Neurog'lia (G. *neuron,* nerve, + *glia,* glue). Supporting tissues of the brain and spinal cord.

Neu'rohypoph'ysis (G. *neuron,* nerve, + *hypophysis,* an undergrowth). Posterior lobe of the hypophysis (pituitary).

Neu'ron (G. *neuron,* nerve). Cellular unit of the nervous system.

Niche (French, an ornamental recess as in a wall). Ecological and functional role of an organism in a community.

No'tochord (G. *notos,* back, + *chorde,* cord, string). Fibrocellular rod constituting the primitive skeletal axis.

Oblique (L. *obliquus,* slanting). Deviating from the perpendicular or horizontal.

Oblonga'ta (L. *oblongatus,* rather long). Relating to the medulla of the brain.

Occip'ital (L. *occipitis,* back of the head). Relating to the skeletal components at the rear of the cranium.

Oc'ulomotor (L. *oculus,* eye, + *motus,* motion). Relating to movements of the eyeball; the third cranial nerve.

Odon'toblast (G. *odont,* tooth, + *blastos,* sprout or bud). One of a layer of cells that line the pulp cavity of a tooth and form dentine.

Odon'tognath'ae (G. *odont,* tooth, + *gnathos,* jaw). Superorder of Cretaceous toothed birds.

Odon'toid (G. *odont,* tooth, + *eidos,* resemblance). Shaped like a tooth.

Olfac'tory (L. *olfactus,* to smell). Relating to the sense of smell.

Ol'igocene (G. *oligos,* small, + *kainos,* new). Third epoch of the Tertiary Period of the Cenzoic Era, characterized by the emergence of modern mammals.

Omen'tum (L. *omentum,* membrane enclosing the bowels). Mesentery passing from the stomach to another abdominal organ.

Ontog'eny (G. *on*, being, + *genesis*, origin). Development of the individual as distinguished from phylogenesis, the evolutionary development of the species.

O'ocyte (G. *oon*, egg, + *kytos*, cell). Primitive ovum in the ovary.

Oogen'esis (L. *oon*, egg, + *genesis*, origin). Process of formation and development of the ovum.

Oogon'ia (G. *oon*, egg, + *gone*, generation). Primitive ova from which the oocytes are developed.

Opaque (L. *opacus*, shady). In reference to a zone or area not translucent or only slightly so.

Oper'culum (L. *operculum*, cover or lid). Any part resembling a lid or cover.

Ophthal'mic (G. *ophthalmos*, eye). Relating to the eye.

Opis'tho- (G. *opisthe*, behind). Prefix denoting behind or to the rear.

Opis'thoneph'ros (G. *opisthe*, behind, + *nephros*, kidney). Adult kidney in the anamniota.

Op'tic (G. *opsis*, sight). Relating to the eye or vision.

O'ral (L. *os*, mouth). Relating to the mouth.

Or'bital (L. *orbita*, a wheel track). Relating to the eye socket.

Ordovic'ian (from the Latin name for an ancient Celtic tribe the Ordovices, from northern Wales). Geologic period of the Paleozoic.

Or'ganogen'esis (G. *organon*, organ, + *genesis*, production). Formation of organs.

Ornithis'chia (G. *ornithos*, bird, + *ischio*, hip). Reptiles with birdlike pelvic girdle.

Os'moregula'tion (G. *osmos*, pushing, + L. *regula*, pattern or rule). Regulatory control of the internal balance of ions in the body fluid.

Os'sicle (L. *ossiculum*, dim. of *os*, bone). Small bone, especially one of the bones of the middle ear.

Os'teich'thyes (G. *osteon*, a bone, + *ichthys*, fish). Fishes possessing a skeleton composed of bone.

Os'teoblast (G. *osteon*, bone, + *blastos*, germ). Bone-forming cell.

Os'teoclast (G. *osteon*, bone, + *klastos*, break). Bone-destroying cell.

Os'teocyte (G. *osteon*, bone, + *kytos*, cell). Bone cell.

Os'teole'pis (G. *osteon*, bone, + *lepis*, scale; named because of the heavy scales). Rhipidistian crossopterygian close to the ancestry of the Amphibia.

Os'tium (L. *os*, mouth). Small opening, especially one of entrance into a hollow organ or canal.

Os'tracoder'mi (G. *ostrakon*, shell, + *derma*, skin). Primitive fishlike vertebrates without jaws, generally encased in armor.

O'tic (G. *otikos*, belonging to the ear). Relating to the ear.

O'tocyst (G. *otos*, ear, + *kystis*, a bladder). Embryonic inner ear.

O'tolith (G. *otos*, ear, + *lithos*, stone). Calcified body in the sacculus of the inner ear.

Ova'le (L. *ovum*, egg). Oval or egg-shaped.

O'vary (L. *ovarium*, egg receptacle). Female reproductive glands containing the ova or germ cells.

O'viduct (L. *ovum*, egg, + *ductus*, duct). Tube serving to transport and/or house the egg and embryo.

Ovip'arous (L. *ovum*, egg, + *pario*, to bear). Animals whose fertilized eggs develop outside the mother.

Ovovivip'arous (L. *ovum*, egg, + *vivus*, alive + *parere*, to bear). Referring to retention of eggs within the body and there developed, but without placental attachment.

O'vum (L. *ovum*, egg). Egg or female sexual cell from which, when fecundated by union with the male element, a new individual is developed.

Oxyto'cin. Trade name of a hormone of the neurophypophysis.

Pal'aeognath'ae (G. *palaios*, ancient + *gnathos*, jaw). Birds exhibiting a primitive type of palate, including the ratites and tinamous; all other living birds are included in the superorder Neognathae.

Pal'aeonis coid (G. *palaios*, ancient). Generalized fish of the Paleozoic Era belonging to the superorder Chondrostei.

Pal'atine (L. *palatinus*, the palate). Referring to the palate.

Pa'leocene (G. *palaios*, ancient, + *kainos*, recent or new). First epoch of the Tertiary Period of the Cenozoic Era.

Paleontol'ogy (G. *palaios*, ancient, + *on* [*ont*], a being, + *logos*, science). Study of life in the most ancient times, as revealed in the fossil remains.

Pa′leopal′lium (G. *palaios*, ancient, + L. *pallium*, cloak). Ancient, primitive cerebral cortex.

Pal′eozo′ic (G. *palaios*, ancient, + *zoe*, life). Geologic era in which the first vertebrates appeared, immediately preceding the Mesozoic Era.

Pal′lium (L. *pallium*, cloak). Cerebral cortex with the subjacent white substance.

Pan′creas (G. *pan*, all, + *kreas*, flesh). Abdominal digestive and endocrine gland.

Pannic′ulus (L. dim. of *pannus*, cloth). Sheet of tissue.

Pan′tothe′ria (G. *pantos*, all, + *therion*, beast). Order of generalized fossil mammals.

Papil′la (L. *papilla*, pimple). Any elevation or nipplelike process.

Par′a- (G. *para*, alongside). Prefix denoting alongside, near, or departure from normal.

Parabron′chus (G. *para*, alongside, + *branchus*, windpipe). Component of avian lung.

Paraph′ysis (G. *para*, alongside, + *physis*, growth). Outgrowth from the roofplate of the telencephalon.

Parapoph′ysis (G. *para*, alongside, + *apophysis*, offshoot). Secondary process in front of the transverse process of a vertebra.

Parap′sida (G. *para*, beside, + *apsis*, arch). Reptiles with single temporal opening on each side, bounded below by postfrontal and supratemporal bones.

Parathy′roid (G. *para*, beside, + *thyreos*, oblong shield, + *eidos*, form). An endocrine gland adjacent to the thyroid.

Pari′etal (L. *paries*, wall). Relating to any sidewall.

Parot′id (G. *para*, beside, + *otos*, ear). Denoting several structures near the ear.

Par′thenogen′esis (G. *parthenos*, virgin, + *genesis*, origin). Development without fertilization.

Pas′serifor′mes (L. *passerinus*, a sparrow). Order containing the advanced perching birds.

Pec′toral (L. *pectoralis*, breastbone). Relating to the breast or chest.

Pellu′cid (L. *pellucidus*; *per*, through, + *lucere*, to shine). In reference to any translucent area or zone.

Pel′vis (L. *pelvis*, a basin). Any basinlike or cup-shaped part, as the pelvis of the kidney or the pelvic girdle.

Pel′ycosau′ria (G. *pelykos*, bowl, + *sauros*,

lizard). Group of Permian reptiles antecedent to mammals.

Pe′nis (L. *penis*). Male organ of copulation.

Per′i- (G. *peri*, around). Prefix denoting around or surrounding.

Pericar′dial (G. *peri*, around, + *kardia*, heart). Surrounding the heart.

Perichon′drium (G. *peri*, around, + *chondros*, cartilage). Fibrous membrane covering cartilage.

Perimys′ium (G. *peri*, around, + *myos*, muscle). Fibrous sheath enveloping each of the primary bundles of muscle fibers.

Perios′teum (G. *peri*, around, + *osteon*, bone). Fibrous membrane covering bone.

Perio′tic (G. *peri*, around, + *otos*, ear). Surrounding the internal ear.

Peris′sodac′tyla (G. *perissos*, odd + *daktylos*, finger). Order of nonruminant ungulate mammals that have an odd number of toes.

Peritone′um (G. *periteino*, I stretch over). Lining of the body cavity and covering of organs.

Per′mian (named for the province of Perm, Russia, where the first sediments of this age were discovered). Last geologic period of the Paleozoic Era.

Pet′romyzon′tia (G. *petros*, stone, + *myzon*, suck). Jawless, suctorial lampreys.

Phag′ocyte (G. *phago*, I eat, + *kytos*, cell). Cell possessing the property of ingesting foreign cells and particles.

Phalan′ges (G. *phalanx*, pl. of line of soldiers). Bones of the fingers or toes.

Phal′lus (G. *phallos*, pertaining to the penis). Shaft of the penis.

Phar′ynx (G. *pharynx*, the throat). Segment of gut between the mouth and the esophagus.

Phylog′eny (G. *phyle, phylon*, a tribe, + *genesis*, origin). Evolutionary development of any plant or animal species.

Pin′eal (L. *pinea*, pine cone). Relating to an outgrowth from the roofplate of the diencephalon.

Pin′na (L. *pinna*, wing). External ear exclusive of the meatus.

Pin′nipe′dia (L. *pinna*, fin, wings, + *pedis*, foot). Suborder of the Carnivora containing the aquatic carnivores such as the seals.

Pi′sces (L. pl. of *piscis*, fish). General term referring to all types of fishes.

Pitu′itary (L. *pituita*, slime). Endocrine gland at the base of the diencephalon.

Placen′ta (L. *placenta*, a cake). Organ of physiological communication between mother and fetus.

Plac′ode (G. *plax*, plate, + *eidos*, like). Denoting any thickened platelike area.

Plac′oder′mi (G. *plakos*, plate, + *derma*, skin). Group of extinct fishes with an armor of large bony plates.

Plac′oid (G. *plax*, plate, + *eidos*, likeness). Relates to a plate.

Plas′ma (G. *plasma*, anything formed). Fluid portion of blood and lymph.

Platys′ma (G. *platysma*, a flat plate). Broad thin muscle of the head and neck.

Pleis′tocene (G. *pleistos*, most, + *kainos*, new). First epoch of the Quaternary Period of the Cenozoic Era, characterized by successive glacial ice sheets in the Northern Hemisphere.

Pleu′ra (G. *pleura*, side). Membrane enveloping the lungs.

Pleu′ro- (G. *pleura*, side). Prefix denoting on the side.

Pleurocen′trum (G. *pleura*, side, + *kentron*, center). One type of centrum.

Plex′us (L. *plexus*, braid). Network of interjoining of parts, for example, blood vessels and nerves.

Pli′ocene (G. *pleion*, more, + *kainos*, new or recent). Last epoch of the Tertiary Period of the Cenozoic Era.

Poiki′lotherm (G. *poikilos*, changeful, + *thermos*, heat). Organisms that do not maintain a constant body temperature.

Pol′y- (G. *polys*, many). Prefix denoting multiplicity.

Poly′odon (G. *poly*, many, + *odon*, tooth). Chondrostean genus constituted by the Mississippi River paddlefish.

Pol′yploid (G. *polys*, many, + *ploos*, equivalent). Term in cytology denoting more than the normal number of chromosomes.

Polyp′terus (G. *poly*, many, + *pteron*, wing, referring to the large, frilled fins). Genus of living chondrostean fish containing the Nile bichirs.

Polysper′my (G. *polys*, many, + *sperma*, seed). Entrance of more than one spermatozoon into the ovum.

Pons (L. *pons*, bridge). Any bridgelike formation connecting two parts; mass in floor of metencephalon.

Pon′tine (L. *pons*, a bridge). Relating to the pons of the brain.

Por′tal (L. *porta*, gate). Communicating part or area of an organism.

Post- (L. *post*, after). Prefix denoting after or behind.

Pre- (L. *prae*, before). Prefix denoting anterior or before, in space and time.

Pri′mates (L. *primas, primat-*, chief). Highest order of mammals, including humans, monkeys, and lemurs.

Pro- (L. *pro*, before). Prefix denoting before or forward.

Pro′boscid′ea (G. *proboskis*, to feed). Order of ungulate mammals consisting of the elephants and their extinct allies.

Proctode′um (G. *prōktos*, anus, + *hodaios*, relating to a way). Terminal portion of rectum formed in the embryo by an ectodermal invagination.

Proes′trus (L. *pro*, before, + *oistros*, mad desire). Period immediately preceding heat in female mammals.

Proges′terone (G. *pro*, in favor of, + L. *gestare*, to bear). Hormone of the corpus luteum.

Prolac′tin (L. *pro*, before, + *lac*, milk). Hormone from the anterior pituitary that stimulates secretion of milk.

Prona′tor (L. *pronare*, to bend forward). Muscle serving to rotate a part downward or backward.

Proneph′ron (G. *pro*, before, + *nephros*, kidney). Tubular unit of a pronephros.

Proneph′ros (G. *pro*, before, + *nephros*, kidney). First stage in the development of the amniote kidney.

Pronu′cleus (G. *pro*, before + L. dim. of *nux*, nut or kernel). Nucleus of the spermatozoon or of the ovum prior to fertilization.

Prosenceph′alon (G. *pros*, before, + *enkephalos*, brain). Most anterior of the three primitive divisions of the brain.

Pros′tate (L. *pro*, in front, + *stare*, to stand). Glandular body that surrounds the beginning of the urethra in the male.

Pro′toplasm (G. *protos*, first, + *plasma*, thing formed). Living matter; substance of which animal and plant tissues are composed.

Pro′tother′ia (G. *protos*, first, + *therion*, wild beast). Primitive group of Mammalia that lay eggs.

Protrac′tor (L. *protrahere*, to draw forth). Muscle serving to draw a part forward.

Prox′imal (L. *proximus*, next). Nearest the trunk or point of origin.

Pseu′do- (G. *pseudes*, false). Prefix denoting deceptive resemblance.

Pteran′odon (G. *pteron*, wing, + *anodontos*, without teeth). Large, toothless pterosaur of the Cretaceous Period.

Pter′obran′chia (G. *pteron*, feather, + *branchion*, gill). Order of sessile hemichordates.

Pter′odac′tyl (G. *pteryx*, wing, + *daktylos*, finger). Common name for the extinct flying reptiles, the pterosaurs.

Pter′osau′ria (G. *pteryx*, wing, + *sauria*, lizard). Extinct flying reptiles.

Pter′ygoid (G. *pteryx*, wing, + *eidos*, likeness). Applied to various anatomical parts in the neighborhood of the sphenoid bone.

Pty′alin (G. *ptyalon*, saliva). Enzyme in the saliva that converts starch into maltose.

Pu′bis (L. *pubis*, mature). Anteroventral bone of pelvic girdle.

Pul′monary (L. *pulmo*, lung). Relating to the lungs.

Py′gostyle (G. *pygo*, rump, + *stylos*, pillar). Plate of bone forming the posterior end of the vertebral column of birds.

Quad′rate (L. *quadratus*, square). Noting an anatomical part more or less square in shape, specifically the quadrate bone.

Quadra′toju′gal (L. *quadratus*, square, + *jugalis*, yolk). Bone connecting the jugal and quadrate bones.

Quadrigem′inal (L. *quadrigeminis*, fourfold). Denoting division into four parts.

Quater′nary (L. *quartus*, fourth). Most recent geologic period of the Cenozoic Era, and the one during which we are now living.

Ra′dius (L. *radius*, spoke). Line from the center to the periphery of a circle, bone of forearm.

Ra′mus (L. *ramus*, a branch). Branch or division.

Ra′tite (L. *ratis*, raft). Group of palaeognathous birds with a flat (unkeeled) sternum; includes flightless ostrich, rhea, and kiwi.

Recep′tor (L. *recipere*, to receive). Sensory nerve ending in skin, or a sense organ.

Rec′tus (L. *rectus*, straight). Referring to straight.

Re′flex (L. *re-*, back, + *flectere*, to bend). Nervous mediated involuntary action.

Re′nal (L. *ren*, kidney). Relating to a kidney.

Reptil′ia (L. *reptilis*, creeping). Class of vertebrates comprising the alligators, crocodiles, lizards, turtles, tortoises, and snakes.

Re′te (L. *retia*, a mesh). Network of nerve fibers or small vessels.

Ret′ina (L. *rete*, net). Light-sensitive layer of the eye.

Retrac′tor (L. *re*, back, + *trahere*, to draw). Muscle serving to withdraw a part.

Rhachit′omous (G. *rachis*, spine, + *temnein*, to cut). Designating vertebrae in which the centrum is composed of separate parts.

Rham′phorhyn′chus (G. *ramphos*, prow, + *rynchos*, break). Tailed, Jurassic pterosaur.

Rhip′idis′tia (G. *rhipidos*, fan, + *histion*, sail). Order of crossopterygian fishes having basal bones of median fins united in one mass.

Rhodop′sin (G. *rhodon*, rose, + *ōps*, eye). Red pigment in the rods of the retina.

Rhombenceph′alon (G. *rhombos*, parallelogram, + *enkephalos*, brain). Posterior of the three primitive brain divisions.

Rhom′boid (G. *rhombos*, a rhomb, + *eidos*, appearance). Having the shape of an oblique parallelogram with unequal sides.

Rhyn′choceph′alia (G. *rhyncos*, snout, + *kephole*, head). Order of reptiles with the general form and appearance of lizards, but having biconcave vertebrae, immovable quadrate bones, and other peculiar osteological characters.

Roden′tia (L. *rodere*, to gnaw). Order of placental mammals that possess one or two pairs of long incisor teeth adapted to gnawing.

Ru′ga (L. *ruga*, a wrinkle). Fold, ridge, or crease.

Sac′culus (L. *sacculus*, little bag). Any small pouch; the smaller of the two sacs in the inner ear.

Sa′crum (L. *sacer*, sacred). Bone formed by welding together of sacral vertebrae.

Sal′ivary (L. *saliva*, spittle). Relating to saliva and the glands producing it.

Sarcolem′ma (G. *sarx*, flesh, + *lemma*, husk). Sheath enclosing a muscle fiber.

Sar′coplasm (G. *sarx*, flesh, + *plasma*, thing formed). Semifluid interstitial substance within a muscle fiber.

Sar′copteryg′ii (G. *sarx*, flesh, + *pterygion*,

wing, fin). Subclass of fishes with fins with fleshy bases.

Sauris′chia (G. *sauria,* lizard, + *ischio,* hip). Reptiles with lizardlike pelvic girdle, with pubis directed forward.

Saur′opod′omor′pha (G. *sauros,* lizard, + *poda,* foot). Suborder of giant herbivorous dinosaurs (order Saurischia) that were quadrupedal.

Saur′opteryg′ia (G. *sauros,* lizard, + *pteron,* wing). Marine reptiles of the Mesozoic (plesiosaurs and relatives) with the limbs transformed into powerful paddles.

Sca′la (L. *scala,* a stairway). Passages in the cochlea.

Sca′phyrhyn′chus (G. *skapho,* boat, + *rhnchos,* nose or beak). Genus of living chondrostean fish comprising the living sturgeons.

Scap′ula (L. *scapula,* blade). Shoulder blade; dorsal component of the pectoral girdle.

Schiz′ocoel (G. *schizo,* split, + *koilos,* hollow). Coelom formed by the splitting of an originally solid layer of mesoderm.

Scle′ra (G. *skleros,* hard). Tough outer coat of the eye.

Scle′rotome (G. *skleros,* hard, + *tomos,* cutting). Segment of the skeleton-producing cells derived from a mesodermal somite.

Scro′tum (L. *scrotum,* hide). Musculocutaneous sac containing the testes.

Seba′ceous (L. *sebum,* tallow, grease). Relating to oil or fat.

Secre′tin (L. *secretus,* to separate). Hormone produced in the intestine that induces pancreatic secretion.

Segmenta′tion (L. *segmentum,* to cut). State or process of being divided into segments.

Sela′chii (G. *selachos,* shark). Order of the subclass Elasmobranchii that contains the sharks.

Semilu′nar (L. *semi,* half, + *luna,* moon). Crescentic or half-moon shape.

Seminif′erous (L. *semen,* seed, + *ferre,* to carry). Carrying or conducting the semen.

Sep′tum (L. *septum,* a partition). Thin wall dividing two cavities or masses of softer tissue.

Sero′sa (L. fem. of *serosus,* a coat). Peritoneal coat of a visceral organ.

Serpen′tes (L. *serpens,* from *serpere,* to creep). Suborder of the Squamata to which the snakes belong.

Seymour′ia (named for Seymour, Texas, where the first specimens were collected). Fossil form structurally intermediate between amphibians and reptiles.

Silur′ian (named after land held by the ancient tribes, the Silures, from southeast Wales). Geologic period of the Paleozoic Era that follows the Ordovician.

Si′nus (L. *sinus,* bay or hollow). Space in some organ or tissue.

Sire′nia (L. *siren,* siren). Order of large aquatic herbivorous mammals.

Skel′eton (G. *skeletos,* dried). Supporting framework of the body.

Skull (early Eng., *skulle,* bowl). Skeletal framework of the head.

Somat′ic (G. *somatikos,* bodily). Relating to the wall of the body.

So′matopleure (L. *somatikos,* bodily, + *plura,* side). Embryonic layer formed by the union of the somatic layer of mesoderm with the ectoderm.

Soma′totro′pin (G. *somatikos,* bodily, + *trophē,* nourishment). Growth hormone produced by the adenohypophysis.

Sper′matid (G. *sperma,* seed). Rudimentary spermatozoon derived from the division of the spermatocyte.

Sper′matocyte (G. *sperma,* seed, + *kytos,* a hollow cell). Cell resulting from the division of the spermatogonium, which in turn by division forms the spermatid.

Sper′matogen′esis (G. *sperma,* seed, + *genesis,* origin). Formation of spermatozoa.

Sper′matogo′nia (G. *sperma,* seed, + *gone,* generation). Primitive sperm cells giving rise, by division, to the spermatocytes.

Sper′matozo′on (G. *sperma,* seed, + *zoon,* animal). Male sexual cell.

Sphen′odon (G. *sphen,* wedge, + *odontos,* tooth). Generic name of the tuatara, the only living rhynchocephalian.

Sphe′noid (G. *sphen,* wedge, + *eidos,* likeness). Relating to any part having wedge-shaped form.

Sphinc′ter (G. *sphinkter,* band). Muscle that serves to close an opening.

Spi′racle (L. *spiraculum,* breathing hole). Aperture or vent for respiration.

Splanch′nic (G. *splanchnon,* viscus). Referring to the viscera.

Splanch′nopleure (G. *splanchnon,* viscus, + *plura,* side). Embryonic layer formed by the union of the visceral layer of mesoderm with the endoderm.

Spleen (G. *splen*, spleen). Large abdominal gland belonging to the circulatory system.

Sple′nial (G. *splenion*, bandage). In anatomy, a structure overlapping another structure; specifically, the splenial bone.

Squama′ta (L. *squamatus*, scaly). Order of reptiles comprising the snakes and lizards.

Squamo′sal (L. *squama*, scale). Relating to a bone in the temple.

Sta′pes (L. *stapes*, stirrup). Smallest of the three middle ear bones.

Steg′osaur′us (G. *stegein*, cover, + *sauros*, lizard). Ornithischian dinosaur named for the armor covering the back.

Ster′num (G. *sternon*, the chest). Breastbone.

Ster′oid (G. *stereos*, solid, + L. *oleum*, oil, + G. *eidos*, form). Resembling a sterol.

Stom′ach (L. *stomachus*). Portion of the gut between the esophagus and the small intestine.

Sto′mochord (G. *stoma*, mouth, + *chorde*, cord). Hemichordate equivalent of the notochord located in the head region.

Stomode′um (G. *stoma*, mouth, + *hodaios*, relating to a way). Ectodermal invagination forming the mouth cavity.

Stra′tum (L. *stratus*, layer). Any given layer of differentiated tissue.

Stria′tum (L. neut. of *striatus*, furrowed). Denoting stripping or furrowing.

Stru′thiomi′mus (G. *strouthion*, ostrich, + *mimos*, an imitator). Coelurosaurian dinosaur that closely resembled an ostrich superficially.

Sub- (L. *sub*, under). Prefix signifying beneath, less than normal.

Subcla′vian (L. *sub*, under, + *clavicula*, small key). Beneath the clavicle or collarbone.

Sul′cus (L. *sulcus*, furrow). Groove or furrow.

Su′pinator (L. *supinus*, lying on the back). Muscle serving to turn a part upward or forward.

Su′pra- (L. *supra*, above). Prefix denoting a position above the part indicated by the word to which it is joined.

Sympathet′ic (G. *syn*, with, + *pathos*, suffering). Division of the nervous system.

Sym′physis (G. *syn*, together, + *physis*, growth). Union or meeting point of any two structures.

Synapse (G. *syn*, together, + *hapstein*, to fasten). Close approximation of the processes of different neurons.

Synap′sida (G. *syn*, together, + *apsis*, arch). Reptiles with a single temporal opening on each side of the skull.

Synap′sis (G. *synapsis*, binding together). Pairing of homologous chromosomes.

Synap′tosau′ria (G. *synaptos*, fastened together, + *sauria*, lizard). Class of reptiles with two dorsal temporal openings.

Syncyt′ium (G. *syn*, with, + *kytos*, cell). Multinucleated protoplasmic mass; aggregation of cells without cell boundaries.

Syndes′mochorial (G. *syndesmos*, ligament, + *chorion*, skin). Placenta type involving contact between chorionic epithelium and connective tissue of uterine mucosa.

Syn′ergist (G. syn, together, + *ergon*, work). That which aids the action of another.

Syntro′phoblast (G. *syn*, together, + *trophe*, nourishment, + *blastos*, germ). Outer, syncytial layer of the trophoblast.

Syr′inx (G. *syrinx*, a tube). Vocal organ of birds.

Tar′sius (G. *tarsus*, sole of the foot). Genus of nocturnal, arboreal, East Indian mammals closely allied to the lemurs.

Tar′sus (G. *tarsus*, wickerwork frame). Ankle, or instep, of the foot.

Tec′tum (L. *tectum*, roof). Any covering or roofing structure.

Tegmen′tum (L. *tegmen*, roof). Any covering structure.

Telenceph′alon (G. *telos*, end, + *enkephalos*, brain). Anterior division of the forebrain.

Tel′eos′tei (G. *teleos*, complete, + *osteon*, bone). Modern bony fishes.

Tel′ic (G. *telikos*, final). Tending to a definite end.

Tel′olec′ithal (G. *telos*, end, + *lekithos*, yolk). Denoting an ovum in which the yolk accumulates at one pole.

Tem′poral (L. *tempora*, temples). Relating to the temple.

Ten′don (L. *tendo*, a cord). Cord serving to attach a muscle to a bone.

Ter′tiary (L. *tertiarius*, of the third rank or formation). First geologic period of the Cenozoic Era, or age of mammals.

Tes′tis (L. *testis*, testicle). Male reproductive gland.

Tet′rapod (G. *tetra*, four, + *pous*, foot). Vertebrate with four legs.

Thal′amus (G. *thalamos*, a bed). Sidewalls of the diencephalon.

The′ca (G. *theke*, a box). Sheath.

The′condon′tia (G. *theke*, case, + *odous*, tooth). Order of reptiles with teeth inserted in sockets.

Therap′sida (G. *theraps*, an attendant). Reptilian ancestors of mammals.

Theria (G. *therion*, wild beast). Mammalian subclass that includes the marsupials and the placentals.

The′ropod (G. *thero*, wild beast, hunter, + *poda*, foot). Suborder of carnivorous dinosaurs (order Saurischia) that walked on their hind limbs.

Thorac′ic (G. *thorax*, breastplate). Relating to the thorax or chest.

Throm′bocyte (G. *thrombos*, clot, + *kytos*, cell). Small platelike blood cell involved in the clotting mechanism.

Thy′mus (G. *thymos*, sweetbread). Ductless gland in the neck.

Thyrogloss′al (G. *thyreos*, a shield, + *glossa*, tongue). Relating to the thyroid gland and tongue.

Thy′roid (G. *thyreos*, oblong shield, + *eidos*, form). Gland or cartilage.

Trabec′ula (L. dim. of *trabs*, beam). Supporting bar or band.

Tra′chea (G. *tracheia*, rough artery). Air tube extending from the larynx to the bronchi.

Trach′odon (G. *trachys*, rough, + *odous* or *odon*, tooth; named because the teeth have a rough pavement). Duck-billed dinosaur of the late Cretaceous Period.

Trape′zium (G. *trapezion*, a table). Four-sided figure in which no two sides parallel.

Trias′sic (named because the sediments are divisible into three divisions). First geologic period of the Mesozoic Era or age of reptiles.

Tri′cer′atops (G. *treis*, three, + *keras*, horn, + *ops*, face). Ceratopsian dinosaur with three horns on the skull.

Tricus′pid (L. *tri*, three, + *cuspis*, point). Having three points or prongs.

Trigem′inal (L. *trigeminis*, triplet). Fifth cranial nerve.

Troch′lear (L. *trochlea*, a pulley). Relating to any structure serving as a pulley; specifically, the fourth cranial nerve.

Tro′phoblast (G. *trophe*, nourishment, + *blastos*, germ). Outer wall of the mammalian blastocyst; concerned with nutrition of embryo within.

Trun′cus (L. *truncus*, stem or trunk). Primary nerve or blood vessel before its division.

Tuber′culum (L. dim. of *tuber*, a nodule). Relating to a proturberance.

Tu′nica (L. *tunica*, a coat). Enveloping layer.

Tupa′ia (from Malay, *tupai*, squirrel). Principal gentus of tree shrews.

Tympan′ic (L. *tympanum*, drum). Relating to the eardrum.

Tyrann′osaur′us (G. *tryannos*, tyrant, + *sauros*, lizard). Largest of the huge, carnivorous therapod dinosaurs of the Cretaceous Period.

Ul′na (L. *ulna*, elbow). Inner and larger of the two bones of the forearm.

Ul′timo- (L. *ultimus*, last). Prefix denoting final or last of a series.

Umbil′ical (L. *umbilicus*, navel). Relating to the umbilicus.

Un′gulate (L. *unguis*, fingernail, toenail). Collective term used to designate hoofed mammals.

Ure′ter (G. *oureter*, urinary canal). Tube conducting the urine from the kidney to the bladder.

Ure′thra (G. *ourethra*, canal). Canal leading from the bladder, discharging the urine externally.

Urinif′erous (L. *urina*, urine, + *ferre*, to carry). Conveying urine, noting the tubules of the kidney.

Ur′ochorda′ta (G. *oura*, tail, + *chorde*, cord). Chordates with notochord and neural tube well developed in tail of larva.

U′rode′la (G. *oura*, tail, + *delos*, visible). Amphibia, including the salamanders and newts that have a long tail.

Uropyg′ial (G. *orros*, end of sacrum, + *pyge*, rump). Relating to a gland at the base of the tail in a bird.

U′rostyle (G. *ouro*, urine, + *stylos*, pillar). Rodlike bone constituting the posterior part of the vertebral column in frogs and toads.

U′terus (L. *uterus*, womb). Womb; the hollow, muscular organ in which the impregnated ovum develops into the fetus.

Utric′ulus (L. *utriculus*, small leather bottle). Larger of the two sacs in the inner ear.

U′vea (L. *uva*, grape). Middle coat of the eyeball.

Vagi′na (L. *vaginae*, sheath). Genital canal in the female extending from the uterus to the vulva.

Va′gus (L. *vagus*, wandering). Tenth cranial nerve.

Vas (L. *vas*, vessel). Tube or vessel.

Vasopres′sin (L. *vas*, vessel, + *pressin*, constriction). Blood pressure-raising hormone of the neurohypophysis.

Veg′etal (L. *vegetare,* to animate). Vegetal pole; denoting the end of a telolecithal egg containing the yolk.

Vein (L. *vena,* a blood vessel). Blood vessel conveying blood toward the heart.

Ven′tral (L. *venter,* belly). Relating to the bottom or belly.

Ven′tricle (L. *ventriculus,* small cavity). Small cavity, especially one in the brain or heart.

Ver′tebra (L. *vertebra,* joint). One of the segments of the spinal column.

Ver′tebra′ta (L. *vertebratus*). Subphylum of chordates characterized by a backbone or spinal column.

Ves′icle (L. *vesicula,* small bladder). Any small sac.

Ves′tibule (L. *vestibulum,* antechamber). Entryway.

Vil′lus (L. *villus,* tuft of hair). Minute projection from the surface of a membrane.

Vis′cus, pl. **viscera** (L. *viscus,* internal organ). Internal organ, especially one of the large abdominal organs.

Vitel′line (L. *vitellus,* yolk). Relating to the yolk of an egg.

Vit′reous (L. *vitreus,* glassy). Having a glassy appearance.

Vivip′arous (L. *vivus,* alive + *parere,* to bear). Giving birth to living young.

Vo′mer (L. *vomer,* ploughshare). Flat bone of trapezoidal shape in roof of mouth.

Xan′thophore (G. *xanthos,* yellow, + *phoros,* carrying). Cell containing yellow pigment.

Zo′na (L. *zona,* girdle). Any encircling area or structure.

Zygapoph′ysis (G. *zygon,* yoke, + *apophysis,* offshoot). Articular process of a vertebra.

Zygomat′ic (G. *zygoma,* cheekbone). Relating to the cheekbone.

Zy′gote (G. *zygotos,* yoked). Fertilized ovum.

Index